List of Sample Documents and Forms

A Topical List of the GUIDELINES Boxes

A Guide to the CHECKLISTS

Technical Communication

Fifteenth Edition

John M. Lannon
University of Massachusetts, Dartmouth

Laura J. Gurak
University of Minnesota

 Pearson

Executive Portfolio Manager: Aron Keesbury
Content Producer: Barbara Cappuccio
Content Developer: Bruce Cantley
Portfolio Manager Assistant: Christa Cottone
Senior Product Marketing Manager: Michael Coons
Product Marketing Manager: Nicholas Bolt
Content Producer Manager: Ken Volcjak
Managing Editor: Cynthia Cox

Digital Studio Course Producer: Elizabeth Bravo
Full-Service Project Management: Integra Software Services
Printer/Binder: LSC Communications, Inc.
Cover Printer: Phoenix Color/Hagerstown
Senior Art Director: Cate Barr
Cover Design: Cadence Design Studio

Acknowledgments of third party content appear on appropriate page within text, which constitutes an extension of this copyright page.

Library of Congress Cataloging-in-Publication Data

Names: Lannon, John M., author. | Gurak, Laura J., author.
Title: Technical communication / John M. Lannon, Laura J. Gurak.
Description: 15e. | New York, NY, USA: Pearson, 2020.
Identifiers: LCCN 2018042238 | ISBN 9780135203224 (0-13-520322-8)
Subjects: LCSH: Technical writing. | Communication of technical information.
Classification: LCC T11 .L24 2020 | DDC 808.06/66–dc23
LC record available at https://lccn.loc.gov/2018042238

Rental Edition
ISBN-10:   0-13-520322-8
ISBN-13: 978-0-13-520322-4

Instructor's Review Copy
ISBN-10:   0-13-520314-7
ISBN-13: 978-0-13-520314-9

Loose-Leaf Edition
ISBN-10:   0-13-520330-9
ISBN-13: 978-0-13-520330-9

Access Code Card
ISBN-10:   0-13-516478-8
ISBN-13: 978-0-13-516478-5

Revel Combo Card
ISBN-10:   0-13-525988-6
ISBN-13: 978-0-13-525988-7

Brief Contents

Detailed Contents

Part 3 Organization, Style, and Visual Design 179

10 Organizing for Readers 180

11 Editing for a Professional Style and Tone 198

Part 4   Specific Documents and Applications   307

14 Email   308

15 Workplace Memos and Letters   325

Preface

Whether digital, face-to-face, handwritten, or printed, workplace communication is more than a value-neutral exercise in "information transfer." Workplace communication is also a complex social transaction. From reports to proposals, job applications to email messages, video chats to oral presentations, every rhetorical situation has its own specific interpersonal, ethical, legal, and cultural demands. Moreover, today's workplace professional needs to be a skilled communicator and a discriminating consumer of information, skilled in methods of inquiry, retrieval, evaluation, and interpretation essential to informed decision making.

Designed in response to these issues, *Technical Communication*, Fifteenth Edition, addresses a wide range of interests for classes in which students from a variety of majors are enrolled. The text explains, illustrates, and applies rhetorical principles to an array of assignments—from memos, résumés, and email to formal reports and proposals. To help students develop awareness of audience and accountability, exercises incorporate the problem-solving demands typical in college and on the job. Self-contained chapters allow for various course plans and customized assignments.

What's New to the Fifteenth Edition?

Technical Communication, Fifteenth Edition, has been thoroughly revised to account for the latest innovations in workplace communication and today's technologically sophisticated, diverse, and global workforce. Students will benefit from a variety of new content and features in this edition:

- **The latest coverage of digital communication and social media.** As in every edition of *Technical*

Communication, the latest innovations in digital communication have been woven throughout the book (for example, using JAWS to make Web pages accessible for visually impaired users; the increasingly common use of video interviews; the use of PDF files for most online instructional material; using collaborative writing apps and spaces such as Google Drive; and more). In this edition, we've placed a special emphasis on the relationships between social media and technical communication, with social media considerations incorporated into nearly every chapter. In addition, the "Social Media" chapter itself (Chapter 25) has been updated and expanded to include new discussion of workplace uses of Instagram and creating instructional videos for YouTube and other social media, including a new "Guidelines for Creating Instructional Videos for Social Media" box.

- **New discussions on the challenges of assessing credible information when using online sources.** In addition to covering the latest in digital technology, we have included content on the challenges writers and everyday citizens face when using the Internet for research, including discussions in Chapters 7, 8, and 9 (on research) about *confirmation bias* and ways that online information tends to reinforce what people already believe or want to believe.

- **Streamlined topical coverage within chapters.** Although we have not changed the overall structure or sequence of chapters in the book, we have done extensive combining of related sections and reorganizing of topics within chapters for improved accessibility and fewer major sections for students to navigate. As in the previous edition, all major sections are tied to the Learning Objectives that appear at the beginning of each chapter.

- **An extensively revised chapter on visuals.** Chapter 12, "Designing Visual Information," now places stronger emphasis on planning visuals by placing the Planning Sheet for Preparing Visuals early in the chapter. In addition, the chapter includes a new section on understanding and creating infographics, as well as a new section on presenting visuals.

- **A revised and expanded chapter on email.** Chapter 14, "Email," now includes a new section on the three major types of email (primary, transmittal, and formatted), as well as revised and expanded coverage of interpersonal issues when using email, updated coverage of global, ethical, and legal issues related to email, and new and revised sample emails throughout the chapter.

- **A thoroughly revised chapter on oral presentations.** Without changing any of the clear and practical advice on planning, preparing, and delivering oral presentations, Chapter 23, "Oral Presentations and Video Conferencing" has been revised to include a new section on when and how to use handouts during presentations, expanded discussion of the cautions of relying too much on presentation apps, more emphasis on considering the needs of off-site audiences when planning and preparing oral presentations, and updated figures throughout the chapter.

- **Extensively revised Projects.** The end-of-chapter Projects—which continue to be organized into General, Team, Digital and Social Media, and Global categories—have been fully revised for this edition.

Hallmarks of *Technical Communication*

Technical Communication, Fifteenth Edition, retains—and enhances or expands—the features that have made it a best-selling text for technical communication over fourteen editions. These include the following:

- **Complete coverage for any course in technical communication, business communication, or professional writing.** The topics move from basic foundational concepts to chapters on research, organization, style, visual characteristics, and document design, and finally to specific documents and applications. The appendix includes thorough coverage of the most recent MLA and APA documentation styles, and a handbook of grammar, mechanics, and usage.

- **A reader-friendly writing style that presents all topics clearly and concisely.** Simple, straightforward explanations of concepts and audience/purpose analyses of specific document types help differentiate technical communication from academic writing.

- **The most current and thorough coverage of workplace technologies, ethics, and global considerations in the workplace.** Always prominent in the book, these three topics have been updated and expanded throughout to keep up with the changes in the contemporary workplace.

- **Strong coverage of information literacy.** According to the American Library Association Presidential Committee on Information Literacy, information-literate people "know how knowledge is organized, how to find information, and how to use information in such a way that others can learn from them." Critical thinking—the basis of information literacy—is covered intensively in Part II and integrated throughout the text, especially in discussions about online information.

- **A focus on applications beyond the classroom.** Clear ties to the workplace have always been a primary feature of this book. This edition includes examples from everyday on-the-job situations and sample documents, as well as dedicated chapters on ethics, teamwork, and global issues. Each chapter opener includes a quote from an on-the-job communicator.

- **Emphasis on the humanistic aspects of technical communication.** Technical communication is ultimately a humanistic endeavor, not just a set of job-related transcription tasks, with broad societal implications. Accordingly, situations and sample documents in this edition address complex technical and societal issues such as climate change, public health issues, environmental and energy topics, digital technology, and genomics.

- **Plentiful model documents and other useful figures throughout the book.** Descriptions and instructions for creating technical documents are accompanied by clear, annotated examples. Graphic illustrations throughout make abstract concepts easy to understand.

- **Highly praised pedagogical features.** Pedagogical features, including chapter-opening Learning Objectives, summary Guidelines boxes, real-world Consider This boxes, Case Studies, annotated figures, summary marginal notes, and end-of-chapter Checklists and Projects reinforce chapter topics. These features are outlined in more detail below.

How this Book is Organized

Technical Communication is designed to allow instructors maximum flexibility. Each chapter is self-contained, and each part focuses on a crucial aspect of the communication process. Following are the five major parts of the book:

- **Part 1: Communicating in the Workplace** treats job-related communication as a problem-solving process. Students learn to think critically about the informative, persuasive, and ethical dimensions of their communications. They also learn how to adapt to the interpersonal challenges of collaborative work, and to address the various needs and expectations of global audiences.

- **Part 2: The Research Process** treats research as a deliberate inquiry process. Students learn to formulate significant research questions; to explore primary and secondary sources in hard copy and digital form; to evaluate and interpret their findings; and to summarize for economy, accuracy, and emphasis. Students are asked to think critically about online information and to consider the credibility and truthfulness of the source.

- **Part 3: Organization, Style, and Visual Design** offers strategies for organizing, composing, and designing messages that readers can follow and understand. Students learn to control their material and develop a readable style. They also learn about the rhetorical implications of graphics and page design—specifically, how to enhance a document's access, appeal, and visual impact for audiences who need to locate, understand, and use the information successfully.

- **Part 4: Specific Documents and Applications** applies earlier concepts and strategies to the preparation of print and electronic documents and oral presentations. Various letters, memos, reports, and proposals offer a balance of examples from the workplace and from student writing. Each sample document has been chosen so that students can emulate it easily. Chapters on email, Web pages, and social media emphasize the ubiquity of digital communication in today's workplace.

- **Part 5: Resources for Writers** includes "A Quick Guide to Documentation," which provides general guidance as well as specific style guides and citation models for MLA and APA styles, and "A Quick Guide to Grammar, Usage, and Mechanics," which provides a handy resource for answering questions about the basic building blocks of writing.

Learning Enhancement Features

This book is written and designed to be a highly accessible document, so that readers can "read to learn and learn to do." *Technical Communication,*

Fifteenth Edition, includes the following learning enhancement features that will help students access the material easily and use the ideas to become effective technical communicators:

- *Chapter opening quotations* demonstrate the real-world applications of each chapter's topic.

- *Learning Objectives* at the beginning of each chapter tie in with the main headed sections of each chapter and provide a set of learning goals for students to fulfill.

- *Guidelines* **boxes** help students prepare specific documents by synthesizing the chapter's information.

- *Cases* **and sample situations** encourage students to make appropriate choices as they analyze their audience and purpose and then compose their document.

- **Sample documents** model various kinds of technical writing, illustrating for students what they need to do. Captions and annotations identify key features in sample documents.

- *Consider This* **boxes** provide interesting and topical applications of the important issues discussed in various chapters, such as collaboration, technology, and ethics.

- *Notes* **callouts** clarify up-to-the-minute business and technological advances and underscore important advice.

- **Marginal notes** summarize larger chunks of information to reinforce key chapter concepts.

- *Checklists* promote careful editing, revision, and collaboration. Students polish their writing by reviewing key criteria for the document and by referring to cross-referenced pages in the text for more information on each point.

- **General, team, global, and digital and social media** *Projects* at each chapter's end help students apply what they have learned.

Revel ™

Revel is an interactive learning environment that deeply engages students and prepares them for class. Media and assessment integrated directly within the authors' narrative lets students read, explore interactive content, and practice in one continuous learning path. Thanks to the dynamic reading experience in Revel, students come to class prepared to discuss, apply, and learn from instructors and from each other.

The Revel features accompanying *Technical Communication* are as follows:

- **Journal Prompts** appear at the end of every major section in each chapter, encouraging hands-on practice through writing. Students are asked to perform brief writing activities that involve reflection, brainstorming, drafting a portion of a document, or analyzing a particular document.

- **Multiple-Choice Quizzes** help reinforce facts and concepts as students move through each major section in each chapter (the end of each major section quiz features three questions directly tied to that section) and then again at the end of the chapter (the end-of-chapter quiz provides five questions covering various sections of the chapter).

- **Table Drag-and-Drop** activities help students remember information by matching terms with their descriptions or placing parts of complex documents in the right order.

- **Fill-in-the-Blank and True/False Quick Check** activities are directly tied to the Guidelines boxes that appear in every chapter and provide a quick means of remembering concepts presented in these important boxes.

- **Shared Writing Activities** at the end of each chapter encourage students to share, discuss, and critique each other's work.

Learn more about Revel
www.pearson.com/revel

Pearson English Assignments Library

Available with your adoption of any © 2019 or © 2020 Pearson English course in Revel is the English Assignments Library comprising 500 essay and Shared Media prompts:

- A series of 300 fully editable essay assignments invite students to write on compelling, wide-ranging writing topics. You can choose from an array of writing prompts in the following genres or methods of development: Argument/Persuasion; Comparison/Contrast; Critique/Review; Definition; Description; Exposition; Illustration; Narration; Process Analysis; Proposal; and Research Project. Assignments can be graded using a rubric based on the WPA Outcomes for First-Year Composition. You can also upload essay prompts and/or rubrics of your own.

- 200 Shared Media assignments ask students to interpret and/or produce various multimedia texts to foster multimodal literacy. Shared Media activities include analyzing or critiquing short professional videos on topics of contemporary interest; posting brief original videos or presentation slides; and sharing original images—such as posters, storyboards, concept maps, or graphs.

Format Options

Below are format options by which Technical Communication is available.

Revel Access Card

Students can purchase a physical Revel access code card at their campus bookstore.

- **INSTANT ACCESS** Students can purchase access directly from Pearson to start their subscription immediately.

- **PRINT UPGRADE** Students can choose to have a printed loose-leaf version sent to them with free shipping

Revel Combo Card

The Revel Combo provides the Revel access code card plus a coupon for the loose-leaf print reference (delivered by mail). This option is perfect for students who need to purchase all of their materials from the campus bookstore.

Print Rental

Students can rent the text from their campus bookstore or directly from Pearson. Barnes & Noble and Follett bookstores are partners in this program.

Rent to Own

If a student has rented the text from either Pearson or their campus bookstore, they can choose to permanently own the text by paying a flat ownership fee.

EBook

Students can choose to purchase or rent the EBook version of the text.

Supplements

Make more time for your students with instructor resources that offer effective learning assessments and classroom engagement. Pearson's partnership with educators does not end with the delivery of course materials; Pearson is there with you on the first day of class and beyond. A dedicated team of local Pearson representatives will work with you to not only choose course materials but also integrate them into your class and assess their effectiveness. Our goal is your goal—to improve instruction with each semester.

Pearson is pleased to offer the following resources to qualified adopters of *Technical Communication*. Several of these supplements are available to instantly download from Revel or on the Instructor Resource Center (IRC); please visit the IRC at www.pearson.com/us to register for access.

- **TEST BANK** Evaluate learning at every level. Reviewed for clarity and accuracy, the Test Bank measures this material's learning objectives with multiple-choice, true/false, and fill-in-the-blank questions. You can easily customize the assessment to work in any major learning management system and to match what is covered in your course. Word and BlackBoard versions are available on the IRC.

- **PEARSON MYTEST** This powerful assessment generation program includes all of the questions in the Test Bank. Quizzes and exams can be easily authored and saved online and then printed for classroom use, giving you ultimate flexibility to manage assessments anytime and anywhere. To learn more, visit www.pearsonmytest.com.

- **INSTRUCTOR'S RESOURCE MANUAL by Lee Scholder, M.S., J.D.** Create a comprehensive roadmap for teaching classroom, online, or hybrid courses. Designed for new and experienced instructors, the Instructor's Resource Manual includes overall teaching strategies (including general teaching ideas, advice on how to use the Revel features accompanying *Technical Communication*, and sample syllabi) and chapter-specific resources (including chapter overviews, Learning Objectives, teaching tips, additional exercises, and quizzes). Available within Revel and on the IRC.

- **POWERPOINT PRESENTATION** Make lectures more enriching for students. The accessible PowerPoint Presentation includes a full lecture outline and figures from the textbook and Revel edition. Available on the IRC.

Acknowledgments

From prior editions, we wish to thank University of Massachusetts colleague Professor Peter Owens for his input on libel law in Chapter 4, Glenn Tarullo for sharing his decisions about the writing process in Chapter 6, and librarians Shaleen Barnes and Ross LaBaugh for their inspirations about the research process in Chapter 7. Also, thank you to Daryl Davis from Northern Michigan University for help in clarifying the descriptive abstract distinctions made in Chapter 9.

Many of the refinements in this and earlier editions were inspired by generous and insightful suggestions from our reviewers. For this edition, we are grateful for the comments of the following reviewers:

- Mikayla Beaudrie, University of Florida
- Mary Faure, The Ohio State University
- William Matter, Richland College
- Ida Patton, Arkansas State University
- Nancy Riecken, Ivy Tech Community College
- Terri Thorson, Arizona State University
- Nicole Wilson, Bowie State University

We thank our colleagues and students at the University of Massachusetts and the University of Minnesota, respectively, for their ongoing inspiration. This edition is the product of much guidance and support from Pearson Education, Ohlinger Publishing Services, and Integra-Chicago. From Aron Keesbury, Cynthia Cox, Maggie Barbieri, Kate Hoefler, Rachel Harbour, Tom Stover, Chris Fegan, Joe Croscup, Carmen Altes, and Valerie Iglar-Mobley, we received outstanding editorial guidance, support, and project management. Many thanks to freelance development editor Bruce Cantley for his generous and unflagging development help and valuable ideas.

From John M. Lannon, special thanks to those who help me keep going: Chega, Daniel, Sarah, Patrick, and Zorro. From Laura J. Gurak, thanks greatly to Nancy, to my friends and family, and to my four-legged companions for the ongoing support and friendship.

—*John M. Lannon and Laura J. Gurak*

Part 1
Communicating in the Workplace

Chapter 1
Introduction to Technical Communication

mama_mia/Shutterstock

❝Writing is essential to my work. Everything we do at my company results in a written product of some kind—a formal technical report, a summary of key findings, recommendations and submissions to academic journals or professional associations. We also write proposals to help secure new contracts. No matter if the document is to be delivered in print or online, writing is the most important skill we seek in potential employees and nurture and reward in current employees. It is very hard to find people with strong writing skills, regardless of their academic background.❞

—*Paul Harder, President, mid-sized consulting firm*

Learning Objectives

1.1 Define technical communication

1.2 Identify the main features of technical communication

1.3 Explain the purposes of technical communication

1.4 Describe the four tasks involved in preparing effective technical documents

What Is Technical Communication?

1.1 Define technical communication

Technical communication is the exchange of information that helps people interact with technology and solve complex problems. Almost every day, we make decisions or take actions that depend on technical information. When we purchase any new device, from a digital camera to a Wi-Ffi range extender, it's the setup information that we look for as soon as we open the box. Before we opt for the latest in advanced medical treatment, we go online and search for all the information we can find about this treatment's benefits and risks. From banking systems to online courses to business negotiations, almost every aspect of daily life involves technology and technical information. Because our technologies are so much a part of our lives, we need information that is technically accurate and, importantly, easy to understand and use.

> Technical communication helps us interact with technology in our daily lives

Technical communication serves various needs in various settings. People may need to perform a task (say, assemble a new exercise machine), answer a question (say, about the safety of a flu shot), or make a decision (say, about suspending offshore oil drilling). In the workplace, we are not only consumers of technical communication but also producers. To be effective and useful, any document or presentation we prepare (memo, letter, report, Web page, PowerPoint presentation) must advance the goals of our readers, viewers, or listeners.

> Technical communication helps us solve complex problems

Figure 1.1 shows a sampling of the kinds of technical communication you might encounter or prepare, either on the job or in the community.

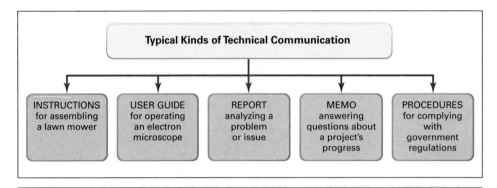

Figure 1.1 **Technical Communication Serves Various Needs**

Technical Communication Is a Digital *and* a Human Activity

Digital communication requires attention to style and tone

In today's workplace, with all of the digital communication available to us, we communicate in writing more than ever. Email, texts, chat sessions, social media and blog posts, document review features (such as Word's track changes when editing group documents): These technologies are a daily part of our workplace communication. Digital technologies make it easy for people to collaborate, especially across different time zones or work schedules. Yet in digital formats, we often communicate with such speed that we forget about basic professional standards for workplace communication. For instance, the informal or even humorous tone we use to text our friends is typically not appropriate for a work-related email. An unclear or inaccurate email sent late in the day when you are rushing to get out of the office could easily cause a safety error or legal problem; an inappropriate tone might result in wasted hours resolving an interpersonal situation instead of working on the project.

Online research is not the same as critical thinking

Digital technology also provides plenty of ways, from simple to sophisticated, to research and find information. Doing a Google or other online search, however, is not a substitute for critical thinking skills. The big questions involved in most workplace projects are questions that require us to take our research findings and make the information meaningful by asking questions such as these:

Questions that only a person can answer

- Which information is relevant to this situation?
- Can I verify the credibility and accuracy of this source?
- What does this information mean?
- What action does it suggest?
- How does this information affect me or my colleagues?
- With whom should I share it?
- How might others interpret this information?

Technical Communication Reaches a Global Audience

Linked as we are through our transportation systems and especially our digital technologies, the global community shares social, political, and financial interests. We can no longer pretend to operate solely within regional or national boundaries. Organizations are increasingly multinational; employees work on teams with colleagues from around the globe. The best collaborations happen when communication is tuned to reflect the diversity of people, countries, and cultures that make up the organization and the project team.

Write to a diverse audience

Understanding the point of view of another culture takes time. Even within specific cultures, people are individuals and can't be lumped together into one stereotype. As researchers in intercultural communication remind us, a key component is the communication's *context* (Collier 9; Martin 6). For instance, people communicate differently in the context of being at home than they do when at work.

Consider the cultural context

Cultures differ over which communication behaviors and approaches are appropriate for business relationships, including contract and other legal negotiations, types of documents (e.g., whether to use email, a memo, or a letter), tone and style, use of visuals, and so forth. An effective and appropriate communication style in one culture may be considered inappropriate or even offensive in another. In the workplace, communication tends to be patterned by a set of norms that have developed over time in different cultures. As one business expert notes,

Understand that communication behaviors differ across cultures

> Every aspect of global communication is influenced by cultural differences. Even the choice of medium used to communicate may have cultural overtones. For example, it has been noted that industrialized nations rely heavily on electronic technology and emphasize written messages over oral or face-to-face communication. Certainly the United States, Canada, the UK and Germany exemplify this trend. But Japan, which has access to the latest technologies, still relies more on face-to-face communications than on the written mode (Goman 1).

This expert goes on to explain how "[i]n some cultures, personal bonds and informal agreements are far more binding than any formal contract. In others, the meticulous wording of legal documents is viewed as paramount" (Goman 2).

The documents you research and write at work need to reflect an understanding and sensitivity to cultural differences and the communication approaches of your teammates at work and your readers (i.e., your customers or clients). Your best bet is to learn as much as you can by listening and observing; asking trusted colleagues; and reading magazines, newspaper articles, blog posts, and other such information (just be sure the information is written by someone with expertise and experience in international communication). You might also try an online short course on international communication. For more on cross-cultural communication, see Chapters 3 and 5 as well as the Global Projects at the end of each chapter.

Take the time to learn about cultural differences

Technical Communication Is Part of Most Careers

Whatever your job description, you should expect to be evaluated at least in part on your written and oral communication skills. Even if you don't anticipate an actual career in writing, every job involves being a technical communicator at some point. You can expect to encounter situations such as the following:

Most professionals serve as part-time technical communicators

- As a medical professional, psychologist, social worker, or accountant, you will keep precise records that are increasingly a basis for legal action.
- As a scientist, you will report on your research and explain its significance.
- As a manager, you will write memos, personnel evaluations, and inspection reports; you will also give oral presentations.
- As a lab or service technician, you will keep daily activity records and help train coworkers in installing, using, or servicing equipment.
- As an attorney, you will research and interpret the law for clients.
- As an engineer or architect, you will collaborate with colleagues as well as experts in related fields before presenting a proposal to your client. (For example, an architect's plans are reviewed by a structural engineer who certifies that the design is sound.)
- As an employee or intern in the nonprofit sector (an environmental group or a government agency), you will research important topics and write brochures, press releases, or handbooks for clients.
- As an employee of any company or organization, you will write status reports, trip reports, memos, proposals, instructions, and many other forms of technical communication.

The more you advance in your field, the more you will need to share information and establish contacts. Managers and executives spend much of their time negotiating, setting policies, and promoting their ideas—often among diverse cultures around the globe.

In addition, most people can expect to work for several different employers throughout their career. Each employer will have questions such as the following:

Employers seek portable skills

- Can you write and speak effectively in a variety of formats and to a range of different people?
- Can you research information, verify its accuracy, figure out what it means, and shape this information for your readers' specific purposes?
- Can you work on a team with people from diverse backgrounds?
- Can you get along with, listen to, and motivate others?
- Are you flexible enough to adapt to rapid changes in business conditions and technology?
- Can you market yourself and your ideas persuasively?
- Are you ready to pursue lifelong learning and constant improvement?

Although technical expertise and experience is important, the above items, most especially the first two (communication and critical thinking), are top among the portable skills employers seek in today's college graduates.

Technical Communicators Play Many Roles

Full-time technical communicators work in many capacities. Job titles include information architect, user experience engineer, technical writer, technical editor, documentation specialist, Web development specialist, and content developer. In the public sector, government agencies (federal, state, and local) hire technical communicators to take technical research and make it accessible to nonexpert readers by writing and designing blog and social media posts, podcasts, Web pages, short reports, and brochures. In the private sector, technical communicators can be found across the spectrum, including in the highly regulated banking, pharmaceutical, and medical device industries, where these skilled communicators create specifications, procedures, and documentation for global audiences. You will also find technical communicators employed at retail companies such as Target and Best Buy to work on websites and technical documentation and in every high-tech company such as Microsoft and Apple where teams of technical communicators are responsible for user manuals, online help, customer support, and much more.

The variety of job titles of technical communicators

Technical communicators also edit reports for punctuation, grammar, style, and logical organization. They may oversee publishing projects, coordinating the efforts of writers, visual artists, graphic designers, content experts, and lawyers to produce a complex manual, report, or proposal. Given their broad range of skills, technical communicators often enter related fields such as technical or scientific publishing, magazine editing, video production (including writing scripts), training, and college teaching.

What technical communicators do

Main Features of Technical Communication

1.2 Identify the main features of technical communication

Almost any form of technical communication displays certain shared features: The communication is reader-centered, accessible and efficient, often produced by teams, and delivered in a variety of digital and hard copy formats.

Reader-Centered

Unlike poetry, fiction, or college essays, a technical document rarely focuses on the writer's personal thoughts and feelings. This doesn't mean your document should have no personality (or voice), but it does mean that the needs of your readers come first.

Focus on the reader, not the writer

Workplace readers typically are interested in "who you are" only to the extent that they want to know what you have done, what you recommend, or how you speak for your company. Reader-centered documents focus on what people need to learn, do, or decide. For example, while the history of how this product was invented may be of interest to the writer, instructions for assembling a new workstation desk should focus on what readers need to do—assemble their desk and start using it. Writing from a reader-centered perspective takes practice and attention (the rest of this text will emphasize reader-centered writing and design to help you get the idea).

Accessible and Efficient

Readers expect to find the information they need and to have questions answered clearly. For instance, the document shown in Figure 1.2 is written and designed so that a nontechnical audience can find and follow the information. Instead of long technical passages, the content is presented in short chunks, answering the main question readers will ask (how to choose the right model).

An accessible and efficient technical document includes elements such as those displayed in Figure 1.2 as well as others listed below.

- **worthwhile content**—includes all (and only) the information readers need
- **sensible organization**—guides the reader and emphasizes important material
- **readable style**—promotes fluid reading and accurate understanding
- **effective visuals**—clarify concepts and relationships, and substitute for words whenever possible
- **effective page design**—provides heads, lists, type styles, white space, and other aids to navigation
- **supplements (abstract, appendix, glossary, linked pages, and so on)**—allow readers to focus on the specific parts of a long document that are relevant to their purpose

Accessible, efficient communication is no mere abstract notion: In the event of a lawsuit, faulty writing is treated like any other faulty product. If your inaccurate, unclear, or incomplete information leads to injury, damage, or loss, you and your company or organization can be held responsible.

NOTE Make sure your message is clear and straightforward—but do not oversimplify. Information designer Nathan Shedroff reminds us that, while clarity makes information easier to understand, simplicity is "often responsible for the 'dumbing down' of information rather than the illumination of it" (280). The "sound bytes" that often masquerade as network news reports serve as a good case in point.

Use a Programmable Thermostat Properly

A programmable thermostat is ideal for people who are away from home during set periods of time throughout the week. Through proper use of pre-programmed settings, a programmable thermostat can save you about $180 every year in energy costs.

← Overview information summarizes the document's main point

How Do You Choose the Right One for You?

← Heading is phrased as the main question readers will ask

To decide which model is best for you, think about your schedule and how often you are away from home for regular periods of time—work, school, other activities—and then decide which of the three different models best fits your schedule:

7-day models are best if your daily schedule tends to change; for example, if children are at home earlier on some days. These models give you the most flexibility and let you set different programs for different days—usually with four possible temperature periods per day.

← Paragraphs and sentences are short

5+2-day models use the same schedule every weekday, and another for weekends.

← Color is used to highlight key items

5-1-1 models are best if you tend to keep one schedule Monday through Friday and another schedule on Saturdays and Sundays.

Programmable Thermostat Settings

You can use the table below as a starting point for setting energy-saving temperatures, and then adjust the settings to fit your family's schedule and stay comfortable.

Setting	Time	Setpoint Temperature (Heat)	Setpoint Temperature (Cool)
Wake	6:00 a.m.	< 70° F	> 78° F
Day	8:00 a.m.	Setback at least 8° F	Setup at least 7° F
Evening	6:00 p.m.	< 70° F	> 78° F
Sleep	10:00 p.m.	Setback at least 8° F	Setup at least 4° F

← Table provides easy-to-read comparative data

Figure 1.2 An Effective Technical Document Language, layout, and PDF format make the information easy for everyday readers to understand and access.

Source: A Guide to Energy-Efficient Heating and Cooling, Energy Star Program, August 2009.

Often Produced by Teams

Prepare for teamwork

Technical documents are often complex. Instead of being produced by a lone writer, complex documents usually are created by teams composed of writers, Web designers, engineers or scientists, managers, legal experts, and other professionals. The teams might be situated at one site or location or distributed across different job sites, time zones, and countries.

Delivered in Paper and Digital Versions

Select the appropriate medium or combination of media

Technical documents can be delivered in a variety of media such as print (hard copy), Web pages, PDF documents, e-books, podcasts, blog and social media posts, tweets, and online videos. In many cases, there is no clear distinction between print and digital communication. Figure 1.2 is a good example: The document is in PDF format and can be read online, downloaded for later reading, or downloaded and printed on paper. Technical communicators must write well but must also be able to think about page design and media choices.

Purposes of Technical Communication

1.3 Explain the purposes of technical communication

What purpose or combination of purposes will your document serve?

Most forms of technical communication address one of three primary purposes: (1) to anticipate and answer questions (inform your readers); (2) to enable people to perform a task or follow a procedure (instruct your readers); or (3) to influence people's thinking (persuade your readers). Often, as in Figure 1.2, these purposes will overlap.

Documents That Inform

Anticipate and answer your readers' questions

Informational documents are designed to inform—to provide information that answers readers' questions clearly and efficiently. Figure 1.2 is primarily informational. It is designed for a wide audience of readers who may have questions but know little about the technical details.

Documents That Instruct

Enable your readers to perform certain tasks

Instructional documents help people do something: assemble a new computer, perform CPR, or, in the case of Figure 1.2, choose and use a programmable thermostat. This page, part of a longer document on energy-efficient heating and cooling, provides basic instructions to help people decide how to choose the most suitable thermostat for their needs. Action verbs and phrases, such as "think about your schedule" and "decide which of the three models best fits," are clear and direct. A simple table provides visual instructions on how and when to set thermostat temperatures.

Documents That Persuade

Persuasion encourages people to take a desired action. While some documents (such as a sales letter) are explicitly persuasive, even the most technical of documents can have an implicitly persuasive purpose. The first paragraph of Figure 1.2, for example, encourages readers to use a programmable thermostat by pointing out how much a person could save in yearly energy bills.

Motivate your readers

Preparing Effective Technical Documents

1.4 Describe the four tasks involved in preparing effective technical documents

Whether you are a full-time communication professional or an engineer, nurse, scientist, technician, legal expert, or anyone whose job requires writing and communicating, the main question you face is this: "How do I prepare the right document for this group of readers and this particular situation?"

A main question you must answer

Other chapters in this book break down the process in more detail. In Chapter 2, for example, you will learn about analyzing the audience and purpose for any document and situation. Later, you will see examples of document types typically used in workplace environments. Regardless of the type, producing an effective document typically requires that you complete the four basic tasks depicted in Figure 1.3 and described below.

- **Deliver information readers can use**—because different people in different situations have different information needs (Chapter 2)

- **Use persuasive reasoning**—because people often disagree about what the information means and what action should be taken (Chapter 3)

- **Weigh the ethical issues**—because unethical communication lacks credibility and could alienate readers (Chapter 4)

- **Practice good teamwork**—in most professions, documents are not produced by one person but by a team of colleagues from different parts of the organization (Chapter 5)

A workplace communicator's four basic tasks

The short cases that follow illustrate how a typical professional confronts these tasks in her own day-to-day communication on the job.

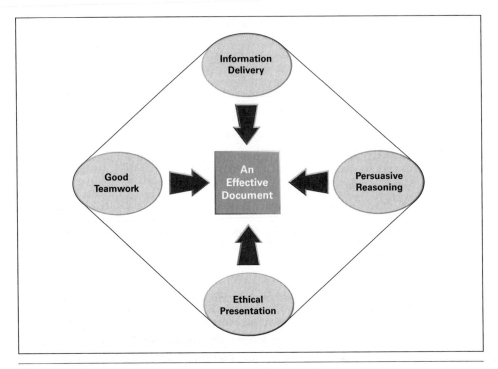

Figure 1.3 **How an Effective Document Is Produced**

Providing Information Readers Can Use

"Can I provide
exactly what
readers need?"

Sarah Burnes was hired two months ago as a chemical engineer for Millisun, a leading maker of film and digital imaging supplies and equipment. Sarah's first major assignment is to evaluate the plant's incoming and outgoing water. (Waterborne contaminants can taint these products during the production process, and the production process itself can pollute outgoing water.) Management wants an answer to this question: How often should we change water filters? The filters are expensive and hard to change, sometimes halting production for up to a day at a time. The company wants as much "mileage" as possible from these filters, without either incurring government fines or tainting its film production.

Sarah will study endless printouts of chemical analysis, review current research and government regulations, do some testing of her own, and consult with her colleagues. When she finally determines what all the data indicate, Sarah will prepare a recommendation report for her bosses.

Later, Sarah will collaborate with the company training manager and the maintenance supervisor to prepare a user manual, instructing employees on how to check and change the filters. To cut down any additional design and publishing costs, the company has asked Sarah to design and produce this manual herself (using Adobe InDesign) and then to make the manual available as a Web page and as a PDF document.

Sarah's report, above all, needs to be accurate; otherwise, the company gets fined or lowers production. Once she has processed all the information, she faces the problem of giving readers what they need: *How much explaining should I do? How will I organize the manual? Do I need visuals?* And so on.

Sarah's next project, described below, also requires research and attention to detail but takes on an even more persuasive quality. In order to resolve the matter, Sarah will seek consensus for *her* view.

Case

Being Persuasive

Millisun and other electronics producers are located on the shores of a small harbor, the port for a major fishing fleet. During the 1980s and 1990s, these companies discharged effluents containing metal compounds, PCBs, and other toxins directly into the harbor. Sarah is on a multicompany team, assigned to work with the U.S. Environmental Protection Agency, as well as state and local environmental agencies, to clean up the harbor. Much of the team's collaboration occurs via email and a shared set of documents (using the company's internal shared document system, similar to Google Docs or Microsoft OneDrive).

Enraged local citizens are demanding immediate action, angry that the process has taken so many years, and the companies themselves are anxious to end what has now become a true public relations nightmare due to the use of Twitter and several Facebook pages that citizens have set up. But the team's analysis reveals that any type of cleanup would stir up harbor sediment, possibly dispersing the solution into surrounding waters and the atmosphere. (Many of the contaminants can be airborne.) Premature action might actually increase danger, but team members disagree on the degree of risk and on how to proceed.

Sarah's communication here takes on a persuasive dimension: She and her team members first have to resolve their own disagreements and produce an environmental impact report that reflects the team's consensus. If the report recommends further study, Sarah will have to justify the delays to her bosses and the public relations office. She will have to make other people understand the dangers as well as she does.

> "Can I influence people to see things my way?"

In the preceding case, the facts are neither complete nor conclusive, and views differ about what these facts mean. Sarah will have to balance the various political pressures and make a case for her interpretation. Also, as company spokesperson, Sarah will be expected to protect her company's interests. Some elements of Sarah's persuasion problem: *Are other interpretations possible? Is there a better way? Can I expect political or legal fallout?*

Case

Considering the Ethical Issues

To ensure compliance with Occupational Safety and Health Administration (OSHA) standards for worker safety, Sarah is assigned to test the air purification system in Millisun's chemical division. After finding the filters hopelessly clogged, she decides to test the air quality and

> "Can I be honest and still keep my job?"

discovers dangerous levels of benzene (a potent carcinogen). She reports these findings in an email memo to the production manager, with an urgent recommendation that all employees be tested for benzene poisoning. The manager phones and tells Sarah to "have the filters replaced," but says nothing at all about her recommendation to test for benzene poisoning. Now Sarah has to decide what to do about this lack of response: Assume the test is being handled? Raise the issue again, and risk alienating her boss? Send copies of her original email to someone else who might take action?

As the preceding case illustrates, Sarah also will have to reckon with the ethical implications of her writing, with the question of "doing the right thing." For instance, Sarah might feel pressured to overlook, sugarcoat, or suppress facts that would be costly or embarrassing to her company.

Situations that compromise truth and fairness present the hardest choices of all: Remain silent and look the other way, or speak out and risk being fired. Some elements of Sarah's ethics problem: *Is this fair? Who might benefit or suffer? What other consequences could this have?*

In addition to solving these various problems, Sarah has to work in a team setting: Much of her writing will be produced in collaboration with others (technical editors, other engineers, project managers, graphic artists), and her audience will extend beyond readers from her own culture.

Case

Working on a Team and Thinking Globally

"Can I connect with all these different colleagues?"

Recent mergers have transformed Millisun into a multinational corporation with branches in 11 countries. Sarah can expect to collaborate with coworkers from diverse cultures on research and development and with government agencies of the host countries on safety issues, patents and licensing rights, product liability laws, and environmental concerns. Also, she can expect to confront the challenges of addressing the unique needs and expectations of people from various cultures across the globe. She will need to be careful about how she writes her daily email status reports, for example, so that these reports convey respect for cultural differences.

In order to standardize the sensitive management of the toxic, volatile, and even explosive chemicals used in film production, Millisun is developing automated procedures for quality control, troubleshooting, and emergency response to chemical leakage. Sarah has been assigned to a team that is preparing Web-based training packages and online instructional videos for all personnel involved in Millisun's chemical management worldwide.

As a further complication, Sarah will have to develop working relationships with people she has never met face to face, people from other cultures, and people she knows only via email and a few conference or video calls.

For Sarah Burnes, or any of us, writing is not just putting words on paper. Writing in the workplace is a process. Throughout this process, we rarely work alone but instead collaborate with others for information, help in writing, and feedback.

Projects

For all projects, check with your instructor about whether to present your findings in class, bring drafts to class for discussion, upload your project to the class learning management system (LMS), and/or use the LMS forum or discussion boards to collaborate and review each activity below.

General

1. Write a memo to your boss, justifying reimbursement for this course. Explain how the course will help you become more effective on the job. (For this and other projects below that request a memo, see Chapter 15 for memo elements and format.)

2. Locate a Web site or Facebook page for an organization that hires graduates in your major. In addition to technical knowledge, what writing and communication skills does this organization seek in job candidates? Discuss your findings in class and create a short presentation for other students, explaining what communication skills they require to find a job in this or a similar organization. If your class is using a learning management system, upload your presentation to the site and refer to it later when working on job search materials (Chapter 16).

Team

Introducing a Classmate

Class members will work together often this semester. To help everyone become acquainted, your task is to introduce to the class the person seated next to you. (That person, in turn, will introduce you.) Follow this procedure:

a. Exchange with your neighbor whatever information you think the class needs: background, major, career plans, kinds of writing and other communication that might be done in that job, and so on. Each person gets 5 minutes to tell her or his story.

b. Take careful notes; ask questions if you need to.

c. Take your notes home and select only the information you think the class will find useful.

d. Prepare a one-page document (a few paragraphs is fine) telling your classmates who this person is.

e. Ask your neighbor to review the document for accuracy; revise as needed.

f. In class, introduce your neighbor using your document as a guide. Turn in your document in whatever format your instructor prefers (uploaded to the learning management system; posted to a class forum or blog; sent as an email attachment; handed in on paper in class).

Digital and Social Media

With a team of 2–3 other students, visit a government Web site, such as the Food and Drug Administration, the Centers for Disease Control, or NASA. Locate documents similar in purpose to Figure 1.2 in this chapter. Analyze these documents, noting whether they conform to one of the three purposes (informative, instructional, persuasive) described in this chapter or whether they are a blend of these purposes. Next, locate the Facebook page of that government agency. How is content presented differently on each site? Does the Facebook page appear to have a different purpose from the Web site? If so, what are the differences?

Global

Look back at the Sarah Burnes case in this chapter. Assume you are about to join a team at work—a team that has members from Ireland, India, China, and the United States. Online, search for information to help you learn what you can about patterns of communication; issues to look for include politeness, turn-taking, use of first names or titles, gender roles, and formal versus informal kinds of language. Describe your findings in a short memo to your instructor. Be sure to include links to the Web sites and other online sources you found.

Chapter 2
Meeting the Needs of Specific Audiences

Ian Lishman/Juice Images/Getty images

“Audience is key. I write for people who have different levels of technical expertise, including service technicians, managers, engineers, and customers. I also work with the marketing team on proposals for commercial clients and with the legal department to write regulatory documents. It's important for me to understand how much background information and technical explanations are required for a particular set of readers. In terms of the document's design and overall size, I need to consider the primary way people will interact with the material: on paper, on a computer, or on a phone.”

—John Bryant, Technical Communication Project Lead

Analyze Your Document's
Audience and Purpose

Assess the Audience's
Technical Background

Anticipate Your
Audience's Preferences

Guidelines for Analyzing
Your Audience and Its
Use of the Document

Develop an Audience and
Use Profile

Checklist: Analyzing
Audience and Purpose

Projects

Learning Objectives

2.1 Ask the right questions to analyze
your audience and purpose

2.2 Assess your audience's technical
background

2.3 Identify the appropriate document
qualities for your audience

2.4 Develop an audience and use profile
to guide your work

All technical communication is intended for people who will use and react to the information. These people are considered to be the *audience* for your document: people who are reading the material in order to do something or learn something.

Definition of audience

Before you start writing, you need to identify with as much precision as possible who will be reading the document and then understand how that particular audience will use your material. For example, you might need to *define* something—as in explaining to insurance clients what the term "variable annuity" means. You might need to *describe* something—as in showing an architectural client what a new office building will look like. You might need to *explain* something—as in instructing an auto repair technician how to reprogram the car's electronic ignition. Or you might need to *propose* something—as in arguing for change in your company's sick-leave policy. Preparing an effective document requires systematic analysis of your audience and the ways in which they will use your document (Figure 2.1).

Why understanding your audience is important

Because people's basic requirements vary, every audience expects a message tailored to its own specific interests, social conventions, ways of understanding problems, and information needs.

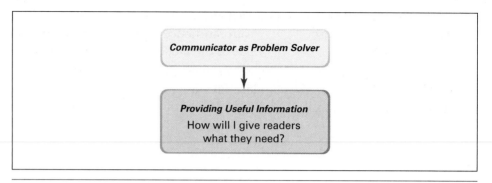

Figure 2.1 **Communicators Begin by Considering Their Audience**

Analyze Your Document's Audience and Purpose

2.1 Ask the right questions to analyze your audience and purpose

Explore all you can about who will use your document, why they will use it, and how they will use it. Begin by analyzing your audience and the background, needs, and preferences of these readers. Among the questions you must answer are these:

Questions for analyzing a document's audience

- Who is the main audience for this document?
- Who else is likely to read it?
- What is your relationship with the audience?
- Are multiple types of relationships involved?
- What information does this audience need?
- How familiar might the audience be with technical details?
- Do these readers have varying levels of expertise?
- What culture or cultures does your audience represent?
- How might cultural differences shape readers' expectations and interpretations?
- How will people interact with the material: in digital formats, on paper, or both?

Answer these questions by considering the suggestions that follow.

Primary and Secondary Audiences

"Who is the main audience for this document?"

When writing a technical document, keep two audiences in mind. Most documents are geared to an immediate audience of readers. This is your primary audience. For instance, a set of instructions for installing and using a new software application for an office network might be directed primarily toward the information technology support staff who would be doing the installing.

But most documents also have a secondary audience, those individuals outside the immediate circle of people who will be needing the information directly. For example, a secondary audience for these instructions might be managers, who will check to see if the instructions comply with company policy, or lawyers, who will make sure the instructions meet legal standards.

"Who else is likely to read it?"

Generally, primary readers are people who will be the main users of the document (often they are also the people who requested the document). Secondary readers are those who will support the project, who will advise any decision makers, or who will be affected by this document in some other way.

Your Relationship to Your Readers

Besides identifying your audience in a general way, you also need to understand your relationship with everyone involved. In your situation, will the readers be superiors, colleagues, or subordinates? Your answer will help you determine the level of formality and authority to use in the document. Are the readers from inside or outside your organization? Answering this question will help you decide how confidential you need to be. Do you know the readers personally? If so, perhaps you can adopt a slightly more informal (but still professional) tone. Are they likely to welcome or to resist your information? Knowing the answer will help you decide how persuasive you need to be. Are they a combination of people from various levels, both inside and outside the company? The answer will help you tailor your document for various readers.

"What is my relationship with this audience?"

Purpose of Your Document

Spell out precisely what you want your document to accomplish and how you expect readers to use it. In other words, determine your purpose. Ask these questions:

- What is the main purpose of the document?
- What other purpose or purposes does the document serve?
- What will readers do with this information?

Questions for deciding on the purpose of your document

Answer these questions by considering the suggestions in the sections that follow.

Primary and Secondary Purposes

Most forms of technical communication fulfill a specific primary purpose. As discussed in Chapter 1, the primary purpose (to inform, to instruct, or to persuade) will affect the document's overall shape and substance.

"What is the main purpose of this document?"

Many documents have one or more secondary purposes. For example, the primary purpose in a typical instruction manual is to instruct, that is, to teach an audience how to assemble or use the product. But for ethical and legal reasons, companies also want people to use the product safely. A user manual for a power tool or a lawnmower, for instance, almost always begins with a page that spells out safety hazards and precautions before instructing readers about how to proceed with the mechanism.

"What other purposes does this document serve?"

Write a clear
audience
and purpose
statement

In planning your document, work from a clear statement that identifies the target audience as well as the document's primary and secondary purposes. For example, "The purpose of my document is *to inform* company employees of the new absentee policy and *to instruct* them on how to follow the procedures properly," or "The purpose of my document is *to inform* my division's programmers about the new antivirus software, as well as *to instruct* them on how to install the software and *to persuade* them of the importance of running weekly virus scans."

Intended Use of the Document

"How will
readers use this
document?"

In addition to determining purposes of a document from your own perspective, also consider how and why it will be used by others. As you plan your document, answer these questions:

Questions for
anticipating how
your document
will be used

- Do my readers simply want to learn facts or understand concepts? Will they use my information in making some type of decision?
- Will people act immediately on the information?
- Do they need step-by-step instructions?
- In my audience's view, what is most important about this document?

Besides answering these questions, try asking members of your audience directly, so you can verify what they want to know.

Assess the Audience's Technical Background

2.2 Assess your audience's technical background

"How much
expertise does
this audience
possess?"

When you write for a close acquaintance (coworker, engineering colleague, chemistry professor who reads your lab reports, or supervisor), you adapt your report to that person's knowledge, interests, and needs. But some audiences are larger and less defined (say, for a journal article, a user manual, a set of first-aid procedures, or an accident report). When you have only a general notion about your audience's background, decide whether your document should be *highly technical, semitechnical,* or *nontechnical,* as depicted in Figure 2.2.

Highly Technical Audience

"Does my
audience
understand
highly technical
information?"

Readers at a specialized level expect to be presented the facts and figures they need—without long explanations. In Figure 2.3, an emergency room physician reports to the patient's doctor, who needs an exact record of symptoms, treatment, and results. In this situation, a highly technical version is both appropriate and important.

For her expert colleague, this physician doesn't need to define the technical terms (*pulmonary edema, sinus rhythm*). Nor does she need to interpret lab findings

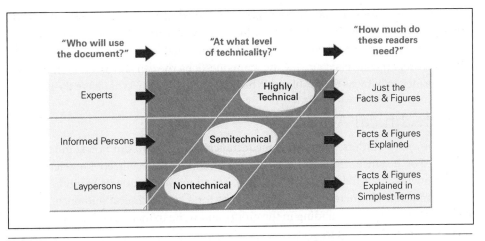

Figure 2.2 **Deciding on a Document's Level of Technicality**

The patient was brought to the ER by ambulance at 1:00 A.M., September 27, 2010. The patient complained of severe chest pains, dyspnea, and vertigo. Auscultation and EKG revealed a massive cardiac infarction and pulmonary edema marked by pronounced cyanosis. Vital signs: blood pressure, 80/40; pulse, 140/min; respiration, 35/min. Lab: wbc, 20,000; elevated serum transaminase; urea nitrogen, 60 mg%. Urinalysis showed 4+ protein and 4+ granular casts/field, indicating acute renal failure secondary to the hypotension.

The patient received 10 mg of morphine stat, subcutaneously, followed by nasal oxygen and 5% D & W intravenously. At 1:25 A.M. the cardiac monitor recorded an irregular sinus rhythm, indicating left ventricular fibrillation. The patient was defibrillated stat and given a 50 mg bolus of Xylocaine intravenously. A Xylocaine drip was started, and sodium bicarbonate administered until a normal heartbeat was established. By 3:00 A.M., the oscilloscope was recording a normal sinus rhythm.

As the heartbeat stabilized and cyanosis diminished, the patient received 5 cc of Heparin intravenously, to be repeated every six hours. By 5:00 A.M. the BUN had fallen to 20 mg% and vital signs had stabilized: blood pressure, 110/60; pulse, 105/min; respiration, 22/min. The patient was now conscious and responsive.

Expert readers need facts and figures, which they can interpret for themselves

Figure 2.3 **A Technical Version of an Emergency Treatment Report** This version is written for medical experts.

(*4+ protein, elevated serum transaminase*). She uses abbreviations that her colleague clearly understands (*wbc, BUN, 5% D & W*). Because her colleague knows all about specific treatments and medications (*defibrillation, Xylocaine drip*), she does not explain their scientific bases. Her report answers concisely the main questions she can anticipate from this particular reader: What was the problem? What was the treatment? What were the results?

Semitechnical Audience

"Does my audience know a little but need further guidance?"

In certain cases, readers will have some technical background but not as much as the experts. For instance, first-year medical students have specialized knowledge, but they know less than the advanced students. Yet all medical students could be considered semitechnical. Therefore, when you write for a semitechnical audience, identify the *lowest* level of understanding in the group, and write to that level. Too much explanation is better than too little.

The partial version of the medical report in Figure 2.4 might appear in a textbook for medical or nursing students, in a report for a medical social worker, or in a monthly report for the hospital administration.

Informed but nonexpert readers need enough explanation to understand what the data mean →

Examination by stethoscope and electrocardiogram revealed a massive failure of the heart muscle along with fluid buildup in the lungs, which produced a cyanotic discoloration of the lips and fingertips from lack of oxygen.

The patient's blood pressure at 80 mm Hg (systolic)/40 mm Hg (diastolic) was dangerously below its normal measure of 130/70. A pulse rate of 140/minute was almost twice the normal rate of 60–80. Respiration at 35/minute was more than twice the normal rate of 12–16.

Laboratory blood tests yielded a white blood cell count of 20,000/cu mm (normal value: 5,000–10,000), indicating a severe inflammatory response by the heart muscle. The elevated serum transaminase enzymes (produced in quantity only when the heart muscle fails) confirmed the earlier diagnosis. A blood urea nitrogen level of 60 mg% (normal value: 12–16 mg%) indicated that the kidneys had ceased to filter out metabolic waste products. The 4+ protein and casts reported from the urinalysis (normal value: 0) revealed that the kidney tubules were degenerating as a result of the lowered blood pressure.

The patient immediately received morphine to ease the chest pain, followed by oxygen to relieve strain on the cardiopulmonary system, and an intravenous solution of dextrose and water to prevent shock.

Figure 2.4 A Semitechnical Version of an Emergency Treatment Report This version is written for readers who are not experts but who have some medical background.

This semitechnical version explains the raw data (highlighted in yellow). Exact dosages are omitted because no one in this audience actually will be treating this patient. Normal values of lab tests and vital signs, however, help readers interpret the report results. (Experts know the normal values.) Knowing what medications the patient received would be especially important in answering this audience's central question: How is a typical heart attack treated?

Nontechnical Audience

People with no specialized training (laypersons) look for the big picture instead of complex details. These readers expect technical data to be translated into words that most people will understand. Laypersons are impatient with abstract theories, but they want enough background to help them make the right decision or take the right action. They are bored or confused by excessive detail but frustrated by raw facts left unexplained or uninterpreted. They expect to understand the document after reading it only once.

"Does my audience have little or no technical background?"

The nontechnical version of the medical report shown in Figure 2.5 might be written for the patient's spouse or other family member, or as part of a script for an online documentary video about emergency-room treatment. Nearly all interpretation (highlighted in yellow), this version mentions no specific medications, lab tests, or normal values. It merely summarizes events and briefly explains what they mean and why these particular treatments were given.

Heart sounds and electrical impulses were both abnormal, indicating a massive heart attack caused by failure of a large part of the heart muscle. The lungs were swollen with fluid and the lips and fingertips showed a bluish discoloration from lack of oxygen.

Blood pressure was dangerously low, creating the risk of shock. Pulse and respiration were almost twice the normal rate, indicating that the heart and lungs were being overworked in keeping oxygenated blood circulating freely.

Blood tests confirmed the heart attack diagnosis and indicated that waste products usually filtered out by the kidneys were building up in the bloodstream. Urine tests showed that the kidneys were failing as a result of the lowered blood pressure.

The patient was given medication to ease the chest pain, oxygen to ease the strain on the heart and lungs, and intravenous solution to prevent the blood vessels from collapsing and causing irreversible shock.

Laypersons need everything translated into terms they understand

Figure 2.5 A Nontechnical Version of an Emergency Treatment Report This version is written for readers who have no medical background.

In a different situation, however (say, a malpractice trial), the layperson jury would require detailed technical information about medication and treatment. Such a report would naturally be much longer—basically a short course in emergency coronary treatment.

Audiences with Varying Technical Backgrounds

The technical background of large and diverse audiences can be variable and hard to pin down. When you must write for audiences at different levels, follow these suggestions:

How to tailor a document to address different technical backgrounds

- If the document is short (a letter, memo, email, or anything less than two pages), rewrite it at different levels for different backgrounds.
- If the document exceeds two pages, address the primary readers. Then provide appendices, glossaries, hyperlinks, or other easily accessible information for secondary readers. Transmittal letters and informative abstracts can also help nonexperts understand a highly technical report. (See Chapter 21 for use and preparation of appendices and other supplements.)

For an illustration of these differences, consider the following case.

Case

Tailoring a Single Document for Multiple Audiences

Different readers have differing information needs

You are a metallurgical engineer in an automotive consulting firm. Your supervisor has asked you to test the fractured rear axle of a 2016 Delphi pickup truck recently involved in a fatal accident. Your assignment is to determine whether the fractured axle *caused* or *resulted from* the accident.

After testing the hardness and chemical composition of the metal and examining microscopic photographs of the fractured surfaces (fractographs), you conclude that the fracture resulted from stress that developed *during* the accident. Now you must report your procedure

"What do these findings mean?"

and your findings to a variety of readers.

Because your report may serve as courtroom evidence, you must explain your findings in meticulous detail. But your primary readers (the decision makers) will be nonspecialists (the attorneys who have requested the report, insurance representatives, possibly a judge and a jury), so you must translate your report, explaining the principles behind the various tests, defining specialized terms such as "chevron marks," "shrinkage cavities," and "dimpled core," and showing the significance of these features as evidence.

"How did you arrive at these conclusions?"

Secondary readers will include your supervisor and outside consulting engineers who will be evaluating your test procedures and assessing the validity of your findings. Consultants will be focusing on various parts of your report, to verify that your procedure has been exact and faultless. For this group, you will have to include appendices spelling out the technical details of your analysis: *how* hardness testing of the axle's case and core indicated that the axle had been properly carburized; *how* chemical analysis ruled out the possibility that the manufacturer had used inferior alloys; *how* light-microscopic fractographs revealed that the origin of the fracture, its direction of propagation, and the point of final rupture indicated a ductile fast fracture, not one caused by torsional fatigue.

In the previous scenario, primary readers need to know *what your findings mean,* whereas secondary readers need to know *how you arrived at your conclusions.* Unless it serves the needs of each group independently, your information will be worthless.

Digital Documents for Multiple Audiences

A great way to address different technical levels of audience is with digital documents, including Web pages, blogs and wikis, PDFs, shared drive documents (such as Google drive), and others. These forms are ideal for providing information for readers from a wide range of backgrounds because you can use hyperlinks, tabs, and other interactive features to direct different audiences to information written and designed to match their interests and backgrounds. The Web site in Figure 2.6 uses tabs and links to accommodate different levels of interest and expertise.

Any document can easily reach across the globe, and digital documents are also useful when readers are from different countries or cultures. The instruction guide you write for a new camera card reader, for example, may well end up on a Web site, where it will be used by customers worldwide. Some international readers may be offended by commands in strongly worded imperative forms, such as "STOP: Do not insert the storage card until you reach Step 3." Or they may be baffled by icons and other visuals that have no meaning in their culture. But with Web-based documents, readers can click on a link to choose from different countries and cultures and will then receive information that is properly translated and localized (adapted) for them.

For more on this topic, see the "Technical Communication Reaches a Global Audience" section in Chapter 1 as well as Chapter 5.

Advantages of digital documents for multiple audiences

Advantages of digital documents for global audiences

Anticipate Your Audience's Preferences

2.3 Identify the appropriate document qualities for your audience

Readers approach any document with certain preferences: its desired length and details, the format and medium in which it should be presented, and the appropriate tone, as well as deadline and budget expectations.

Length and Details

The length and amount of detail in your document depends on what you can learn about your audience's needs. Were you asked to "keep it short" or to "be comprehensive"? Are people more interested in conclusions and recommendations, or do they want everything spelled out?

Give readers only what they need and want

Format and Medium

Does your audience expect a letter, a memo, an email, a short report, or a long, formal report with supplements (title page, table of contents, appendixes, and so on—see Chapter 21, "Front Matter and End Matter Supplements")? Can visuals and page

Decide how your document will look and will be distributed

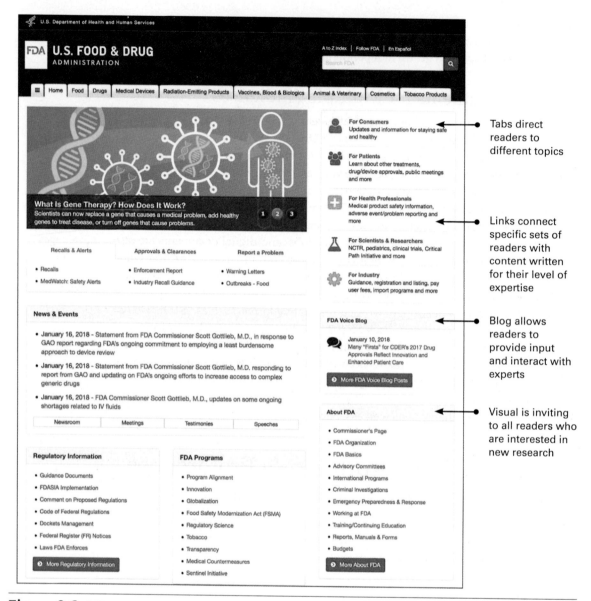

Figure 2.6 A Web Page Designed for Multiple Audiences This page uses links and tabs to provide information for people from a range of backgrounds, including consumers, patients, health professionals, and scientists. A blog allows readers to share their views. The site is available in both English and Spanish.

Source: U.S. Food and Drug Administration.

layout (charts, graphs, drawings, headings, lists) make the material more accessible? Should the document be available as a PDF, as a Web site with links, in hard copy, as a social media post, or in some combination of the above?

Tone

The tone of your writing conveys an image of who you are: your *persona*—the image that comes through between the lines. Tone can range from formal (as in a business letter to a client) to semiformal (as in a memo announcing a change in company dress policy) to informal (as in a quick email to colleagues announcing the upcoming company picnic). Workplace readers expect a tone that reflects both the importance or urgency of the topic and the relationship between writer and reader. For example, the letter to a client that begins with "We are pleased to forward your annual investment statement" is probably appropriate. But a similar tone used in the memo about the company picnic would seem stuffy and pretentious ("I am pleased to announce...").

> Decide on the appropriate tone for your situation

At the same time, the tone of your writing can range from friendly and encouraging to distant and hostile. For example, a bossy tone in a memo to your employees ("It would behoove you to...") would make them feel demeaned and resentful. In short, your tone is effective when you sound like a likable person talking to people in a workplace setting. The notion of *workplace setting* is key here: Always avoid the kind of unprofessional free-for-all tone that is common in tweets, text messages, and emails among casual friends outside of work.

Due Date and Timing

Does your document have a deadline? Workplace documents almost always do. Is there a best time to submit it? Do you need to break down the deadline into a schedule of milestones? Will any of your information become outdated if you wait too long to complete the document?

> Know when to submit the document

Budget

Does your document have a production budget? If so, how much? Where can you save money? How much time can your company afford to allot you for creating the document? How much money can you spend obtaining permission to use materials from other sources? How much can you spend on Web design, page layout, and, if in hard copy, on printing, binding, and distributing your document?

> Calculate the financial costs

> NOTE *Although a detailed analysis can tell you a great deal, rarely is it possible to pin down an audience with certainty, especially if your audience is large and diverse. Before you circulate or submit a final document, ask selected readers for feedback. For task-oriented documents, such as instructions or procedures, you can also conduct a usability test (see Chapter 19).*

Guidelines

for Analyzing Your Audience and Its Use of the Document

➤ **Picture exactly what these readers need and how they expect to use your document.** Whether it's the company president or the person next to you in class, that person has specific concerns and information needs. Your readers may need to complete a task, solve a problem, make a decision, evaluate your performance, or take a stand on an issue. Think carefully about exactly what you want your readers to be able to do.

➤ **Learn all you can about who will use your document.** Are your primary readers superiors, colleagues, or subordinates? Are they inside or outside your organization? Who else might be interested or affected? What do readers already know about this topic? How much do they care? Are they likely to welcome or reject your information?

➤ **In planning your document, work from a clear statement of audience and purpose.** For example, "The purpose of my document is to [describe using verbs: persuade, instruct, inform] the target audience [identify precisely: colleagues, superiors, clients]."

➤ **Consider your audience's technical background.** Colleagues who speak your technical language will understand raw data. Managers who have limited technical knowledge expect interpretations and explanations. Clients with little or no technical background want to know what this information means to them, personally (to their health, pocketbook, safety). However, none of these generalizations might apply to your situation. When in doubt, aim for low technicality.

➤ **When you don't know exactly who will be reading your document, picture the "general reader."** A nontechnical audience will expect complex information to be explained in ways that have meaning for them, personally, and insofar as possible in everyday language. (For example, refer to "heart and lungs" instead of "cardiopulmonary system." Instead of "A diesel engine generates 10 BTUs per gallon of fuel compared with 8 BTUs generated by a conventional gasoline engine," write "A diesel engine yields 25 percent better gas mileage than its gas-burning counterpart.")

➤ **Consider readers' cultural backgrounds.** Identify as closely as possible your audience's specific customs and values. How might cultural differences play a role in readers' interpretation of your presentation?

➤ **Anticipate your audience's reactions.** If the topic is controversial or the news is bad, will some people resist your message? Will some feel threatened or offended? Should you be bold and outspoken or tread lightly? No matter how accurate your information or how sensible your ideas, an alienated audience will reject them out of hand.

➤ **Anticipate your audience's questions.** Based on their needs and concerns, readers have questions such as these: What is the purpose of this document? Why should I read it? What happened, and why? Who was involved? How do I perform this task? How did you perform it? What action should be taken, and why? How much will it cost? What are the risks? Give readers what they need to know, instead of what they already know. Give them enough material to understand your position and to react appropriately.

➤ **Anticipate your audience's preferences.** Try to pinpoint the length, detail, format, medium, tone, timing, and budget preferred by this audience. As the situation allows, adjust your document accordingly.

Develop an Audience and Use Profile

2.4 Develop an audience and use profile to guide your work

In order to focus sharply on your audience, purpose, and the many factors discussed in this chapter, develop your own version of the Audience and Use Profile Sheet shown in Figure 2.7 for any document you prepare. Modify this sheet as needed to suit your own situation, as shown in the following case.

Audience and Purpose

Primary audience: _____ *(name, title)*

Secondary audience(s): _____ *(technicians, managers, other)*

Relationship with audience: _____ *(colleague, employer, other)*

Purpose of document: _____ *(inform, instruct, persuade)*

Audience and purpose statement: _____

_____

_____

Intended use of document: _____ *(perform tasks, solve a problem, other)*

Information needs: _____ *(background, basic facts, other)*

Technical background: _____ *(layperson, expert, other)*

Cultural considerations: _____ *(level of detail or directness, other)*

Probable questions: _____ ?

_____ ?

_____ ?

_____ ?

Probable reaction: _____ *(resistance, approval, anger, other)*

Audience Preferences about the Document

Length and detail: _____ *(comprehensive, conscise, other)*

Format and medium: _____ *(letter, memo, email, other)*

Tone: _____ *(businesslike, confident, informal, other)*

Due date and timing: _____ *(meet deadline, wait for the best time, other)*

Budget: _____ *(what can be spent on what)*

Figure 2.7 Audience and Use Profile Depending on your situation, you can adapt this sheet, as shown in the case earlier in this section. For a completed profile in a persuasive situation, see Chapter 3, Figure 3.5.

Case

Developing an Audience and Use Profile

Assume you face this situation: First-year students increasingly are dropping out of your major because of low grades or stress or inability to keep up with the workload. As part of your work–study duties, your department chairperson asks you to prepare a "Survival Guide" for next year's incoming students to the major. This one- or two-page memo should focus on the challenges and the pitfalls of the major and should include a brief motivational section along with whatever additional information you decide readers need.

Adapt Figure 2.7 to develop your audience and use profile. Here are some possible responses:

Audience and Use Profile

Audience and Purpose

➤ *Who is my primary audience?* Incoming students to the major

➤ *Any secondary audiences?* Department faculty.

➤ *What is my relationship with everyone involved?* Primary audience: student colleagues who don't know me very well; secondary audience: major faculty, who must approve the final document.

➤ *What is the purpose of the document?* This document has multiple purposes: to inform, instruct, and persuade.

➤ *Audience and purpose statement:* The purpose of this document is to explain to incoming students the challenges and pitfalls of year 1 of our major. I will show how the number of dropouts has increased, describe what seems to go wrong and explain why, suggest steps for avoiding common mistakes, and emphasize the benefits of enduring the first year.

➤ *Intended use of this document:* To enable students to craft their own survival plan based on the information, advice, and encouragement provided in the document.

➤ *Information needs:* Incoming students know very little about this topic. They need everything spelled out.

➤ *Technical background:* In regard to this topic, the primary audience can be considered laypersons.

➤ *Cultural considerations:* The document will refer readers from other countries and cultures (exchange students, nonnative speakers of English, and so on) to designated advisors for additional assistance.

➤ *Probable questions (along with others you anticipate):* "How big is the problem?" "How can this problem affect me personally?" "How much time will I need to devote to homework?" "How should I budget my time?" "Can I squeeze in a part-time job?" "Why do so many students drop out?" "Whom should I see if I'm having a problem?"

➤ *Probable reaction to document:* Most readers should welcome this information and take it seriously. However, some students who don't know the meaning of failure might feel patronized or offended. Some faculty might resent any suggestions that courses are too demanding.

Audience Preferences about the Document

➤ *Length and detail:* Because the document was requested by the department and not by the primary audience, I can't expect students to tolerate more than a page or two.

> ➤ *Format and medium:* Social media post (to the student Facebook page and Twitter) that provides a link to a PDF attachment. The social media post will be brief but very welcoming, encouraging students to click on the link and read the document. For ease of access, the document will also be available on the department's Web site.
>
> ➤ *Tone:* Since we students are all in this situation together, a friendly, informal, and positive (to avoid panic) but serious tone seems best.
>
> ➤ *Due date and timing:* This document must be available before students arrive next fall—but not so early that it gets forgotten or overshadowed by other registration paperwork.
>
> ➤ *Budget:* No printing costs are involved, but I will need someone to proofread my document and check that the PDF conversion looks good.

Checklist

Analyzing Audience and Purpose

Use the following Checklist as you analyze a document's audience and purpose:

- ❑ Have I identified primary and secondary audiences? (See "Primary and Secondary Audiences" in this chapter.)
- ❑ Have I identified my relationship to these readers? (See "Your Relationship to Your Readers" in this chapter.)
- ❑ Have I identified primary and secondary purposes? (See "Primary and Secondary Purposes" in this chapter.)
- ❑ Have I assessed how readers will use the document? (See "Intended Use of the Document" in this chapter.)
- ❑ Have I researched the technical background of my audience? (See "Assess the Audience's Technical Background" in this chapter.)
- ❑ Have I considered the cultural backgrounds of my audience? (See "Digital Documents for Multiple Audiences" in this chapter.)
- ❑ Have I considered other audience preferences, including media format? (See "Anticipate Your Audience's Preferences" in this chapter.)
- ❑ Have I developed an audience and use profile? (See "Develop an Audience and Use Profile" in this chapter.)

Projects

For all projects, check with your instructor about whether to present your findings in class, bring drafts to class for discussion, upload your project to the class learning management system (LMS), and/or use the LMS forum or discussion boards to collaborate and review each activity below.

General

1. Find a short article from your field (or part of a long article or a selection from your text book for an advanced course). Choose a piece written at the highest level of technicality you can understand, and then rewrite a version of that piece suitable for a layperson, as in the example in Figure 2.5. Exchange your document with a classmate from a different major, and see if he or she can understand your version. Read that student's version, and use a document comment system (such as track changes in Word) to provide comments evaluating the

level of technicality. Submit to your instructor a copy of the original, your new version, and your evaluation of your neighbor's translation.

2. Assume that a new employee is taking over your job because you have been promoted. Identify a specific problem in the job that could cause difficulty for the new employee. Assume that you will need to write instructions for the employee to help him or her avoid or cope with the problem. Create an audience and use profile based on Figure 2.7. Use the Case in the "Develop an Audience and Use Profile" section in this chapter as a model for your responses.

Team

For this project, you will work on creating a document to help students in your major understand the job market. Form teams of 3 to 4 people who are in the same or closely related majors (electrical engineering, biology, graphic design). As a team, perform an audience analysis of your potential readers, using the Audience and Use profile in Figure 2.7. Next, have each person individually research the job market for graduates in this major, including specific types of skills that employers seek beyond the required courses. Start a shared document (Google Drive or another app that is available through your school or your class learning management system), and contribute your ideas to this document. Appoint a team member to present your completed Audience and Purpose profile along with your key findings. Ask classmates to let you know if your findings match the needs of your audience.

Digital and Social Media

Locate a Facebook page or Web site that accommodates various readers at different levels of technicality. Sites for state and federal government agencies that deal with the space, environment, public health, or food safety are good sources of both general and specialized information. Search on "government agencies environment" or the like; add in the name of your state if you want to be more specific. Examine one of these sites and find an example of (a) material aimed at a general audience and (b) material on the same topic aimed at a specialized or expert audience. First, list the specific features that enabled you to identify each piece's level of technicality. Next, using the Audience and Use Profile Sheet (see "Develop an Audience and Use Profile" in this chapter), record the assumptions about the audience made by the author of the nontechnical version. Finally, evaluate how well that piece addresses a nontechnical reader's information needs. Be prepared to discuss your evaluation in class or to post to the forum of your class learning management system.

Global

The World Health Organization's Web site is designed for a global audience. Visit the site, then answer these questions:

• What topics are covered on the main page? Do these topics appeal to readers from different parts of the world?

• How many different languages is the site available in? Does the site make it easy or difficult to change languages?

Present your findings in class.

Chapter 3
Persuading Your Audience

Monkey Business Images/Shutterstock

❝For me, persuasion is mostly about getting along with coworkers. What I didn't understand when I started working after college is the whole idea of organizational behavior. I assumed that, as long as I worked hard and did a good job, I would get my raises and promotions. But it's not that simple. I had to leave my first job because I didn't learn soon enough how people react in certain situations, for instance, that 'constructive' suggestions from the new person are not always appreciated.

My advice: Learn about the people you're working for and with. Spend a lot of time observing, listening, and asking questions about how the organization works at the person-to-person level.❞

—*Ryan Donavan, Software Developer*

Learning Objectives

3.1 Define persuasion and claims

3.2 Determine your persuasive approach

3.3 Anticipate how your audience may react

3.4 Choose a strategy to connect with your audience

3.5 Respect various limitations when making an argument

3.6 Support your argument using evidence and reason

3.7 Understand how social media influences persuasion

3.8 Prepare a convincing argument

What Is Persuasion?

3.1 Define persuasion and claims

Definition of persuasion

Persuasion means trying to influence someone's actions, opinions, or decisions (Figure 3.1). In the workplace, we rely on persuasion daily: to win coworker support, to attract clients and customers, to request funding. But changing someone's mind is never easy, and sometimes impossible. Your success will depend on what you are requesting, whom you are trying to persuade, and how entrenched those people are in their own views.

Implicit versus explicit persuasion

Almost all workplace documents, to some extent, have an *implicitly* persuasive goal: namely, to assure readers the information is accurate; the facts are correct; and the writer is fluent, competent, and knowledgeable. But the types of documents featured in this chapter have an *explicitly* persuasive goal: namely, to win readers over to a particular point of view about an issue that is in some way controversial.

Explicit persuasion is required whenever you tackle an issue about which people disagree. Assume, for example, that you are Manager of Employee Relations at

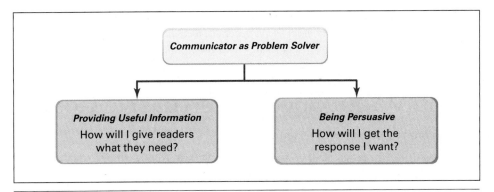

Figure 3.1 Informing and Persuading Require Audience Awareness

Softbyte, a software developer whose recent sales have plunged. To avoid layoffs, the company is trying to persuade employees to accept a temporary cut in salary.

As you plan various memos and presentations on this volatile issue, you must first identify your major *claims*. (A claim is a statement of the point you are trying to prove.) For instance, in the Softbyte situation, you might first want employees to recognize and acknowledge facts they've ignored:

> Because of the global recession, our software sales in two recent quarters have fallen nearly 30 percent; earnings should remain flat all year.

Even when a fact is obvious, people often disagree about what it means or what should be done about it. And so you might want to influence their interpretation of the facts:

> Reduced earnings mean temporary layoffs for roughly 25 percent of our staff. But we could avoid layoffs entirely if each of us at Softbyte would accept a 10 percent salary cut until the market improves.

And eventually you might want to ask for direct action:

> Our labor contract stipulates that such an across-the-board salary cut would require a two-thirds majority vote. Once you've had time to examine the facts, we hope you'll vote "yes" on next Tuesday's secret ballot.

As you present your case, you will offer support for your claims before you finally ask readers to take the action you favor. Whenever people disagree about what the facts are or what the facts mean or what should be done, you need to make the best case for your own view.

On the job, your emails, memos, reports, and proposals advance claims like these (Gilsdorf, "Executives' and Academics' Perception" 59–62):

| We can't possibly meet this production deadline without sacrificing quality.
| We're doing all we can to correct your software problem.
| Our equipment is exactly what you need.
| I deserve a raise.

Such claims, of course, are likely to be rejected—unless they are backed up by a convincing argument.

Margin notes:
Definition of claims

A claim about what the facts are

A claim about what the facts mean

A claim about what should be done

Claims require support

NOTE "Argument," in this context, means "a process of careful reasoning in support of a particular claim"—it does not mean "a quarrel or dispute." People who "argue skillfully" are able to connect with others in a rational, sensible way, without causing animosity. But people who are merely "argumentative," on the other hand, simply make others defensive.

Identify Your Specific Persuasive Goal

3.2 Determine your persuasive approach

What do you want people to be doing or thinking? Arguments differ considerably in the level of involvement they ask from people.

Types of persuasive goals

- **Arguing to influence people's opinions.** Some arguments ask for minimal audience involvement. Maybe you want people to agree that the benefits of bioengineered foods outweigh the risks or that your company's monitoring of employee email is hurting morale. The goal here is merely to move readers to change their thinking, to say "I agree."

- **Arguing to enlist people's support.** Some arguments ask people to take a definite stand. Maybe you want readers to support a referendum that would restrict cloning experiments or to lobby for a daycare center where you work. The goal is to get people actively involved, to get them to ask "How can I help?"

- **Submitting a proposal.** Proposals offer plans for solving problems. The proposals we examine in Chapter 22 typically ask audiences to take—or approve—some form of direct action (say, a plan for improving your firm's computer security or a Web-based orientation program for new employees). Your proposal goal is achieved when people say "Okay, let's do this project."

- **Arguing to change people's behavior.** Getting people to change their behavior is a huge challenge. Maybe you want a coworker to stop dominating your staff meetings or to be more open about sharing information that you need to do your job. People naturally take such arguments personally. And the more personal the issue, the greater people's resistance. After all, you're trying to get them to admit, "I was wrong. From now on, I'll do it differently."

The above goals can and often do overlap, depending on the situation. But never launch an argument without a clear view of exactly what you want to see happen.

Try to Predict Audience Reaction

3.3 Anticipate how your audience may react

Audience questions about your attempts to persuade

Any document can evoke different reactions depending on a reader's temperament, interests, fears, biases, ambitions, or assumptions. Whenever peoples' views are challenged, they react with defensive questions such as these:

- Says who?
- So what?

- Why should I?
- What's in this for me?
- What will it cost?
- What are the risks?
- What are you up to?
- What's in it for you?
- Will it mean more work for me?
- Will it make me look bad?

People read between the lines. Some might be impressed and pleased by your suggestions for increasing productivity or cutting expenses; some might feel offended or threatened. Some might suspect you of trying to undermine your boss, while others might question your logic or reasoning.

Reactions to persuasive messages can differ greatly

As one researcher puts it, "Delivering bad news is tough" (A. Gallo 2). No one wants bad news; some people prefer to ignore it. To anticipate any potential reactions, you will need to find out as much as possible about your audience and the organization you work for. Will your supervisor or manager back you up? Does the organization have an employee assistance program to help people get the support they might need? For more on conveying bad news messages, see Chapter 15, "Memo Tone" and "Conveying Bad or Unwelcome News."

Expect Audience Resistance

People who haven't made up their minds about what to do or think are more likely to be receptive to persuasive influence.

> We need others' arguments and evidence. We're busy. We can't and don't want to discover and reason out everything for ourselves. We look for help, for short cuts, in making up our minds. (Gilsdorf, "Write Me" 12)

People rely on persuasion to make up their minds

People who *have* decided what to think, however, naturally assume they're right, and they often refuse to budge even when presented with solid evidence to the contrary. In fact, once people have made up their minds, you can expect resistance, due in part to something known as *confirmation bias*—"where people selectively seek evidence that is consistent with their prior beliefs and expectations" (Hernandez and Preston). Confirmation bias is especially strong when people get information from online sources, because such information tends to confirm what people already know and want to know (see "Digital Persuasion" later in this chapter). Getting people to admit you might be right means getting them to admit they might be wrong. The more strongly they identify with their position, the more resistance you can expect.

When people do yield to persuasion, they may respond grudgingly, willingly, or enthusiastically (Figure 3.2). A classic work on this idea, still widely

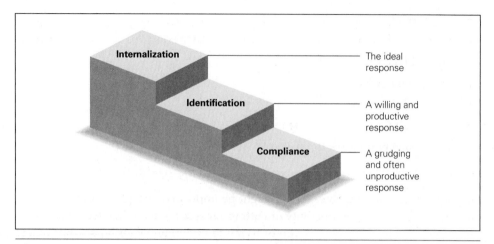

Figure 3.2 **The Levels of Response to Persuasion**

cited, categorized these responses as compliance, identification, or internalization (Kelman 51–60):

- **Compliance:** "I'm yielding to your demand in order to get a reward or to avoid punishment. I really don't accept it, but I feel pressured, and so I'll go along to get along."

- **Identification:** "I'm going along with your appeal because I like and believe you, I want you to like me, and I feel we have something in common."

- **Internalization:** "I'm yielding because what you're suggesting makes good sense and it fits my goals and values."

Although achieving compliance is sometimes necessary (as in military orders or workplace safety regulations), nobody likes to be coerced. If readers merely comply because they feel they have no choice, you probably have lost their loyalty and goodwill—and as soon as the threat or reward disappears, you will also lose their compliance.

Know How to Connect with the Audience

3.4 Choose a strategy to connect with your audience

Persuasive people know when to simply declare what they want, when to reach out and create a relationship, when to appeal to reason and common sense, or when to employ some combination of these strategies (Kipnis and Schmidt 40–46). These three strategies, respectively, can be labeled the *power connection*, the *relationship connection*, and the *rational connection* (Figure 3.3). To get a better understanding of these three different strategies, picture the scenario in the case study below.

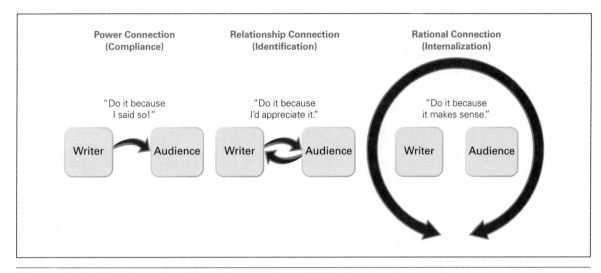

Figure 3.3 **Three Strategies for Connecting with an Audience** Instead of intimidating your audience, try to appeal to the relationship or—better yet—appeal to people's intelligence.

Case

Connecting with the Audience

Your Company, XYZ Engineering, has just developed a fitness program, based on findings that healthy employees work better, take fewer sick days, and cost less to insure. This program offers clinics for smoking, stress reduction, and weight loss, along with group exercise. In your second month on the job, you receive the following email:

To: Employees at XYZ.com
From: GMaximus@XYZ.com
Date: June 6, 20XX
Subject: Physical Fitness

On Monday, June 10, all employees will report to the company gymnasium at 8:00 A.M. for the purpose of choosing a walking or jogging group. Each group will meet for 30 minutes three times weekly during lunch time.

Power connection: Orders readers to show up

How would you react to the previous notice? Despite the reference to "choosing," the recipients of the memo are given no real choice. They are simply ordered to show up at the gym. Typically used by bosses and other authority figures, this type of *power connection* does get people to comply, but it almost always alienates them, too!

Suppose, instead, you receive this next version of the email. How would you react in this instance?

To: Employees at XYZ.com
From: GMaximus@XYZ.com
Date: June 6, 20XX
Subject: An Invitation to Physical Fitness

Relationship connection:
Invites readers to participate

Leaves choice to readers

> I realize most of you spend lunch hour playing cards, reading, or just enjoying a bit of well-earned relaxation in the middle of a hectic day. But I'd like to invite you to join our lunchtime walking/jogging club.
>
> We're starting this club in hopes that it will be a great way for us all to feel more healthy. Why not give it a try?

This second version conveys the sense that "we're all in this together." Instead of being commanded, readers are invited to participate. Someone who seems likable and considerate is offering readers a real choice.

Often the biggest variable in a persuasive message is the reader's perception of the writer. Readers are more open to people they like and trust. The *relationship connection* often works for this reason, and it is especially vital in cross-cultural communication—as long as it does not sound too "chummy" and informal to carry any real authority. (For more on tone, see Chapter 11, "Adjusting Your Tone.") In short, people need to find the claim believable ("Exercise will help me feel more healthy") and relevant ("I personally need this kind of exercise"); the claim is enhanced if it comes from a trusted source.

Here is a third version of the email. As you read, think about the ways in which its approach differs from those of the first two examples:

> To: Employees at XYZ.com
> From: GMaximus@XYZ.com
> Date: June 6, 20XX
> Subject: Invitation to Join One of Our Jogging or Walking Groups

Rational connection:
Presents authoritative evidence

> I want to share a recent study from the *New England Journal of Medicine,* which reports that adults who walk two miles a day could increase their life expectancy by three years.
>
> Other research shows that 30 minutes of moderate aerobic exercise, at least three times weekly, has a significant and long-term effect in reducing stress, lowering blood pressure, and improving job performance.

Offers alternatives

> As a first step in our exercise program, XYZ Engineering is offering a variety of daily jogging groups: The One-Milers, Three-Milers, and Five-Milers. All groups will meet at designated times on our brand new, quarter-mile, rubberized clay track.

Offers a compromise

> For beginners or skeptics, we're offering daily two-mile walking groups. For the truly resistant, we offer the option of a Monday–Wednesday–Friday two-mile walk.
>
> Coffee and lunch breaks can be rearranged to accommodate whichever group you select.

Leaves choice to readers

Offers incentives

> Why not take advantage of our hot new track? As small incentives, XYZ will reimburse anyone who signs up as much as $100 for running or walking shoes and will even throw in an extra fifteen minutes for lunch breaks. And with a consistent turnout of 90 percent or better, our company insurer may be able to eliminate everyone's $200 yearly deductible in medical costs.

This version conveys respect for the reader's intelligence and for the relationship. With any reasonable audience, the rational connection stands the best chance of success.

NOTE Keep in mind that no cookbook formula exists, and in many situations even the best persuasive attempts may be rejected.

In addition to these three strategies, you can connect with your audience by allowing for give and take, being specific about the response you wish to receive, and being reasonable in your request.

Allow for Give-and-Take

Reasonable people expect a balanced argument, with both sides of the issue considered evenly and fairly. Persuasion requires flexibility on your part. Instead of merely pushing your own case forward, consider other viewpoints. In advocating your position, for example, you need to do these things (Senge 8):

- explain the reasoning and evidence behind your stance
- invite people to find weak spots in your case, and to improve on it
- invite people to challenge your ideas (say, with alternative reasoning or data)

How to promote your view

When others offer an opposing view, you need to do these things:

- try to see the issue their way, instead of insisting on your way
- rephrase an opposing position in your own words, to be sure you understand it accurately
- try reaching agreement on what to do next, to resolve any insurmountable differences
- explore possible compromises others might accept

How to respond to opposing views

Perhaps some XYZ employees (see the previous case), for example, have better ideas for making the exercise program work for everyone.

Ask for a Specific Response

Unless you are giving an order, diplomacy is essential in persuasion. But don't be afraid to ask for what you want:

> The moment of decision is made easier for people when we show them what the desired action is, rather than leaving it up to them.... No one likes to make decisions: there is always a risk involved. But if the writer asks for the action, and makes it look easy and urgent, the decision itself looks less risky. (Cross 3)

Spell out what you want

Let people know what you want them to do or think.

> NOTE *Overly direct communication can offend audiences from some other cultures. Don't mistake bluntness for clarity.*

Never Ask for Too Much

People never accept anything they consider unreasonable. And the definition of "reasonable" varies with the individual. Employees at XYZ Engineering (see the case study in "Know How to Connect with the Audience" in this chapter), for example, differ as

Stick with what is achievable

to which walking/jogging option they might accept. To the runner writing the memo, a daily five-mile jog might seem perfectly reasonable, but to most people this would seem outrageous. XYZ's program, therefore, has to offer something most of its audience (except, say, couch potatoes and those in poor health) would accept as reasonable.

Recognize All Constraints

3.5 Respect various limitations when making an argument

Constraints are limits or restrictions imposed by the situation:

Communication constraints in persuasive situations

- What can I say around here, to whom, and how?
- Should I say it in person, by phone, in print, online?
- Could I be creating any ethical or legal problems?
- Is this the best time to say it?
- What is my relationship with the audience?
- Who are the personalities involved?
- Is there any peer pressure to overcome?
- How big an issue is this?

Organizational Constraints

Constraints based on company rules

Organizations announce their own official constraints: deadlines; budgets; guidelines for organizing, formatting, and distributing documents; personnel policies and procedures; and so forth. There are also unstated rules, such as what person(s) to speak with if you have a complaint or suggestion. These rules, stated and unstated, vary among organizations, but anyone who ignores those rules (say, by going over your supervisor's head) invites disaster.

Airing even a legitimate gripe in the wrong way through the wrong medium to the wrong person can be fatal to your work relationships. The following email, for instance, is likely to be interpreted by the executive officer as petty and whining behavior, and by the maintenance director as a public attack:

To: CEO@XYZ.com

Cc: MaintenanceDirector@XYZ.com

From: Middle Manager@XYZ.com

Date: May 13, 20XX

Re: Trash Problem

Wrong way to the wrong person

Please ask the Maintenance Director to get his people to do their job for a change. I realize we're all understaffed, but I've gotten dozens of complaints this week about the filthy restrooms and overflowing wastebaskets in my department. If he wants us to empty our own wastebaskets, why doesn't he let us know?

Instead, why not address the message directly to the key person—or better yet, pick up the phone and have a conversation (or suggest a meeting)? The following email invites discussion and collaboration:

> To: MaintenanceDirector@XYZ.com
>
> From: MiddleManager@XYZ.com
>
> Date: May 13, 20XX
>
> Re: Staffing Shortage
>
> I wonder if we could meet to exchange some ideas about how our departments might be able to help one another during these staff shortages.

A better way to the right person

Can you identify the unspoken rules in companies where you have worked? What happens when such rules are ignored?

Legal Constraints

In any attempts at persuasion, keep in mind that what you are allowed to say may be limited by contract or by laws protecting confidentiality and customer rights and/or laws on product liability.

Constraints based on the law

- In many states, collection agencies are not allowed to use misleading or deceptive statements in any verbal or written communication.

Major legal constraints on communication

- States have differing laws about what you can and can't say about an employee during a reference check without permission from the employee. Companies may also have their own internal policies about such matters.

- When writing and designing sales literature or manuals, training videos, instructional podcasts, or the like, you and your company are typically liable for faulty information that leads to injury or damage.

Whenever you prepare a document, be aware of possible legal problems. For instance, suppose an employee of XYZ Engineering (see the case study in "Know How to Connect with the Audience" in this chapter) is injured or dies during the new exercise program you've marketed so persuasively. Could you and your company be liable? Should you require physical exams and stress tests (at company expense) for participants? When in doubt, always consult an attorney.

Public relations and legal liabilities

Ethical Constraints

While legal constraints are defined by federal and state law, ethical constraints are defined by honesty and fair play. For example, it may be perfectly legal to promote a new pesticide by emphasizing its effectiveness while downplaying its carcinogenic effects; whether such action is *ethical*, however, is another issue entirely. To earn people's trust, you will find that "saying the right thing" involves more than legal considerations. (See Chapter 4 for more on ethics.)

Constraints based on honesty and fair play

> NOTE Persuasive skills carry tremendous potential for abuse. "Presenting your best case" does not mean deceiving others—even if the dishonest answer is the one people want to hear.

Time Constraints

Constraints
based on the
right timing

Persuasion often depends on good timing. Should you wait for an opening, release the message immediately, or what? Let's assume you're trying to "bring out the vote" among members of your professional society on some hotly debated issue, say, whether to refuse to work on any project related to biological warfare. You might prefer to wait until you have all the information you need or until you've analyzed the situation and planned a strategy. But if you delay, rumors or paranoia could cause people to harden their positions—and their resistance.

Social and Psychological Constraints

Constraints
based on
audience

"What is our
relationship?"

Too often, what we say can be misunderstood or misinterpreted because of constraints such as these:

- **Relationship with the audience:** Is your reader a superior, a subordinate, a peer? (Try not to dictate to subordinates nor to shield superiors from bad news.) How well do you know each other? Can you joke around, or should you be serious? Do you get along, or do you have a history of conflict or mistrust? What you say and how you say it—and how it is interpreted—will be influenced by the relationship.

"How receptive is
this audience?"

- **Individual and group personality:** Some people are just naturally more open to hearing many points of view and potentially changing their stance. Other people are more stubborn. And when individuals are considering different perspectives as part of a larger group of people, group dynamics (peer pressure) come into play. Does the group have a strong sense of identity (union members, conservationists, engineers)? Will group loyalty or pressure to conform prevent certain appeals from working? Address the group's collective concerns. (See Chapter 4, "Mistaking Groupthink for Teamwork" for more on the idea of "groupthink.")

"How entrenched
is this audience
in their exisiting
views?"

- **Confirmation bias:** As discussed earlier in this chapter, once people have a strong and set opinion, they often become more fixed in their views even when presented with extremely strong evidence and facts that points to a different conclusion. On social media and other online spaces, this phenomenon is quite strong due to the way most social media platforms cater to our individual preferences, often presenting us only with information that confirms what we already believe. To counter this phenomenon, have conversations in person, on video, or by phone, preferably with smaller groups of people. Also, structure your documents so that people read more slowly (Hernandez and Preston). People read print copies more slowly than digital, for example.

"Where are most
people coming
from on this
issue?"

- **Perceived size and urgency of the issue:** Does the audience see this as a cause for fear or for hope? Is trouble looming or has a great opportunity emerged? Has the issue been understated or overstated? Big problems often cause people to exaggerate their fears, loyalties, and resistance to change—or to seek quick solutions. Assess the problem realistically. Don't downplay a serious problem, but don't cause panic, either.

Consider This

People Often React Emotionally to Persuasive Appeals

We've all been on the receiving end of attempts to influence our thinking:

➤ *You need this product!*
➤ *This candidate is the one to vote for!*
➤ *Try doing things this way!*

How do we decide which appeals to accept or reject? One way is by evaluating the argument itself, by asking *Does it make good sense? Is it balanced and fair?* But arguments rarely succeed or fail merely on their own merits. Emotions play a major role. In fact, our desire to separate emotions from logic does not fully reflect the way the human brain works or the nature of how people make decisions. Emotions and logic work in tandem, but emotions, which are based to a large extent in the "flight or fight" parts of our brain, often rise to the surface, especially in difficult or personal decisions.

Why We Say No

Management expert Edgar Schein outlines various fears that prevent people from trying or learning something new (34–39):

➤ **Fear of the unknown:** *Why rock the boat?* (Change can be scary, and so we cling to old, familiar ways of doing things, even when those ways aren't working.)

➤ **Fear of disruption:** *Who needs these headaches?* (We resist change if it seems too complicated or troublesome.)

➤ **Fear of failure:** *Suppose I screw up?* (We worry about the shame or punishment that might result from making errors.)

To overcome these basic fears, Schein explains, people need to feel "psychologically safe":

> They have to see a manageable path forward, a direction that will not be catastrophic. They have to feel that a change will not jeopardize their current sense of identity and wholeness. They must feel that... they can... try out new things without fear of punishment. (59)

Why We Say Yes

Social psychologist Robert Cialdini pinpoints six subjective criteria that move people to accept a persuasive appeal (76–81):

➤ **Reciprocation:** *Do I owe this person a favor?* (We feel obligated—and we look for the chance—to reciprocate, or return, a good deed.)

➤ **Consistency:** *Have I made an earlier commitment along these lines?* (We like to perceive ourselves as behaving consistently. People who have declared even minor support for a particular position [say by signing a petition] will tend to accept requests for major support of that position [say, a financial contribution].)

➤ **Social validation:** *Are other people agreeing or disagreeing?* (We often feel reassured by going along with our peers.)

➤ **Liking:** *Do I like the person making the argument?* (We are far more receptive to people we like—and often more willing to accept a bad argument from a likable person than a good one from an unlikable person!)

> ➤ **Authority:** *How knowledgeable does this person seem about the issue?* (We place confidence in experts and authorities.)
>
> ➤ **Scarcity:** *Does this person know (or have) something that others don't?* (The scarcer something seems, the more we value it [say, a hot tip about the stock market].)
>
> A typical sales pitch, for example, might include a "free sample of our most popular brand, which is nearly sold out" offered by a chummy salesperson full of "expert" details about the item itself.

Support Your Claims Convincingly

3.6 Support your argument using evidence and reason

Persuasive claims are backed up by reasons that have meaning for the audience

The most persuasive argument will be the one that presents the strongest case—from the audience's perspective:

> When we seek a project extension, argue for a raise, interview for a job … we are involved in acts that require good reasons. Good reasons allow our audience and ourselves to find a shared basis for cooperating…. [Y]ou can use marvelous language, tell great stories, provide exciting metaphors, speak in enthralling tones, and even use your reputation to advantage, but what it comes down to is that you must speak to your audience with reasons they understand. (Hauser 71)

Imagine yourself in the following situation: As documentation manager for Bemis Lawn and Garden Equipment, a rapidly growing company, you supervise preparation and production of all user manuals. The present system for producing manuals is inefficient because three respective departments are involved in (1) providing the technical content, (2) taking the technical content and writing it in language suitable to the consumer, and (3) designing and publishing the final document in various formats (print, PDF, and Web). Much time and energy are wasted as Microsoft Word files go back and forth among the engineering and product testing group, the technical writing group, and the publications group. After studying the problem and calling in a consultant, you decide that the process could be far more efficient and the final user manuals more accurate if Bemis was using a collaborative writing and tracking tool, such as content management software. This software would reduce problems with trying to keep track of various email attachments and versions—a process that currently introduces errors into the material. To sell this plan to supervisors and coworkers you will need good reasons, in the form of *evidence* and *appeals to readers' needs and values* (Rottenberg 104–06).

Offer Convincing Evidence

Criteria for worthwhile evidence

Evidence (factual support from an outside source) is a powerful persuader—as long as it measures up to readers' standards. Discerning readers evaluate evidence by using these criteria (Perloff 157–58):

- **The evidence has quality.** Instead of sheer quantity, people expect evidence that is strong, specific, new, different, and verifiable (provable).

- **The sources are credible.** People want to know where the evidence comes from, how it was collected, and who collected it.
- **The evidence is considered reasonable.** When the evidence is of high quality and the sources are credible, most people will consider the evidence to be reasonable.

Common types of evidence include factual statements, statistics, data from credible sources, examples that are representative of many situations, and expert testimony.

FACTUAL STATEMENTS.   A *fact* is something that can be demonstrated by observation, experience, research, or measurement—and that your audience is willing to recognize:

> Most of our competitors already have content management systems in place.

Offer the facts

Be selective. Decide which facts best support your case. In the above example, you might also provide a list of the competitors and the content management systems they are using.

STATISTICS.   Numbers can be highly convincing. Many readers focus on the "bottom line": costs, savings, losses, profits:

> After a cost/benefit analysis, our accounting office estimates that an integrated content management system will save Bemis 30 percent in production costs and 25 percent in production time—savings that will enable the system to pay for itself within one year.

Cite the numbers

But numbers can mislead. Your statistics must be accurate, trustworthy, and easy to understand and verify (see Chapter 8, "Faulty Statistical Analysis"). Always cite your source and make sure the source is considered credible by the majority of people in your field.

EXAMPLES.   Examples help people visualize and remember the point. For example, the best way to explain what you mean by "inefficiency" in your company is to show "inefficiency" occurring:

> The figure illustrates the inefficiency of Bemis's present system for producing manuals:

Show what you mean

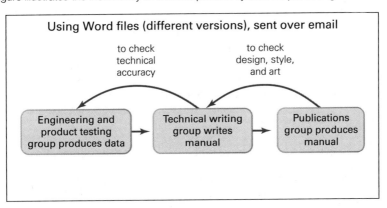

> A manual typically goes back and forth through this cycle numerous times, wasting time in all three groups. In addition, the use of Microsoft Word files (email attachments) causes confusion about who has the latest version.

Always explain how each example fits the point it is designed to illustrate.

EXPERT TESTIMONY.   Expert opinion—if it is unbiased and if people recognize the expert—lends authority and credibility to any claim. Note, however, that in difficult cases, each side will find their own "experts." Experts are considered as such when they are recognized by the highest level of professional society, when they have degrees from legitimate colleges and universities, and when their work has been subject to peer review:

Cite the experts

> Ron Catabia, nationally recognized expert in document control systems, has studied our needs and strongly recommends we move ahead with a content management system.

> NOTE *Finding evidence to support a claim often requires that we go beyond our own experience by doing some type of research. (See Part 2, "The Research Process.")*

Appeal to Common Goals and Values

"What makes these people tick?"

Evidence alone may not be enough to change a person's mind. At Bemis, for example, the bottom line might be very persuasive for company executives, but managers and employees will be asking: Does this threaten my authority? Will I have to work harder? Will I fall behind? Is my job in danger? These readers will have to perceive some benefit beyond company profit.

Appeal to shared goals

If you hope to create any kind of consensus, you have to identify at least one goal you and your audience have in common: "What do we all want most?" Bemis employees, like most people, share these goals: job security and control over their jobs and destinies. Any persuasive recommendation will have to take these goals into account:

Appeal to shared goals

> I'd like to show how content management skills, instead of threatening anyone's job, would only increase career mobility for all of us.

People's goals are shaped by their values (qualities they believe in, ideals they stand for): friendship, loyalty, honesty, equality, fairness, and so on (Rokeach 57–58).

At Bemis, you might appeal to the commitment to quality and achievement shared by the company and individual employees:

Appeal to shared values

> None of us needs reminding of the fierce competition in our industry. The improved collaboration among departments will result in better manuals, keeping us on the front line of quality and achievement.

Give your audience reasons that have real meaning for *them* personally. For example, in a recent study of teenage attitudes about the hazards of smoking, respondents listed these reasons for not smoking: bad breath, difficulty concentrating, loss of friends, and trouble with adults. No respondents listed dying of cancer—presumably because this last reason carries little meaning for young people personally (Baumann et al. 510–30).

> NOTE *We are often tempted to emphasize anything that advances our case and to ignore anything that impedes it. But any message that prevents readers from making their best decision is unethical, as discussed in Chapter 4.*

Consider the Cultural Context

Chapters 1 and 2 discussed the relationship of culture to technical communication, describing how technical communication reaches audiences from different countries and cultures. In terms of persuasion, you need to consider that appeals to common goals and values are much easier when all parties share those goals or values. But cultures may differ in their willingness to debate, criticize, or express disagreement or emotion. There are also differences in tone and style (direct or indirect), the importance of relationship building, the value of evidence and logic, and other areas that are key to a successful negotiation. International business professor Erin Meyer reminds us that when it comes to persuasion,

How cultural differences govern a persuasive situation

> "[t]he many theories about negotiation may work perfectly when you're doing a deal with a company in your own country. But in today's globalized economy you could be negotiating a joint venture in China, an outsourcing agreement in India, or a supplier contract in Sweden. If so, you might find yourself working with very different norms of communication. What gets you to 'yes' in one culture gets you to 'no' in another." (E. Meyer 1)

She recommends several ways to be attentive to your audience, including an important suggestion that you "adapt the way you express disagreement." She quotes from several international business people who explain the vast differences between cultures in how one expresses disagreement: up front and direct ("I disagree") versus more indirect and polite ("I do not quite understand your point") (E. Meyer 1–2). In addition, in all situations, people want to feel respected and not be embarrassed or made to look foolish, especially in front of colleagues and superiors. This idea, known as "saving face" (Oetzel et al.) is important across all countries and cultures but according to research is particularly important among Asian cultures (Brill 1; Oetzel et al. 237–238).

How to navigate cultural differences in persuasive situations

Take time to know your audience, to appreciate their frame of reference, and to establish common ground. Get to know people and listen carefully. Use the Internet to search for credible information on the countries and cultures you will be working with. (See Chapters 1, 2, and 5 for more on cultural considerations.)

> NOTE *Violating a person's cultural frame of reference is offensive, but so is reducing individual complexity to a laundry list of cultural stereotypes. Any generalization about a group presents a limited picture and in no way accurately characterizes even one much less all members of the group.*

Digital Persuasion and Social Media

3.7 Understand how social media influences persuasion

Positive and negative effects of the Internet on persuasion

The Internet has a powerful effect on how we research, receive, and process information. On the plus side, more people than ever before have access to vast amounts of information, giving us the ability to consider many points of view. On the negative side, however, as our technologies get smarter about our individual likes and dislikes, people become isolated from viewpoints that are in conflict with their own. As discussed earlier, confirmation bias enhances what people already believe, often despite factual information to the contrary.

How social media can confirm our biases

Social media plays a powerful role in how and why we are persuaded to believe various ideas, theories, or concepts. Because social media combines text, images, sound, and video, messages come at us in multiple formats, triggering our brains to remember a composite of the idea but not necessarily the details. Furthermore, social media is tailored to our own interests and belief systems, and the more we click on various memes, articles, and links, the smarter these systems become in terms of pointing us toward information that reinforces what we already believe or want to believe. The speed and reach of social media (Gurak 29), combined with the constant checking on our phones and desire to respond immediately (Bosker), creates situations where we might be persuaded by a story that is entirely untrue. As one computer science expert notes, "[t]his algorithmic bias toward engagement over truth reinforces our social and cognitive biases" (Menczer 1).

How to avoid confirmation bias

One suggestion for correcting this problem is to read and consider what you find online with more care, taking time to check the source, and to read more slowly (Hernandez and Preston) and not jump to any immediate conclusions. In other words, to be persuasive means you need to understand the other person's point of view; to do that, you need to look carefully at all forms of digital media before taking any point as factual and to consider that the people you are working with are also being influenced by social and digital persuasion. (For more on this topic, see Part 2, "The Research Process," as well as Chapters 24 and 25).

Guidelines

for Persuasion

Later chapters offer specific guidelines for various persuasive documents such as sales letters and proposals. Beyond attending to the unique requirements of a particular document, remember this principle:

No matter how brilliant, any argument rejected by its audience is a failed argument.

If readers find cause to dislike you or conclude that your argument has no meaning for them personally, they usually reject *anything* you say. Connecting with an audience means being able to see things from their perspective. The following guidelines can help you make that connection.

Analyze the Situation

➤ **Assess the political climate.** Who will be affected by your document? How will they react? How will they interpret your motives? Can you be outspoken? Could the argument cause legal problems? The better you assess readers' political feelings, the less likely your document will backfire. Do what you can to earn confidence and goodwill:

- Be aware of your status in the organization; don't overstep.
- Do not expect anyone to be perfect—including yourself.
- Never overstate your certainty or make promises you cannot keep.
- Be diplomatic; don't make anyone look bad or lose face.
- Ask directly for support: "Is this idea worthy of your commitment?"
- Ask your intended readers to review early drafts.

When reporting company negligence, dishonesty, incompetence, or anything else that others do not want to hear, expect fallout. Decide beforehand whether you want to keep your job (or status) or your dignity (more in Chapter 4).

➤ **Learn the unspoken rules.** Know the constraints on what you can say, to whom you can say it, and how and when you can say it. Consider the cultural context.

➤ **Decide on a connection (or combination of connections).** Does the situation call for you to merely declare your position, appeal to the relationship, or appeal to common sense and reason?

➤ **Anticipate your audience's reaction.** Will people be surprised, annoyed, angry? Try to address their biggest objections beforehand. Express your judgments ("We could do better") without making people defensive ("It's all your fault").

Develop a Clear and Credible Plan

➤ **Define your precise goal.** Develop the clearest possible view of what you want to see happen.

➤ **Do your homework.** Be sure your facts are straight, your figures are accurate, and that the evidence supports your claim.

➤ **Think your idea through.** Are there holes in this argument? Will it stand up under scrutiny?

➤ **Never make a claim or ask for something that people will reject outright.** Consider how much is *achievable* in this situation by asking what people are thinking. Invite them to share in decision making. Offer real choices.

➤ **Consider the cultural context.** Will some audience members feel that your message ignores their customs? Will they be offended by a direct approach or by too many facts and figures without a relationship connection? Remember that, beyond racial and ethnic distinctions, cultural groups also consist of people who share religious or spiritual views, sexual orientations, political affiliations, and so on.

Prepare Your Argument

➤ **Be clear about what you want.** Diplomacy is always important, but people won't like having to guess about your purpose.

➤ **Avoid an extreme persona.** Persona is the image or impression of the writer's personality suggested by the document's tone. Resist the urge to "sound off" no matter how strongly you feel; audiences tune out aggressive people no matter how sensible the argument. Admit the imperfections in your case. Invite people to respond. A little humility never hurts. Don't hesitate to offer praise when it's deserved.

➤ **Find points of agreement with your audience.** "What do we *all* want?" Focus early on a shared value, goal, or experience. Emphasize your similarities.

➤ **Never distort the opponent's position.** A sure way to alienate people is to cast the opponent in a more negative light than the facts warrant.

➤ **Try to concede something to the opponent.** Reasonable people respect an argument that is fair and balanced. Admit the merits of the opposing case before arguing for your own. Show empathy and willingness to compromise. Encourage people to air their own views.

➤ **Do not merely criticize.** If you're arguing that something is wrong, be sure you can offer realistic suggestions for making it right.

➤ **Stick to claims you can support.** Show people what's in it for them—but never distort the facts just to please the audience. Be honest about the risks.

➤ **Stick to your best material.** Not all points are equal. Decide which material—from your audience's view—best advances your case.

Present Your Argument

➤ **Before releasing the document, seek a second opinion.** Ask someone you trust and who has no stake in the issue at hand. If possible, have your company's legal department review the document.

➤ **Get the timing right.** When will your case most likely fly—or crash and burn? What else is going on that could influence people's reactions? Look for a good opening in the situation.

➤ **Decide on the proper format.** Does this audience and topic call for a letter, a memo, or some type of report? Your decision will affect how positively your message is received. Can visuals and page layout (charts, graphs, drawings, headings, lists) make the material more accessible?

➤ **Decide on the appropriate medium.** In today's workplace, email has become the default method for communicating. But consider your argument carefully, and decide if it is too complex for email. Most persuasive situations require readers to take time and consider the facts carefully; email, on the other hand, tends to be read very quickly. Consider starting the discussion with a phone call, meeting, or video chat. Create a shared document where everyone can contribute ideas; only use email to summarize the conversations.

➤ **Be sure everyone involved receives the documents and is included in meetings and other discussions.** People hate being left out of the loop—especially when any change that affects them is being discussed.

➤ **Invite responses.** After people have had a chance to consider your argument, gauge their reactions by asking them directly.

➤ **Do not be defensive about negative reactions.** Admit mistakes, invite people to improve on your ideas, and try to build support.

➤ **Know when to back off.** If you seem to be "hitting the wall," don't push. Try again later, or drop the whole effort. People who feel they have been bullied or deceived will likely become your enemies.

Shaping Your Argument

3.8 Prepare a convincing argument

To understand how our guidelines are employed in an actual persuasive situation, see Figure 3.4. The letter is from Rosemary Garrido of Energy Empowerment, Inc., a consulting firm that works with contractors to maximize the energy efficiency of offices and retail locations. Garrido's letter is a persuasive answer to her potential customer's main question: "Is it worthwhile to make energy-efficient changes to the storefront we've bought?" As you read the letter, notice the evidence and appeals Rosemary uses to support her opening claim and how she focuses on her reader's needs. Rosemary used the Audience and Use Profile Sheet (Figure 3.5) to help formulate her approach to the letter.

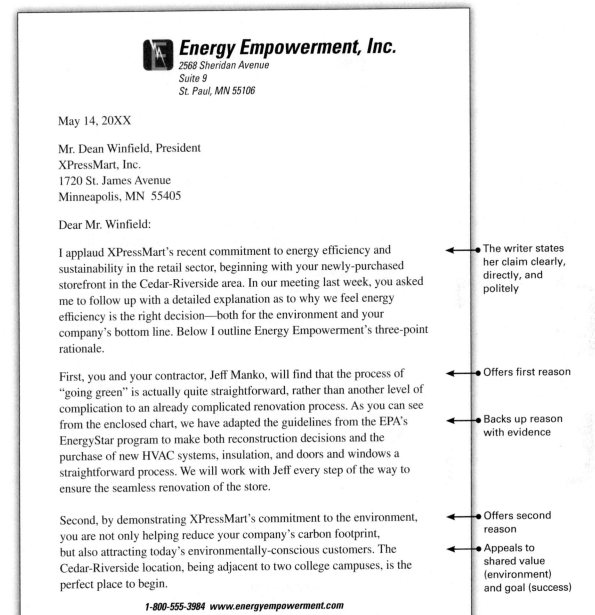

Energy Empowerment, Inc.
2568 Sheridan Avenue
Suite 9
St. Paul, MN 55106

May 14, 20XX

Mr. Dean Winfield, President
XPressMart, Inc.
1720 St. James Avenue
Minneapolis, MN 55405

Dear Mr. Winfield:

I applaud XPressMart's recent commitment to energy efficiency and sustainability in the retail sector, beginning with your newly-purchased storefront in the Cedar-Riverside area. In our meeting last week, you asked me to follow up with a detailed explanation as to why we feel energy efficiency is the right decision—both for the environment and your company's bottom line. Below I outline Energy Empowerment's three-point rationale.

First, you and your contractor, Jeff Manko, will find that the process of "going green" is actually quite straightforward, rather than another level of complication to an already complicated renovation process. As you can see from the enclosed chart, we have adapted the guidelines from the EPA's EnergyStar program to make both reconstruction decisions and the purchase of new HVAC systems, insulation, and doors and windows a straightforward process. We will work with Jeff every step of the way to ensure the seamless renovation of the store.

Second, by demonstrating XPressMart's commitment to the environment, you are not only helping reduce your company's carbon footprint, but also attracting today's environmentally-conscious customers. The Cedar-Riverside location, being adjacent to two college campuses, is the perfect place to begin.

1-800-555-3984 www.energyempowerment.com

The writer states her claim clearly, directly, and politely

Offers first reason

Backs up reason with evidence

Offers second reason

Appeals to shared value (environment) and goal (success)

Figure 3.4 **Supporting a Claim with Good Reasons** Give your audience a clear and logical path.

Dean Winfield, May 14, 20XX, page 2

Cites statistics)

Currently the storefront rates only a 42 on EnergyStar's performance scale. By improving that rating to 75 or above, this location will qualify for an EnergyStar display sticker, which, according to the EPA's Annual Report last year, increases retails sales in urban areas. Between the EnergyStar rating and your focus on sustainable products, expect to attract the interest of all residents of this forward-thinking community.

Closes with best reason

Appeals to shared goal (cost)

Finally, and perhaps most importantly, rest assured that the costs you put into reconstruction, systems, and other materials will pay for themselves in less than two years. Jeff and I have assembled a preliminary proposal itemizing costs, to which I will add my estimates regarding cost recuperation. However, know that in Energy Empowerment's 12-year history, every store and office renovation project has paid for itself remarkably quickly. As a recent example, consider our recent small office renovation in Columbia Heights, which recouped its costs in only 14 months.

Offers example

If I can answer any further questions, please do not hesitate to email me at rgarrido@esi.com or call me (extension 646). Again, we applaud your commitment to the environment and look forward to working with you and Jeff.

Best regards,

Rosemary Garrido
Executive Manager

cc: Jeff Manko, Manko Construction
Encl. EnergyStar's performance chart

Figure 3.4 Continued

Audience and Purpose

Primary audience: *Dean Winfield, President, XPressMart, Inc.*

Secondary audience(s): *Jeff Manko, Owner, Manko Construction*

Relationship with audience: *A possible customer for an energy-efficient renovation*

Purpose of document: *To help in gaining an important consulting project*

Audience and purpose statement: *To pave the way for a potential customer to hire us for our consulting services*

Intended use of document: *To provide information/rationale for a hiring decision*

Technical background: *Novice to moderate*

Prior knowledge about this topic: *Is new to the process of energy-efficiency renovation costs and results*

Information needs: *Needs costs, benefits, and statistics before proceeding*

Cultural considerations: *None in particular*

Probable questions: *How complicated is this process going to be?*

Will these green renovations impact business in any real way?

Will the cost of making these major renovations pay for themselves?

Audience's Probable Attitude and Personality

Attitude toward topic: *Highly interested but somewhat skeptical*

Probable objections: *These renovations may be too expensive for our bottom line*

Probable attitude toward this writer: *Receptive but cautious*

Organizational climate: *Open and flexible*

Persons most affected by this document: *Winfield and other decision makers*

Temperament: *Winfield takes a conservative approach to untested innovations*

Probable reaction to document: *Readers should feel somewhat reassured*

Audience Expectations about the Document

Material important to this audience: *Evidence that the renovations will be cost effective*

Potential problems: *Readers may have further questions I haven't anticipated*

Length and detail: *A concise argument that gets right to the point*

Format and medium: *A formal letter delivered via overnight mail*

Tone: *Encouraging, friendly, and confident*

Due date and timing: *ASAP—to illustrate our responsiveness to customer concerns*

Figure 3.5 Audience and Use Profile Sheet Notice how this profile sheet expands on the one shown in Figure 2.7 (see Chapter 2, "Develop an Audience and Use Profile"), to account for specific considerations in preparing an explicitly persuasive document.

> NOTE *People rarely change their minds quickly or without good reason. A truly resistant audience will dismiss even the best arguments and may end up feeling threatened and resentful. Even with a receptive audience, attempts at persuasion can fail. Often, the best you can do is avoid disaster and allow people to ponder the merits of the argument.*

Checklist

Persuasion

Use the following Checklist as you prepare to persuade an audience.

Planning Your Argument

❑ Do I understand the difference between explicit and implicit persuasion? (See "What Is Persuasion?" in this chapter.)

❑ Can I identify the different types of claims? (See "What Is Persuasion?" in this chapter.)

❑ Have I identified my precise goal in this situation? (See "Identify Your Specific Persuasive Goal" in this chapter.)

❑ Have I attempted to predict audience reaction? (See "Try to Predict Audience Reaction" in this chapter.)

❑ Am I prepared for audience resistance? (See "Try to Predict Audience Reaction" in this chapter.)

❑ Have I appealed to the rational audience connection? (See "Know How to Connect with the Audience" in this chapter.)

❑ Have I promoted my own view while planning how to respond to opposing views? (See "Allow for Give-and-Take" in this chapter.)

❑ Have I spelled out what I want? (See "Ask for a Specific Response" in this chapter.)

❑ Have I stuck with what is achievable? (See "Never Ask for Too Much" in this chapter.)

❑ Have I considered the various constraints in this situation? (See "Recognize All Constraints" in this chapter.)

❑ Have I filled out an Audience and Use Profile? (See "Shaping Your Argument" in this chapter.)

Presenting Your Argument

❑ Do I provide convincing evidence to support my claims? (See "Offering Convincing Evidence" in this chapter.)

❑ Will my appeals have personal meaning for this audience? (See "Appeal to Common Goals and Values" in this chapter.)

❑ Have I successfully navigated cultural differences in my argument? (See "Consider the Cultural Context" in this chapter.)

❑ Have I avoided confirmation bias in my argument? (See "Digital Persuasion and Social Media" in this chapter.)

Projects

For all projects, check with your instructor about whether to present your findings in class, bring drafts to class for discussion, upload your project to the class learning management system (LMS), and/or use the LMS forum or discussion boards to collaborate and review each activity below.

General

1. Find a persuasive letter that you feel is effective. In a memo (Chapter 15) to your instructor, explain how the message succeeds. Base your evaluation on the"Guidelines for Persuasion" and "Checklist: Persuasion" in this chapter. Attach the letter to your memo. Now, evaluate a letter or document that you feel is ineffective,

explaining how and why it fails. If your instructor agrees, upload your work to the class learning management system and provide feedback on each others' work.

2. Think about some change you would like to see on your campus or at work. Perhaps you would like to promote something new such as changes in course offerings or requirements, an off-campus shuttle service, or a daycare center. Or perhaps you would like to improve something such as the grading system, campus lighting, the system for student evaluation of teachers, or the promotion system at work. Or perhaps you would like to stop something from happening such as noise in the library or conflict at work.

Decide whom you want to persuade and write a memo (Chapter 15) to that audience. Anticipate your audience's questions, such as:

- Do we really have a problem or need?
- If so, should we care enough about it to do anything?
- Can the problem be solved?
- What are some possible solutions?
- What benefits can we anticipate? What liabilities?

Can you think of additional audience questions? Do an audience and use analysis based on the "Audience and Use Profile Sheet" in this chapter.

Don't think of this memo as the final word but as a consciousness-raising introduction that gets the reader to acknowledge that the issue deserves attention. At this early stage, highly specific recommendations would be premature and inappropriate.

Team

As a class, select a topic that involves persuasion. Topics might include childhood obesity, climate change, nutritional supplements, or other. In teams of 2–3 students, find a document that you feel makes a persuasive case about the topic. Using a shared document system if possible (such as Google Drive or your class learning management system), have each member contribute a short summary of what techniques these documents use to make a persuasive case. Are you able to identify the document's specific persuasive goal (see "Identify Your Specific Persuasive Goal" in this chapter)? Combine your ideas into a simple PowerPoint (or other slide software) presentation and share with class.

Digital and Social Media

At work, most communication takes place online, via email, video conferencing, chats, and collaborative writing tools. In teams of 2–3 students, look at the sample email in Chapter 14 (Figure 14.2), where Frank tries to persuade his manager to reconsider Frank's recent performance review. Write an email to your instructor explaining the strengths and weaknesses of Frank's attempt at persuasion, using the first set of items (Planning Your Argument) in "Checklist Persuasion" in this chapter to structure your ideas.

Global

Effective persuasive techniques in one culture may not work in another culture. Do a Web search on "intercultural communication" related to a specific country or culture and locate a topic that is important for technical communication. For instance, you might locate information about the different ways in which certain types of visuals that are persuasive in one culture are not effective in another or ways in which politeness is used in negotiations. Write a short summary of your findings, cite your sources, and present your information in class as a presentation or a shared document.

Chapter 4
Weighing the Ethical Issues

"Most of my writing is for clients who will make investment decisions based on their understanding of complex financial data, presented in a concise, *nontechnical* way. These people are not experts. They want to know, 'What do I do next?' While I can never guarantee the certainty of any stock or mutual fund investment, my advice has to be based on an accurate and honest assessment of all the facts involved.**"**

— Chris Fernandez, Certified Financial Planner

 # Learning Objectives

4.1 Understand the major causes of unethical workplace communication

4.2 Recognize common types of workplace communication abuses

4.3 Understand ethical issues related to digital communication

4.4 Use critical thinking to solve ethical dilemmas

4.5 Know the limitations of legal guidelines and avoid plagiarism

4.6 Determine when and how to report ethical violations on the job

As discussed in Chapter 3, most forms of workplace communication involve some level of persuasion. Yet being persuasive should not mean being dishonest, stretching the facts, or omitting key information just to win over your audience. Even in cases where claims are technically accurate, the way these claims are written can be misleading. For example, companies may win customers with a claim such as "Our artificial sweetener is composed of proteins that occur naturally in the human body [amino acids]" or "Our Krunchy Cookies contain no cholesterol!" Such claims are technically accurate but misleading: Amino acids in certain sweeteners can alter body chemistry to cause headaches, seizures, and possibly brain tumors; processed food snacks often contain saturated fat and trans fats from which the liver produces cholesterol.

At work, people are often tempted to emphasize any information or perspective that advances their case while downplaying or even ignoring ideas that might impede it. Communication is considered unethical if it leaves recipients at a disadvantage or prevents them from making their best decision. Ethical communication should provide the most useful, accurate information while always recognizing that choices about persuasion need to be guided by this question: "How can I be sure of doing the right thing?" (Figure 4.1).

Definition of unethical workplace communication

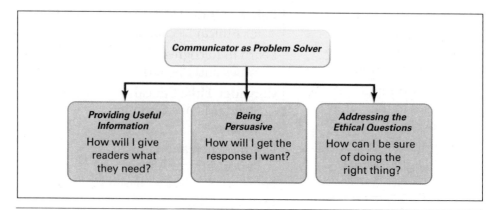

Figure 4.1 **In Addition to Being Informative and Persuasive, Communicators Must Be Ethical**

Recognize Examples and Causes of Unethical Workplace Communication

4.1 Understand the major causes of unethical workplace communication

Unethical communication in the workplace is all too common

Examples of unethical workplace behavior are in the news regularly and include financial scandals (where executives ignore warning signs in favor of maximum profits); corporate irresponsibility (where long-term environmental concerns take a back seat to short-term gains); and bad engineering decisions (based on getting a product to market or launching a rocket before proper safety checks are completed). While these big corporate and other scandals make for dramatic headlines, more routine examples of deliberate miscommunication are rarely publicized:

Routine instances of unethical communication

- A person lands a great job by exaggerating his credentials, experience, or expertise
- A marketing specialist for a chemical company negotiates a huge bulk sale of its powerful new pesticide by downplaying its carcinogenic hazards
- A manager writes a strong recommendation to get a friend promoted, while overlooking someone more deserving
- An engineer ignores the results of a safety test, sending the memo back to the team and asking them to find a different interpretation

Other instances of unethical communication, however, are less black and white. Here is one engineer's description of the gray area in which issues of product safety and quality often are decided:

Ethical decisions are not always "black and white"

> The company must be able to produce its products at a cost low enough to be competitive....To design a product that is of the highest quality and consequently has a high and uncompetitive price may mean that the company will not be able to remain profitable, and be forced out of business. (Burghardt 92)

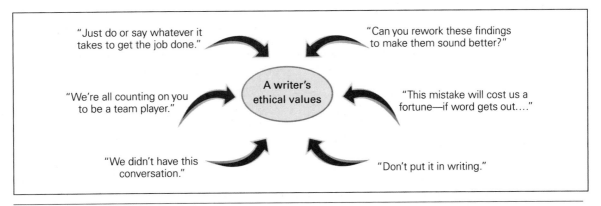

Figure 4.2 How Workplace Pressures Can Influence Ethical Values A decision more efficient, profitable, or better for the company might overshadow a person's sense of what is right.

When creating a technical marketing brochure, do you emphasize the need for very careful maintenance of a highly sensitive piece of lab equipment and risk losing a sale? Or do you downplay maintenance requirements, focusing instead on the lab equipment's positive features? The decisions we make in these situations are often influenced by the pressures we feel.

To save face, escape blame, or get ahead, anyone might be tempted to say what people want to hear or to suppress or downplay bad news. But normally honest people usually break the rules only when compelled by an employer, coworkers, or their own bad judgment. Figure 4.2 depicts how workplace pressures to "succeed at any cost" can influence ethical values. *Recognizing the major causes of unethical communication*

Yielding to Social Pressure

Sometimes, you may have to choose between doing what you know is right and doing what your employer or organization expects, as in this next example:

> Just as your automobile company is about to unveil its new pickup truck, your safety engineering team discovers that the reserve gas tanks (installed beneath the truck but outside the frame) may, in rare circumstances, explode on impact from a side collision. You know this information should be included in the owner's manual or, at a minimum, in a letter to the truck dealers, but the company has spent a fortune building this truck and does not want to hear about this problem. *Looking the other way*

Companies often face the contradictory goals of *production* (producing a product and making money on it) and *safety* (producing a product but spending money to avoid accidents that may or may not happen). When production receives first priority, safety concerns may suffer (Wickens 434–36). In these circumstances, you need to rely on your own ethical standards. In the case of the reserve gas tanks, if you decide to publicize the problem, expect to be fired for taking on the company.

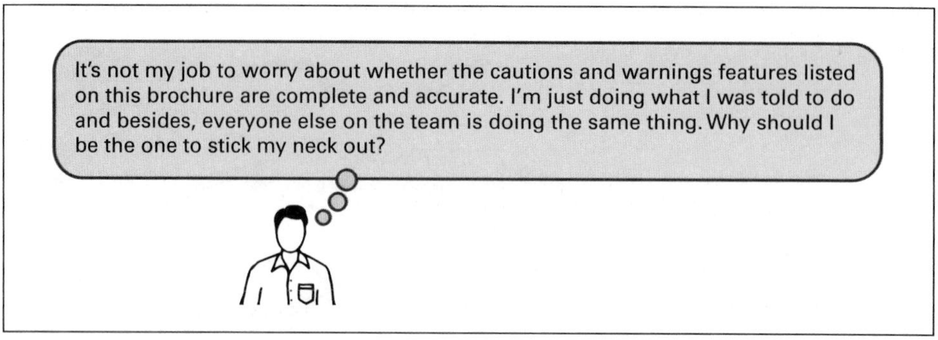

> It's not my job to worry about whether the cautions and warnings features listed on this brochure are complete and accurate. I'm just doing what I was told to do and besides, everyone else on the team is doing the same thing. Why should I be the one to stick my neck out?

Figure 4.3 **Hiding behind Groupthink**

Mistaking Groupthink for Teamwork

Blindly following the group

Organizations rely on teamwork and collaboration to get a job done; technical communicators often work as part of a larger team of writers, editors, designers, engineers, and production specialists. Teamwork is important in these situations, but teamwork should not be confused with *groupthink*, which occurs when group pressure prevents individuals from questioning, criticizing, reporting bad news, or "making waves" (Janis 9). Group members may feel a need to be accepted by the team, often at the expense of making the right decision. Anyone who has ever given in to adolescent peer pressure has experienced a version of groupthink.

Groupthink also can provide a handy excuse for individuals to deny responsibility (see Figure 4.3). For example, because countless people work on a complex project (say, a new airplane), identifying those responsible for an error is often impossible—especially in errors of omission, that is, when something that should have been done was overlooked (Unger 137).

One source offers this observation:

> People commit unethical acts inside corporations that they never would commit as individuals representing only themselves (Bryan 86).

Once their assigned task has been completed, employees might mistakenly assume that their responsibility has been fulfilled.

Types of Communication Abuses in the Workplace

Telling the truth versus meeting workplace expectations

4.2 **Recognize common types of workplace communication abuses**

On the job, you write in the service of your employer. Your effectiveness is judged by how well your documents speak for the company and advance its interests and agendas (Ornatowski 100–01). You walk the proverbial line between telling the truth and doing what your employer expects (Dombrowski 79).

Workplace communication influences the thinking, actions, and welfare of numerous people—customers, investors, coworkers, the public, policy makers—to name a few. These people are victims of communication abuse whenever we give them information that is less than the truth as we know it, as in the following situations.

Suppressing Knowledge the Public Needs

Pressures to downplay concerns about science and technology can result in censorship.

- Despite a federal law requiring the labeling of food containing genetically modi-fied ingredients, companies make this information difficult to access by hiding it in a QR code.

- Certain U.S. government agencies require technical writers and research scientists to avoid using the phrases "climate change" or "global warming."

- For example, information about airline safety lapses and near accidents is often suppressed by air traffic controllers because of "a natural tendency not to call attention to events in which their own performance was not exemplary" or their hesitation to "squeal" about pilot error (Barnett, qtd. in Ball 13).

Examples of suppressed information

Hiding Conflicts of Interest

Can scientists and engineers who have a financial stake in a particular issue or experi-ment provide fair and impartial information about the topic?

- In one analysis of 800 scientific papers, Tufts University's Sheldon Krimsky found that 34 percent of authors had "research-related financial ties," but none had been disclosed (King B1).

- Medical researchers at most U.S. research universities are required to reveal any potential conflicts of interest (such as serving as a consultant for a pharmaceutical company that manufacturers a drug being studied by this researcher). But some find ways around this requirement.

- Analysts on a popular TV financial program have recommended certain company stocks (thus potentially inflating the price of that stock) without disclosing that their investment firms hold stock in these companies (Oxfeld 105).

Examples of hidden conflicts of interest

Exaggerating Claims about Technology

Organizations that have a stake in a particular technology (for example, bioengineered foods) may be especially tempted to exaggerate its benefits, potential, or safety and to downplay the technology's risks. Especially if your organization depends on outside funding (as in the defense or space industry), you might find yourself pressured to make unrealistic promises.

Examples of exaggerated claims about technology

Falsifying or Fabricating Data

When researchers are under pressure to obtain grant funding, research data might be manipulated or even invented to support specific agendas that will appeal to the funding agency. Or if researchers have a high stake in the outcome of the analysis—because, for example, they hope the new results will confirm prior research—researchers might be tempted to make up or modify the new data set.

Using Visual Images That Conceal the Truth

Pictures can be more powerful than words and can easily distort the real meaning of a message. For example, as required by law, commercials for prescription medications must identify a drug's side effects—which can often be serious. But the typical drug commercial lists the side effects in small type or reads the information quickly, all the while showing images of smiling, healthy people. The happy images eclipse the sobering verbal message. (See Chapter 12 for more on ethics and visual communication.)

Stealing or Divulging Proprietary Information

Information that originates in a specific company is typically the exclusive intellectual property of that company. Proprietary information may include company records, product formulas, test and experiment results, surveys financed by clients, market research, plans, specifications, and minutes of meetings (Lavin 5). In theory, such information is legally protected, but it remains vulnerable to sabotage, theft, or leaks to the press or on social media. Fierce competition among rival companies for the very latest intelligence gives rise to measures like these:

> Companies have been known to use business school students to garner information on competitors under the guise of conducting "research." Even more commonplace is interviewing employees for slots that don't exist and wringing them dry about their current employer. (Gilbert 24)

Withholding Information People Need for Their Jobs

Nowhere is the adage that "information is power" more true than among coworkers. One sure way to sabotage a colleague is to deprive that person of information about the task at hand. Studies show that employees also withhold information for more benign reasons such as fear that someone else might take credit for their work or might "shoot them down" (Davenport 90).

Exploiting Cultural Differences

Based on its level of business experience or its particular social values or financial need, a given culture might be especially vulnerable to manipulation or deception. Some cultures, for example, place greater reliance on interpersonal trust than on lawyers or legal wording, and a handshake can be worth more than the fine print of a legal contract. Other cultures may tolerate abuse or destruction of their natural resources in order to generate much-needed income. If you know something about a culture's habits or business practices and then use this information unfairly to get a sale or make a profit, you are behaving unethically. (For more on cultural considerations, see Chapters 3 and 5.)

Ethical Issues with Social and Digital Media

4.3 Understand ethical issues related to digital communication

As discussed in Chapter 3, social media provides an efficient way to target information to specific audiences. Yet this same feature also allows plenty of opportunities for ethical breaches, including playing on people's fears using distorted visuals and inflammatory language; creating posts and tweets that look real but are actually based on incomplete or out of date information; starting a "tweet storm" by writing a meme or post that goes viral before any of the facts can be checked. Keep in mind the speed with which social media posts can be distributed, and always check to be sure the post is in keeping with the ethical standard of "doing the right thing."

Ethics and social media

Web sites can also be used in ways that are ethical or unethical. Like social media, Web sites are read quickly and can circulate widely. People often believe what they read on the Web simply because a friend recommended the site.

Ethics and Web sites

Examples of ethical issues with social media and the Web include the following:

- Offering inaccurate or unsubstantiated medical advice online
- Leaving out any cautions or warnings in an FAQ or online help system about digital information used by the product or company
- Purposely using unclear language in a Web site privacy statement or making the privacy information difficult to find (or leaving it out altogether)
- Tweeting or posting anonymous attacks, or smear campaigns, against people, products, or organizations
- Creating a Facebook post or tweet that is only partially accurate in order to get people to repost it

Along with the potential communication abuses involved with social media and Web sites, the ever-increasing amount of personal information stored in digital formats requires anyone working as part of a communication team (technical writer, engineer, scientist) to consider ethical issues. Schools, governments, credit card companies, insurance companies, pharmacies, and many other organizations house vast databases of personal, often private information about patients, customers, citizens, and employees. As we are all aware, this information can be breeched by hackers or others intent on identity theft or worse. Private information is sometimes left open for stealing on nonsecure servers; in addition, organizations often share information without regard for accuracy or to be used for unethical purposes.

Ethics and digital data

The proliferation of "big data" and digital information creates broad opportunities for communication abuse, as in these examples:

- Failing to safeguard the privacy of personal information about a Web site visitor's health, finances, buying habits, or affiliations
- Providing customers with little to no information about how their personal data will be used, either on a Web site or in any other situation (e.g., medical, insurance, at work)
- Ignoring any data breaches observed in the workplace

Rely on Critical Thinking for Ethical Decisions

4.4 Use critical thinking to solve ethical dilemmas

Because of their effects on people and on your career, ethical decisions challenge your critical thinking skills:

Ethical decisions require critical thinking

- How can I know the "right action" in this situation?
- What are my obligations, and to whom, in this situation?
- What values or ideals do I want to represent in this situation?
- What is likely to happen if I do X—or Y?

Ethical issues resist simple formulas, but the following criteria offer a limited form of guidance.

Reasonable Criteria for Ethical Judgment

Reasonable criteria (standards that most people consider acceptable) take the form of *obligations*, *ideals*, and *consequences* (Christians et al. 17–18; Ruggiero 33–34). *Obligations* are the responsibilities you have to everyone involved:

Our obligations are varied and often conflicting

- **Obligation to yourself** to act in your own self-interest and according to good conscience
- **Obligation to clients and customers** to stand by the people to whom you are bound by contract—and who pay the bills
- **Obligation to your company** to advance its goals, respect its policies, protect confidential information, and expose misconduct that would harm the organization
- **Obligation to coworkers** to promote their safety and well-being
- **Obligation to the community** to preserve the local economy, welfare, and quality of life
- **Obligation to society** to consider the national and global impact of your actions and choices

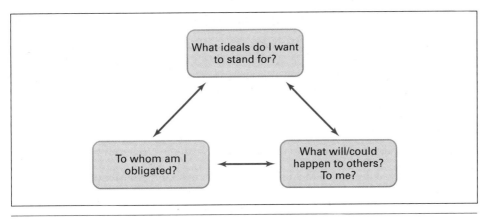

Figure 4.4 **Reasonable Criteria for Ethical Judgment**

When the interests of these parties conflict—as they often do—you have to decide where your primary obligations lie.

Ideals are the values you believe in or stand for: loyalty, friendship, compassion, dignity, fairness, and whatever qualities make you who you are. *Consequences* are the beneficial or harmful results of your actions. Consequences may be immediate or delayed, intentional or unintentional, obvious or subtle. Some consequences are easy to predict; some are difficult; some are impossible. Figure 4.4 depicts the relations among these three criteria.

The criteria for ethical judgment above help us understand why even good intentions can produce bad judgments, as in the following situation:

> Someone observes…that waste from the local mill is seeping into the water table and polluting the water supply…. But before [a remedy] can be found, extremists condemn the mill for lack of conscience and for exploiting the community. People get upset and clamor for the mill to be shut down and its management tried on criminal charges. The next thing you know, the plant does close, 500 workers are without jobs, and no solution has been found for the pollution problem. (Hauser 96)

What seems like the "right action" might be the wrong one

Because of their zealous dedication to the *ideal* of a pollution-free environment, the extremist protestors failed to anticipate the *consequences* of their protest or to respect their *obligation* to the community's economic welfare.

Ethical Dilemmas

Ethics decisions are especially frustrating when no single choice seems acceptable (Ruggiero 35). For example, the proclaimed goal of "welfare reform" is to free people from lifelong economic dependence. One could argue that dedication to this *consequence* would violate our *obligations* (to the poor and the sick) and our *ideals* (of compassion or fairness). On the basis of our three criteria, how else might the welfare-reform issue be considered?

Ethical questions often resist easy answers

Anticipate Some Hard Choices

Communicators' ethical choices basically are concerned with honesty in choosing to reveal or conceal information:

What to reveal and what to conceal is at the root of every hard choice

- What exactly do I report and to whom?
- How much do I reveal or conceal?
- How do I say what I have to say?
- Could misplaced obligation to one party be causing me to deceive others?

The following case illustrates the difficulty of making sound ethical decisions.

Case

A Hard Choice

You are an assistant structural engineer working on the construction of a nuclear power plant in a developing country. After years of construction delays and cost overruns, the plant finally has received its limited operating license from the country's Nuclear Regulatory Commission (NRC).

During your final inspection of the nuclear core containment unit, on February 15, you discover a ten-foot-long hairline crack in a section of the reinforced concrete floor, within twenty feet of the area where the cooling pipes enter the containment unit. (The especially cold and snowless winter likely has caused a frost heave under a small part of the foundation.) The crack has either just appeared or was overlooked by NRC inspectors on February 10.

The crack could be perfectly harmless, caused by normal settling of the structure; and this is, after all, a "redundant" containment system (a shell within a shell). But, then again, the crack might also signal some kind of serious stress on the entire containment unit, which ultimately could damage the entry and exit cooling pipes or other vital structures.

You phone your boss, who is just about to leave on vacation and who tells you, "Forget it; no problem," and hangs up.

You may have to choose between the goals of your organization and what you know is right

You know that if the crack is reported, the whole start-up process scheduled for February 16 will be delayed indefinitely. More money will be lost; excavation, reinforcement, and further testing will be required—and many people with a stake in this project (from company executives to construction officials to shareholders) will be furious—especially if your report turns out to be a false alarm. All segments of plant management are geared up for the final big moment. Media coverage will be widespread. As the bearer of bad news—and bad publicity—you suspect that, even if you turn out to be right, your own career could be damaged by your apparent overreaction.

On the other hand, ignoring the crack could compromise the system's safety, with unforeseeable consequences. Of course, no one would ever be able to implicate you. The NRC has already inspected and approved the containment unit—leaving you, your boss, and your company in the clear. You have very little time to decide. Start-up is scheduled for tomorrow, at which time the containment system will become intensely radioactive.

What would you do? Justify your decision on the basis of the obligations, ideals, and consequences involved.

Working professionals commonly face choices similar to the one depicted above. They must often make these choices alone or on the spur of the moment, without the luxury of contemplation or consultation.

Learn to Recognize Legal Issues and Plagiarism

4.5 Know the limitations of legal guidelines and avoid plagiarism

Can the law tell you how to communicate ethically? Sometimes. If you stay within the law, are you being ethical? Not always—as illustrated in this chapter's earlier section on communication abuses. Legal standards "sometimes do no more than delineate minimally acceptable behavior." In contrast, ethical standards "often attempt to describe ideal behavior, to define the best possible practices for corporations" (Porter 183).

Communication can be legal without being ethical

Except for the instances listed below, lying is rarely illegal. Common types of legal lies are depicted in Figure 4.5. Later chapters cover other kinds of lying that are often legal, such as page design that distorts the real emphasis of the content or words that are deliberately unclear, misleading, or ambiguous.

What, then, are a communicator's legal guidelines? Workplace communication is regulated by the types of laws described below.

- **Laws against deception** prohibit lying under oath, lying to a federal agent, lying about a product so as to cause injury, or breaking a contractual promise.

Laws that govern workplace communication

- **Laws against libel** prohibit any false written statement that attacks or ridicules anyone. A statement is considered libelous when it damages someone's reputation, character, career, or livelihood or when it causes humiliation or mental suffering. Material that is damaging but *truthful* would not be considered libelous unless it were used intentionally to cause harm. In the event of a libel suit, a writer's ignorance is no defense; even when the damaging material has been obtained from a source presumed reliable, the writer and publisher are accountable.

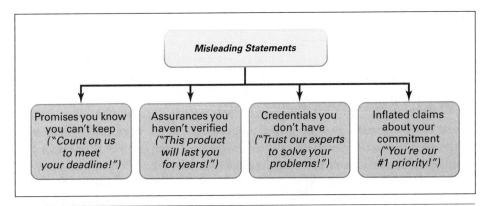

Figure 4.5 **Statements That Are Misleading but May Be Legal**

- **Laws protecting employee privacy** impose strict limits on information employers are allowed to give out about an employee.

- **Copyright laws** protect the rights of authors—or of their employers, in cases where the content was produced by a regular, salaried employee. The **No Electronic Theft Act (NET Act, passed in 1997)** covers copyright even in cases where no one is making a profit (such as the sharing of music files among students). Other laws include those related to software theft, hacking, and file sharing.

- **Laws against stealing or revealing trade secrets.** According to many government agencies and private firms, billions of dollars of proprietary information (trade secrets and other intellectual property) are stolen annually. The Economic Espionage Act makes such theft a federal crime; this law classifies as "trade secret" not only items such as computer source code or the recipe for our favorite cola, but even a listing of clients and contacts brought from a previous employer (Farnham 114, 116).

- **Laws against deceptive or fraudulent advertising** prohibit false claims or suggestions, for example, implying that a product or treatment will cure disease, or representing a used product as new. Fraud is defined as lying that causes another person monetary damage (Harcourt 64). Even a factual statement such as "our cigarettes have fewer additives" is considered deceptive because it implies that a cigarette with fewer additives is safer than other cigarettes (Savan 63).

- **Liability laws** define the responsibilities of authors, editors, and publishers for damages resulting from incomplete, unclear, misleading, or otherwise defective information. The misinformation might be about a product (such as failure to warn about the toxic fumes from a spray-on oven cleaner) or a procedure (such as misleading instructions in an owner's manual for using a tire jack).

Legal standards for product literature vary from country to country. A document must satisfy the legal standards for safety, health, accuracy, or language for the country in which it will be distributed. For example, instructions for any product requiring assembly or operation have to carry warnings as stipulated by the country in which the product will be sold. Inadequate documentation, as judged by that country's standards, can result in a lawsuit (Caswell-Coward 264–66; Weymouth 145).

> *NOTE Large companies have legal departments to consult about various documents. Most professions have ethics guidelines (as in Figure 4.7). If your field has its own formal code, obtain a copy.*

Learn to Recognize Plagiarism

What is plagiarism?

Ethical communication includes giving proper credit to the work of others. In both workplace and academic settings, plagiarism—representing the words, ideas, or perspectives of others as your own—is a serious breach of ethics. Even when your use of a source may be perfectly legal, you will still be violating ethical standards if you fail to cite the information source or fail to identify material that is directly quoted. Figure 4.6 shows some ways in which plagiarism can occur.

Blatant versus Unintentional Plagiarism

Blatant cases of plagiarism occur when a writer consciously lifts passages from another work (print or online) and incorporates them into his or her own work without quoting or documenting the original source. As most students know, this can result in a failing grade and potential disciplinary action. More often, writers will simply fail to cite a source being quoted or paraphrased, often because they misplaced the original source and publication information, or forgot to note it during their

Plagiarism can be both intentional and unintentional

Original Source. To begin with, language is a system of communication. I make this rather obvious point because to some people nowadays it isn't obvious: they see language as above all a means of "self-expression." Of course, language is one way that we express our personal feelings and thoughts—but so, if it comes to that, are dancing, cooking, and making music. Language does much more: it enables us to convey to others what we think, feel, and want. Language-as-communication is the prime means of organizing the cooperative activities that enable us to accomplish as groups things we could not possibly do as individuals. Some other species also engage in cooperative activities, but these are either quite simple (as among baboons and wolves) or exceedingly stereotyped (as among bees, ants, and termites). Not surprisingly, the communicative systems used by these animals are also simple or stereotypes. Language, our uniquely flexible and intricate system of communication, makes possible our equally flexible and intricate ways of coping with the world around us: In a very real sense, it is what makes us human (Claiborne 8).

Plagiarism Example 1 One commentator makes a distinction between language used as a means of self-expression and language-as-communication. It is the latter that distinguishes human interaction from that of other species and allows humans to work cooperatively on complex tasks (8).

What's wrong? The source's name is not given, and there are no quotation marks around words taken directly from the source (highlighted in yellow in the example).

Plagiarism Example 2 Claiborne notes that language "is the prime means of organizing the cooperative activities." Without language, we would, consequently, not have civilization.

What's wrong? The page number of the source is missing. Parenthetical references should immediately follow the material being quoted, paraphrased, or summarized. You may omit a parenthetical reference only if the information that you have included in your attribution is sufficient to identify the source in your Works Cited list and no page number is needed.

Plagiarism Example 3 Other animals also engage in cooperative activities. However, these actions are not very complex. Rather they are either the very simple activities of, for example, baboons and wolves or the stereotyped activities of animals such as bees, ants, and termites (Claiborne 8).

What's wrong? A paraphrase should capture a specific idea from a source but must not duplicate the writer's phrases and words (highlighted in yellow in the example). In the example, the wording and sentence structure follow the source too closely.

Figure 4.6 A Few Examples of Plagiarism Can you spot the plagiarism in the examples above that follow the original source?

research (Anson and Schwegler 633–36). Whereas this more subtle, sometimes unconscious, form of misrepresentation is less blatant, it still constitutes plagiarism and can undermine the offender's credibility, or worse. Whether the infraction is intentional or unintentional, people accused of plagiarism can lose their reputation and be sued or fired.

Plagiarism and the Internet

Digital considerations

The Internet makes is easy to search and find thousands of sites with information on any topic imaginable. Yet some people mistakenly assume that because material posted on a Web site, Facebook page, Twitter feed, or blog is free, it can be paraphrased or copied without citation. Despite the ease of cutting and pasting digital information, the fact remains: Any time you borrow someone else's words, ideas, perspectives, or images, you need to document the original source accurately, even if that source is public domain (see "Consider This: Frequently Asked Questions about Copyright" in Chapter 7).

Plagiarism and Your Career

Career considerations

Whatever your career plans, learning to gather, incorporate, and document authoritative source material is an absolutely essential job skill. By properly citing a range of sources in your work, you bolster your own credibility and demonstrate your skills as a researcher and a writer. (For more on documenting sources and avoiding plagiarism, see "Appendix A: A Quick Guide to Documentation.")

> NOTE *Plagiarism and copyright infringement are not the same. You can plagiarize someone else's work without actually infringing copyright. These two issues are frequently confused, but plagiarism is primarily an ethical issue, whereas copyright infringement is a legal and economic issue. (For more on copyright and related legal issues, see Chapter 7.)*

Consider This

Ethical Standards are Good for Business

To be sustainable, businesses need to make money. Yet earning a profit does not have to come at the expense of ethics; profits and ethical standards can and should coexist. As one business writer notes, "CEOs do have a responsibility to keep businesses profitable. Their success often depends on profit-and-loss statements. But some businesses also value having high ethical standards" (Baldelomar 1).

Social media can spread news quickly
Consumers also value companies that have clear, visible ethical standards and are responsive to individual and societal concerns. With social media, word can spread quickly about organizations with poor ethical standards such as companies that stretch the truth, don't deliver on promises, or engage in shady workplace practices. At the same

time, social media can also spread the word about companies that are highly valued by customers. Here is how a writer for *Forbes* magazine describes the relationship between social media and business ethics:

> With the instant feedback loop that is driven by technology, businesses quickly learn when customers are happy...and when they're not. Angie's List, Yelp, TripAdvisor and...other on-line services expose all to the public. This leaves little room for a business to run roughshod over customers. Bad behavior leads to bad reviews. And that's bad for business.
>
> On the flip side, when customers are treated ethically, word now gets around quickly. (Parish)

High standards earn customer trust

Describing a potentially profitable business proposal, which he decided to walk away from due to ethical concerns, one business owner says this: "Trust and ethical behavior go hand in hand. Ethics comprise the foundation of your character and cannot be compromised without major consequences. There are no shortcuts when it comes to being ethical: Either you do the right thing or you don't" (Gensler 1).

Companies that take a clear and visible ethical stance and demonstrate these standards by their actions will create long-term trust with customers over time. Numerous examples exist, including companies that follow through on promises (such as lifetime warranties) and are upfront and take immediate corrective actions when problems arise. Long-term trust means repeat customers and customer referrals.

Workplace cultures that encourage sharing can pay off

Information expert Keith Devlin describes how one company's "strong culture of sharing ideas paid a handsome dividend":

> The invention of Post-it Notes by 3M's Art Fry came about as a result of a memo from another 3M scientist who described the new glue he had developed. The new glue had the unusual property of providing firm but very temporary adhesion. As a traditional bonding agent, it was a failure. But Fry was able to see a novel use for it, and within a short time, Post-it Notes could be seen adorning every refrigerator door in the land. (179–80)

Decide When and How to Report Ethical Abuses

4.6 Determine when and how to report ethical violations on the job

Suppose your employer asks you to cover up fraudulent Medicare charges or a violation of federal pollution standards. If you decide to resist, your choices seem limited: resign or go public (i.e., blow the whistle).

What is whistle-blowing?

Walking away from a job isn't easy, and whistle-blowing can spell career disaster. Many organizations refuse to hire anyone blacklisted as a whistle-blower. Even if you do not end up being fired, expect your job to become hellish. Consider, for example, the Research Triangle Institute's study of consequences for whistle-blowers in 68 different instances. Following is an excerpt:

> More than two-thirds of all whistle-blowers reported experiencing at least one negative outcome.... Those most likely to experience adverse consequences were "lower ranking [personnel]." Negative consequences included pressure to drop their allegations,

Consequences of whistle-blowing

[ostracism] by colleagues, reduced research support, and threatened or actual legal action. Interestingly…three-fourths of these whistle-blowers experiencing "severe negative consequences" said they would definitely or probably blow the whistle again. (qtd. in "Consequences of Whistle Blowing in Scientific Misconduct Reported" 2)

Despite the retaliation they suffered, few people surveyed regretted their decision to go public.

Employers are generally immune from lawsuits by employees who have been dismissed unfairly but who have no contract or union agreement specifying length of employment (Unger 94). Federal and state laws, however, offer some protection for whistle-blowers.

Limited legal protections for whistle-blowers

- The Whistleblower Protection Act (and the subsequent Whistleblower Protection Enhancement Act) protects federal government employees who disclose information about illegal activities, public safety issues, abuse of authority, and other unethical activities in any government organization.

- The Federal False Claims Act allows an employee to sue, in the government's name, a contractor who defrauds the government (say, by overcharging for military parts). Also, this law allows employees of government contractors to sue when they are punished for whistle-blowing (Stevenson A7).

- Anyone punished for reporting employer violations to a regulatory agency (Federal Aviation Administration, Nuclear Regulatory Commission, Occupational Health and Safety Administration, and so on) can request a Labor Department investigation.

- Federal as well as some state laws prohibit employers from punishing job applicants or employees who file a complaint or are part of an Equal Employment Opportunity (EEO) complaint or investigation. (EEO complaints typically involve issues such as sexual harassment, religious practices, and disability accommodations.)

- Beyond requiring greater accuracy and clarity in the financial reports of publicly traded companies, The Sarbanes-Oxley Act imposes criminal penalties for executives who retaliate against employees who blow the whistle on corporate misconduct. This legislation also requires companies to establish confidential hotlines for reporting ethical violations.

Even with such protections, an employee who takes on a company without the backing of a labor union or other powerful group can expect lengthy court battles, high legal fees (which may or may not be recouped), and disruption of life and career.

Before accepting a job offer, do some online and library research about the company's reputation. Of course, you can learn only so much before actually working there, but with the wide reach of social media and with most public documents available online, there is a lot you can find out in advance of accepting a job.

Learn whether the company has *ombudspersons*, who help employees file complaints, or hotlines for advice on ethics problems or for reporting misconduct. Ask whether the company or organization has a formal code for personal and organizational behavior. For example, Figure 4.7 is the code of ethics for the Institute of

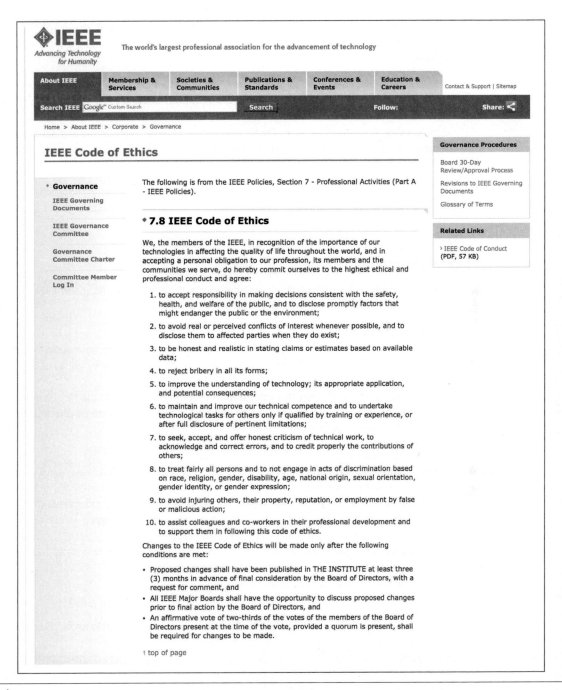

Figure 4.7 A Sample Code of Ethics Notice the many references to ethical communication in this engineering association's code of professional conduct, available on its Web site.

Source: Reprinted by permission of the Institute of Electrical and Electronic Engineers, Inc.

Electrical and Electronics Engineers (IEEE). Finally, assume that no employer, no matter how ethical, will tolerate any public statement that makes the company look bad.

> NOTE *Sometimes the right choice is obvious, but often it is not. No one has any sure way of always knowing what to do. This chapter is only an introduction to the inevitable hard choices that, throughout your career, will be yours to make and to live with. For further guidance and case examples, go to The Online Ethics Center for Engineering and Science, part of the U.S. National Academy of Engineering.*

Guidelines

for Ethical Communication

How do we balance self-interest with the interests of others—our employers, the public, our customers? Listed below are guidelines:

Satisfying the Audience's Information Needs

➤ **Give your readers everything they need to know.** To accurately see things as you do, your audience needs more than just a partial view. Don't bury readers in needless details, but do make sure they get all of the facts and get them straight. If you're at fault, admit it and apologize immediately.

➤ **Give people a clear understanding of what the information means.** Even when all the facts are known, they can be misinterpreted. Do all you can to ensure that your readers understand the facts as you do. If you're not certain about your own understanding, say so.

Taking a Stand versus the Company

➤ **Get your facts straight, and keep a record.** Don't blow matters out of proportion, but do keep a trail (email or print or both) in case of possible legal proceedings.

➤ **Appeal your case in terms of the company's interests.** Instead of being pious and judgmental ("This is a racist and sexist policy, and you'd better get your act together"), focus on what the company stands to gain or lose ("Promoting too few women and minorities makes us vulnerable to legal action").

➤ **Aim your appeal toward the right person.** If you have to go to the top, find someone who knows enough to appreciate the problem and who has enough clout to make something happen.

➤ **Get legal advice.** Contact a lawyer and your professional society.

Leaving the Job

➤ **Make no waves before departure.** Discuss your departure only with people "who need to know." Say nothing negative about your employer to clients, coworkers, or anyone else.

➤ **Leave all proprietary information behind.** Take no hard-copy documents or digital files prepared on the job—except for those records tracing the process of your resignation or termination.

"Checklist: Ethical Communication" in this chapter incorporates additional guidelines from other chapters.

Source: Adapted from G. Clark 194; Lenzer and Shook 102; Unger 127–30.

Checklist

Ethical Communication

Use the following Checklist as you weigh ethical issues.

Accuracy
❑ Have I explored all sides of the issue and all possible alternatives? (See Chapter 3, "Allow for Give and Take.")
❑ Do I provide enough information and interpretation for recipients to understand the facts as I know them? (See Chapter 1, "Reader-Centered.")
❑ Do I avoid exaggeration, understatement, sugarcoating, or any distortion or omission that would leave readers at a disadvantage? (See "Understand the Potential for Communication Abuse" in this chapter.)
❑ Do I state the case clearly instead of hiding behind jargon and euphemism? (See Chapter 11, "Avoid Useless Jargon.")

Honesty
❑ Do I make a clear distinction between "certainty" and "probability"? (See Chapter 8, "Evaluate the Evidence.")
❑ Are my information sources valid, reliable, and relatively unbiased? (See Chapter 8, "Evaluate the Sources")
❑ Do I actually believe what I'm saying, instead of being a mouthpiece for groupthink or advancing some hidden agenda? (See "Mistaking Groupthink for Teamwork" in this chapter.)
❑ Would I still advocate this position if I were held publicly accountable for it? (See "Yielding to Social Pressure" in this chapter.)
❑ Do I inform people of all the consequences or risks (as I am able to predict) of what I am advocating? (See "Suppressing Information the Public Needs" in this chapter.)

Fairness
❑ Am I reasonably sure this document will harm no innocent persons or damage their reputations? (See "Learn to Recognize Legal Issues and Plagiarism" in this chapter.)
❑ Am I respecting all legitimate rights to privacy and confidentiality? (See "Consider Ethical Issues and Digital Communication" in this chapter.)
❑ Do I credit all contributors and sources of ideas and information? (See "Learn to Recognize Legal Issues and Plagiarism" in this chapter.)

Source: Adapted from Brownell and Fitzgerald 18; Bryan 87; Johannesen 21–22; Larson 39; Unger 39–46; Yoos 50–55.

Projects

For all projects, check with your instructor about whether to present your findings in class, bring drafts to class for discussion, upload your project to the class learning management system (LMS), and/or use the LMS forum or discussion boards to collaborate and review each activity below.

General

1. Visit a Web site or the Facebook page for a professional association in your field (American Psychological Association, Society for Technical Communication, American Nursing Association). See if you can locate this association's code of ethics. How often are communication-related issues mentioned? Save a copy of the code (upload to your class learning management system if your instructor agrees) for a class discussion of the role of ethical communication in different fields.

2. Review the plagiarism material in this chapter, then locate additional resources on plagiarism that will be useful for students in class. Begin by checking with your school's library or writing center to see what materials they have available (for instance, online resources or brochures). Next, go online and do a search, paying particular attention to credible sites such as college and university libraries and recognized writing centers. Create a two-page document that provides a practical, working definition of plagiarism, a list of strategies for avoiding it, and a list of online resources. See Appendix A for further information on how to avoid plagiarism and how to document sources. When you've completed the handout, upload it to the class learning management system as directed by your instructor, and share it with classmates or coworkers.

Team

Assume you are a training manager for ABC Corporation, which is in the process of overhauling its policies on company ethics. Developing the company's official Code of Ethics will require months of research and collaboration with attorneys, ethics consultants, editors, and company officers. Meanwhile, your boss has asked you to develop a brief but practical set of "Guidelines for Ethical Communication," as a quick and easy reference for all employees until the official code is finalized. Using the material in this chapter, prepare a two-page memo (see Chapter 15) for employees, explaining how to avoid ethical pitfalls in corporate communication.

Digital and Social Media

The Internet allows us to find more information then ever before on any possible topic we can think of. Yet the power, speed, and research potential of online information must be balanced against the wide range of ideas presented, some of which are not factual and many of which could even be considered unethical. Pick a controversial topic that has ethical implications: for example, individual choice about vaccinating children; issues about the safety of nuclear energy; the science of bioengineered food and crops. Review Web sites and social media posts and compare the competing claims being made. Do you find possible examples of unethical communication, such as conflicts of interest or exaggerated claims? Refer to "Types of Communication Abuses in the Workplace" and "Checklist: Ethical Communication" in this chapter to help you evaluate the various claims. Report your findings in a memo (see Chapter 15) to your instructor and classmates.

Global

Find an example of a document designed to "sell" an item, an idea, or a viewpoint (a sales brochure for a new automobile, a pharmaceutical brochure for some popular prescription medication, a Web page for an environmental or political organization, or something similar). Write a brief description of places in the document where you think the writers considered (or ignored) the importance of writing for a cross-cultural audience. For each item you describe, explain why this item is an important ethical consideration.

Chapter 5
Teamwork and Global Considerations

kantver/123RF

❝My work in preparing user manuals is almost entirely collaborative. The actual process of writing takes maybe 30 percent of my time. I spend more time consulting with subject matter experts (we call them "SMEs"), such as the software engineers and field support people. I then meet with the publication and graphics departments to plan the manual's structure and online and paper formats. As I prepare the drafts, I have to keep track of which reviewer has which draft. Because I rely on others' feedback, I circulate materials often. One major challenge is getting everyone to agree on a specific plan of action and then stay on schedule so we can meet our publication deadline.❞

—*Pam Herbert, Technical Writer for a mobile app company*

Learning Objectives

5.1 Manage a team project

5.2 Participate in virtual teams and
conduct face-to-face meetings

5.3 Recognize and negotiate conflicts in
collaborative groups

5.4 Brainstorm using face-to-face and
digital methods

5.5 Review and edit the work of your
peers

5.6 Avoid unethical behavior as a team
member

5.7 Understand how to work
productively on a global team

Teamwork and Project Management

5.1   Manage a team project

Technical writers
often collaborate

Complex documents (especially long reports, proposals, and manuals) are rarely
produced by one person working alone. For example, instruction manuals (for products ranging from medical devices to lawn care equipment to digital cameras) are typically produced not just by one lone writer but with input from a team that includes
technical writers, technical editors, engineers, graphic artists, subject matter experts,
reviewers, marketing personnel, and lawyers.

Teams are often
distributed

At work, teams are increasingly distributed across different job sites, time zones, and
countries. Digital technology—email, texting, collaborative writing spaces, video conferencing, webinars, and online conferencing tools—offer many ways for distributed teams
to collaborate and stay in touch across distances and time zones. But whether the team
is onsite or distributed (or both), members have to find ways of expressing their views

persuasively, accepting constructive criticism, working with digital apps effectively, and getting along and reaching agreement with others who hold different views.

Teamwork is successful only when there is strong cooperation and trust, a recognized team structure, and clear communication. Collaboration is key to a good outcome for any project. The following guidelines explain how to manage a collaborative team project in a systematic way. Following these guidelines, Figure 5.1 provides a template for a basic project planning form, which you can modify to fit your specific team, project, and deadlines.

Qualities of successful teamwork

Project Planning Form

Project title:

Audience:

Project manager:

Team members:

Purpose of the project:

Type of document required:

Specific Assignments	**Due Dates**	**Person(s) Responsible**
Research:		
Planning:		
Drafting:		
Revising:		
Preparing progress report:		
Preparing oral briefing for the team:		
Final document first draft:		
Final document:		

Work Schedule

Team meetings:	*Date*	*Place*	*Time*	*Note taker*
#1				
#2				
#3				
etc.				
Mtgs. w/instructor				
#1				
#2				
etc.				

Miscellaneous

How will disputes and grievances be resolved?

How will performances be evaluated?

Other matters (use of technology: email, Google Drive, other?)

Figure 5.1 Project Planning Form for Managing a Collaborative Project To manage a team project you need to (a) spell out the project goal, (b) break the entire task down into manageable steps, (c) create a climate in which people work well together, and (d) keep each phase of the project under control.

Guidelines

for Managing a Collaborative Project

➤ **Appoint a group manager.** The manager assigns tasks, enforces deadlines, conducts meetings, consults with supervisors, and "runs the show."

➤ **Define a clear and definite goal.** Compose an audience and purpose statement (see Chapter 2, "Primary and Secondary Purposes") that spells out the project's goal and the plan for achieving the goal. Be sure each member understands the goal.

➤ **Identify the type of document required.** Is this a report, a proposal, a manual, a brochure? Are visuals and supplements (abstract, appendices, and other front and end matter) needed? Will the document be in hard copy or digital form or both?

➤ **Divide the tasks.** Who will be responsible for which parts of the document or which phases of the project? Who is best at doing what (writing, editing, layout and graphics, oral presentation)? Which tasks will be done individually and which collectively? Spell out, in writing, clear expectations for each team member.

➤ **Establish a timetable.** Gantt and PERT charts (see Chapter 12, "Gantt and PERT Charts") help the team visualize the whole project as well as each part, along with start-up and completion dates for each phase.

➤ **Decide on a meeting schedule.** How often will the group meet, and where and for how long?

➤ **Establish a procedure for responding to the work of other members.** How will the group review and respond to document drafts? Can all aspects of the review be conducted through document comment systems (see next item), or will the team need some regular meeting, too?

➤ **Decide on the most suitable digital writing apps.** Will you work on a shared collaborative document (via Google docs or Microsoft cloud or similar), or will you share Microsoft Word files using Track Changes (see Figure 5.2)? If using Microsoft Word files, develop a file naming system (version 1, 2, and so forth), so that no one accidentally saves a previous version and loses the most recent file.

➤ **Establish procedures for dealing with interpersonal problems.** How will gripes and disputes be aired and resolved (by vote, by the manager, or by some other means)? How will irrelevant discussion be curtailed?

➤ **Select a group decision-making style.** Will decisions be made alone by the group manager, or by group input or majority vote?

➤ **Decide how to evaluate each member's contribution.** Will the manager assess each member's performance and in turn be evaluated by each member? Will members evaluate each other? What are the criteria? Members might keep a journal of personal observations for overall evaluation of the project.

➤ **Prepare a project management plan.** Figure 5.1 shows a sample planning form. Keep this document in a shared space and update it only if all team members agree.

➤ **Submit regular progress reports.** These reports (see Chapter 20, "Progress Reports") track activities, problems, and rate of progress.

Source: Adapted from Debs, "Collaborative Writing" 38–41; Duin 45–50; Hulbert, "Developing" 53–54; McGuire 467–68; Morgan 540–41.

Teamwork: Virtual and Face to Face

5.2 Participate in virtual teams and conduct face-to-face meetings

In most of today's workplace settings, employees are spread out over different time zones and countries. And even within the same time zone, organizations often allow employees to work from home (telecommute) or to work different shifts or hours. Even in college, students working on team projects may need to meet in person and online, due to different class and work schedules.

Teamwork happens online and face to face

At work, you can expect to be involved in meetings where some team members are in the same room and others are connected via conference call, webinar, Skype, Google Hangouts, or other connections. Note that most organizations prefer secure apps versus public ones such as Skype.

- **Email:** Although the most popular tool for general workplace communication, email is the least effective way to hold a virtual meeting or deal with a complicated decision. Email does not allow for facial expressions, voice, or other social cues. (For more on workplace email, see Chapter 14.)

Apps for virtual meetings

- **Blogs:** Similar to email, blogs do not allow for social cues or the give and take that make meetings so effective. Yet blogs offer a single location to keep track of ideas. Instead of sorting through old emails, team members can go to the blog site and read through a chronological discussion. (For more on blogs, see Chapter 24.)

- **Conference calls:** Phone calls provide more cues (vocal inflection; pauses) than written text, allowing for the give and take necessary to come to consensus on complex issues. Many organizations use conference call services, where participants dial in and join the discussion. Ask one person to volunteer to open a shared document and take notes during the phone call.

- **Video conferencing:** Many organizations have options for team members to connect via video conference. As noted earlier in this chapter, companies generally prefer internal apps that provide security, although third-party programs such as GoToMeeting, WebEx, or Adobe Connect offer secure connections that create a meeting-like experience, with windows that display each participant and allow people to share presentations, documents, and other files.

- **Webinars:** A particular kind of video conference, webinars are Web-based seminars for giving presentations, conducting trainings, and the like via the Internet. (For more on video conferencing and Webinars, see Chapter 23.)

- **Digital whiteboards:** Useful as part of an Internet video conference, webinar, or conference call, digital whiteboards offer a large screen that allows participants to write, sketch, and revise in real time on their own computers.

- **Shared folders and collaborative writing spaces:** Applications such as Google Drive, Dropbox, and many others allow teams to set up shared folders. Inside these folders, you can store important project documents, including spread-sheets and presentations. These programs allow teams to "meet in the document" (log in to the document at the same time), write collaboratively, and use a chat window to discuss ideas. Although these sites are reasonably secure (one person, the document owner, sets up the people who can access the site, edit, or comment), most large organizations prefer to use similar shared spaces on their own internal file servers. (See Chapter 6 for more on digital technology and the writing process.)

- **Project management software:** To keep track of tasks, due dates, and respon-sibilities, software such as Microsoft Project, or even a simple spreadsheet, can ensure that after the meeting, everyone knows his or her role and what's due when.

Face-to-face meetings remain vital in the workplace

Despite the many digital tools for collaboration, face-to-face meetings are still a fact of life because they provide vital *personal contact.* Meetings are usually scheduled for two purposes: to convey or exchange information or to make deci-sions. Informational meetings tend to run smoothly because there is less cause for disagreement. But decision-making meetings often fail to reach clear resolution about various debatable issues. Such meetings often end in frustration because the leader has never managed to take charge. Even more frustrating is when a meeting is partially face to face and partially virtual. In these situations, the team leader needs to be especially sure to send out an agenda in advance, stick to the topics at hand, and make sure all participants—in the room and online—have a chance to contribute.

Taking charge doesn't mean imposing one's views or stifling opposing views. Taking charge *does* mean moving the discussion along and keeping it centered on the issue, as explained in the guidelines below.

Importance of meeting minutes

Someone should take minutes at each meeting. This process can be accomplished by assigning one person to be the minute-taker for all meetings or by doing a round-robin approach (everyone takes a turn). Meeting minutes can be brief and should be distributed (via email or on the company file server) right after the meeting, with opportunities for other team members to make corrections. Minutes help ensure that everyone has the same understanding of the meeting goals and outcomes. To ensure that the content stays intact, save the meeting minutes as a PDF file. (For more on meeting minutes, see Chapter 20.)

Guidelines

for Running a Meeting

➤ **Set an agenda.** Distribute copies to members beforehand (via email, a shared file system, on paper, or some combination most appropriate for your team). Be clear: "Our 10 a.m. Monday meeting will cover the following items:..." Spell out each item and set a strict time limit for discussion, and stick to this plan.

➤ **Ask each person to prepare as needed.** Meetings works best when each member has a role to play (e.g., prepares a specific contribution).

➤ **Appoint a different "observer" for each meeting.** At Charles Schwab & Co., the observer keeps a list of what worked well during the meeting and what didn't. The list is added to the meeting's minutes (Matson 31).

➤ **Begin by summarizing the minutes of the last meeting.** This step need not take long; you can provide bullet points in advance and then review quickly.

➤ **Give all members a chance to speak.** Don't allow anyone to monopolize. Be sure to include the people who are connected via conference call, video, or other app.

➤ **Stick to the issue.** Curb irrelevant discussion. Politely nudge members back to the original topic.

➤ **Keep things moving.** Don't get hung up on a single issue; work toward a consensus by highlighting points of agreement; push for a resolution.

➤ **Observe, guide, and listen.** Don't lecture or dictate.

➤ **Summarize major points before calling for a vote.**

➤ **End the meeting on schedule.** This is not a hard-and-fast rule. If you feel the issue is about to be resolved, continue.

NOTE For detailed advice on motions, debate, and voting, consult Robert's Rules of Order, *the classic guide to meetings, available online or in your library.*

Identifying and Managing Conflicts in Collaborative Groups

5.3   Recognize and negotiate conflicts in collaborative groups

Workplace surveys show that people view meetings as "their biggest waste of time" (Schrage 232). This fact alone accounts for the boredom, impatience, or irritability that might crop up in any meeting. But even the most dynamic group setting can produce conflict because of differences such as the following. (adapted from Bogert and Butt 51; Debs, "Collaborative Writing" 38; Duin 45–46; Nelson and Smith 61).

Interpersonal Differences

People might clash because of differences in personality, working style, commitment, standards, or ability to take criticism. Some might disagree about exactly what or how much the group should accomplish, who should do what, or who should have the

How personality influences communication

final say. Some might feel intimidated or hesitant to speak out. These interpersonal conflicts can actually worsen when the group interacts exclusively online: Lack of personal contact makes it hard for trust to develop.

Gender Differences

Gendered communication styles may influence team dynamics

Collaboration involves working with peers—those of equal status, rank, and expertise. But gender differences in communication styles can sometimes create perceptions of inequality. Studies over the years have shown that in certain workplace settings, gendered communication styles may influence behavior (Kelley-Reardon). For example, some research has shown that women may be more likely than men to take as much time as needed to explore an issue, build consensus and relationship among members, use tact in expressing view, and consider the listener's feelings.

Similarly, women may be more likely to make requests (*Could I have this report by Friday?*) versus give commands (*Have this ready by Friday.*) or to preface assertions in ways that avoid offending (*I don't want to seem disagreeable here, but...*). Any woman who breaches the gender code by being too assertive, for example, may be perceived as "too controlling" (Kelley-Reardon 6).

On the other hand, a different study on gender and communication conducted at a large Danish corporation found that both men and women preferred "an indirect, normatively feminine management style." Yet, despite this similarity in how men and women communicated, the authority of female managers was challenged more frequently than that of male managers (Ladegaard).

These studies show that it is not possible to generalize about gender differences in communication and that such traits are not always gender specific. People of either gender can be soft spoken and reflective. But these traits are often attributed to the "feminine" stereotype and, in this way, perceived gender differences in communication styles may influence team dynamics.

Cultural Differences

How culture influences communication

Another source of conflict in collaborative groups is the potential for misunderstandings based on cultural differences. Issues such as the use of humor, ways of expressing politeness, or cultural references (to sports or television shows, for example) could cause members of the group to understand ideas and meanings differently. See "Global Considerations When Working in Teams" in this chapter as well as information on intercultural communication in Chapters 2 and 3 for more on this topic.

Managing Group Conflict

No team will agree about everything. Before any group can reach final agreement, conflicts must be addressed openly. Management expert David House has this advice for overcoming personal differences (Warshaw 48):

- Give everyone a chance to be heard
- Take everyone's feelings and opinions seriously
- Don't be afraid to disagree
- Offer and accept constructive criticism
- Find points of agreement with others who hold different views
- When the group makes a decision, support it fully

How to manage group conflict

Business etiquette expert Ann Marie Sabath offers the following suggestions for reducing animosity (108–10):

- If someone is overly aggressive or keeps wandering off track, try to politely acknowledge valid reasons for such behavior: "I understand your concern about this, and it's probably something we should look at more closely." If you think the point has value, suggest a later meeting: "Why don't we take some time to think about this and schedule another meeting to discuss it?"
- Never attack or point the finger by using "aggressive 'you' talk": "You should," "You haven't," or "You need to realize." (See Chapter 11, "Using Passive Voice Selectively," for ways to avoid a blaming tone.)

On email and with social media, animosity and irritation often flair up quickly; people write and post with such speed that they often give little consideration to how their message will affect readers (and how it will affect the team and the project). Also, writing from the safety of a computer screen provides a kind of shield from the reaction the writer might receive if the same idea were expressed in person.

How to reduce animosity

Ultimately, collaboration requires compromise and consensus: Each person must give a little. Before your meeting, review Chapter 3, "Guidelines for Persuasion;" also, try really *listening* to what other people have to say. Set some ground rules about acceptable and unacceptable behaviors for the online components of future meetings and the project as a whole. For instance, if email is getting too confusing and team members seem frustrated, you might decide to call a meeting to clear the air and give everyone a chance to work out their differences in person.

Overcoming Differences by Active Listening

Listening is key to getting along, building relationships, and learning. In one manager survey, the ability to listen was ranked second (after the ability to follow instructions) among thirteen communication skills sought in entry-level graduates (cited in Goby and Lewis 42). Many of us seem more inclined to speak, to say what's on our minds, than to listen. We often hope someone else will do the listening. Effective listening requires *active* involvement instead of merely passive reception, as explained in the following guidelines. (Note that these guidelines apply mainly to face-to-face settings, but most of these ideas can and should be adapted to digital communication, including email, video conferencing, and the other technologies and apps described earlier in this chapter).

Listen actively to avoid conflict

Guidelines

for Active Listening

➤ **Don't dictate.** If you are the group moderator, don't express your view until everyone else has had a chance.

➤ **Be receptive.** Instead of resisting different views, develop a "learner's" mind-set: take it all in first, and evaluate it later.

➤ **Keep an open mind.** Reserve judgment until everyone has had their say.

➤ **Be courteous.** Don't smirk, roll your eyes, whisper, fidget, or wisecrack. Online, avoid angry, rude, accusatory language.

➤ **Show genuine interest.** Eye contact is vital, and so is body language (nodding, smiling, leaning toward the speaker). Make it a point to remember everyone's name. In digital writing, you can demonstrate interest by acknowledging the points made by others.

➤ **Hear the speaker out.** Instead of "tuning out" a message you find disagreeable, allow the speaker to continue without interruption (except to ask for clarification). Delay your own questions, comments, and rebuttals until the speaker has finished. Instead of blurting out a question or comment, raise your hand and wait to be recognized. With email, don't respond immediately (unless there is a deadline): take time to let the writer's intent really sink in before you compose your reply.

➤ **Focus on the message.** Instead of thinking about what you want to say or email next, try to get a clear understanding of the speaker's position.

➤ **Ask for clarification.** If anything is unclear, say so: "Can you run that by me again?" To ensure accuracy, paraphrase the message: "So what you're saying is. . . . Did I understand you accurately?" Whenever you respond, try repeating a word or phrase that the other person has just used. If email isn't helping clarify the situation, pick up the phone and give the person a call.

➤ **Observe the 90/10 rule.** You rarely go wrong spending 90 percent of your time listening, and 10 percent speaking. President Calvin Coolidge claimed that "Nobody ever listened himself out of a job." Some historians would argue that "Silent Cal" listened himself right into the White House.

Source: Adapted from Armstrong 24+; Cooper 78–84; Pearce, Johnson, and Barker 28–32; Sittenfeld 88; Smith 29.

Thinking Creatively

5.4 Brainstorm using face-to-face and digital methods

Team projects benefit when members are encouraged to think creatively, bringing innovative ideas into the mix before any final decisions are made. Instead of starting with a top-down project plan, teams that spend time engaged in creative thinking may come up with ideas that might never have made it into the mix if their process was constrained by linear thinking. Creative thinking is very productive in group settings, often by using one or more of the following techniques.

Brainstorm as a Way of Getting Started

The more ideas, the better

When we begin working with a problem, we search for useful material—insights, facts, statistics, opinions, images—anything that sharpens our view of the audience ("Who here needs what?"), the problem ("How can we increase market share for our new health care app?"), and potential solutions ("Which of these ideas might work best?"). *Brainstorming* is a creative technique for generating useful material well before any final decisions are made. The aim of brainstorming is to produce as many ideas as possible (on paper, screen, whiteboard, or the like) from team members' personal inventories.

1. **Choose a quiet setting and agree on a time limit.**
2. **Decide on a clear and specific goal for the session.** For instance, "We need at least five good ideas about why we are losing top employees to other companies."
3. **Focus on the issue or problem.**
4. **As ideas begin to flow, record every one.** Don't stop to judge relevance or worth, and don't worry about spelling or grammar. This step is very important in terms of allowing creative ideas to flow, regardless of how these ideas sound at the outset.
5. **If ideas are still flowing at session's end, keep going.**
6. **Take a short break.**
7. **Now confront your list.** Strike out what is useless and sort the remainder into categories. Include any new ideas that crop up.

> A procedure for brainstorming

Brainstorming with Digital Technologies

Brainstorming is effective in many types of team settings. Face-to-face groups work best, but with team members often scattered across different time zones and countries, digital technologies provide a range of options for brainstorming online. Shared documents (via Google Drive or similar applications) allow writers to contribute to the same document and use a chat window to discuss ideas at the same time. You can see ideas as they are being typed by others, comment on these ideas, and use the chat window to keep up a conversation. Apps such as Google Hangouts allow video conferencing and uploading and real-time editing of documents. Track Changes (see Figure 5.2), a feature available in most word-processing programs, also allows for a back-and-forth brainstorming of ideas.

> Tools for digital brainstorming

Mind-Mapping

A more structured version of brainstorming, *mind-mapping* helps group members visualize relationships. Team members begin by drawing a circle around the main issue or concept, centered on the paper or whiteboard. Related ideas are then added, each in its own box, connected to the circle by a ruled line (or "branch"). Other branches are then added, as lines to some other distinct geometric shape containing supporting ideas. Unlike a traditional outline, a mind-map does not require sequential thinking: As each idea pops up, it is connected to related ideas by its own branch. Mind-mapping software is used widely in many industries; apps such as Mindjet automate this process of visual thinking. You can search online examples of mind-mapping and related apps and processes.

> How mind-mapping works

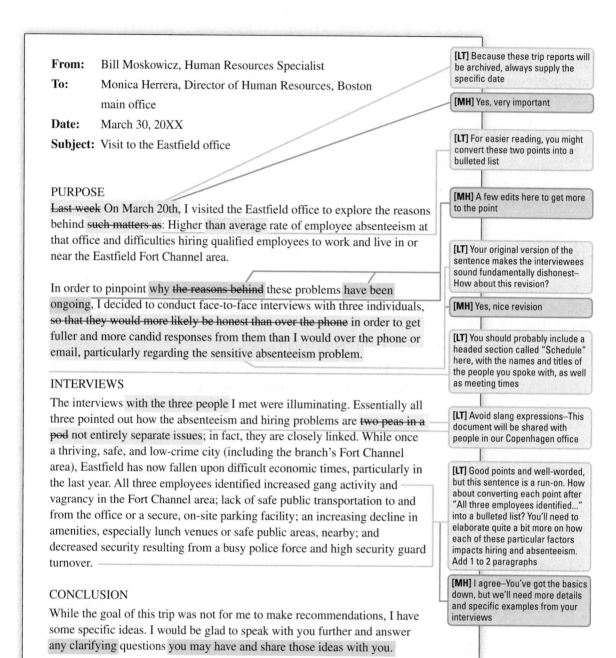

From: Bill Moskowicz, Human Resources Specialist

To: Monica Herrera, Director of Human Resources, Boston main office

Date: March 30, 20XX

Subject: Visit to the Eastfield office

PURPOSE

~~Last week~~ On March 20th, I visited the Eastfield office to explore the reasons behind ~~such matters as~~: Higher than average rate of employee absenteeism at that office and difficulties hiring qualified employees to work and live in or near the Eastfield Fort Channel area.

In order to pinpoint why ~~the reasons behind~~ these problems have been ongoing, I decided to conduct face-to-face interviews with three individuals, ~~so that they would more likely be honest than over the phone~~ in order to get fuller and more candid responses from them than I would over the phone or email, particularly regarding the sensitive absenteeism problem.

INTERVIEWS

The interviews with the three people I met were illuminating. Essentially all three pointed out how the absenteeism and hiring problems are ~~two peas in a pod~~ not entirely separate issues; in fact, they are closely linked. While once a thriving, safe, and low-crime city (including the branch's Fort Channel area), Eastfield has now fallen upon difficult economic times, particularly in the last year. All three employees identified increased gang activity and vagrancy in the Fort Channel area; lack of safe public transportation to and from the office or a secure, on-site parking facility; an increasing decline in amenities, especially lunch venues or safe public areas, nearby; and decreased security resulting from a busy police force and high security guard turnover.

CONCLUSION

While the goal of this trip was not for me to make recommendations, I have some specific ideas. I would be glad to speak with you further and answer any clarifying questions you may have and share those ideas with you.

Comment boxes:

[LT] Because these trip reports will be archived, always supply the specific date

[MH] Yes, very important

[LT] For easier reading, you might convert these two points into a bulleted list

[MH] A few edits here to get more to the point

[LT] Your original version of the sentence makes the interviewees sound fundamentally dishonest–How about this revision?

[MH] Yes, nice revision

[LT] You should probably include a headed section called "Schedule" here, with the names and titles of the people you spoke with, as well as meeting times

[LT] Avoid slang expressions–This document will be shared with people in our Copenhagen office

[LT] Good points and well-worded, but this sentence is a run-on. How about converting each point after "All three employees identified..." into a bulleted list? You'll need to elaborate quite a bit more on how each of these particular factors impacts hiring and absenteeism. Add 1 to 2 paragraphs

[MH] I agree–You've got the basics down, but we'll need more details and specific examples from your interviews

Figure 5.2 A Document That Has Been Edited Using a Track Changes System Notice how two different people, signified by their initials and different text colors, have read and edited Bill Moskowicz's first draft of a trip report. Compare this edited draft with the final version in Chapter 20 (Figure 20.4).

Storyboarding

A technique for visualizing the shape of an entire process, document, video, Web site, or other item is *storyboarding*. Group members write each idea and sketch each visual on a large index card. Cards are then displayed on a wall or bulletin board so that others can comment on or add, delete, refine, or reshuffle ideas, topics, and visuals (Kiely 35–36).

Many digital technologies are available to help you with storyboarding. Presentation software (such as PowerPoint or Google Slides) allows you to arrange images and ideas in a linear sequence that can be viewed and edited by the entire team. You can also use video and movie software, such as Apple's iMovie, to create a storyboard that can be played in real time.

How storyboarding works

Reviewing and Editing Others' Work

5.5 Review and edit the work of your peers

Documents produced collaboratively should always be reviewed and edited extensively. *Reviewing* means evaluating a document for audience and purpose considerations as well as technical accuracy. Reviewers typically examine a document to make sure it includes these features:

- Appropriate, useful content, suited for the audience and purpose
- Technical accuracy of the document's content
- Material organized for the reader's understanding
- Clear, easy-to-read, and engaging style
- Effective visuals and page design

What reviewers look for

When you are in the role of reviewer, you explain to the writer how you respond as a reader; you point out what works or doesn't work. This commentary helps a writer think about ways of revising. (Criteria for reviewing various documents appear in Checklists throughout this text.)

Editing means actually "fixing" the piece by making it more precise and readable. Editors typically suggest improvements like these:

- Rephrasing or reorganizing sentences
- Clarifying a topic sentence
- Choosing a better word or phrase
- Correcting spelling, usage, or punctuation, and so on

Ways in which editors "fix" writing

(Criteria for editing appear in Chapter 11 and in Appendix B: "A Quick Guide to Grammar, Usage, and Mechanics.")

> NOTE *Your job as a reviewer or editor is to help clarify and enhance a document—but without altering its original meaning.*

Guidelines

for Peer Reviewing and Editing

➤ **Read the entire piece at least twice before you comment.** Develop a clear sense of the document's purpose and audience. Try to visualize the document as a whole before you evaluate specific parts.

➤ **Remember that mere mechanical correctness does not guarantee effectiveness.** Poor usage, punctuation, or mechanics distract readers and harm the writer's credibility; however, a "correct" piece of writing might still contain faulty rhetorical elements (inferior content, confusing organization, or unsuitable style).

➤ **Understand the acceptable limits of editing.** In the workplace, editing can range from fine-tuning to an in-depth rewrite (in which case editors are cited prominently as consulting editors or coauthors). In school, however, rewriting a piece to the extent that it ceases to belong to its author may constitute plagiarism. (See Chapter 4, "Learn to Recognize Legal Issues and Plagiarism" and "Claiming Credit for Others' Work" in this chapter.)

➤ **Be honest but diplomatic.** Begin with something positive before moving to suggested improvements. Be supportive instead of judgmental.

➤ **Focus first on the big picture.** Begin with the content and the shape of the document. Is the document appropriate for its audience and purpose? Is the supporting material relevant and convincing? Is the discussion easy to follow? Does each paragraph do its job? Then discuss specifics of style and correctness (tone, word choice, sentence structure, and so on).

➤ **Always explain why something doesn't work.** Instead of "this paragraph is confusing," say "because this paragraph lacks a clear topic sentence, I had trouble discovering the main idea." Help the writer identify the cause of the problem.

➤ **Make specific recommendations for improvements.** Write out suggestions in enough detail for the writer to know what to do. Provide brief reasons for your suggestions.

➤ **Be aware that not all feedback has equal value.** Even professional editors can disagree. If different readers offer conflicting opinions of your own work, seek your instructor's advice.

Ethical Issues in Workplace Collaboration

5.6   Avoid unethical behavior as a team member

Teamwork versus survival of the fittest

Our "lean" and "downsized" corporate world sends coworkers a conflicting message, encouraging teamwork while "rewarding individual stars, so that nobody has any real incentive to share the glory" (Fisher, "Is My Team Leader" 291). The resulting mistrust may promote unethical behavior such as the following.

Intimidating One's Peers

A dominant personality may intimidate peers into silence or agreement (Matson 30). Intimidated employees resort to "mimicking"—merely repeating what the boss says (Haskin 55). In email, people often resort to intimidating, bullying language, saying things they might not have the courage to say to someone's face.

Claiming Credit for Others' Work

Workplace plagiarism occurs when the team or project leader claims all the credit. Even with good intentions, "the person who speaks for a team often gets the credit, not the people who had the ideas or did the work" (Nakache 287–88). Team expert James Stern describes one strategy for avoiding plagiarism among coworkers:

> Some companies list "core" and "contributing" team members, to distinguish those who did most of the heavy lifting from those who were less involved. (qtd. in Fisher, "Is My Team Leader" 291)

Stern advises groups to decide beforehand—and in writing—exactly who will be given what credit.

How to ensure that the deserving get the credit

Hoarding Information

Surveys reveal that the biggest obstacle to workplace collaboration is people's "tendency to hoard their own know-how" (Cole-Gomolski 6) when confronted with questions like these:

- Whom do we contact for what?
- Where do we get the best price, the quickest repair, the best service?
- What's the best way to do X?

Information people need to do their jobs

People hoard information when they think it gives them power or self-importance, or when having exclusive knowledge might provide job security (Devlin 179). In a worse case, they withhold information when they want to sabotage peers.

Global Considerations When Working in Teams

5.7   Understand how to work productively on a global team

In today's workplace, teams are often composed of people from many different cultures and countries who are spread throughout different time zones and different continents. Cultures and countries have different norms of communication, and all members of any team need to understand the cultures and personalities of other team members.

Teamwork is an increasingly global activity

> NOTE *Each team member is an individual person, not a stereotype, and so both the culture and the person need to be taken into consideration.*

In many global organizations, teams meet virtually, using a variety of technologies (see list of digital collaboration technologies and apps earlier in this chapter).

Interpersonal Issues in Global Teams

Whether working virtually or face to face, teams often experience unique interpersonal issues when members are from different countries and cultures. Many cultures value the social (or relationship) function of communication as much as the informative function (Archee 41). Some of these issues are discussed below.

Virtual meetings may lack important social cues

DIGITAL COMMUNICATION AND SOCIAL CUES. For global teams, face-to-face meetings may be impossible, Yet digital communication—even video—may leave out important social cues that would otherwise be visible in person. Such cues include age, gender, appearance, ethnicity, team status, facial features, seating position, and more (Sproull and Kiesler 40–54; Wojahn 747–48). These cues provide important information, such as who is sitting at the head of the table, how people are reacting to what one is saying, and whether the team is made up of international members.

In email, people may adopt a writing style that sounds friendly and conversational to make up for lack of eye contact and speaking tone. But excessive informality may be interpreted differently by people in different cultures. (See Chapter 14, "Global Considerations When Using Email," for more on global considerations when using email.)

Become familiar with cultural codes

MISUNDERSTANDING CULTURAL CODES. International business expert David A. Victor describes cultural codes that influence group interactions: Some cultures value silence more than speech, intuition and ambiguity more than hard evidence or data, and politeness and personal relations more than business relationships. Cultures also differ in their perceptions of time. Some are "all business" and in a big rush; others take as long as needed to weigh the issues, engage in small talk and digressions, and chat about family, health, and other personal matters (233). In Anthony's email (Figure 5.3), the writer's hurried approach ("Let's get right down to business"; "We need to get moving right away") and directness ("Marcus, can you handle this?") might be offensive in cultures that value a more personal and subtle approach.

Individuals from various cultures may differ in their willingness to express disagreement; to question or be questioned; to leave things unstated; or to touch, shake hands, kiss, hug, or backslap. Also, direct eye contact is not always a good indicator of listening; some cultures find it offensive. Other eye movements, such as squinting, closing the eyes, staring away, staring at legs or other body parts, are acceptable in some cultures but insulting in others (Victor 206). Finally, some cultures value formality over informality. Anthony's email (Figure 5.3) might be perceived as impolite and overly informal by members of some cultures.

Be careful with humor, slang, and idioms

MISUSING HUMOR, SLANG, AND IDIOMS. Humor often can relieve tension, especially in a team meeting where tempers might flair or personalities might clash. But humor is culturally dependent. Its timing and use differ greatly among cultures; moreover, the examples used in a joke are often specific to one culture. For instance, a joke that references a sport such as U.S. football may fall flat for people from countries

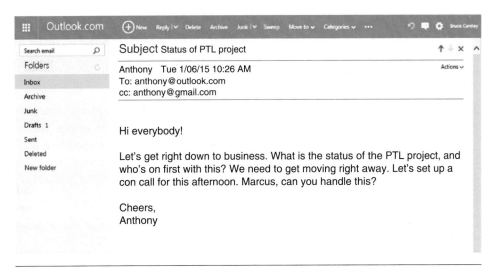

Figure 5.3 An Inappropriate Email Message for a Global Audience Notice the variety of ways in which Anthony's email fails to successfully address international and culturally diverse readers. *Source:* Microsoft Outlook 2013.

where soccer is "football." Also, what some people consider funny may insult others. Avoid, for example, jokes at the expense of others (even if meant affectionately).

Slang means informal words or phrases, such as *bogus* or *cool*. Slang makes sense to those inside the culture but not to those outside. Idioms are phrases that have a meaning beyond their literal meaning, such as "all bark and no bite," which would make no sense to anyone unfamiliar with the meaning behind this phrase. Anthony's email in Figure 5.3 uses a slang phrase ("con call") and an idiom ("who's on first with this?"). (See Chapter 11 for more on culture and writing style.)

MISUSING CULTURALLY SPECIFIC REFERENCES. References to television shows, movie stars, sports, politics, locations, and so forth are geographically as well as culturally dependent and may not make sense to everyone on the team. In Anthony's email (Figure 5.3), "who's on first with this?" is both an idiom and a culturally specific reference to baseball.

> Avoid culturally dependent references

FAILING TO ALLOW FOR EASY TRANSLATION. Often, the documents your team produces will need to be translated into a number of different languages. Be careful to use English that is easy to translate. Idioms, humor, and analogies are often difficult for translators. Also, certain grammatical elements are important for translation. The lack of an article (*a, the*) or of the word *that* in certain crucial places can cause a sentence to be translated inaccurately. Consider the following examples (Kohl 151):

> Use language that can be translated easily

Programs **that are** currently running in the system are indicated by icons in the lower part of the screen.

Programs currently running in the system are indicated by icons in the lower part of the screen.

The first sentence contains the phrase "that are," which might ordinarily be left out by native English writers, as in the second sentence. This second sentence is harder to translate because the phrase "that are" provides the translator with important clues about the relationship of the words *programs, currently,* and *running.*

Active "listening" occurs virtually as well as face-to-face

FAILING TO LISTEN.   Active listening, described earlier in this chapter, is an excellent way to learn more about your colleagues. As noted in that section, active "listening" can also occur when you read email or text messages; instead of responding with your usual style and tone, take note of the approach of each team member, tuning your eye and ear to that person's particular way of analyzing problems, presenting information, and persuading.

Guidelines
for Communicating on a Global Team

➤ **Use the right technology for the situation.** Email and shared documents are great for keeping track of discussions and the project's status, but video or phone conferencing are important to keep the lines of communication open and clear.

➤ **Remember that social cues are not conveyed well in digital communication.** Email and other text-based systems leave out crucial information such as body language, facial expressions, and gender. Even video chats do not always convey social cues in full.

➤ **Consider how cultural codes might influence team dynamics.** Some team members may want to get right down to business, whereas others may value small talk or forms of politeness before getting started.

➤ **Avoid humor, slang, idioms, and cultural references.** These items are not understood identically by people from different cultures and countries, especially if different languages are involved.

➤ **Write with translation in mind.** Keep your writing simple, with short sentences and nothing too complex in structure that might confuse a translator. (If your company does a lot of translation, there will usually be a techncial editor on hand who can give you more tips.)

➤ **Create a glossary so that everyone is using identical vocabulary.** Team members likely will be using specific terminology and abbreviations. A glossary (which can be kept updated on a wiki or on a shared collaborative document) ensures that everyone understands and uses terms identically. (For more on glossaries, see Chapter 17, "Placing Definitions in a Document.")

➤ **Agree in advance on technical standards.** For instance, will your team use the metric system (meters, liters) or the Imperial unit system (feet, gallons)? Remember that dates and times are expressed differently in different countries (e.g., 12-20-2018 in the United States would be written as 20-12-2018 in many European countries).

➤ **Be polite and professional.** All around the globe, respectful behavior and professional communication is always appreciated. Phrases such as *thank you, I appreciate it,* and *you're welcome* are always valued.

➤ **Be respectful of the rank and status of all team members, even if you are the team leader.** Many cultures are quite sensitive to rank.

➤ **Be an active listener.** Remember that listening, and learning about others, can occur in a meeting or over email.

➤ **Show respect for differences by choosing words carefully.** *Third world country* is disrespectful because the phrase implies inequality; instead, say "emerging markets" or "developing nations." Spell the names of people and their countries and cities using the correct accent marks and other symbols. Learn to pronounce the names of all team members.

➤ **Use visual information carefully.** Not all symbols have the same meaning across cultures. When possible, use internationally identifiable symbols such as those endorsed by the International Standards Organization (ISO). (See Chapter 12 for more information.)

Checklist

Teamwork and Global Considerations

Use the following Checklist when working in a team.

Teamwork

❑ Have we appointed a team manager? (See "Guidelines for Managing a Collaborative Project" in this chapter.)

❑ Do we have a plan for how to divide the tasks? (See "Guidelines for Managing a Collaborative Project" in this chapter.)

❑ Have we established a timetable and decided on a meeting schedule? (See "Guidelines for Managing a Collaborative Project" in this chapter.)

❑ Do we have an agenda for our first meeting? (See "Guidelines for Running a Meeting" in this chapter.)

❑ Do we have a clear understanding of how we will share drafts of the document and name our files? (See "Guidelines for Managing a Collaborative Project" in this chapter.)

❑ Have we decided what tools to use (Track Changes, Google docs, other)? (See Figure 5.2 in this chapter.)

❑ Are we using the Project Planning Form? (See Figure 5.1 in this chapter.)

Running a Meeting

❑ Has the team manager created an agenda and circulated it in advance? (See "Guidelines for Running a Meeting" in this chapter.)

❑ Do members understand their individual roles on the team so they can be prepared for the meeting? (See "Guidelines for Managing a Collaborative Project" in this chapter.)

❑ Has someone been appointed to take meeting minutes? (See "Guidelines for Running a Meeting" in this chapter.)

❑ Does the team manager keep discussion focused on agenda items? (See "Guidelines for Running a Meeting" in this chapter.)

❑ Does the meeting end on schedule? (See "Guidelines for Running a Meeting" in this chapter.)

Active Listening

❑ Are team members receptive to each other's viewpoints? (See "Guidelines for Active Listening" in this chapter.)

❑ Does everyone communicate with courtesy and respect? (See "Guidelines for Active Listening" in this chapter.)

❑ In face-to-face settings, are all team members allowed to speak freely? (See "Guidelines for Active Listening" in this chapter.)

❑ Are interruptions discouraged? (See "Guidelines for Active Listening" in this chapter.)

Continued

❏ In email, do people take time to reflect on ideas before responding? (See Chapter 14, "Email Style and Tone.")

❏ Do people observe the 90/10 rule (listen 90% of the time; speak 10% of the time)? (See "Guidelines for Active Listening" in this chapter.)

Peer Review and Editing

❏ Have I read the entire document twice before I make comments? (See "Guidelines for Peer Review and Editing" in this chapter.)

❏ Have I focused on content, style, and logical flow of ideas before looking at grammar, spelling, and punctuation? (See "Guidelines for Peer Review and Editing" in this chapter.)

❏ Do I know what level of review and editing is expected of me (focus only on the content, or focus on style, layout, and other factors)? (See "Guidelines for Peer Review and Editing" in this chapter.)

❏ Am I being honest but polite and diplomatic in my response? (See "Guidelines for Peer Review and Editing" in this chapter.)

❏ Do I explain exactly why something doesn't work? (See "Guidelines for Peer Review and Editing" in this chapter.)

❏ Do I make specific recommendations for improvements? (See "Guidelines for Peer Review and Editing" in this chapter.)

Global Considerations

❏ Do I understand the communication customs of the international audience for my document? (See "Global Considerations When Working in Teams" and "Guidelines for Communicating on a Global Team" in this chapter.)

❏ Have I avoided humor, idioms, and slang? (See "Global Considerations When Working in Teams" and "Guidelines for Communicating on a Global Team" in this chapter.)

❏ Have I avoided stereotyping of different cultures and groups of people? (See "Global Considerations When Working in Teams" and "Guidelines for Communicating on a Global Team" in this chapter.)

❏ Does my document avoid cultural references (such as TV shows and sports), which may not make sense to a global audience? (See "Global Considerations When Working in Teams" and "Guidelines for Communicating on a Global Team" in this chapter.)

Projects

For all projects, check with your instructor about whether to present your findings in class, bring drafts to class for discussion, upload your project to the class learning management system (LMS), and/or use the LMS forum or discussion boards to collaborate and review each activity below.

General

Describe the role of collaboration in a company, organization, or campus group where you have worked or volunteered. Among the questions: What types of projects require collaboration? How are teams organized? Who manages the projects? How are meetings conducted? Who runs the meetings? How is conflict managed? Summarize your findings in a one- or two-page memo.

Hint: If you have no direct experience, interview a school administrator, faculty member, or editor of the campus newspaper for input on how they collaborate and work with teams. (See Chapter 7, "Guidelines for Informational Interviews" for interview advice.)

Team

At school and in the workplace, much emphasis is placed on writing and speaking clearly. But not enough focus is given to the importance of active listening. As noted in this chapter, listening is key to getting along with others and to building strong relationships. In teams of 3–4, review the "Guidelines for Active Listening" in this chapter, then, practice by having a discussion about the Digital and Social Media project on this page. Take turns listening to each person describe his or her experience with collaborative writing tools. Pay special attention to showing genuine interest (eye contact, body language), letting the speaker finish without interruption, and asking clarification questions. When each person has had a turn, appoint one person to present your observations to the entire class.

Digital and Social Media

Use a collaborative writing app (such as Google Drive or Microsoft Office Online or a writing space that is part of your class learning management system) to work together as a team on a group memo such as the memo described in the Global project on this page. Notice how the collaborative writing tool affects team interactions. If using Google Drive's document app, experiment with using the real-time chat window while drafting your memo. When your project is completed, prepare a brief presentation for class about the use of collaborative writing tools.

Global

Teams increasingly comprise people from different countries and cultures. Go online and search for "international teams communication best practices" and carefully review your findings. Look for articles and ideas from credible sources (business sections of major newspapers; business magazines; journal articles; blogs from international companies). Compare your findings with those of 2–3 other students and write a 2–3 page summary (include references; see Chapter 9 for more on summaries). Share with your instructor and class via your class learning management system (a discussion forum or whatever is most appropriate for your class).

Chapter 6

An Overview of the Technical Writing Process

Weekend Images Inc/E+/Getty Images

❝Deadlines affect how our team approaches the writing process. With plenty of time, we can afford the luxury of the whole process: careful decisions about audience, purpose, content, organization, and style—and plenty of revisions. With limited time, we have limited revisions—so we have to think and work quickly. But even when deadlines are tight, we try our best to understand how our audience thinks: 'How can we make this information logical to our audience? Will readers understand what we want them to understand?'❞

—*Blair Cordasco, Training Specialist for an international bank*

Learning Objectives

6.1 Apply critical thinking to the four
stages of the writing process

6.2 Identify the gathering, planning,
drafting, and revising stages of the
writing process

6.3 Understand why proofreading is an
important final step

6.4 Consider the use of digital and
collaborative writing tools and
applications

Although the writing process (researching, planning, drafting, and revising) is similar
across all disciplines, the process for technical writing differs from the process for essay and other forms of college writing in ways such as the following:

Factors that influence the technical writing process

- Research often involves discussions with technical experts.
- Analysis of audience needs and expectations is critical.
- Complex organizational settings and "office politics" play an important role.
- Colleagues frequently collaborate in preparing a document.
- Collaboration usually takes place via digital writing and collaboration apps.
- Many workplace documents are carefully reviewed before being released.
- Proper format (letter, memo, report, brochure, and so on) for a document is
  essential.
- Proper distribution medium (hard copy, digital, or both) is essential.
- Deadlines often limit the amount of time that can be spent preparing a document.

In order to navigate these types of complex decisions, technical communicators rely on
the critical thinking strategies discussed and illustrated throughout this chapter.

In *critical thinking,* you test the strength of your ideas or the quality of your information. Instead of accepting an idea at face value, you examine, evaluate, verify,
analyze, weigh alternatives, and consider consequences—at every stage of that idea's

Definition of critical thinking

development. You use critical thinking to examine your evidence and your reasoning, to discover new connections and new possibilities, and to test the effectiveness and the limits of our solutions.

Critical Thinking in the Writing Process

6.1 Apply critical thinking to the four stages of the writing process

Whether you are working alone or as part of a team, you apply critical thinking throughout the four stages in the technical writing process:

Stages in the writing process

1. Gather and evaluate ideas and information.
2. Plan the document.
3. Draft the document.
4. Revise the document.

As the arrows in Figure 6.1 indicate, no single stage of the writing process is complete until all stages are complete.

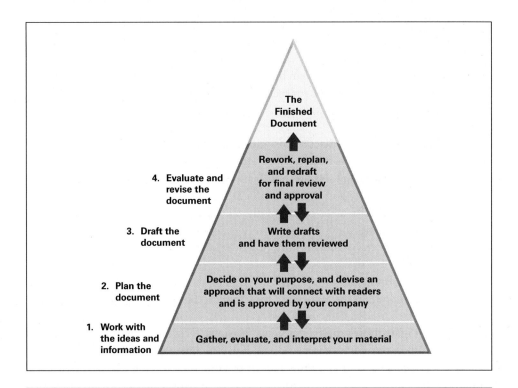

Figure 6.1 **The Writing Process for Technical Documents** Like the exposed tip of an iceberg, the finished document provides the only visible evidence of your (or your team's) labor in preparing it.

Figure 6.2 lists the kinds of questions you answer at various stages. On the job, you must often complete these stages under deadline pressure.

This next section will follow one working writer through an everyday writing situation. You will see how he approaches his unique informational, persuasive, and ethical considerations and how he collaborates to design a useful and efficient document.

---

1. Work with the ideas and information:

- Have I defined the issue accurately?
- Is the information I've gathered complete, accurate, reliable, and unbiased?
- Can it be verified?
- How much of it is useful?
- Is a balance of viewpoints represented?
- What do these facts mean?
- What conclusions seem to emerge?
- Are other interpretations possible?
- What, if anything, should be done?
- What are the risks and benefits?
- What other consequences might this have?
- Should I reconsider?

2. Plan the document:

- What do I want it to accomplish?
- Who is my audience, and why will they use this document?
- What do they need to know?
- What are the "political realities" (feelings, egos, cultural differences, and so on)?
- How will I organize?
- What format and visuals should I use?
- Whose help will I need?
- When is it due?

3. Draft the document:

- How do I begin, and what comes next?
- How much is enough?
- What can I leave out?
- Am I forgetting anything?
- How will I end?
- Who needs to review my drafts?
- What are the best tools for collaboration in this case (Word document using Track Changes? Shared Google or other shared document system?)

4. Evaluate and revise the document:

- Does this draft do what I want it to do?
- Is the content useful?
- Is the organization sensible?
- Is the style readable?
- Is everything easy to find?
- Is the format appealing?
- Is everything accurate, complete, appropriate, and correct?
- Is the information honest and fair?
- Who needs to review and approve the final version?
- Does it advance my organization's goals?
- Does it advance my audience's goals?
- What is the best way to deliver the document: As an email attachment? As a shared document? On paper?

Figure 6.2 **Critical Thinking in the Technical Writing Process** The actual "writing" (putting words on the page) is only a small part of the overall process.

A Sample Writing Situation

6.2 **Identify the gathering, planning, drafting, and revising stages of the writing process**

The setting

The company is Microbyte, developer of security software. The writer is Glenn Tarullo (BS, Management; Minor, Computer Science). Glenn has been on the job three months as Assistant Training Manager for Microbyte's Marketing and Customer Service Division.

The assignment

For three years, Glenn's boss, Marvin Long, has periodically offered a training program for new managers. Long's program combines an introduction to the company with instruction in management skills (time management, motivation, communication). Long seems satisfied with his two-week program but has asked Glenn to evaluate it and write a report as part of a company move to upgrade training procedures.

The audience

Glenn knows his report will be read by Long's boss, George Hopkins (Assistant Vice President, Personnel) and Charlotte Black (Vice President, Marketing, the person who devised the upgrading plan). Copies will go to other division heads, to the division's chief executive, and to Long's personnel file.

This writer's problem

Glenn spends two weeks (Monday, October 3 to Friday, October 14) attending and taking notes in Long's classes. On October 14, the trainees evaluate the program. After reading these evaluations and reviewing his notes, Glenn concludes that the program was successful but could use some improvement. How can Glenn be candid without harming or offending anyone (instructors, his boss, or guest speakers)? Figure 6.3 depicts Glenn's problem.

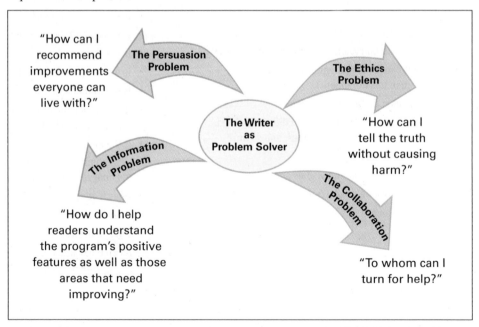

Figure 6.3 **Glenn's Fourfold Problem** Everyday writing situations typically pose similar problems.

Glenn is scheduled to present his report in conference with Long, Black, and Hopkins on Wednesday, October 19. Right after the final class (1 P.M., Friday, the 14th), Glenn begins work on his report.

The deadline

Working with the Information

Glenn spends half of Friday afternoon fretting over the details of his situation, the readers and other people involved, the political realities, constraints, and consequences. (He knows about interpersonal communication issues between Long and Black, and he wants to steer clear of their ongoing conflict.) By 3 P.M., Glenn hasn't written a word. Desperate, he decides to write whatever comes to mind:

Getting started

> Although the October Management Training Session was deemed quite successful, several problems have emerged that require our immediate attention.
>
> - Too many of the instructors had poor presentation skills. A few never arrived on time. One didn't stick to the topic but rambled incessantly. Jones and Wells seemed poorly prepared. Instructors in general seemed to lack any clear objectives. Also, because instructors all used the same PowerPoint format, with no visuals, many presentations seemed to bore the trainees.
> - The trainees (all new people) were not at all cognizant of how the company was organized or functioned, so the majority of them often couldn't relate to what the speakers were talking about.
> - It is my impression that this was a weak session due to the fact that there were insufficient members (only five trainees). Such a small class makes the session a waste of time and money. For instance, Lester Beck, Senior Vice President of Personnel, came down to spend over an hour addressing only a handful of trainees. Another factor is that with fewer trainees in a class, less dialogue occurs, with people tending to just sit and get talked at.
> - Last but not least, executive speakers generally skirted the real issues, saying nothing about what it was really like to work here. They never really explained how to survive politically (e.g., never criticize your superior; never complain about the hard work or long hours; never tell anyone what you *really* think; never observe how few women are in executive or managerial positions or how disorganized things seem to be). New employees shouldn't have to learn these things the hard way.
>
> In the final analysis, if these problems can be addressed immediately, it is my opinion we can look forward in the future to effectuating management training sessions of even higher quality than those we now have.

Glenn's first draft

Glenn completes this draft at 5:10 P.M. Displeased with the results but not sure how to improve the piece, he asks an experienced colleague for advice and feedback. Blair Cordasco, a senior project manager, has collaborated with Glenn on several earlier projects. Blair agrees to study Glenn's draft over the weekend. Because of this document's sensitive and complex topic, they agree to a phone call to discuss the document on Monday morning. Using the secure company server, Glenn uploads the draft and shares it with Blair.

The request for help

At 8:05 A.M. Monday, Blair reviews the document with Glenn, looking over some ideas that Blair inserted using Track Changes. First, she points out obvious style problems: wordiness ("due to the fact that"), jargon ("effectuating"), triteness ("in the final analysis"), implied bias ("weak presentation," "skirted"), among others.

The colleague's peer review

NOTE See Chapter 5, Figure 5.2 for an example of how tracking is used to share comments.

Blair points out other problems. The piece is disorganized, and even though Glenn is being honest, he isn't being particularly fair. The emphasis is too critical (making Glenn's boss look bad to his superiors), and the views are too subjective (no one is interested in hearing Glenn gripe about the company's political problems). Moreover, the report lacks persuasive force because it contains little useful advice for solving the problems he identifies. The tone is bossy and judgmental. Glenn is in no position to make this kind of *power connection* (see Chapter 3, "Know How to Connect with the Audience"). In its current form, the report will only alienate people and harm Glenn's career. He needs to be more fair, diplomatic, and reasonable.

Planning the Document

Glenn realizes he needs to begin by focusing on his writing situation. His audience and use analysis goes like this (note that throughout this section, Glenn's analysis will address the areas illustrated in the Audience and Use Profile Sheet in Figure 3.5):

I'd better decide *exactly* what my primary reader wants.

Long requested the report, but only because Black developed the scheme for division-wide improvements. So I really have two primary readers: my boss and the big boss.

My major question here: Am I including enough detail for all the bosses? The answer to this question will require answers to more specific questions:

Anticipated readers' questions

What are we doing right, and how can we do it better?

What are we doing wrong, and does it cost us money?

Have we left anything out, and does it matter?

How, specifically, can we improve the program, and how will those improvements help the company?

Because all readers have participated in these sessions (as trainees, instructors, or guest speakers), they don't need background explanations.

I should begin with the *positive* features of the last session. Then I can discuss the problems and make recommendations. Maybe I can eliminate the bossy and judgmental tone by *suggesting improvements* instead of *criticizing weaknesses*. Also, I could be more persuasive by describing the *benefits* of my suggestions.

Glenn realizes that if he wants successful future programs, he can't afford to alienate anyone. After all, he wants to be seen as a loyal member of the company yet preserve his self-esteem and demonstrate he is capable of making objective recommendations:

Now, I have a clear enough sense of what to do.

Audience and purpose statement

The purpose of my document is to provide my supervisor and interested executives with an evaluation of the workshop by describing its strengths, suggesting improvements, and explaining the benefits of these changes.

From this plan, I should be able to revise my first draft, but that first draft lacks important details. I should brainstorm to get *all* the details (including the *positive* ones) I want to include.

Glenn's first draft touched on several topics. Incorporating them into his brainstorming (see Chapter 5, "Thinking Creatively"), he comes up with the following list.

Glenn's brainstorming list

1. Better-prepared instructors and more visuals
2. On-the-job orientation *before* the training session
3. More members in training sessions
4. Executive speakers should spell out qualities needed for success
5. Beneficial emphasis on interpersonal communication
6. Need follow-up evaluation (in six months?)
7. Four types of training evaluations:
   a. Trainees' reactions
   b. Testing of classroom learning
   c. Transference of skills to the job
   d. Effect of training on the organization (high sales, more promotions, better-written reports)
8. Videotaping and critiquing of trainee speeches worked well
9. Acknowledge the positive features of the session
10. Ongoing improvement ensures quality training
11. Division of class topics into two areas was a good idea
12. Additional trainees would increase classroom dialogue
13. The more trainees in a session, the less time and money wasted
14. Instructors shouldn't drift from the topic
15. On-the-job training to give a broad view of the division
16. Clear course objectives to increase audience interest and to measure the program's success
17. Marvin Long has done a great job with these sessions over the years

By 9:05 A.M., the office is hectic. Glenn puts his list aside to spend the day on work that has been piling up. He returns to his report at 4 P.M.

Now what? I should delete whatever my audience already knows or doesn't need, or whatever seems unfair or insincere: Item 7 can go (this audience needs no lecture in training theory); 14 is too negative and critical—besides, the same idea is stated more positively in 4; 17 is obvious brown-nosing, and I'm in no position to make such grand judgments.

Maybe I can unscramble this list by arranging items within categories (strengths, suggested changes, and benefits) from my statement of purpose.

Notice here how Glenn discovers additional *content* (see italics type) while he's deciding about *organization.*

Strengths of the Workshop

Glenn's brainstorming list rearranged

- Division of class topics into two areas was useful
- Emphasis on interpersonal communication
- Videotaping of trainees' oral reports, followed by critiques

Well, that's one category done. Maybe I should combine *suggested changes* with *benefits* since I'll wantw to cover them together in the report.

Suggested Changes/Benefits

- More members per session would increase dialogue and use resources more efficiently.
- Varied on-the-job experiences before the training sessions would give each member a broad view of the marketing division.
- Executive speakers should spell out qualities required for success, and *future sessions should cover professional behavior to provide trainees with a clear guide.*
- Follow-up evaluation in 6 months *by both supervisors and trainees would reveal the effectiveness of this training and suggest future improvements.*
- Clear course objectives and more visual aids would increase *instructor efficiency* and audience interest.

Now that he has a fairly sensible arrangement, Glenn can get this list into report form, even though he will probably think of more material to add as he works. Since this is *internal* correspondence, he uses a memo format.

Drafting the Document

Glenn produces a usable draft—one containing just about everything he wants to cover. (Note: Sentences are numbered for our later reference. These numbers are not footnotes.)

A later draft

[1]In my opinion, the Management Training Session for the month of October was somewhat successful. [2]This success was evidenced when most participants rated their training as "very good." [3]But improvements are still needed.

[4]First and foremost, a number of innovative aspects in this October session proved especially useful. [5]Class topics were divided into two distinct areas. [6]These topics created a general-to-specific focus. [7]An emphasis on interpersonal communication skills was the most dramatic innovation. [8]This change helped class members develop a better attitude toward things in general. [9]Videotaping of trainees' oral reports, followed by critiques, helped clarify strengths and weaknesses.

[10]A detailed summary of the trainees' evaluations is attached. [11]Based on these and my past observations, I have several suggestions.

- [12]All management training sessions should have a minimum of 10–15 members. [13]This would better utilize the larger number of managers involved and the time expended in the implementation of the training. [14]The quality of class interaction with the speakers would also be improved with a larger group.
- [15]There should be several brief on-the-job training experiences in different sales and service areas. [16]These should be developed prior to the training session. [17]This would provide each member with a broad view of the duties and responsibilities in all areas of the marketing division.
- [18]Executive speakers should take a few minutes to spell out the personal and professional qualities essential for success with our company. [19]This would provide trainees with a concrete guide to both general company and individual supervisors' expectations. [20]Additionally, by the next training session, we should develop a presentation dealing with appropriate attitudes, manners, and behavior in the business environment.

- [21]Do a 6-month follow-up. [22]Get feedback from supervisors as well as trainees. [23]Ask for any new recommendations. [24]This would provide a clear assessment of the long-range impact of this training on an individual's job performance.
- [25]We need to demand clearer course objectives. [26]Instructors should be required to use more visual aids and improve their course structure based on these objectives. [27]This would increase instructor quality and audience interest.
- [28]These changes are bound to help. [29]Please contact me if you have further questions.

Although now developed and organized, this version still is not near the finished document. Glenn has to make further decisions about his style, content, arrangement, audience, and purpose.

Blair Cordasco offers to review the piece once again and to work with Glenn on a thorough edit.

Revising the Document

At 8:15 A.M. Tuesday, Blair and Glenn set up a phone call and, with the shared document on the screen, begin a sentence-by-sentence revision for worthwhile content, sensible organization, and readable style. Their discussion goes something like this:

Sentence 1 begins with a needless qualifier, has a redundant phrase, and sounds insulting ("somewhat successful"). Sentence 2 should be in the passive voice, to emphasize the training—not the participants. Also, 1 and 2 are choppy and repetitious and should be combined. Here are your original two sentences followed by a one-sentence revision:

In my opinion, the Management Training Session for the month of October was somewhat successful. This success was evidenced when most participants rated their training as "very good." (28 words)	Original
The October Management Training Session was successful, with training rated "very good" by most participants. (15 words)	Revised

Sentence 3 is too blunt. An orienting sentence should forecast content diplomatically. This statement can be candid without being so negative. Here is your blunt sentence, followed by a more diplomatic revision:

But improvements are still needed.	Original
A few changes—beyond the recent innovations—should result in even greater training efficiency.	Revised

In sentence 4, "First and foremost" is trite, "aspects" only adds clutter, and word order needs changing to improve the emphasis (on innovations) and to lead into the examples. Here is your original sentence followed by a revised version:

First and foremost, a number of innovative aspects in this October session proved especially useful.	Original
Especially useful in this session were several program innovations.	Revised

In collaboration with his colleague, Glenn continues this editing and revising process on his report. Wednesday morning, after much revising and proofreading, Glenn puts the final draft on letterhead and creates a PDF, shown in Figure 6.4.   The final version

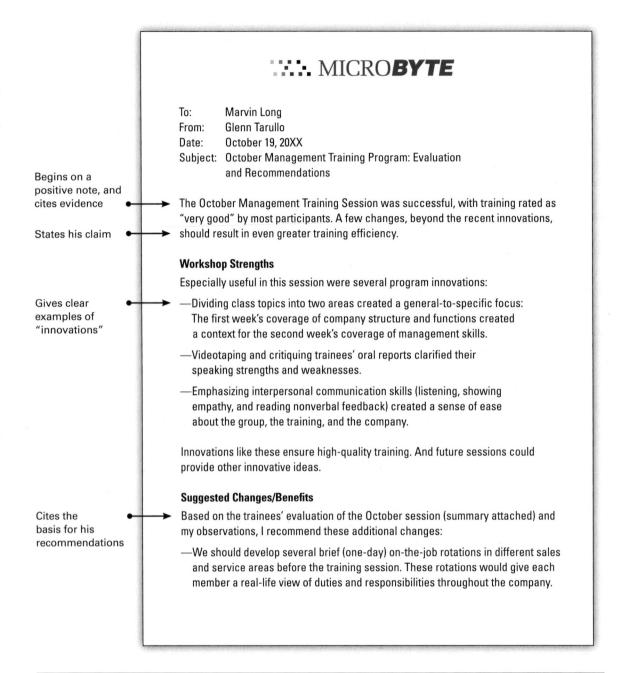

Begins on a positive note, and cites evidence

States his claim

Gives clear examples of "innovations"

Cites the basis for his recommendations

MICROBYTE

To: Marvin Long
From: Glenn Tarullo
Date: October 19, 20XX
Subject: October Management Training Program: Evaluation and Recommendations

The October Management Training Session was successful, with training rated as "very good" by most participants. A few changes, beyond the recent innovations, should result in even greater training efficiency.

Workshop Strengths

Especially useful in this session were several program innovations:

—Dividing class topics into two areas created a general-to-specific focus: The first week's coverage of company structure and functions created a context for the second week's coverage of management skills.

—Videotaping and critiquing trainees' oral reports clarified their speaking strengths and weaknesses.

—Emphasizing interpersonal communication skills (listening, showing empathy, and reading nonverbal feedback) created a sense of ease about the group, the training, and the company.

Innovations like these ensure high-quality training. And future sessions could provide other innovative ideas.

Suggested Changes/Benefits

Based on the trainees' evaluation of the October session (summary attached) and my observations, I recommend these additional changes:

—We should develop several brief (one-day) on-the-job rotations in different sales and service areas before the training session. These rotations would give each member a real-life view of duties and responsibilities throughout the company.

Figure 6.4 Glenn's Final Draft Note how this seemingly routine document is the product of a complex—but vital—process. (For a vivid example of how critical thinking pays off, compare this final draft with Glenn's first draft, in "Working with the Information" in this chapter.)

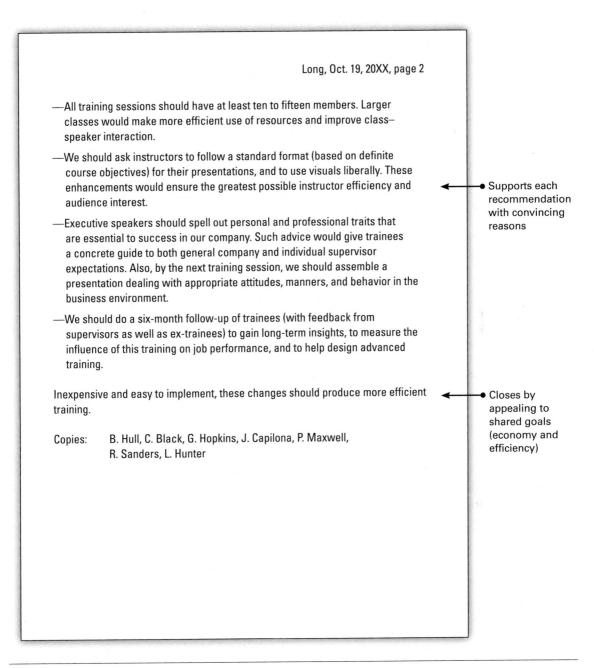

—All training sessions should have at least ten to fifteen members. Larger classes would make more efficient use of resources and improve class–speaker interaction.

—We should ask instructors to follow a standard format (based on definite course objectives) for their presentations, and to use visuals liberally. These enhancements would ensure the greatest possible instructor efficiency and audience interest.

→ Supports each recommendation with convincing reasons

—Executive speakers should spell out personal and professional traits that are essential to success in our company. Such advice would give trainees a concrete guide to both general company and individual supervisor expectations. Also, by the next training session, we should assemble a presentation dealing with appropriate attitudes, manners, and behavior in the business environment.

—We should do a six-month follow-up of trainees (with feedback from supervisors as well as ex-trainees) to gain long-term insights, to measure the influence of this training on job performance, and to help design advanced training.

Inexpensive and easy to implement, these changes should produce more efficient training.

→ Closes by appealing to shared goals (economy and efficiency)

Copies: B. Hull, C. Black, G. Hopkins, J. Capilona, P. Maxwell,
          R. Sanders, L. Hunter

Figure 6.4 **Continued**

Glenn's final report is both informative and persuasive. But this document did not appear magically. Glenn made deliberate decisions about purpose, audience, content, organization, and style. He sought advice and feedback on every aspect of the document. Most importantly, he *spent time revising and then proofreading.*

> NOTE *Writers work in different ways. Some begin by brainstorming. Some begin with an outline. Others simply write and rewrite. Some sketch a quick draft before thinking through their writing situation. Introductions and titles are often written last. Whether you write alone or collaborate in preparing a document, whether you are receiving feedback or providing it, no one step in the process is complete until the whole is complete. Notice, for instance, how Glenn sharpens his content and style while he organizes. Every document you write will require all these decisions, but you rarely will make them in the same sequence.*
>
> *No matter what the sequence, revision is a fact of life. It is the one constant in the writing process. When you've finished a draft, you have in a sense only begun. Sometimes you will have more time to submit a document than Glenn did, sometimes much less. Whenever your deadline allows, leave time to revise.*

> NOTE *Revising a draft doesn't always guarantee that you will improve it. Save each draft and then compare them to select the best material from each one.*

The final version is in PDF format, to be sent to Glenn's manager as an email attachment and cc'd to the people listed on the letter's next page.

Make Proofreading Your Final Step

6.3   Understand why proofreading is an important final step

No matter how engaging and informative the document, basic errors distract the reader and make the writer look bad (including on various drafts being reviewed by colleagues). In Glenn Tarullo's situation, all of his careful planning, drafting, and revising could be lost on his readers if the final document contained spelling or other mechanical errors. Proofreading detects easily correctable errors such as these:

Errors we look for during proofreading

- **Sentence errors** such as fragments, comma splices, or run-ons
- **Punctuation errors** such as missing apostrophes or excessive commas
- **Usage errors** such as "it's" for "its," "lay" for "lie," or "their" for "there"
- **Mechanical errors** such as misspelled words, inaccurate dates, or incorrect abbreviations
- **Format errors** such as missing page numbers, inconsistent spacing, or incorrect form of documenting sources
- **Typographical errors** (typos) such as repeated or missing words or letters, missing word endings (say, *-s* or *-ed* or *-ing*), or a missing quotation mark

Guidelines

for Proofreading

➤ **Save it for the final draft.** Proofreading earlier drafts might cause writer's block and distract you from the document's "rhetorical features" (content, organization, style, and design).

➤ **Take a break before proofreading your final document.**

➤ **Work from hard copy.** Research indicates that people read more perceptively (and with less fatigue) from a printed page than from a computer screen.

➤ **Keep it slow.** Read each word—don't skim. Slide a ruler under each line or move backward through the document, sentence by sentence. For a long document, read only small chunks at one time.

➤ **Be especially alert for problem areas in your writing.** Do you have trouble spelling? Do you get commas confused with semicolons? Do you make a lot of typos? Make one final pass to check on any problem areas.

➤ **Proofread more than once.** The more often you do it, the better.

➤ **Ask a trusted colleague to proofread, if time permits.** If doing so won't violate confidentiality, you can ask a colleague to proof your document.

➤ **Don't rely solely on autocorrect or grammar and spell check.** A synonym found in an electronic thesaurus may distort your meaning. Spell check cannot differentiate among correctly spelled words such as "their," "they're," and "there" or "it's" versus "its;" grammar checkers don't always catch everything either. In the end, use these apps to do a first run through the document, but remember that nothing can replace your own careful proofreading. (Chapter 11, "Using Digital Editing Tools Effectively," summarizes the limitations of digital editing tools such as autocorrect.)

Digital Technology and the Writing Process

6.4   Consider the use of digital and collaborative writing tools and applications

A variety of digital tools and programs provide support for the technical writing process (consult this book's index), including:

Digital tools for the technical writing process

- The outline feature in Microsoft Word and similar programs, which allows you to outline

- Brainstorming and storyboarding software, which lets you and team members collaborate on the early stages of the writing process (see Chapter 5)

- Social media such as Facebook, LinkedIn, and Twitter, where you can research your topic and connect with other experts (see Chapter 25)

- Customer review sites, such as Yelp and Angie's List, where you can find information on how customers feel about a product or service

- Programs such as PowerPoint and Microsoft Word (or similar apps for the Mac or on Google), which provide templates and formatting options for your presentations and document

- Software such as Visio, which allows flowcharting and mapping
- Apps that turn word-processing documents into Web pages, PDF documents, and more
- Wikis or tools such as Google Drive, which let you and others contribute to a common document
- Tracking systems (Track Changes), available in most word-processing programs, which record all suggested edits and comments from team members
- Email, texting, and chat programs, which provide quick turnarounds needed during the writing process

In Glenn Tarullo's situation, some of these resources helped him with certain parts of the project such as brainstorming, outlining, drafting, collaborating, and document formatting. Glenn and Blair Cordasco used Track Changes to share and discuss comments within Glenn's draft (see "Working with the Information" in this chapter), although due to the document's sensitive content, they shared the document only with each other via a secure server at work, not via email.

No matter how efficient the writing tools, many situations require you to set up a meeting, video chat, or phone call (as Glenn did with Blair). Documents with lots of tracked changes can be difficult to read through; more importantly, the ideas or sentences being considered may be so sensitive or complex that they require a conversation, not more Track Changes bubbles.

In the end, effective communication depends on a deliberative process, balancing smart choices about writing technologies with appropriate choices about audience, purpose, language, document length, method of delivery, timing, and the other items illustrated by Glenn's situation. For all types of writing—from a report emailed to the boss to a research paper uploaded for an instructor—you should continue to rely on proven strategies that comprise the writing process. The human brain remains our ultimate tool for navigating the critical thinking and decisions that produce effective writing and communication.

Checklist

Proofreading

Use the following Checklist when proofreading any document.

Sentences
- ❑ Are all sentences complete (no unacceptable fragments)? (See "Sentence Fragments" in Appendix B.)
- ❑ Is the document free of comma splices and run-on sentences? (See "Comma Splices" and "Run-on Sentences" in Appendix B.)
- ❑ Does each verb agree with its subject? (See "Faulty Agreement—Subject and Verb" in Appendix B.)
- ❑ Does each pronoun refer to and agree with a specific noun? (See "Faulty Agreement—Pronoun and Referent" in Appendix B.)

❏ Are ideas of equal importance coordinated? (See "Faulty Coordination" in Appendix B.)

❏ Are ideas of lesser importance subordinated? (See "Faulty Subordination" in Appendix B.)

❏ Is each pronoun in the correct case (nominative, objective, possessive)? (See "Faulty Pronoun Case" in Appendix B.)

❏ Is each modifier positioned to reflect the intended meaning? (See "Dangling and Misplaced Modifiers" in Appendix B.)

❏ Are items of equal importance expressed in equal (parallel) grammatical form? (See "Faulty Parallelism" in Appendix B.)

Punctuation

❏ Does each sentence conclude with appropriate end punctuation? (See "Punctuation" in Appendix B.)

❏ Are semicolons and colons used correctly as a break between items? (See "Semicolon" and "Colon" in Appendix B.)

❏ Are commas used correctly as a pause between items? (See "Comma" in Appendix B.)

❏ Do apostrophes signal possessives, contractions, and certain plurals? (See "Apostrophe" in Appendix B.)

❏ Do quotation marks set off direct quotes and certain titles? (See "Quotation Marks" in Appendix B.)

❏ Is each quotation punctuated correctly? (See "Quotation Marks" in Appendix B.)

❏ Do ellipses indicate material omitted from a quotation? (See "Ellipses" in Appendix B.)

❏ Do italics indicate certain titles or names, or emphasis or special use of a word? (See "Italics" in Appendix B.)

❏ Are brackets, parentheses, and dashes used correctly and as needed? (See "Brackets," "Parentheses," and "Dashes" in Appendix B.)

Mechanics

❏ Are abbreviations used correctly and without confusing the reader? (See "Abbreviation" in Appendix B.)

❏ Are hyphens used correctly? (See "Hyphenation" in Appendix B.)

❏ Are words capitalized correctly? (See "Capitalization" in Appendix B.)

❏ Are numbers written out or expressed as numerals as needed? (See "Numbers and Numerals" in Appendix B.)

❏ Have autocorrect and spell check been supplemented by actual proofreading for words spelled correctly but used incorrectly (as in "there" for "their")? (See "Spelling" in Appendix B.)

Format and Usage

❏ Are commonly confused words used accurately? (See Table B.1 in Appendix B.)

❏ Are pages numbered correctly? (See Chapter 13, "Shaping the Page.")

❏ Are sources cited in a standard form of documentation? (See Appendix A.)

❏ Have typographical errors been corrected? (See "Make Proofreading Your Final Step" in this chapter.)

Projects

For all projects, check with your instructor about whether to present your findings in class, bring drafts to class for discussion, upload your project to the class learning management system (LMS), and/or use the LMS forum or discussion boards to collaborate and review each activity below.

General

Compare Glenn's second draft (see "Working with the Information" in this chapter) with his final draft (see Figure 6.4). Identify all improvements in content, arrangement, and style besides those already discussed.

Team

Working in groups, assume you are a training team for XYZ Corporation. After completing the first section of this text and the course, what advice about the writing process would you have for a beginning writer who will frequently need to write reports on the job? In a one- or two-page memo to new employees, explain the writing process briefly, and give a list of guidelines these beginning writers can follow. Include a suggestion that new employees explore the company's policy on using email and other digital communication technologies within the company and with clients/customers. Post to your class learning management system based on your instructor's requirements.

Digital and Social Media

In groups of two students, imagine that you have been asked to create a job description for an entry-level job at your company. Use an internship or job you've had to help you think of a job that would be suitable for this exercise, or look at some online job sites like Monster.com (see Chapter 16). Before writing the ad, you'll need to generate some ideas, similar to the process involved with Glenn's first draft (see "Working with the Information" in this chapter). Decide on your approach: Should each of you write individual ideas in separate document files, then compare ideas and start drafting? Or should you try using a collaborative writing app (like Google Docs) where you can actually write at the same time? Write a brief memo to your instructor discussing your experience drafting a document with whichever approach you chose. If your instructor prefers, you can create your memo as an email, or you can upload a document to the learning management system for your class.

Global

Imagine that Blair, who works out of the Palo Alto (CA) office, and three other colleagues, based in London (UK), Shanghai (China), and Melbourne (Australia), all attended the same training program, but online, via a webinar. All four employees report to the same manager, Marvin Long, and have been asked by Marvin to write a team evaluation similar to the one Blair wrote individually in this chapter. Look back at Chapter 5, "Checklist: Teamwork and Global Considerations," and make a list of the most important items for the team to agree on before getting started.

Part 2
The Research Process

Chapter 7
Thinking Critically about the Research Process

Geber86/E+/Getty Images

" As a freelance researcher, I search online databases, Web sites, and social media for any type of specialized information needed by my clients. For example, yesterday I did a search for a corporate attorney who needed the latest information on some specific product-liability issues, plus any laws or court decisions involving specific products. For the legal research I accessed *LexisNexis*, the legal database that offers full-text copies of articles and cases. For the liability issue I began with *Factiva* and then double-checked by going into the *Dialog* database. "

—*Martin Casamonte, Freelance Researcher*

Learning Objectives

7.1 Ask the right questions, explore a balance of views, and achieve adequate depth in your research

7.2 Evaluate and interpret your sources

7.3 Differentiate between primary and secondary research

7.4 Conduct secondary research using online and traditional sources

7.5 Perform primary research using interviews, surveys, and other techniques

Major decisions in the workplace are based on careful research, with the findings recorded in a written report. Some parts of the research process follow a recognizable sequence (Figure 7.1A). But research is not merely a numbered set of procedures. The procedural stages depend on the many decisions that accompany any legitimate inquiry (Figure 7.1B). These decisions require you to *think critically* about each step of the process and about the information you gather for your research.

The importance of critical thinking at every stage of the research process

Three Essential Approaches to Research

7.1 Ask the right questions, explore a balance of views, and achieve adequate depth in your research

When approaching a research assignment, first be sure you have defined and refined your topic by asking the right questions. Then, as you locate sources, be sure to

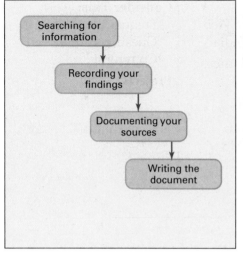

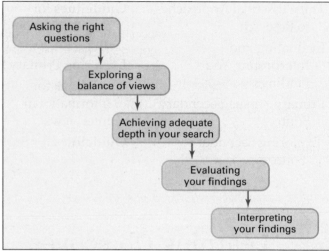

Figure 7.1A The Procedural Stages of the Research Process

Figure 7.1B Stages of Critical Thinking in the Research Process

explore a balance of views by considering your topic from all angles. Finally, achieve adequate depth in your research by exploring beyond the surface level.

Asking the Right Questions

The answers you uncover will only be as good as the questions you ask. Suppose, for instance, you face the following scenario:

Case

Defining and Refining a Research Question

You are the public health manager for a small, New England town in which high-tension power lines run within 100 feet of the elementary school. Parents are concerned about danger from electromagnetic radiation (EMR) emitted by these power lines in energy waves known as electromagnetic fields (EMFs). Town officials ask you to research the issue and prepare a report to be distributed at the next town meeting in six weeks.

Defining the
questions

First, you need to identify your exact question or questions. Initially, the major question might be: Do the power lines pose any real danger to our children? After phone calls around town and discussions at the coffee shop, you discover that townspeople actually have three main questions about electromagnetic fields: What are they? Do they endanger our children? If so, then what can be done?

Refining the
questions

To answer these questions, you need to consider a range of subordinate questions, like those in the Figure 7.2 tree chart. Any one of those questions could serve as subject of a worthwhile research report on such a complex topic. As research progresses, this chart will grow. For instance, after some preliminary reading, you learn that electromagnetic fields radiate not only from power lines but from *all* electrical equipment, and even from the Earth itself. So you face this additional question: Do power lines present the greatest hazard as a source of EMFs?

You now wonder whether the greater hazard comes from power lines or from other sources of EMF exposure. Critical thinking, in short, has helped you to define and refine the essential questions.

Let's say you've chosen this question: Do electromagnetic fields from various sources endanger our children? Now you can consider sources to consult (journals, interviews, reports, Internet sites, database searches, and so on).

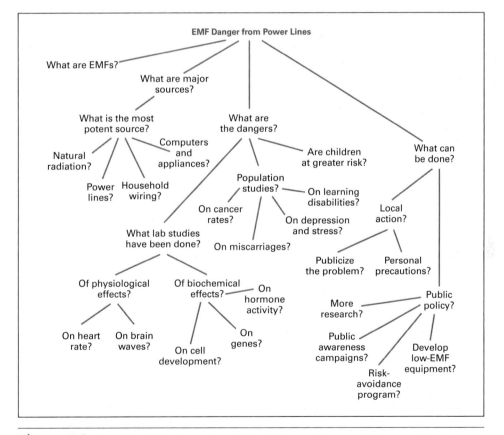

Figure 7.2 How the Right Questions Help Define a Research Problem You cannot begin to solve a problem until you have defined it clearly.

Exploring a Balance of Views

Instead of settling for the most comforting or convenient answer, pursue the *best* answer. Even "expert" testimony may not be enough, because experts can disagree or be mistaken. To answer fairly and accurately, consider a balance of perspectives from up-to-date and reputable sources (see Figure 7.3). Remember that online, many sources may *look* reputable (anyone can make a credible looking Web site); you need to check the source, examine where the information comes from, and determine if the information is based on reliable, factual information that is generally agreed to be sound. Exploring a balance of views means that you should look carefully at what you find, but it does not mean that you need to consider all ideas as equally weighted (for instance, conclusions based on extreme or highly politicized interpretations).

Try to consider all the angles

- What do informed sources have to say about this topic?
- On which points do sources agree?
- On which points do sources disagree?

> NOTE *Recognize the difference between "balance" (sampling a full range of opinions) and "accuracy" (getting at the facts). Government or power industry spokespersons, for example, might present a more positive view (or "spin") of the EMF issue than the facts warrant. Not every source is equal, nor should we report points of view as though they were equal (Trafford 137).*

Achieving Adequate Depth in Your Search

Balanced research examines a broad *range* of evidence; thorough research, however, examines that evidence in sufficient *depth*. Different sources of information about any topic occupy different levels of detail and dependability (Figure 7.4).

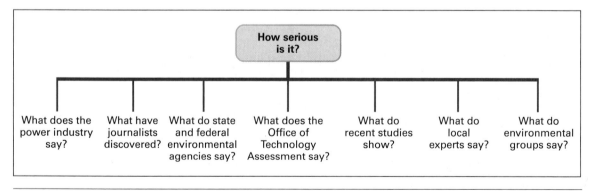

How serious is it?

What does the power industry say? — What have journalists discovered? — What do state and federal environmental agencies say? — What does the Office of Technology Assessment say? — What do recent studies show? — What do local experts say? — What do environmental groups say?

Figure 7.3 A Range of Essential Viewpoints No single source is likely to offer "the final word." Ethical researchers rely on evidence that represents a fair balance of views.

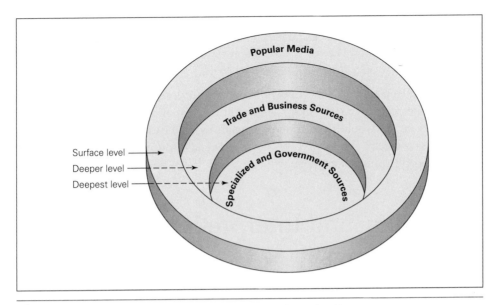

Figure 7.4 **Effective Research Achieves Adequate Depth**

1. The surface level offers items from the popular media (print and online news-papers and magazines, radio, TV, Web sites, Twitter feeds, Facebook pages, blogs). Designed for general consumption, this layer of information often merely skims the surface of an issue. *The depth of a source often determines its quality*

2. At the next level are trade, business, and technical publications (*Frozen Food World, Publisher's Weekly,* and so on). Often available in both print and digital formats, these publications are designed for readers who range from moderately informed to highly specialized. This layer of information focuses more on practice than on theory, on issues affecting the field, and on public relations. While the informa-tion is usually accurate, the general viewpoints may reflect a field's particular biases.

3. At a deeper level is the specialized literature (peer-reviewed journals from professional associations—academic, medical, legal, engineering). Designed for practicing professionals, this layer of information focuses on theory as well as on practice; on descriptions of the latest studies (written by the researchers them-selves and scrutinized by peers for accuracy and objectivity); on debates among scholars and researchers; and on reviews, critiques, and refutations of prior studies and publications.

Also at this deepest level are government sources and corporate documents available through the Freedom of Information Act. Designed for anyone willing to investigate its complex resources, this information layer offers hard facts and detailed discussion, and (in many instances) relatively impartial views.

NOTE *Web pages and social media feeds offer links to increasingly specific levels of detail. But the actual "depth" and quality of such information depend on the sponsorship and reliability of the organization behind the information (see "Guidelines for Researching on the Web and with Social Media" in this chapter).*

How deep is deep enough? This depends on your purpose, your audience, and your topic. But the real story most likely resides at deeper levels. Research on the EMF issue, for example, would need to look beneath media headlines and biased special interests (say, electrical industry or environmental groups), focusing instead on studies by a wide range of experts.

Evaluating and Interpreting Your Findings

7.2 Evaluate and interpret your sources

Not all findings have equal value. Some information might be distorted, incomplete, or misleading. Information might be tainted by *source bias*, in which a source understates or overstates certain facts, depending on whose interests that source represents (say, power company, government agency, parent group, or a reporter seeking headlines). Information might also be skewed based on *confirmation bias,* where new ideas that run counter to what a person already believes tend not to change minds but to reinforce existing beliefs. (See Chapter 3, "Expect Audience Resistance" and "Social and Psychological Constraints" for more on confirmation bias.) To evaluate a particular finding, ask these questions:

Questions for evaluating a particular finding

- Is this information accurate, reliable, and relatively unbiased?
- Do the facts verify the claim?
- How much of the information is useful?
- Is this the whole or the real story?
- Do I need more information?

Instead of merely emphasizing findings that support their own biases or assumptions, ethical researchers seek out and report the most *accurate* answer.

Once you have decided which of your findings seem legitimate, you need to decide what they all mean by asking these questions:

Questions for interpreting your findings

- What are my conclusions and do they address my original research question?
- Do any findings conflict?
- Are other interpretations possible?
- Should I reconsider the evidence?
- What, if anything, should be done?

For more advice on evaluating and interpreting data, see Chapter 8.

> NOTE *Never force a simplistic conclusion on a complex issue. Sometimes the best you can offer is an indefinite conclusion: "Although controversy continues over the extent of EMF hazards, we all can take simple precautions to reduce our exposure." A wrong conclusion is far worse than no definite conclusion at all.*

Primary versus Secondary Sources

7.3 Differentiate between primary and secondary research

Primary research means getting information directly from the source by conducting interviews and surveys and by observing people, events, or processes in action. *Secondary research* is information obtained second hand by reading what other researchers have compiled in books and articles in print or online. Most information found on the Internet would be considered a secondary source. Some Web-based information is more accurate than others; for instance, a Web page created by a high school student might be interesting but not overly reliable, whereas a Web site that is the equivalent of a traditional secondary source (encyclopedia, research index, newspaper, journal) would be more reliable for your research.

How primary and secondary research differ

Whenever possible, combine primary and secondary research. Typically, you would start by using secondary sources because they are readily available and can help you get a full background understanding of your topic. However, don't neglect to add your own findings to existing ones by doing primary research.

Why you should combine primary and secondary research

Working with primary sources can help you expand on what other people have already learned and add considerable credibility to your work. For instance, assume that your boss asks you to write a report about how well your company's new product is being received in the marketplace: You might consult sales reports and published print and online reviews of the product (secondary research), but you might also survey people who use the product and interview some of them individually (primary research).

How primary research adds credibility to your work

Exploring Secondary Sources

7.4 Conduct secondary research using online and traditional sources

Secondary sources include a wide range of publications including Web sites; news outlets (newspapers both print and online, radio, TV, Internet); blogs and wikis; social media sites; books; magazines; peer-reviewed journals; government documents and other public records; trade publications; and more. Your research assignments will begin more effectively when you first uncover and sort through what is already known about your topic before adding to that knowledge yourself. Many times, starting with secondary research can help you narrow down the interview questions and

Examples of secondary sources

other methods you might use for primary research (discussed in more detail later in this chapter).

Locating secondary sources and ideas

To begin exploring secondary sources, start by doing some general searching to see what's out there, what themes arise, what authors and organizations are at the forefront of research, and so forth. Take notes on what you find, using a Word file or a shared document (if yours is a team project). Keep careful records of the Web address (URL) and all source information (author, date, title, page numbers if applicable) so you can look up specific articles later. You might want to use a citation manger such as Zotero or Endnote to help you keep track. See the first few pages of Appendix A for information on citations management apps and for models on how to take careful notes and record your information.

> NOTE *Most of us engage in online research throughout the day, looking up everything from weather trends to the price of a new phone to restaurant reviews. Consider this activity as "everyday research," which differs from research for a school or workplace project. Workplace research, as discussed previously in this chapter, requires you to achieve adequate depth by reviewing numerous sources, keeping careful records, and using critical thinking skills to evaluate and report your findings.*

Searching for Secondary Sources

A systematic approach to secondary research

When it comes to secondary sources, don't just ask your phone to search for you; you have no idea what information your phone's digital assistant may be accessing. Use a systematic approach by searching via the options listed below, and be sure to assess the outcomes of these searches based on what is discussed within each option and also on the "Guidelines for Research on the Web and Social Media" later in this chapter.

Start with a general Google search

GOOGLE AND OTHER SEARCH ENGINES. From students to professionals, most of us begin our research of secondary sources by doing a Google search. Google, the most popular of the search engines, searches Web pages, government documents, online news sites, and other sources. Google also has a large collection of books and journal articles that it makes available through agreements with publishers or by digitizing works that either are in the public domain or are out of copyright.

Refine your Google search

It's fine to start with a Google search to see what's out there and to develop approaches to get started. But you quickly will need to narrow down your findings and do some deeper digging. For instance, a search on "electromagnetic radiation" will yield thousands or even millions of results. Look for sites from reliable sources such as universities and government research labs. Once you have some basic information, try searching on Google Scholar for links to peer-reviewed journal articles, which you may be able to obtain online or through the library.

Try using other search engines

Also, keep in mind that different search engines use different algorithms (ways of searching) and will yield different results. Try a few different search engines to see what you find. A few other popular ones are Bing, Yahoo!, and Ask. WolframAlpha calls itself an "answer engine" and searches through external data sources to answer specific questions in a range of categories.

WIKIPEDIA. Most Google and other searches quickly lead to Wikipedia, the popular online encyclopedia. Wikipedia's content is provided and edited by countless people worldwide. Although these pages can provide a good starting point, the content may not be completely accurate. Use a Wikipedia entry to get an overview of the topic, and to help you locate other sources.

Use Wikipedia as a starting point

The Wikipedia page on electromagnetic radiation contains footnotes to other sources. You can track down these sources at the library or over the Internet. Think of Wikipedia as a place to get your research started, but not as your final destination.

Use Wikipedia footnotes to find additional sources

DIGITAL LIBRARIES. Explore the Internet Public Library, the Haithi Trust Digital Library, and the Internet Archive. These online libraries are entirely searchable via the Internet, and their holdings are almost completely in digital format (either digital content itself, like online news sites, or print books and articles that have been digitized). Some of these sites also include ways to ask a question (usually via an online form) of an actual librarian. The Internet Archive offers its "WayBackMachine," which searches through all current and past versions of Web pages and other online materials.

What digital libraries offer

Keep in mind that due to copyright restrictions, these sites may not include current books or other documents and materials still under copyright. (See the "Consider This" box at the end of this chapter for more on copyright.) Supplement any digital library research with research at an actual library either at your school or in your community.

Limitations of digital libraries

LIBRARIES AND LIBRARY WEB SITES. Your college or university library offers a wide range of search tools, most of which you can use from your home or dorm room computer. School libraries and public libraries provide access to a host of digital materials (digital versions of newspapers, magazines, peer-reviewed articles, and so forth) as well as access to printed books, historic material (such as old maps), and specialized archives (such as a collection on the history of medicine or the history of aerospace engineering).

The value of traditional libraries

After you have gathered some ideas and potential sources from your Google and other online searching, start with the library's main search engine, then refine your search via subject directories, indexes, databases, and other tools. Libraries offer a general search tool of all holdings, called their online public access catalog (OPAC).

Using the OPAC

Some of the most popular general periodical databases (for magazines, newspapers, and other more general information) include InfoTrac, NewsBank, ProQuest, and EBSCOHost. Your library may also subscribe to more specific services such as PsychINFO (for articles in psychology), Business Source Premier (for articles on business), Ovid Medline (for articles on biomedical research), as well as indexes and other resources for engineering, the humanities, world affairs, and other areas. For more information on the range of subject and other indexes available through the library, see the sections on bibliographies and indexes under "Reference Works," later in this chapter.

Using periodical databases

Before initiating a periodical database search, meet with your reference librarian for a tour of the various databases and instructions for searching effectively. Or see if the library offers a webinar or other online tutorial. You can also communicate with a

The value of consulting a librarian

librarian via email or via a chat app if you need assistance. Libraries are an excellent resource, and you should become comfortable and familiar with using your school and public library. Many large technical organizations have company libraries, too.

Types of Secondary Sources

Varieties of hard copy and digital secondary sources

The following is a listing of the many types of secondary sources available to you for workplace and school-related research. Except for those items that are only available in digital format (a Web site or social media app, for example), most of these sources are available in both hard copy and online.

WEB SITES (GENERAL).   Search engines pull up a wide variety of hits, most of which will be commercial (.com), organizational (.org), and academic (.edu) Web sites. If a commercial site looks relevant to your search, by all means take a look at it, as long as you think critically about the information presented. Does the company's effort to sell you something affect the content? Be careful also of organizational Web sites, which are likely to be well-researched, but may have a particular social or political agenda. Academic Web sites tend to be credible, especially if the content is based on research studies and peer-reviewed publications.

GOVERNMENT WEB SITES.   Search engines will also locate government Web sites, including local, state, and federal. Most government organizations offer online access to research and reports. Examples at the federal level include the Food and Drug Administration's site (FDA), for information on food recalls, clinical drug trials, and countless related items; the Centers for Disease Control (CDC), for data on recent flu outbreaks, vaccines, and more. State and county sites provide information on auto licenses, tax laws, and local property issues. From some of these sites you can link to specific government-sponsored research projects, and you can often download the data directly. For more on locating government documents, see the next section of this chapter.

> NOTE *Be sure to check the dates of reports or data you locate on a government Web site, and find out how often the site is updated.*

BLOGS.   Blogs offer ideas, opinions, research discussions, and more. On most blogs, readers can post comments. The postings and attached discussions are displayed in reverse chronological order. Links that the owner has selected also supply ways to connect to other blogs on similar topics. Blogs are great for finding current information about a specific topic from individuals, companies, and nonprofit organizations. But be sure to evaluate the information on individual blogs carefully and decide which ones are most relevant and reliable based on the author's credentials, cited sources (does the author only cite him or herself?), and reader feedback.

Blogs nearly always represent the particular views of the blog author (whether an individual, company, organization, or academic institution) and of those who reply to the postings. Check any information you find on a blog against a professionally edited or peer-reviewed source.

WIKIS (INCLUDING WIKIPEDIA). Wikis are community encyclopedias that allow anyone to add to or edit the content of a listing. The most popular wiki is Wikipedia (discussed earlier in this chapter). The theory of a wiki is that if the information from one posting is wrong, someone else will correct it, and over time the site will reach a high level of accuracy and reliability.

Many wikis have no oversight. Aside from a few people who determine whether to delete articles based on requests from readers, the content on a wiki is not checked by editors for accuracy. Always check the information against other peer-reviewed or traditional sources. Remember that most of what is posted on a wiki has not been evaluated objectively. See Chapter 24 for more on blogs and wikis.

SOCIAL MEDIA. Almost every organization (company, government, nonprofit) and every topic imaginable has a Facebook page as well as Twitter, Instagram, and other social media feeds. The posting on these sites can provide you with ideas as well as sources of information to explore more deeply. For instance, your local electric utility company's Twitter feed might contain a link to a new study about EMFs, and that study could turn out to be very useful for your research. Or, their Facebook page might announce a local citizen advisory committee that's being formed; you might want to attend that meeting to learn more. As with all secondary sources, especially online, keep in mind that the material being presented comes with the biases of its particular organization; you need to search more deeply for the original source of the material. (See Figure 7.3 about balancing a range of viewpoints.)

ONLINE GROUPS. The Internet is home to thousands of online forums, some of which are affiliated with specific organizations and others that are run by individuals. These forums might be on Yahoo! Groups or Google Groups, or Reddit. These forums are popular for discussions about news, technology, science, politics, and almost any other topic imaginable. Also, many Web sites, such as health-related sites like WebMD or the Mayo Clinic have forums where patients and family members can post questions and ideas. If membership is open to all, you can join one of these groups and see if any of the information seems useful. Keep in mind that material from these groups may be insightful but biased. Visit a variety of groups to get a broad perspective on the issue. Groups with a moderator tend to have less "noise," while unmoderated groups could be less reliable.

NEWS OUTLETS (INCLUDING MAGAZINES AND OTHER PERIODICALS). Most major news organizations offer online versions of their broadcasts and print publications. Examples include the *New York Times,* the *Wall Street Journal,* CNN, and National Public Radio. Magazines such as *Time, Newsweek,* and *Forbes,* also offer both print and online versions. Some news is available in online formats only, as in the online magazines *Slate* and *Salon*. Increasingly, groups with extreme positions have begun to set up online news sites that "look and feel" like mainstream news but are not. These sites often hand-pick certain facts and then manipulate this information to an extreme, contributing to the confirmation bias (discussed previously in Chapter 3). When conducting serious research, look for a range of news information sources, but stick with news outlets that are well recognized, have serious editorial policies, and are committed to well-established norms of journalistic excellence and integrity.

BOOKS. Books have been and remain a significant source of information for research. Look for books that are published by well-regarded publishing houses; these books will typically have been put through a systematic review process, where other experts as well as an in-house editorial staff read various drafts, check research sources, and provide feedback. For time-sensitive topics (such as science and technology), more recent publication dates may provide you with up-to-date information. But books from past years may offer useful insights into previous theories or ideas. Use your library's search tool to locate books; you can also look on the Internet Archive and the Haithi Trust for books that have been digitized. The larger or more specialized the library you visit, the more likely you are to find books by specialist publishers and periodicals that delve into more specific subject areas. When consulting books and periodicals, always check the copyright date and supplement the source with additional information from more recent sources, if necessary.

PEER-REVIEWED JOURNAL ARTICLES.   Peer-reviewed journal articles represent the highest standard in research. As discussed earlier in this chapter (see item 3 under "Achieving Adequate Depth in Your Search"), specialized literature in medical, legal, engineering, business, humanities, and other fields provides theoretical and practical information written by researchers and reviewed rigorously for accuracy and objectivity by peers in the field. Journals are staffed by editorial boards representing the best and brightest minds in that particular field. Studies are based on past research and must conform to the highest of research standards. Often, you will find a citation or mention of a peer-reviewed article when looking at a news clip. For instance, the *New York Times* science section or magazines such as *Scientific American* contain stories geared toward general audiences about newly published research. You can then search for the research article itself.

Many scientific articles are readily available to the National Institutes of Health's PubMed Web site. Keep in mind that unless you are an expert in this particular field, you may not be able to fully grasp the nuances of the research. Ask a professor in that field or a librarian to help you. Or, you can always email the paper's author(s) if you have specific questions (see "Exploring Primary Sources" later in this chapter).

GRAY LITERATURE.   Some useful printed information may be unavailable at any library. These documents are known as "gray literature," or materials that are unpublished or not typically catalogued. Examples include pamphlets published by organizations or companies (such as medical pamphlets or company marketing materials), unpublished government documents (available under the Freedom of Information Act), papers presented at certain professional conferences, or self-published works.

The only way to track down gray literature is to contact those who produce such literature and request anything available in your subject area. For instance, you could contact a professional organization and request any papers on your topic that were delivered at their recent annual conference, or contact a government agency for statistics relevant to your topic. Before doing so, be knowledgeable about your topic and know specifically whom to contact. Don't make vague, general requests.

Keep in mind that gray literature, like some material found on the Web, is often not carefully scrutinized for content by editors. Therefore, it may be unreliable and should be backed up by information from other sources.

REFERENCE WORKS. Reference works are general information sources that provide background and can lead to more specific information.

- **Bibliographies.** Bibliographies are lists of books and/or articles categorized by subject. To locate bibliographies in your field, begin by consulting the *Bibliographic Index Plus,* a list (by subject) of major bibliographies, which indexes more than 500,000 bibliographies worldwide. You can also consult such general bibliographies as *Books in Print* or the *Readers' Guide to Periodical Literature.* Or, examine subject area bibliographies such as *Bibliography of World War II History,* or highly focused bibliographies, such as *Chemical Engineering.*

- **Indexes.** Book and article bibliographies may also be referred to as "indexes." Yet there are other types of indexes that collect information not likely found in standard bibliographies. One example is the *Index to Scientific and Technical Proceedings,* which indexes conference proceedings in the sciences and engineering. While limited versions of some of these indexes may be available for free on the Internet, most are only available via a library subscription. Other types of indexes that may be useful for your research include *newspaper indexes* (which in some cases index articles from the entire history of the paper), *periodical indexes* (which list articles from magazines and journals, for example the *Reader's Guide to Periodical Literature* and the periodical databases InfoTrack, NewsBank, ProQuest, and EBSCOHost), *citation indexes* (which track publications in which original material has been cited), *technical report indexes* (which allow you to look for government and private-sector reports), and *patent indexes* (which allow you to search for both U.S. and international patents).

- **Encyclopedias.** Encyclopedias are alphabetically arranged collections of articles. Like Wikipedia, encyclopedias are a good starting point for your research, but you should use these to guide you to other material (such as journal articles or government reports). The *Encyclopedia Britannica* is the most popular of the general encyclopedias. You should also examine more subject-focused encyclopedias, such as *Encyclopedia of Nutritional Supplements, Encyclopedia of Business and Finance,* or *Illustrated Encyclopedia of Aircraft.*

- **Government documents.** The federal government publishes maps, periodicals, books, pamphlets, manuals, research reports, and other information. These publications may be available in digital as well as hard-copy formats. To help you find what you are looking for, you will need to use an access tool such as the following. A librarian can teach you to use tools such as the *Monthly Catalog of the United States Government* (a pathway to government publications and reports), the *Government Reports Announcements and Index* (a listing of more than one million federally-sponsored and published research reports and patents issues since 1964) and the *Statistical Abstract of the United States* (an annual update of statistics on population, health, employment, and many other areas).

- **Dictionaries.** Dictionaries are alphabetically arranged lists of words, including definitions, pronunciations, and word origins. If you can't locate a particular word in a general dictionary (e.g., a highly specialized term or jargon specific to a certain field), consult a specialized dictionary, such as *Dictionary of Engineering and Technology, Dictionary of Psychology,* or *Dictionary of Media and Communication Studies.*

- **Handbooks.** Handbooks offer condensed facts (formulas, tables, advice, examples) about particular fields. Examples include the *Civil Engineering Handbook* and *The McGraw-Hill Computer Handbook.*

- **Almanacs.** Almanacs are collections of factual and statistical data, usually arranged by subject area and published annually. Examples include general almanacs, such as the *World Almanac and Book of Facts,* or subject-specific almanacs, such as the *Almanac for Computers* or *Baer's Agricultural Almanac.*

- **Directories.** Directories provide updated information about organizations, companies, people, products, services, or careers, often listing addresses and phone numbers. Examples include *The Career Guide: Dun's Employment Opportunities Directory* and the *Directory of American Firms Operating in Foreign Countries.* Ask your librarian about *Hoover's Company Capsules* (for basic information on thousands of companies) and *Hoover's Company Profiles* (for detailed information).

- **Abstracts.** Abstracts are collected summaries of books and/or articles. Reading abstracts can help you decide whether to read or skip an article and can save you from having to track down a journal you may not need. Abstracts usually are titled by discipline: *Biological Abstracts, Computer Abstracts,* and so on. For some current research, you might consult abstracts of doctoral dissertations in *Dissertation Abstracts International.*

Most reference works are now available in both print and digital formats. When using a reference work, check the copyright date to make sure you are accessing the most current information available.

Guidelines

for Researching on the Web and with Social Media

➤ **Expect different results from different search engines and subject directories.** Search engines use algorithms (a set of rules) based on where you are located, what items you've searched on in the past, what sites are most popular, and many other factors. No single search engine can index all of the material available on the Internet; different search engines and subject directories will list different findings.

➤ **When using a search engine, select keywords or search phrases that are varied and technical rather than general.** Some search terms generate more useful information than others. In addition to "electromagnetic radiation," for example, try "electromagnetic fields," "power lines and health," or "electrical fields." Specialized terms (say, "vertigo" versus "dizziness") offer the best access to reliable sites. However, if you are not able to locate much by using a specialized term, widen your search somewhat. Try different search engines (Bing; Google; Yahoo!; others as discussed in this chapter).

➤ **When using Wikipedia, look carefully at the footnotes and other citations.** These references can direct you to other sources, such as government documents, books in the library, or published journal articles.

➤ **Consider the domain type (where the site originates).** Standard domain types in the United States include .com (commercial organization), .edu (educational institution), .gov or .mil (government or military organization), .net (general usage), and .org (organizational).

➤ **Identify the Web site or social media feed's purpose and sponsor.** Is the intent merely to relay information, to sell something, or to promote an ideology or agenda? For Web sites, the domain type might alert you to bias or a hidden agenda. A .com site might provide accurate information but also some type of sales pitch. Sites that have the ".org" extension might reflect a political or ideological bias. Looking for a site's sponsor can also help you evaluate its postings. For example, a Web site about the dangers of bioengineered foods that is sponsored by an advocacy organization may be biased. On social media, do some digging so you can determine what organization or person(s) is sponsoring these feeds and how the purpose and sponsor may influence the quality of information you find. Facebook, Twitter, Reddit, and other social media sites may provide some credible secondary source content (for instance, if a company representative makes a statement about a new product release), but think of these sources as starting points. You should then use your insights to research a balance of solid, credible secondary sources.

➤ **Look beyond the style of a site.** Research shows that the visual "look and feel" of a Web site is the primary item influencing whether readers think a site is credible or not (Fogg). Anyone can make a Web site that looks professional and "real;" be sure to look more closely at the items listed in these guidelines before making a decision about a site's credibility. The same is true with social media: it takes only seconds to create a professional looking Facebook and Twitter account. What you see is not always what you get!

➤ **Assess the creation date and updates of the site's materials.** When was the material created, posted, and updated? Many Web sites have not been updated in months or years. On social media, old news stories are often recirculated, and people are all too willing to share the post before checking to discover that the story is actually from many years ago. Less-than credible online news sites bury the date at the bottom of the page, or in small print, so that readers are drawn into the headline and pictures and do not notice that the story is too old to be useful or current.

➤ **Assess the author's credentials and assertions.** Check the author's reputation, expertise, and institutional affiliation (university, company, environmental group). Do not confuse the author (the person who wrote the material) with the Webmaster (the person who created and maintains the site). Follow links to other sites that mention the author. Where, on the spectrum of expert opinion and accepted theory, does this author fall? Is each assertion supported by solid evidence? Verify any extreme claim through other sources, such as a professor or expert in the field. Consider whether your own biases might predispose you to accept certain ideas. On social media, do some checking to see who is responsible for writing and posting to Facebook, Instagram, Twitter, and other feeds.

➤ **Use bookmarks and URL apps for quick access to favorite Web sites.** It is always frustrating when you can't find a helpful Web site that you accessed earlier but didn't bookmark. Most Web browsers (Chrome, Firefox, Internet Explorer, Safari) offer apps that let you save and organize your bookmarks. You can also use citation management apps such as Zotero or Endnote (see Appendix A for more on citations management apps and uses).

➤ **Save what you need before it changes or disappears.** Even if you save the URL, remember that Web sites often change their content or "go dead." You can also use a screen capture tool (Apple Grab, Microsoft Snipping Tool, or Snagit) to take a snapshot of the Web site or social media post. On Twitter, you can search on specific hashtags to capture and research certain trends and discussions.

➤ **Download only what you need, use it ethically, obtain permission, and credit your sources.** Unless they are crucial to your research, omit graphics, sound, and video files. Do not use material created by others in a way that harms the material's creator. For any type of commercial use of material that's on social media, a Web site, or any other online source, you should obtain written permission from the material's owner and credit the source exactly as directed by its owner. For more information on copyright, see the Consider This box later in this chapter.

Exploring Primary Sources

7.5 **Perform primary research using interviews, surveys, and other techniques**

Types of primary sources

Once you have explored your research topic in depth by finding out what others have uncovered (secondary sources), supplement that knowledge with information you discover yourself by doing primary research. Primary sources include unsolicited inquiries, informational interviews, surveys, and observations or experiments.

Unsolicited Inquiries

Unsolicited inquiries uncover basic but important information

The most basic form of primary research is a simple, unsolicited inquiry. Letters, phone calls, or email inquiries to experts you have identified in your reading and initial research can clarify or supplement information you already have. Try to contact the right individual instead of a company or department. Also, ask specific questions that cannot be answered elsewhere. Be sure what you ask about is not confidential or otherwise sensitive information.

Unsolicited inquiries, especially by phone, can be intrusive or even offensive. Therefore, initiate your request for an interview with a short email.

Informational Interviews

Informational interviews can lead to original, unpublished material

An excellent primary source of information is the informational interview. Expert information may never be published. Therefore, you can uncover highly original information by spending time with your respondent and asking pertinent questions. In addition, an interviewee might refer you to other experts or sources of information.

Expert opinion is not always reliable

Of course, an expert's opinion can be just as mistaken or biased as anyone else's. Like patients who seek second opinions about medical conditions, researchers must seek a balanced range of expert opinions about complex problems or controversial issues. In researching the effects of electromagnetic fields (EMFs), for example, you would seek opinions not only from a company engineer and environmentalist, but also from presumably more objective third parties such as a professor or journalist who has studied the issue. Figure 7.5 provides a partial text of an interview about persuasive challenges faced by a corporation's manager.

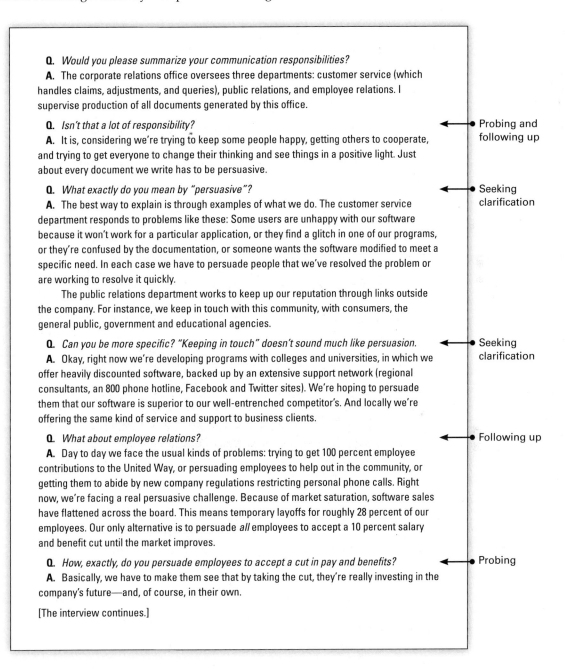

Q. *Would you please summarize your communication responsibilities?*
A. The corporate relations office oversees three departments: customer service (which handles claims, adjustments, and queries), public relations, and employee relations. I supervise production of all documents generated by this office.

Q. *Isn't that a lot of responsibility?*   ← Probing and
A. It is, considering we're trying to keep some people happy, getting others to cooperate,   following up
and trying to get everyone to change their thinking and see things in a positive light. Just about every document we write has to be persuasive.

Q. *What exactly do you mean by "persuasive"?*   ← Seeking
A. The best way to explain is through examples of what we do. The customer service   clarification
department responds to problems like these: Some users are unhappy with our software because it won't work for a particular application, or they find a glitch in one of our programs, or they're confused by the documentation, or someone wants the software modified to meet a specific need. In each case we have to persuade people that we've resolved the problem or are working to resolve it quickly.

The public relations department works to keep up our reputation through links outside the company. For instance, we keep in touch with this community, with consumers, the general public, government and educational agencies.

Q. *Can you be more specific? "Keeping in touch" doesn't sound much like persuasion.* ← Seeking
A. Okay, right now we're developing programs with colleges and universities, in which we   clarification
offer heavily discounted software, backed up by an extensive support network (regional consultants, an 800 phone hotline, Facebook and Twitter sites). We're hoping to persuade them that our software is superior to our well-entrenched competitor's. And locally we're offering the same kind of service and support to business clients.

Q. *What about employee relations?*   ← Following up
A. Day to day we face the usual kinds of problems: trying to get 100 percent employee contributions to the United Way, or persuading employees to help out in the community, or getting them to abide by new company regulations restricting personal phone calls. Right now, we're facing a real persuasive challenge. Because of market saturation, software sales have flattened across the board. This means temporary layoffs for roughly 28 percent of our employees. Our only alternative is to persuade *all* employees to accept a 10 percent salary and benefit cut until the market improves.

Q. *How, exactly, do you persuade employees to accept a cut in pay and benefits?* ← Probing
A. Basically, we have to make them see that by taking the cut, they're really investing in the company's future—and, of course, in their own.

[The interview continues.]

Figure 7.5 Partial Text of an Informational Interview This page from an informational interview shows you how to use clear, specific questions and how to follow up and seek clarification to answers.

Guidelines
for Informational Interviews

Planning the Interview

➤ **Know exactly what you're looking for from whom.** Write out your plan.

Audience
and purpose
statement

> I will interview Anne Hector, Chief Engineer at Northport Electric, to ask about the company's approaches to EMF (electromagnetic field) risk avoidance—in the company as well as in the community.

➤ **Do your homework.** Learn all you can. Be sure the information this person might provide is unavailable in print.

➤ **Make arrangements by phone, letter, or email.** (See Karen Granger's letter in Figure 15.10.) Ask whether this person objects to being quoted or taped. If possible, submit your questions beforehand.

Preparing the Questions

➤ **Make each question clear and specific.** Avoid questions that can be answered "yes" or "no":

An unproductive
question

> In your opinion, can technology find ways to decrease EMF hazards?

Instead, phrase your question to elicit a detailed response:

A clear and
specific
question

> Of the various technological solutions being proposed or considered, which do you consider most promising?

➤ **Avoid loaded questions.** A loaded question invites or promotes a particular bias:

A loaded
question

> Wouldn't you agree that EMF hazards have been overstated?

Ask an impartial question instead:

An impartial
question

> In your opinion, have EMF hazards been accurately stated, overstated, or understated?

➤ **Save the most difficult, complex, or sensitive questions for last.**

➤ **Write out each question on a separate notecard.** Use the notecard to summarize the responses during the interview.

Conducting the Interview

➤ **Make a courteous start.** Express your gratitude, explain why you believe the respondent can be helpful, and explain exactly how you will use the information.

➤ **Respect cultural differences.** Consider the level of formality, politeness, directness, and other behaviors appropriate in the given culture. (See Chapters 3 and 5.)

➤ **Let the respondent do most of the talking.**

➤ **Be a good listener.** For listening advice, see Chapter 5, "Overcoming Differences by Active Listening" and "Guidelines for Active Listening."

➤ **Stick to your interview plan.** If the conversation wanders, politely nudge it back on track (unless the peripheral information is useful).

➤ **Ask for clarification if needed.** Keep asking until you understand.

| —Could you go over that again?
| —What did you mean by [word]?

Clarifying
questions

➤ **Repeat major points in your own words and ask if your interpretation is correct.** But do not put words into the respondent's mouth.
➤ **Be ready with follow-up questions.**

| —Why is it like that?
| —Could you say something more about that?
| —What more needs to be done?

Follow-up
questions

➤ **Keep note taking to a minimum.** Record statistics, dates, names, and other precise data, but don't record every word. Jot key terms or phrases that can refresh your memory later.

Concluding the Interview

➤ **Ask for closing comments.** Perhaps these can point to additional information.

| —Would you care to add anything?
| —Is there anyone else I should talk to?
| —Can you suggest other sources that might help me better understand this issue?

Concluding
questions

➤ **Request permission to contact your respondent again, if new questions arise.**
➤ **Invite the respondent to review your version for accuracy.** If the interview is to be published, ask for the respondent's approval of your final draft. Offer to provide copies of any document in which this information appears.
➤ **Thank your respondent and leave promptly.**
➤ **As soon as possible, write a complete summary (or record one verbally).**

Surveys

Surveys help you form impressions of the concerns, preferences, attitudes, beliefs, or perceptions of a large, identifiable group (a *target population*) by studying representatives of that group (a *sample*). While interviews allow for greater clarity and depth, surveys offer an inexpensive way to get the viewpoints of a large group. Respondents can answer privately and anonymously—and often more candidly than in an interview.

Surveys provide multiple, fresh viewpoints on a topic

The tool for conducting surveys is the questionnaire. See Figures 7.6 and 7.7 for a sample cover email and questionnaire.

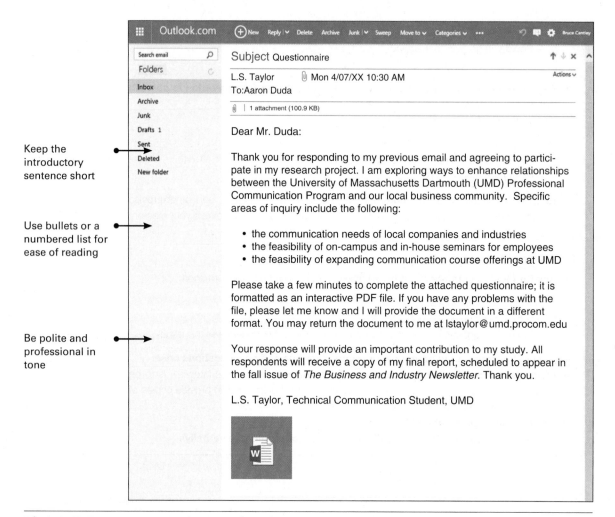

Keep the introductory sentence short

Use bullets or a numbered list for ease of reading

Be polite and professional in tone

Figure 7.6 A Cover Email for a Questionnaire Use a brief email to explain your questionnaire; be sure to include the attachment and provide contact information.
Source: Microsoft Outlook 2013.

Observations and Experiments

Observations and experiments offer proof to back up assumptions about a topic

Observations or experiments should be your final step because you now know exactly what to look for.

When you make observations, have a plan in place. Know how, where, and when to look, and jot down or record your observations immediately. You might even take photos or draw sketches of what you observe.

Experiments are controlled forms of observations designed to verify assumptions (e.g., the role of fish oil in preventing heart disease) or to test something untried (e.g., the relationship between background music and productivity). Each field has

Communication Questionnaire

1. Describe your type of company (e.g., manufacturing, high tech).

   _____

2. Number of employees. (Please check one.)

   | _____ 0–4 | _____ 26–50 | _____ 101–150 | _____ 301–450 |
   | _____ 5–25 | _____ 51–100 | _____ 151–300 | _____ 451+ |

3. What types of written communication occur in your company? (Label by frequency: daily, weekly, monthly, never.)

_____ memos	_____ letters	_____ advertising
_____ manuals	_____ reports	_____ newsletters
_____ procedures	_____ proposals	_____ other (Specify.)
_____ email	_____ catalogs	_____

4. Who does most of the writing? (Pls. give titles.) _____

5. Please characterize your employees' writing effectiveness.

   _____ good   _____ fair   _____ poor

6. Does your company have formal guidelines for writing?

   _____ no   _____ yes (Pls. describe briefly.) _____

7. Do you offer in-house communication training?

   _____ no   _____ yes (Pls. describe briefly.) _____

8. Please rank the usefulness of the following areas in communication training (from 1–10, 1 being most important).

_____ organizing information	_____ audience awareness
_____ summarizing information	_____ persuasive writing
_____ editing for style	_____ grammar
_____ document design	_____ researching
_____ email etiquette	_____ web page design
_____ other (specify) _____	_____ social media management

9. Please rank these skills in order of importance (from 1–6, 1 being most important).

   | _____ reading | _____ listening | _____ speaking to groups |
   | _____ writing | _____ collaborating | _____ speaking face-to-face |

10. Do you provide tuition reimbursement for employees?

    _____ no   _____ yes

11. Would you consider having UMD communication interns work for you part-time?

    _____ no   _____ yes

12. Should UMD offer Saturday seminars in communication?

    _____ no   _____ yes

    Additional comments/suggestions: _____

    _____

Annotations:
- Open-ended question (allows people to respond as they choose) → points to question 1
- Closed-ended question (provides a limited choice of responses) → points to question 3
- Questions and sentences are short and to the point → points to question 7
- Page layout is clean and easy to read → points to question 9

Figure 7.7 **A Questionnaire** A questionnaire will help you gather answers to specific questions and topics.

Guidelines

for Surveys

➤ **Define the survey's purpose and target population.** Ask yourself, "Why is this survey being performed?" "What, exactly, is it measuring?" "How much background research do I need?" "How will the survey findings be used?" and "Who is the exact population being studied?"

➤ **Identify the sample group.** Determine how many respondents you need. Generally, the larger the sample surveyed the more dependable the results (assuming a well-chosen and representative sample). Also determine how the sample will be chosen. Will they be randomly chosen? In the statistical sense, random does not mean "haphazard": A random sample means that each member of the target population stands an equal chance of being in the sample group.

➤ **Define the survey method.** How will the survey be administered—by phone, by mail, or online? Each method has benefits and drawbacks: Phone surveys yield fast results and high response rates; however, they take longer than written surveys. Also, many people find them annoying and tend to be less candid when responding in person. Mail surveys promote candid responses, but many people won't bother returning the survey, and results can arrive slowly. Surveys via the Web (using apps such as SurveyMonkey) or email yield quick results, but people may have technical problems that you can't anticipate. To anticipate technical problems, try a short pilot test first: create a small survey and ask a few colleagues to try it and give you feedback.

➤ **Decide on types of questions.** Questions can be open- or closed-ended. Open-ended questions allow respondents to answer in any way they choose. Measuring the data gathered from such questions is more time-consuming, but they do provide a rich source of information. An open-ended question is worded like this:

Open-ended question

| How much do you know about electromagnetic radiation at our school?

Closed-ended questions give a limited number of choices, and the data gathered are easier to measure. Here are some types of closed-ended questions:

Closed-ended questions

Are you interested in joining a group of concerned parents?

YES _____   NO _____

Rate your degree of concern about EMFs at our school.

HIGH _____   MODERATE _____   LOW _____   NO CONCERN _____

Circle the number that indicates your view about the town's proposal to spend $20,000 to hire its own EMF consultant.

1. ... 2. ... 3. ... 4. ... 5. ... 6. ... 7

Strongly   No   Strongly

Disapprove   Opinion   Approve

How often do you ...?

ALWAYS _____   OFTEN _____   SOMETIMES _____   RARELY _____   NEVER _____

To measure exactly where people stand on an issue, choose closed-ended questions.

➤ **Develop an engaging introduction and provide appropriate information.** Persuade respondents that the questionnaire relates to their concerns, that their answers matter, and that their anonymity is ensured:

> Your answers will help our school board to speak accurately for your views at our next town meeting. All answers will be kept confidential. Thank you.

A survey introduction

Researchers often include a cover letter or email with the questionnaire, as in Figure 7.7. Begin with the easiest questions, usually the closed-ended ones. Respondents who commit to these are likely to answer later, more difficult questions.

➤ **Make each question unambiguous.** All respondents should be able to interpret identical questions identically. An ambiguous question allows for misinterpretation:

> Do you favor weapons for campus police? YES____ NO____

An ambiguous question

"Weapons" might mean tear gas, clubs, handguns, tasers, or some combination of these. The limited "yes/no" format reduces an array of possible opinions to an either/or choice. Here is an unambiguous version:

> Do you favor (check all that apply):
> _____ Having campus police carry mace and a club?
> _____ Having campus police carry nonlethal "stun guns"?
> _____ Having campus police store handguns in their cruisers?
> _____ Having campus police carry handguns?
> _____ Having campus police carry large-caliber handguns?
> _____ Having campus police carry no weapons?
> _____ Don't know

A clear and incisive question

To account for all possible responses, include options such as "Other," "Don't know," or an "Additional Comments" section.

➤ **Avoid biased questions:**

> Should our campus tolerate the needless endangerment of innocent students by lethal weapons? YES____ NO____

A loaded question

Avoid emotionally loaded and judgmental words ("endangerment," "innocent," "needless"), which can influence a person's response (Hayakawa 40).

➤ **Make it brief, simple, and inviting.** Long questionnaires usually get few replies. And people who do reply tend to give less thought to their answers. Limit the number and types of questions. Include a stamped, return-addressed envelope, and stipulate a return date.

➤ **Have an expert review your questionnaire before use, whenever possible.**

its own guidelines for conducting experiments (e.g., you must use certain equipment, scrutinize your results in a certain way); follow those guidelines to the letter when conducting your own experiments.

Remember that observations and experiments are not foolproof. During observation or experimentation, you may be biased about what you see (focusing on the wrong events, ignoring something important). In addition, if you are observing people or experimenting with human subjects, they may be conscious of being observed and may alter their normal behaviors.

Consider This

Frequently Asked Questions about Copyright

Research often involves working with copyrighted materials. Copyright laws have an ethical purpose: to balance the reward for intellectual labors with the public's right to use information freely.

1. *What is a copyright?*

   A copyright is the exclusive legal right to reproduce, publish, and sell a literary, dramatic, musical, or artistic work that is fixed in a tangible medium (digital or print). Written permission must be obtained to use all copyrighted material except where fair use applies or in cases where the copyright holder has stated other terms of use. For example, a musician might use a Creative Commons "attribution" license as part of a song released on the Internet. This license allows others to copy, display, and perform the work without permission, but only if credit is given.

2. *What are the limits of copyright protection?*

   Copyright protection covers the exact wording of the original, but not the ideas or information it conveys. For example, consider a newspaper article about climate change. The actual wording and images in the article have copyright protection, but the idea (climate change) does not. Others can write articles or books about climate change, too. But when paraphrasing or quoting from another article, you need to cite the source.

3. *How long does copyright protection last?*

   Works published before January 1, 1978, are protected for 95 years. Works published on or after January 1, 1978, are copyrighted for the author's life plus 70 years.

4. *Must a copyright be officially registered in order to protect a work?*

   No. Protection begins as soon as a work is created.

5. *What is "fair use" and how is it determined?*

   "Fair use" is the legal and limited use of copyrighted material without permission. The source should, of course, be acknowledged.

   In determining fair use, the courts ask these questions:

   ➤ *Is the material being used for commercial or for nonprofit purposes?* For example, nonprofit educational use is viewed more favorably than for-profit use.

   ➤ *Is the copyrighted work published or unpublished?* Use of published work is viewed more favorably than use of unpublished essays, correspondence, and so on.

   ➤ *How much, and which part, of the original work is being used?* The smaller the part, the more favorably its use will be viewed.

   ➤ *How will the economic value of the original work be affected?* Any use that reduces the potential market value of the original will be viewed unfavorably.

6. *What is the exact difference between copyright infringement and fair use?*

   Courts differ on when and how to apply the four fair use questions, especially in cases related to digital media and material used from the Internet. Check the source to see if the author has added any language about copyright or terms of use (including any of the Creative Commons licensing statements). When in doubt, obtain written permission.

7. *What is material in the "public domain"?*

   "Public domain" refers to material not protected by copyright or material on which copyright has expired. Works published in the United States 95 years before the current year are in the public domain. Most government publications and commonplace information, such as height and weight charts or a metric conversion table, are in the public domain. These works might contain copyrighted material (used with permission and properly acknowledged). If you are not sure whether an item is in the public domain, request permission.

8. *What about international copyright?*
Copyright protection varies among individual countries, and some countries offer little or no protection for foreign works.

9. *Who owns the copyright to a work prepared as part of one's employment?*
Typically, materials prepared by an employee are considered "works for hire"; in these cases, the employer holds the copyright. For instance, a user manual researched, designed, and written by an employee would be in this category. Contract employees may be asked to sign over the copyright to materials they produce on the job.

Projects

For all projects, check with your instructor about whether to present your findings in class, bring drafts to class for discussion, upload your project to the class learning management system (LMS), and/or use the LMS forum or discussion boards to collaborate and review each activity below.

General

Begin researching for the analytical report (Chapter 22) due at semester's end.

Phase One: Preliminary Steps
a. Choose a topic that affects you, your workplace, or your community directly.
b. Develop a tree chart (See Figure 7.2 in this chapter) to help you ask the right questions.
c. Complete an audience and use profile. (See Chapter 2, "Develop and Audience and Use Profile.")
d. Narrow your topic, checking with your instructor for approval and advice.
e. Make a working bibliography to ensure sufficient primary and secondary sources.
f. List what you already know about your topic.
g. Write an audience and purpose statement (see Chapter 2, "Primary and Secondary Purposes") and submit it in a research proposal (see Chapter 22, "Research Proposals").
h. Make a working outline.

Phase Two: Collecting, Evaluating, and Interpreting Data (Read Chapters 8–9 in preparation for this phase.)
a. In your research, begin with general works for an overview, and then consult more specific sources.
b. Skim the sources, looking for high points.
c. Take notes selectively (see Appendix A, "Taking Notes"), summarize (see Chapter 9), and record each source.
d. Plan and administer questionnaires, interviews, and inquiries.
e. Try to conclude your research with direct observation.
f. Evaluate each finding for accuracy, reliability, fairness, and completeness.
g. Decide what your findings mean.
h. Use "Checklist: The Research Process" in Chapter 8 to reassess your methods, interpretation, and reasoning.

Phase Three: Organizing Your Data and Writing the Report

a. Revise your working outline as needed.

b. Document each source of information. (See Appendix A).

c. Write your final draft according to "Checklist: The Research Process" in Chapter 8.

d. Proofread carefully. Add front and end matter supplements (Chapter 22).

Due Dates: To Be Assigned by Your Instructor

List of possible topics due:

Final topic due:

Proposal memo due:

Working bibliography and working outline due:

Notecards (or note files) due:

Copies of questionnaires, interview questions, and inquiry letters due:

Revised outline due:

First draft of report due:

Final version with supplements and full documentation due:

Team

Divide into groups according to majors. Assume that several employers in your field are holding a job fair on campus next month and will be interviewing entry-level candidates. Each member of your group is assigned to develop a profile of one of these companies or organizations by researching its history, record of mergers and stock value, management style, financial condition, price/earnings ratio of its stock, growth prospects, products and services, multinational affiliations, ethical record, environmental record, employee relations, pension plan, employee stock options or profit-sharing plans, commitment to equal opportunity, number of women and minorities in upper management, or any other features important to a prospective employee. The entire group will then edit each profile and assemble them in one single document to be used as a reference for students in your major.

Digital and Social Media

Since so many people start and end their research by looking only at Wikipedia, use this project to explore Wikipedia not as the end of the research process but as the beginning. Go to Wikipedia and search on a topic for this or another class. Select a topic that is "big" (for example, climate change). Look carefully at the footnotes (citations) on the Wikipedia page. Carefully organize these footnotes into subcategories (you can cut and paste into a Word document, using headings to create your subcategories). For instance, of the 100+ footnotes, you might discover several related to articles about global warming and biodiversity or global warming and its effects on different economies. Select one of these subtopics for your report. Next, use a library and online resources to locate the articles, reports, and other publications cited in the footnotes. From these articles, you can check the references pages and find even more articles. Write a short report (3–4 pages) describing your research process and what you learned about using Wikipedia as a starting point for research.

Global

Using the Guidelines for Informational Interviews in this chapter, write an email to an interviewee who speaks fluent English but comes from another country. How might you approach the request for an interview differently? In addition to research about your topic and your respondent's background prior to the interview, what other research might you do? How might you compose your interview questions with your interviewee's nationality and/or culture in mind? (For more on global considerations, see Chapters 3 and 5.) Compare the different approaches.

Chapter 8
Evaluating and Interpreting Information

Alexander Raths/Shutterstock

"Our clients make investment decisions based on feasibility and strategy for marketing new ideas and products. Our job is to research consumer interest and experiences with products and services such as a new secure phone app and Web site where patients can review their medical test results. We use surveys to help understand what works and what doesn't. In designing surveys, I have to translate the client's information needs into precise questions. I have to be certain that the respondents are answering *exactly* the question I had in mind and not inventing their own version of the question. Then I have to take these data and translate them into accurate interpretations and recommendations for our clients.**"**

—*Jessica North, Senior Project Manager, technical marketing research firm*

Learning Objectives

8.1 Evaluate your sources, with special attention to online sources

8.2 Assess the quality of your evidence

8.3 Interpret your findings accurately and without bias

8.4 Recognize common errors in reasoning and statistical analysis

8.5 Understand that even careful research can be limited and imperfect

Determine if your research sources are valid and reliable

Not all information is equal. Not all interpretations are equal either. For instance, if you really want to know how well the latest innovation in robotic surgery works, you need to check with other sources besides, say, the device's designer (from whom you could expect an overly optimistic or insufficiently critical assessment).

Whether you work with your own findings or the findings of other researchers, you need to decide if the information is valid and reliable. Then you need to decide what your information means. Figure 8.1 outlines your critical thinking decisions, and the potential for error at any stage in this process.

Evaluate the Sources

8.1 Evaluate your sources, with special attention to online sources

Not all sources are equally dependable. A source might offer information that is out-of-date, inaccurate, incomplete, mistaken, or biased.

"Is the source up-to-date?"

- **Determine the currency of the source.** Even newly published books contain information that can be more than a year old, and journal articles typically undergo a lengthy process of peer review before they are published.

  NOTE *The most recent information is not always the most reliable—especially in scientific research, a process of ongoing inquiry where new research findings may enhance, modify, or invalidate previous studies.*

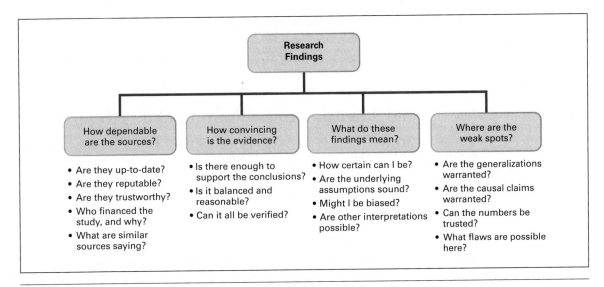

Figure 8.1 Critical Thinking Decisions in Evaluating and Interpreting Information Collecting information is often the easiest part of the research process. Your larger challenge is in getting the exact information you need, making sure it's accurate, figuring out what it means, and then double-checking for possible errors along the way.

- **Assess the reputation of the source.** Check the publication's copyright page. Is the work published by a university, professional society, trusted research organization, or respected news organization? Is the publication *refereed* (all submissions reviewed by experts before acceptance)? Has the author created a well-researched and thorough bibliography? Also check citation indexes (see Chapter 7, "Reference Works") to see what other experts have said about this particular source. Many periodicals also provide brief biographies or descriptions of authors' earlier publications and achievements.

  "Is the source reputable?"

- **Consider the possible motives of those who have funded the study.** Much of today's research is paid for by private companies or special-interest groups that have their own agendas (Crossen 14, 19). Medical research may be financed by drug or tobacco companies; nutritional research by food manufacturers; environmental research by oil or chemical companies. Instead of a neutral and balanced inquiry, this kind of "strategic research" is designed to support one special interest or another (132–34). Research financed by opposing groups can produce opposing results (234). Try to determine exactly what those who have funded a particular study stand to gain or lose from the results.

  "Who financed the study, and why?"

NOTE Keep in mind that any research ultimately stands on its own merits. Thus, funding by a special interest should not automatically discredit an otherwise valid and reliable study. Also, financing from a private company often sets the stage for beneficial research that might otherwise be unaffordable, as when research funded by Quaker Oats led to other studies proving that oats can lower cholesterol (Raloff 189).

"What are similar sources saying?"

- **Cross-check the source against other, similar sources.** Most studies have some type of flaw or limitation (see "Flaws in Research Studies" in this chapter). Instead of relying on a single source or study, you should seek a consensus among various respected sources.

Pay Special Attention to Evaluating Online Sources

"Does this online source reflect only one particular point of view?"

As noted above, you should evaluate all sources; however, for online sources, special scrutiny is needed. Many peer-reviewed, edited publications are available on the Internet (either through your school's library Web interface or on the site of publication itself). High quality, credible documents that were once only available in print (books, journal articles, government studies, newspaper articles) are now available in both print and online. But keep in mind that online, anyone can become a publisher, so you will also find information that is not subject to the scrutiny of an editorial board. For instance, many Web sites, blogs, and Facebook pages may contain valuable ideas, but the material might only reflect the perspective of the one person or group that created and maintains the page. As just one example, Web sites about genetically modified organisms (GMOs) will contain a range of various kinds of useful information. But depending on the sponsorship, each site may be biased toward a particular point of view (some in favor, some against, some taking a more middle of the road approach). That bias can actually provide you with useful clues about the different sides of the issue. However, the burden falls to you, as the researcher, to weigh this information against other perspectives and to check the facts as carefully as you can.

So, when evaluating online information, be sure to dig deeper and find out more about the organization behind the Web site, Twitter feed, Facebook page, Instagram account, or other site. Then, look for information from a range of sources—but make sure these sources are credible—so you can assess a balance of viewpoints. Throughout this chapter, refer to Chapter 7, "Guidelines for Researching on the Web and with Social Media," for more information on how to determine if sources are credible and how to use your own critical thinking skills to evaluate online information.

Evaluate the Evidence

8.2 Assess the quality of your evidence

Evidence is any finding used to support or refute a particular claim. Although evidence can serve the truth, it can also distort, misinform, and deceive. For example:

Questions that invite distorted evidence

| How much money, material, or energy does recycling really save?
| How well are public schools educating children?
| Which investments or automobiles are safest?
| How safe and effective are herbal medications?

Competing answers to such questions often rest on evidence that has been chosen to support a particular view or agenda.

- **Determine the sufficiency of the evidence.** Evidence is sufficient when nothing more is needed to reach an accurate judgment or conclusion. Say you are researching the stress-reducing benefits of low-impact aerobics among employees at a fireworks factory. You would need to interview or survey a broad sample: people who have practiced aerobics for a long time; people of both genders, different ages, different occupations, different lifestyles before they began aerobics; and so on. But responses even from hundreds of practitioners might be insufficient unless those responses were supported by laboratory measurements of metabolic and heart rates, blood pressure, and so on.

  NOTE Although anecdotal evidence ("This worked great for me!") may offer a starting point, personal experience rarely provides enough evidence from which to generalize.

  "Is there enough evidence?"

- **Differentiate hard evidence from soft evidence.** "Hard evidence" consists of facts, expert opinions, or statistics that can be verified. "Soft evidence" consists of uninformed opinions or speculations, data obtained or analyzed unscientifically, and findings that have not been replicated or reviewed by experts.

  "Can the evidence be verified?"

- **Decide whether the presentation of evidence is balanced and reasonable.** Evidence may be overstated, such as when overzealous researchers exaggerate their achievements without revealing the limitations of their study. Or vital facts may be omitted, as when acetaminophen pain relievers are promoted as "safe," even though acetaminophen is the leading cause of U.S. drug fatalities (Easton and Herrara 42–44).

  "Is this claim too good to be true?"

- **Consider how the facts are framed.** A *frame of reference* is a set of ideas, beliefs, or views that influences our interpretation or acceptance of other ideas. In medical terms, for example, is a "90 percent survival rate" more acceptable than a "10 percent mortality rate"? Framing sways our perception (Lang and Secic 239–40). For instance, what we now call a financial "recession" used to be a "depression"—a term that was coined as a euphemism for "panic" (Bernstein 183). For more on euphemisms, see Chapter 11, "Avoid Misleading Euphemisms."

  Is the glass "half full" or "half empty"?

Whether the language is provocative ("rape of the environment," "soft on terrorism"), euphemistic ("teachable moment" versus "mistake"), or demeaning to opponents ("bureaucrats," "tree huggers"), deceptive framing—all too common in political "spin" strategies—obscures the real issues.

Interpret Your Findings

8.3 Interpret your findings accurately and without bias

Interpreting means trying to reach the truth of the matter: an overall judgment about what the findings mean and what conclusion or action they suggest.

"What does this all mean?"

Unfortunately, research does not always yield answers that are clear or conclusive. Instead of settling for the most *convenient* answer, we must pursue the most *reasonable* answer by critically examining a full range of possible meanings.

Identify Your Level of Certainty

Research can yield three distinct and very different levels of certainty:

1. The ultimate truth—the *conclusive answer:*
   "Truth," as one expert notes, "is *what is so* about something, as distinguished from what people wish, believe, or assert to be so." Citing the philosopher Israel Scheffler, this author goes on to explain that "truth is the view 'which is fated to be ultimately agreed to by all who investigate' [and that] [t]he word *ultimately* is important" because, as many examples from history illustrate, "[i]nvestigation may produce a wrong answer for years, even for centuries" (emphases original) (Ruggiero 21). The word *investigate* is also important: in scientific and technical fields, a working theory, backed by evidence but not yet fully proven, may exist for years. However, the scientific method provides a system for leading us to the truth. The careful process of scientific inquiry is always open to new ideas but must be evidence-based using the highest standards of data collection, analysis, and peer review. In this way, we determine evidence-based truths about the situations and systems that surround us.

2. The *probable answer:* the answer that stands the best chance of being true or accurate, given the most we can know at this particular time. Probable answers are subject to revision in light of new information. This is especially the case with *emergent science,* such as gene therapy, certain kinds of new cancer treatments, or discoveries made in outer space.

3. The *inconclusive answer:* the realization that the truth of the matter is more elusive, ambiguous, or complex than we expected.

We need to decide what level of certainty our findings warrant. For example, we are *certain* about the perils of smoking and sunburn, *reasonably certain* about the health benefits of exercise, but *less certain* about the perils of genetically modified food or the benefits of taking fiber supplements.

Examine the Underlying Assumptions

Assumptions are notions we take for granted, ideas we often accept without proof. The research process rests on assumptions such as these: a sample group accurately represents a larger target group, survey respondents remember facts accurately, and mice and humans share many biological similarities. For a study to be valid, the underlying assumptions have to be accurate.

Consider this example: You are an education consultant evaluating the accuracy of IQ testing as a predictor of academic performance. Reviewing the evidence, you perceive an association between low IQ scores and low achievers. You then verify your statistics by examining a cross section of reliable sources. Can you justifiably conclude that IQ tests do predict performance accurately? This conclusion might be invalid unless you verify the following assumptions:

1. No one—parents, teachers, or children—had seen individual test scores, which could produce biased expectations.
2. Regardless of score, each child had completed an identical curriculum instead of being "tracked" on the basis of his or her score.

> NOTE *Assumptions can be easier to identify in someone else's thinking than our own. During team discussions, ask members to help you identify your own assumptions.*

Be Alert for Personal Bias

To support a particular version of the truth, our own bias might cause us to overestimate (or deny) the certainty of our findings.

Personal bias is a fact of life

> Unless you are perfectly neutral about the issue, an unlikely circumstance, at the very outset…you will believe one side of the issue to be right, and that belief will incline you to…present more and better arguments for the side of the issue you prefer. (Ruggiero 134)

Because personal bias is hard to transcend, *rationalizing* often becomes a substitute for *reasoning:*

Reasoning versus rationalizing

> You are reasoning if your belief follows the evidence—that is, if you examine the evidence first and then make up your mind. You are rationalizing if the evidence follows your belief—if you first decide what you'll believe and then select and interpret evidence to justify it. (Ruggiero 44)

Personal bias is often unconscious until we examine our attitudes long held but never analyzed, assumptions we've inherited from our backgrounds, and so on. The *confirmation bias* makes the situation even more complicated; when presented with solid evidence that disproves their beliefs, people will often cling even more tightly to what they want to be true. Recognizing our own biases and being willing to listen to new evidence is a crucial first step in managing personal bias.

Consider Other Possible Interpretations

Settling on a final meaning can be difficult—and sometimes impossible. For example, issues such as the need for defense spending or the causes of inflation are always controversial and will never be resolved. Although we can get verifiable data and can reason persuasively on many subjects, no close reasoning by any expert and no supporting statistical analysis will "prove" anything about a controversial subject. Some problems are simply more resistant to solution than others, no matter how dependable the sources.

"What else could this mean?"

> NOTE *Not all interpretations are equally valid. Never assume that any interpretation that is possible is also allowable—especially in terms of its ethical consequences.*

Consider This

Standards of Proof Vary for Different Audiences

How much evidence is enough to "prove" a particular claim? This often depends on who is making the inquiry:

- **The scientist** demands at least 95 percent certainty. A scientific finding must be evaluated and replicated by other experts. Good science looks at the entire picture. Findings are reviewed before they are reported. Even then, answers in science are never "final," but open-ended and ongoing.

- **The juror** in a civil case requires evidence that indicates 51 percent certainty (a "preponderance of the evidence"). In a criminal case, the standard is higher ("beyond a reasonable doubt"). Either way, jurors rarely see the big picture—just the information given by lawyers and witnesses. Based on such evidence, courts must make final decisions (Monastersky 249; Powell 32+).

- **The executive** demands immediate (even if insufficient) evidence. In a global business climate of overnight developments (data breaches, political strife, natural disasters), business decisions are often made on the spur of the moment. On the basis of incomplete or unverified information—or even hunches—executives must react to crises and try to seize opportunities (Seglin 54).

Avoid Distorted or Unethical Reasoning

8.4 Recognize common errors in reasoning and statistical analysis

Finding the truth, especially in a complex issue or problem, often is a process of elimination, of ruling out or avoiding errors in reasoning. As we interpret, we make *inferences:* We derive conclusions about what we don't know by reasoning from what we do know (Hayakawa 37). For example, we might infer that a drug that boosts immunity in laboratory mice will boost immunity in humans or that a rise in campus crime statistics is caused by the fact that young people have become more violent. Whether a particular inference is on target or dead wrong depends largely on our answers to one or more of these questions:

Questions for testing inferences

- To what extent can these findings be generalized?
- Is *Y* really caused by *X*?
- To what extent can the numbers be trusted, and what do they mean?

Three major reasoning errors that can distort our interpretations are faulty generalization, faulty causal reasoning, and faulty statistical analysis.

Faulty Generalization

Consider, for example, the numerous times we hear about a crisis of some sort—a natural disaster, a missing airplane, a terrible accident—and all we know initially is based on limited news reports, random Twitter posts, and cell phone videos that go viral. Based only on this information, people will often make general claims and jump to conclusions that later prove to be false after more evidence is uncovered.

We engage in faulty generalization when we jump from a limited observation to a sweeping conclusion. Even "proven" facts can invite mistaken conclusions, as in the following examples:

1. "Some studies have shown that gingko [an herb] improves mental functioning in people with dementia [mental deterioration caused by maladies such as Alzheimer's Disease]" (Stix 30).

2. "[I]n some cases, a two-year degree or technical certificate can offer students a better return on investment than a four-year degree" (Sheehy).

3. "Adult female brains are significantly smaller than male brains—about 8% smaller, on average" (Seligman 74).

> Factual observations

1. Gingko is food for the brain!
2. Four-year degrees are a waste of money.
3. Women are less intelligent than men.

> Invalid conclusions

When we accept findings uncritically and jump to conclusions about their meaning (as in points 1 and 2, above) we commit the error of *hasty generalization*. When we overestimate the extent to which the findings reveal some larger truth (as in point 3, above) we commit the error of *overstated generalization*.

> "How much can we generalize from these findings?"

> *NOTE We often need to generalize, and we should. For example, countless studies support the generalization that fruits and vegetables help lower cancer risk. But we ordinarily limit general claims by inserting qualifiers such as "usually," "often," "sometimes," "probably," "possibly," or "some."*

Faulty Causal Reasoning

Causal reasoning tries to explain why something happened or what will happen, often in very complex situations. Sometimes a *definite cause is apparent* ("The engine's overheating is caused by a faulty radiator cap"). We reason about definite causes when we explain why the combustion in a car engine causes the wheels to move, or why the moon's orbit makes the tides rise and fall. However, causal reasoning often explores *causes that are not so obvious, but only possible or probable*. In these cases, much analysis is needed to isolate a specific cause.

> "Did *X* possibly, probably, or definitely cause *Y*?"

Suppose you ask: "Why are there no children's daycare facilities on our college campus?" Brainstorming yields these possible causes:

- lack of need among students
- lack of interest among students, faculty, and staff
- high cost of liability insurance
- lack of space and facilities on campus
- lack of trained personnel
- prohibition by state law
- lack of government funding for such a project

Assume that you proceed with interviews, surveys, and research into state laws, insurance rates, and availability of personnel. As you rule out some items, others appear as probable causes. Specifically, you find a need among students, high campus interest, an abundance of qualified people for staffing, and no state laws prohibiting such a project. Three probable causes remain: high insurance rates, lack of funding, and lack of space. Further inquiry shows that high insurance rates and lack of funding *are* issues. You think, however, that these obstacles could be eliminated through new sources of revenue such as charging a modest fee per child, soliciting donations, and diverting funds from other campus organizations. Finally, after examining available campus space and speaking with school officials, you conclude that one definite cause is lack of space and facilities. In reporting your findings, you would follow the sequence shown in Figure 8.2.

The persuasiveness of your causal argument will depend on the quality of evidence you bring to bear, as well as on your ability to explain the links in the chain of your reasoning. Also, you must convince audiences that you haven't overlooked important alternative causes.

> NOTE *Any complex effect is likely to have more than one cause. You have to make sure that the cause you have isolated is the right one. In the daycare scenario, for example, you might argue that lack of space and facilities somehow is related to funding. And the college's inability to find funds or space might be related to student need or interest, which is not high enough to exert real pressure. Lack of space and facilities, however, does seem to be the immediate cause.*

Here are common errors that distort or oversimplify cause–effect relationships:

Ignoring other causes	Investment builds wealth. [*Ignores the roles of knowledge, wisdom, timing, and luck in successful investing.*]
Ignoring other effects	Running improves health. [*Ignores the fact that many runners get injured and that some even drop dead while running.*]
Inventing a causal sequence	Right after buying a rabbit's foot, Felix won the state lottery. [*Posits an unwarranted causal relationship merely because one event follows another.*]

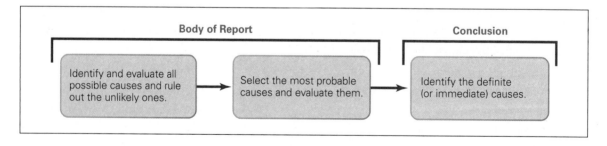

Figure 8.2 The Reporting Sequence in a Causal Analysis Be sure readers can draw conclusions identical to your own on the basis of the reasoning you present.

Women in Scandinavian countries drink a lot of milk. Women in Scandinavian countries have a high incidence of breast cancer. Therefore, milk must be a cause of breast cancer. [*The association between these two variables might be mere coincidence and might obscure other possible causes, such as environment, fish diet, and genetic predisposition* (Lemonick 85).]

Confusing correlation with causation

My grades were poor because my exams were unfair. [*Denies the real causes of one's failures.*]

Rationalizing

Media researcher Robert Griffin identifies three criteria for demonstrating a causal relationship:

> Along with showing correlation [say, an association between smoking and cancer], evidence of causality requires that the alleged causal agent occurs prior to the condition it causes (e.g., that smoking precedes the development of cancers) and—the most difficult task—that other explanations are discounted or accounted for (240).

For example, epidemiological studies found this correlation: People who eat lots of broccoli, cauliflower, and other cruciferous vegetables have lower rates of some cancers. But other explanations (say, that big veggie eaters might also have many other healthful habits) could not be ruled out until lab studies showed how a special protein in these vegetables actually protects human cells (Wang 182).

Faulty Statistical Analysis

The purpose of statistical analysis is to determine the meaning of a collected set of numbers. In primary research, our surveys and questionnaires often lead to some kind of numerical interpretation ("What percentage of respondents prefer X?" "How often does Y happen?"). In secondary research, we rely on numbers collected by survey researchers.

How numbers can mislead

Numbers seem more precise, more objective, more scientific, and less ambiguous than words. They are easier to summarize, measure, compare, and analyze. But numbers can be totally misleading. For example, radio or television phone-in surveys often produce distorted data: Although "90 percent of callers" might express support for a particular viewpoint, people who bother to respond tend to have the greatest anger or extreme feelings—representing only a fraction of overall attitudes (Fineman 24). Mail-in or Internet surveys can produce similar distortion. Before relying on any set of numbers, we need to know exactly where they come from, how they were collected, and how they were analyzed.

Faulty statistical reasoning produces conclusions that are unwarranted, inaccurate, or deceptive. Following are typical fallacies.

Common statistical fallacies

THE SANITIZED STATISTIC. Numbers can be manipulated (or "cleaned up") to obscure the facts. For instance, when creating a chart or graph to represent sales trends for a product, marketers might leave out numbers at the very low end of the range, or start with the years when sales began to pick up, in order to show an upward trend more favorable to investors. (See Chapter 12, "Ethical Considerations," for more about ethics and visual communication.)

"Exactly how well are we doing?"

THE MEANINGLESS STATISTIC. Exact numbers can be used to quantify something so inexact or vaguely defined that it should only be approximated (Huff 247; Lavin 278): "Boston has 3,247,561 rats." "Super brand detergent makes laundry 10 percent brighter." An exact number looks impressive, but it can hide the fact that certain subjects (child abuse, cheating in college, drug and alcohol abuse, eating habits) cannot be quantified exactly because respondents don't always tell the truth (on account of denial or embarrassment or guessing). Or they respond in ways they think the researcher expects.

THE UNDEFINED AVERAGE. The mean, median, and mode can be confused in representing an "average" (Huff 244; Lavin 279): (1) The *mean* is the result of adding up the values of items in a set of numbers, and then dividing that total by the number of items in the set. (2) The *median* is the result of ranking all the values from high to low, then identifying the middle value (or the 50th percentile, as in calculating SAT scores). (3) The *mode* is the value that occurs most often in a set of numbers.

Each of these three measurements represents some kind of average. But unless we know which "average" (mean, median, or mode) is being presented, we cannot possibly interpret the figures accurately.

Assume, for instance, that we want to determine the average salary among female managers at XYZ Corporation (ranked from high to low):

Manager	Salary
"A"	$90,000
"B"	$90,000
"C"	$80,000
"D"	$65,000
"E"	$60,000
"F"	$55,000
"G"	$50,000

In the above example, the mean salary (total salaries divided by number of salaries) is $70,000; the median salary (middle value) is $65,000; the mode (most frequent value) is $90,000. Each is, legitimately, an "average," and each could be used to support or refute a particular assertion (for example, "Women managers are paid too little" or "Women managers are paid too much").

Research expert Michael R. Lavin sums up the potential for bias in the reporting of averages:

> Depending on the circumstances, any one of these measurements may describe a group of numbers better than the other two....[But] people typically choose the value which best presents their case, whether or not it is the most appropriate to use. (279)

Although the mean is the most commonly computed average, this measurement is misleading when one or more values on either end of the scale (*outliers*) are extremely high or low. Suppose, for instance, that manager "A" (above) was paid a $200,000 salary. Because this figure deviates so far from the normal range of salary figures for "B" through "G," it distorts the average for the whole group, increasing the mean salary by more than 20 percent (Plumb and Spyridakis 636).

THE DISTORTED PERCENTAGE FIGURE.   Percentages are often reported without explanation of the original numbers used in the calculation (Adams and Schvaneveldt 359; Lavin 280): "Seventy-five percent of respondents prefer our brand over the competing brand"—without mention that, say, only four people were surveyed.

> NOTE *In small samples, percentages can mislead because the percentage size can dwarf the number it represents: "In this experiment, 33% of the rats lived, 33% died, and the third rat got away" (Lang and Secic 41). When your sample is small, report the actual numbers: "Five out of ten respondents agreed . . . ."*

"Is 51 percent really a majority?"

Another fallacy in reporting percentages occurs when the *margin of error* is ignored. This is the margin within which the true figure lies, based on estimated sampling errors in a survey. For example, a claim that "most people surveyed prefer Brand X" might be based on the fact that 51 percent of respondents expressed this preference; but if the survey carried a 2 percent margin of error, the real figure could be as low as 49 percent or as high as 53 percent. In a survey with a high margin of error, the true figure may be so uncertain that no definite conclusion can be drawn.

"How large is the margin of error?"

THE BOGUS RANKING. This distortion occurs when items are compared on the basis of ill-defined criteria (Adams and Schvaneveldt 212; Lavin 284). For example, the statement "Last year, the Batmobile was the number-one selling car in America" does not mention that some competing car makers actually sold *more* cars to private individuals and that the Batmobile figures were inflated by hefty sales—at huge discounts—to rental-car companies and corporate fleets. Unless we know how the ranked items were chosen and how they were compared (the *criteria*), a ranking can produce a seemingly scientific number based on a completely unscientific method.

"Which car should we buy?"

CONFUSION OF CORRELATION WITH CAUSATION. *Correlation* is a numerical measure of the strength of the relationship between two variables (say smoking and increased lung cancer risk, or education and income). *Causation* is the demonstrable production of a specific effect (smoking causes lung cancer). Correlations between smoking and lung cancer or between education and income signal a causal relationship that has been demonstrated by many studies. But not every correlation implies causation. For instance, a recently discovered correlation that people who eat apples daily have a decreased heart disease risk offers no sufficient proof that eating apples actually *causes* less heart disease.

"Does *X* actually cause *Y*?"

In any type of causal analysis, be on the lookout for *confounding factors*, which are other possible reasons or explanations for a particular outcome. For instance, people who eat apples daily might also have other healthy habits, such as taking daily walks or avoiding saturated fats, that contribute to lower instances of heart disease.

"Could something else have caused *Y*?"

Many highly publicized correlations are the product of *data mining:* In this process, computers randomly compare one set of variables (say, eating habits) with another set (say, range of diseases). From these countless comparisons, certain relationships or associations are revealed (say, between coffee drinking and pancreatic cancer risk). Data mining has gotten quite sophisticated over the past few years, as companies work to understand the relationships between consumers and spending habits online; these tools can then be used to understand other relationships. Despite the power of data mining, it is important for researchers to evaluate and consider the results.

"Who selected which studies to include?"

THE BIASED META-ANALYSIS. In a meta-analysis, researchers examine a whole range of studies that have been done on one topic (say, high-fat diets and cancer risk). The purpose of this "study of studies" is to decide the overall meaning of the collected findings. Because results ultimately depend on which studies have been included and which omitted, a meta-analysis can reflect the biases of the researchers who select the material. Also, because small studies have less chance of being published than large ones, they may get overlooked (Lang and Secic 174–76). That said, meta-analyses can be quite helpful if done properly and by a credible agency or group of scientists; by examining a range of studies that took place over a long period of time, for example, in 2009 the U.S. Preventive Services Task Force changed it recommendations for breast cancer screening.

"How have assumptions influenced this computer model?"

THE FALLIBLE COMPUTER MODEL. Computer models process complex *assumptions* (see "Examine the Underlying Assumptions" in this chapter) to predict or estimate costs, benefits, risks, and probable outcomes. But answers produced by any computer model depend on the assumptions (and data) programmed in. Assumptions might be influenced by researcher bias or the sponsors' agenda. For example, a prediction of human fatalities from a nuclear reactor meltdown might rest on assumptions about the availability of safe shelter, evacuation routes, time of day, season, wind direction, and the structural integrity of the containment unit. But these assumptions could be manipulated to overstate or understate the risk (Barbour 228). For computer-modeled estimates of accident risk (oil spill, plane crash) or of the costs and benefits of a proposed project or policy (international space station, health care reform), consumers rarely know the assumptions behind the numbers.

"Do we all agree on what these terms mean?"

MISLEADING TERMINOLOGY. The terms used to interpret statistics sometimes hide their real meaning. For instance, the widely publicized figure that people treated for cancer have a "50 percent survival rate" is misleading in two ways: (1) *survival* to laypersons means "staying alive," but to medical experts, staying alive for only five years after diagnosis qualifies as survival; (2) the "50 percent" survival figure covers *all* cancers, including certain skin or thyroid cancers that have extremely high *cure rates* as well as other cancers (such as lung or ovarian) that are rarely curable and have extremely low *survival rates* ("Are We" 6).

"Is this news good, bad, or insignificant?"

Even the most valid and reliable statistics require that we interpret the reality behind the numbers. For instance, the overall cancer rate today is "higher" than it was in 1910.

What this may mean is that people are living longer and thus are more likely to die of cancer and that cancer today rarely is misdiagnosed—or mislabeled because of stigma ("Are We" 4). The finding that rates for certain cancers "double" after prolonged exposure to electromagnetic waves may really mean that cancer risk actually increases from 1 in 10,000 to 2 in 10,000.

The numbers may be "technically accurate" and may seem highly persuasive in the interpretations they suggest. But the actual "truth" behind these numbers is far more elusive. Any interpretation of statistical data carries the possibility that other, more accurate interpretations have been overlooked or deliberately excluded (Barnett 45).

Acknowledge the Limits of Research

8.5 Understand that even careful research can be limited and imperfect

Legitimate researchers live with uncertainty. They expect to be wrong far more often than right. Following is a brief list of items that go wrong with research and interpretation.

Obstacles to Validity and Reliability

Validity and *reliability* determine the dependability of any research (Adams and Schvaneveldt 79–97; Crossen 22–24). *Valid research* produces correct findings. A survey, for example, is valid when (1) it measures what you want it to measure, (2) it measures accurately and precisely, and (3) its findings can be generalized to the target population. Valid survey questions enable each respondent to interpret each question exactly as the researcher intended; valid questions also ask for information respondents are qualified to provide.

What makes a survey valid

Survey validity depends largely on trustworthy responses. Even clear, precise, and neutral questions can produce mistaken, inaccurate, or dishonest answers. People often see themselves as more informed, responsible, or competent than they really are. Respondents are likely to suppress information that reflects poorly on their behavior, attitudes, or will power when answering such leading questions as "How often do you take needless sick days?" "Would you lie to get ahead?" "How much time do you spend on Facebook?" They might exaggerate or invent facts or opinions that reveal a more admirable picture when answering the following types of questions: "How much do you give to charity?" "How many books do you read?" "How often do you hug your children?" Even when respondents don't know, don't remember, or have no opinion, they often tend to guess in ways designed to win the researcher's approval.

Why survey responses can't always be trusted

Reliable research produces findings that can be replicated. A survey is reliable when its results are consistent; for instance, when a respondent gives identical answers to the same survey given twice or to different versions of the same questions. Reliable survey questions can be interpreted identically by all respondents.

What makes a survey reliable

Much of your communication will be based on the findings of other researchers, so you will need to assess the validity and reliability of their research as well as your own.

Flaws in Research Studies

Although some types of studies are more reliable than others, each type has limitations (Cohn 106; Harris 170–72; Lang and Secic 8–9; Murphy 143):

Common flaws in epidemiologic studies

- **Epidemiological studies.** Epidemiologists study various populations (human, animal, or plant) to find correlations (say, between computer use and cataracts). Conducted via observations, interviews, surveys, or records review, these studies are subject to faulty sampling techniques (see Chapter 7, "Surveys" and "Guidelines for Surveys") and observer bias (seeing what one wants to see). Even with a correlation that is 99 percent certain, an epidemiological study alone doesn't "prove" anything. (The larger the study, however, the more credible.)

Common flaws in laboratory studies

- **Laboratory studies.** Although a laboratory offers controlled conditions, these studies also carry limitations. For example, the reactions of experimental mice to a specific treatment or drug often are not generalizable to humans. Also, the reaction of an isolated group of cells does not always predict the reaction of the entire organism.

Common flaws in clinical trials

- **Human exposure studies (clinical trials).** These studies compare one group of people receiving medication or treatment with an untreated group, the *control group*. Limitations include the possibility that the study group may be nonrepresentative or too different from the general population in overall health, age, or ethnic background. (For example, even though gingko may slow memory loss in sick people, that doesn't mean it will boost the memory of healthy people.) Also, anecdotal reports are unreliable. Respondents often invent answers to questions such as "How often do you eat ice cream?" or "Do you sometimes forget to take your medication?"

Social Media and Research Reporting

Social media can promote inaccurate reporting

The sheer number of today's social media outlets allows for dozens if not hundreds of versions of any particular story. Deceptive reporting takes place when writers or media outlets purposely skew the facts, leave out important information, or create just enough confusion to keep people wondering. Often, after spending so much time online, the public comes away with a distorted picture about the research. For instance, a peer-reviewed study about the potential for a particular herbal medication to treat anxiety and depression might be accompanied by several cautionary notes, such as the need for more studies to address the variations in this herbal compound across different brands or the need to conduct future studies using different age groups.

Yet on social media feeds run by companies that sell this product or by people who have a particular agenda against traditional pharmaceutical medications, the research may be reported in glowing terms, promoting the use of the product, leaving out any cautionary notes, and using eye-catching headlines and visuals ("An herbal cure for depression!") to promote reposting on social media feeds.

> *NOTE All this potential for error doesn't mean we should believe nothing. But we need to be discerning about what we do choose to believe. Critical thinking is essential. Review Chapter 7 for more on critical thinking and evaluating research.*

Guidelines

for Evaluating and Interpreting Information

Evaluate the Sources

➤ **Check the posting or publication date.** The latest information is not always the best, but keeping up with recent developments is vital.

➤ **Assess the reputation of each source.** Check the copyright page for background on the publisher, the bibliography for the quality of research, and (if available) the author's brief biography.

➤ **Assess the quality of your source material.** If using material from the Internet and social media, see "Evaluate Online Information" in this chapter as well as Chapter 7, "Guidelines for Researching on the Web and with Social Media."

➤ **Don't let looks deceive you.** Several studies have found that people judge Web sites and other online information based in large part on the look of the page (Fogg). Professional formatting can disguise a special-interest or biased point of view.

➤ **Identify the study's sponsor.** If a study proclaiming the crashworthiness of the Batmobile has been sponsored by the Batmobile Auto Company, be skeptical about the study's findings.

➤ **Look for corroborating sources.** A single study rarely produces definitive findings. Learn what other sources say, why they agree or disagree, and where most experts stand.

Evaluate the Evidence

➤ **Decide whether the evidence is sufficient.** Evidence should surpass personal experience, anecdote, or media reports. Reasonable and informed observers should be able to agree on its credibility.

➤ **Look for a fair and balanced presentation.** Suspect any claims about "breakthroughs" or "miracle cures" or the like.

➤ **Try to verify the evidence.** Examine the facts that support the claims. Look for replication of findings.

Interpret Your Findings

➤ **Don't expect "certainty."** Complex questions are mostly open-ended, and a mere accumulation of facts doesn't "prove" anything. Even so, the weight of evidence usually suggests some reasonable conclusion.

➤ **Examine the underlying assumptions.** As opinions taken for granted, assumptions are easily mistaken for facts.

➤ **Identify your personal biases.** Examine your own assumptions. Don't ignore evidence simply because it contradicts your original assumptions.

➤ **Consider alternative interpretations.** What else might this evidence mean?

Check for Weak Spots

➤ **Scrutinize all generalizations.** Decide whether the evidence supports the generalization. Suspect any general claim not limited by a qualifier such as "often," "sometimes," or "rarely."

➤ **Treat causal claims skeptically.** Differentiate correlation from causation, as well as possible from probable or definite causes. Consider confounding factors (other explanations for the reported outcome).

➤ **Look for statistical fallacies.** Determine where the numbers come from, and how they were collected and analyzed—information that legitimate researchers routinely provide. Note the margin of error.

➤ **Consider the limits of computer analysis.** Data mining often produces intriguing but random correlations; a computer model is only as accurate as the assumptions and data that were programmed in.

➤ **Look for misleading terminology.** Examine terms that beg for a precise definition in their specific context: "survival rate," "success rate," and so on.

➤ **Interpret the reality behind the numbers.** Consider the possibility of alternative, more accurate interpretations of the data.

➤ **Consider the study's possible limitations.** Small, brief studies are less reliable than large, extended ones; epidemiological studies are less reliable than laboratory studies (which have their own flaws); and animal exposure studies are often not generalizable to human populations.

➤ **Look for the whole story.** Consider whether bad news is underreported or good news is exaggerated. Also consider whether bad science is camouflaged and sensationalized and whether promising but unconventional topics are ignored.

Checklist

The Research Process

Use the following Checklist as you conduct research.

Methods

❏ Did I ask the right questions? (See Chapter 7, "Asking the Right Questions.")

❏ Is each source appropriately up to date, reputable, trustworthy, relatively unbiased, and borne out by other, similar sources? (See "Evaluate the Sources" in this chapter.)

❏ For digital sources, do I have a clear idea of the author or organization behind the information? (See "Evaluate Online Information" in this chapter.)

❏ Does the evidence clearly support all of the conclusions? (See Chapter 7, "Evaluating and Interpreting Your Findings.")

❏ Is a fair balance of viewpoints represented? (See Chapter 7, "Exploring a Balance of Views.")

❏ Can all the evidence be verified? (See "Evaluate the Evidence" in this chapter.)

❏ Has my research achieved adequate depth? (See Chapter 7, "Achieving Adequate Depth in Your Search.")

❏ Has the entire research process been valid and reliable? (See "Obstacles to Validity and Reliability" in this chapter.)

Interpretation and Reasoning

❏ Am I reasonably certain about the meaning of these findings? (See Chapter 7, "Evaluating and Interpreting Your Findings" and "Interpret Your Findings" in this chapter.)

❏ Can I discern assumption from fact and reasoning from rationalizing? (See "Be Alert to Personal Bias" in this chapter.)

❏ Can I discern correlation from causation? (See "Confusion of Correlation with Causation" in this chapter.)

❏ Can I rule out other possible interpretations or conclusions? (See "Consider Other Possible Interpretations" in this chapter.)

❏ Have I accounted for all sources of bias, including my own? (See "Examine the Underlying Assumptions" and "Be Alert to Personal Bias" in this chapter.)

❏ Are my generalizations warranted by the evidence? (See "Faulty Generalization" in this chapter.)

❏ Am I confident that my causal reasoning is accurate? (See "Faulty Causal Reasoning" in this chapter.)

❏ Can I rule out confounding factors? (See "Confusion of Correlation with Causation" in this chapter.)

❏ Can all of the numbers, statistics, and interpretations be trusted? (See "Faulty Statistical Analysis" in this chapter.)

❏ Have I resolved (or at least acknowledged) any conflicts among my findings? (See Chapter 7, "Evaluating and Interpreting Your Findings.")

❏ Can I rule out any possible error or distortion in a given study? (See "Flaws in Research Studies" in this chapter.)

❏ Am I getting the whole story, and getting it straight? (See "Social Media and Research Reporting" in this chapter.)

Documentation

❏ Have I taken good notes in order to document my sources accurately? (See Appendix A, "Taking Notes.")

❏ Is my documentation consistent, complete, and correct? (See Appendix A, "MLA Documentation Style" or "APA Documentation Style.")

❏ Have I used quotations sparingly, marked them accurately, and integrated them properly? (See Appendix A, "Quoting the Work of Others" and "Guidelines for Quoting the Work of Others.")

❏ Are all paraphrases accurate and clear? (See Appendix A, "Paraphrasing the Work of Others" and "Guidelines for Paraphrasing.")

❏ Have I documented all sources not considered common knowledge? (See Appendix A, "What You Should Document.")

Projects

For all projects, check with your instructor about whether to present your findings in class, bring drafts to class for discussion, upload your project to the class learning management system (LMS), and/or use the LMS forum or discussion boards to collaborate and review each activity below.

General

Based on your own personal experience or on a social media post, Facebook page, or Web site, identify an example of each of the following sources of distortion or of interpretive error:

- a study with questionable sponsorship or motives
- reliance on insufficient evidence
- unbalanced presentation
- deceptive framing of facts
- overestimating the level of certainty
- biased interpretation
- rationalizing
- unexamined assumptions

- faulty causal reasoning
- hasty generalization
- overstated generalization
- sanitized statistic
- meaningless statistic
- undefined average
- distorted percentage figure
- bogus ranking
- fallible computer model
- misinterpreted statistic
- deceptive reporting

Submit your examples to your instructor, along with a memo explaining each error.

Team

Projects from the previous or following section may be done as team projects. See the Global Project, below, for one such activity.

Digital and Social Media

Uninformed opinions are usually based on assumptions we've never really examined. With social media, one post or tweet often leads to another one (and to other sites) with the same point of view, reinforcing assumptions rather than challenging them. Consider one of these two popular assumptions and go online to see what you can find that challenges the accuracy of the statement.

- "Bottled water is safer than most tap water in the United States."
- "The fewer germs in their environment, the healthier our children will be."

Keep track of the Web sites, online publications, tweets, social media posts, government documents, research reports, and other sources of online information you find. Pay attention to whether the source directs you to similar information or provides a range of ideas. Write up your findings in a short memo. Or, create a class wiki where you can share your findings.

Global

Look for studies from a credible global organization, such as the World Health Organization (WHO) or the International Standards Organization (ISO). Select one study or report and examine the way it evaluates and interprets information. Look especially at areas where causal reasoning is employed; for example, claims made that unsanitary drinking water causes childhood illness or that improper labeling of technical products causes injuries. Are these claims supported by the evidence? What steps do the writers take to make the report accessible and clear to a global audience? For the Team Project, do this activity in teams of 2–3 people.

Chapter 9

Summarizing Research Findings and Other Information

robophobic/Shutterstock

❝Every time I run a training session in corporate communication, participants tell horror stories about working weeks or months on a report, only to have it disappear somewhere up the management chain. We use copies of those 'invisible' reports as case studies, and invariably, the summary turns out to have been poorly written, providing readers few or no clues as to the report's significance. I'll bet companies lose millions because new ideas and recommendations get relegated to that stockpile of reports unread yearly in corporate America.❞

— *Frank Sousa, Communications Consultant*

Learning Objectives

9.1 Describe the audience and purpose
   of summaries in the workplace

9.2 Identify the elements of a usable
   summary

9.3 Follow the step-by-step process for
   creating a summary

9.4 Differentiate among four special
   types of summaries

9.5 Exercise caution when summarizing
   for social media

A *summary* is a restatement of the main ideas in a longer document. Summaries are
used to convey the general meaning of the ideas in the original source without all the
details or examples that may appear in the original. When you write a summary, provide only the essential information clearly and concisely in your own words, leaving
out anything that isn't central to an understanding of the original.

Considering Audience and Purpose

9.1 Describe the audience and purpose of summaries in the workplace

Summaries as a
research aid

Chapter 7 shows how abstracts (a type of summary) aid our research by providing an
encapsulated glimpse of an article or other long document (see Chapter 7, "Reference
Works"). Also, as we record our research findings, we summarize to capture the main
ideas in a compressed form. In addition to this dual role as a research aid, summarized information is vital in day-to-day workplace transactions.

Summaries in
the workplace

On the job, you have to write concisely about your work. You might report on
meetings or conferences, describe your progress on a project, or propose a money-
saving idea. A routine assignment for many new employees is to provide superiors
(decision makers) with summaries of the latest developments in their field.

Audience
considerations

Formal reports and proposals (discussed in Chapters 21 and 22) and other long
documents are typically submitted to busy people: researchers, developers, managers, vice presidents, customers, and so on. For readers who must act quickly or who
only need to know the "big picture," reading an entire long report may not only be
too time-consuming but also irrelevant. As a result, most long reports, proposals, and
other complex documents are commonly preceded by a summary.

The purpose of summaries, then, is to provide only an overview and the essential facts. Whether you summarize your own writing or someone else's, your summary should do three things for readers: (1) describe, in short form, what the original document is all about; (2) help readers decide whether to read the entire document, parts of it, or none of it; and (3) give readers a framework for understanding the full document that will follow if they do plan to read it.

An effective summary communicates the *essential message* accurately and in the fewest possible words. To get a basic idea of how summaries fulfill audience needs, consider the examples in Figures 9.1 and 9.2.

In Figure 9.1, three ideas make up the essential message: (1) scientists are certain that greenhouse gases largely produced by human activities are warming the planet; (2) rising temperatures worldwide during the 20th century have been demonstrated; and (3) this warming trend almost certainly will continue.

Figure 9.2 shows a summary of the complex passage in Figure 9.1. Note how it captures the original's main ideas but in a compressed and less technical form that busy readers and general audiences would appreciate. The summary does not, however, change the essential meaning of the original—it merely boils the original down to its basic message.

> NOTE *For letters, memos, or other short documents that can be read quickly, the only summary needed is usually an opening thesis or topic sentence that previews the contents.*

Purpose of summaries

Scientists know with certainty that human activities are changing the composition of Earth's atmosphere. Increasing levels of greenhouse gases like carbon dioxide (CO_2) since pre-industrial times are well-documented and understood. The atmospheric buildup of CO_2 and other greenhouse gases is largely the result of human activities such as the burning of fossil fuels. Increasing greenhouse gas concentrations tend to warm the planet. A warming trend of about 0.7 to 1.5°F occurred during the 20th century in both the Northern and Southern Hemispheres and over the oceans. The major greenhouse gases remain in the atmosphere for periods ranging from decades to centuries. It is therefore virtually certain that atmospheric concentrations of greenhouse gases will continue to rise over the next few decades.

Figure 9.1 **A Passage to Be Summarized**
Source: Adapted from *State of Knowledge*, U.S. Environmental Protection Agency.

Scientists are certain that greenhouse gases largely produced by human activities are warming the planet. Temperatures have risen worldwide during the 20th century and undoubtedly will continue.

Figure 9.2 **A Summarized Version of Figure 9.1**
Source: From *State of Knowledge*, U.S. Environmental Protection Agency.

What Readers Expect from a Summary

9.2 Identify the elements of a usable summary

Whether you summarize your own documents (like the sample abstract page of Figure 21.3) or someone else's, readers will have these expectations:

Elements of a usable summary

- **Accuracy:** Readers expect a precise sketch of the content, emphasis, and line of reasoning from the original.

- **Completeness:** Readers expect to consult the original document only to find more detail—but not to have to make sense of the main ideas and their relationships as these appear in the summary.

- **Readability:** Readers expect a summary to be clear and straightforward—easy to follow and understand.

- **Conciseness:** Readers expect a summary to be informative yet brief, and they may stipulate a word limit (say, 200 words).

- **Nontechnical style:** Unless they are all experts on this particular subject, readers prefer a document that uses nontechnical terms and simplifies complex ideas—without distorting those ideas.

Although the summary is written last, it is what readers of a long document turn to first. Take the time to do a good job.

How to Create a Summary

9.3 Follow the step-by-step process for creating a summary

The step-by-step nature of summaries

More so than for other types of documents, writing summaries involves a straightforward process: reading and rereading the original document, marking the key information, and writing and revising your summary while checking it against the original. The Guidelines for Summarizing Information following this section provide an overview of these steps, along with audience and purpose considerations.

Do an initial read to get the big picture

- **Step 1: Read the Original Document.** Before you can effectively summarize a document, you need a solid understanding of what it says. Read the original from start to finish, without highlighting or taking notes. An initial read will give you a general understanding of the document's main point and key subpoints.

Reread and highlight for a sharper understanding

- **Step 2: Reread and Mark Essential Material.** Read the document again, this time using a highlighter or pen (if working on hard copy) or your computer's highlighting feature. Pay close attention to key words or phrases to help you identify the essential points.

Create a new document from the highlighted information

- **Step 3: Cut and Paste the Key Information.** If you are working from a hard copy of the original, type or scan the material you highlighted into a new document. If you are working from an electronic version of the original, copy and paste only the highlighted portions into a new file. Don't worry about the overall organization or sentence construction at this point; just gather everything you need.

- **Step 4: Redraft the Information into Your Own Organizational Pattern and Words.** At this point, reorganize the material from the original source into patterns that work well in a smaller space. Use your own wording to express the original concepts. Don't worry about length just yet. You just want to get your own wording down on paper.

  Turn the cut and paste version into your own version

- **Step 5: Edit Your Draft.** In the previous step, the goal was to get your own wording written down. Now, edit your version. Keeping sentences clear and grammatical, cross out needless words; get rid of any asides; combine related concepts; look for any repetitions.

  Focus on shortening your version

- **Step 6: Compare Your Version with the Original Document.** As a final step, double-check that you have not altered the meaning, intent, or emphasis of the original document. Reread the original and then immediately reread your own version.

  Make sure you preserve the intent of the original

The following case study illustrates the application of this step-by-step process.

Case

Assume that you work in the communications unit of your state's public health agency. You and your team have been assigned to look through published research from the U.S. Food and Drug Administration (FDA) and write summaries of new research, to be posted on your state's public health Web site and linked to from the agency's Facebook page and weekly Twitter feed. The FDA material is too long for your purposes; you want to post short summaries that get right to the point and don't require too much time to read.

This week, it's your turn to select a topic. The FDA article you find most useful, about different types of dietary fats, is shown in Figure 9.3. It's available in PDF format, so you download it. You then go through the article, underline key phrases (see underlining in Figure 9.3), and make other notes using the Adobe Acrobat comments tool.

Because potential readers in this situation represent a broad cross section of interests and backgrounds, your summary has to be accessible to a general audience. However, the summary must be sufficiently informative to enable people to understand the most important parts of the research and to make informed decisions about their own dietary choices. Readers here need a balanced and accurate representation of the original article, without any hint of bias on the part of the writer. Also, because the summary will be read on the computer or a phone or tablet, the content must be brief but can contain links so readers can read in more depth if they wish to.

The summary version in Figure 9.4 remains true to the original but is much shorter—only about 25 percent of the original's length. Notice that the essential message remains intact. Related ideas are combined and fewer supporting details are included. But ideas familiar to a general audience (dairy products, snack foods) keep the content real. Transitions ("Although," "In fact," "To make wise choices") provide flow and help guide readers through the content. A link at the bottom of the summary allows interested readers to access the full report and learn more.

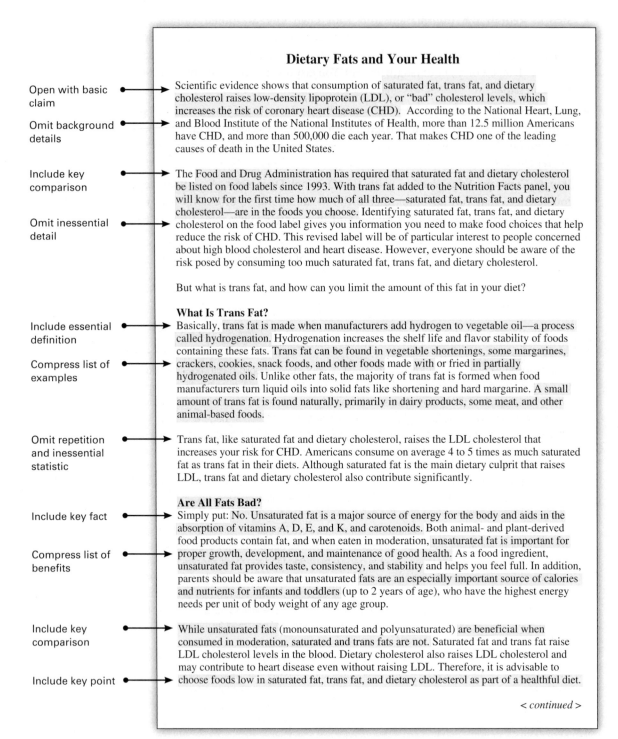

Open with basic claim

Omit background details

Include key comparison

Omit inessential detail

Include essential definition

Compress list of examples

Omit repetition and inessential statistic

Include key fact

Compress list of benefits

Include key comparison

Include key point

Dietary Fats and Your Health

Scientific evidence shows that consumption of saturated fat, trans fat, and dietary cholesterol raises low-density lipoprotein (LDL), or "bad" cholesterol levels, which increases the risk of coronary heart disease (CHD). According to the National Heart, Lung, and Blood Institute of the National Institutes of Health, more than 12.5 million Americans have CHD, and more than 500,000 die each year. That makes CHD one of the leading causes of death in the United States.

The Food and Drug Administration has required that saturated fat and dietary cholesterol be listed on food labels since 1993. With trans fat added to the Nutrition Facts panel, you will know for the first time how much of all three—saturated fat, trans fat, and dietary cholesterol—are in the foods you choose. Identifying saturated fat, trans fat, and dietary cholesterol on the food label gives you information you need to make food choices that help reduce the risk of CHD. This revised label will be of particular interest to people concerned about high blood cholesterol and heart disease. However, everyone should be aware of the risk posed by consuming too much saturated fat, trans fat, and dietary cholesterol.

But what is trans fat, and how can you limit the amount of this fat in your diet?

What Is Trans Fat?
Basically, trans fat is made when manufacturers add hydrogen to vegetable oil—a process called hydrogenation. Hydrogenation increases the shelf life and flavor stability of foods containing these fats. Trans fat can be found in vegetable shortenings, some margarines, crackers, cookies, snack foods, and other foods made with or fried in partially hydrogenated oils. Unlike other fats, the majority of trans fat is formed when food manufacturers turn liquid oils into solid fats like shortening and hard margarine. A small amount of trans fat is found naturally, primarily in dairy products, some meat, and other animal-based foods.

Trans fat, like saturated fat and dietary cholesterol, raises the LDL cholesterol that increases your risk for CHD. Americans consume on average 4 to 5 times as much saturated fat as trans fat in their diets. Although saturated fat is the main dietary culprit that raises LDL, trans fat and dietary cholesterol also contribute significantly.

Are All Fats Bad?
Simply put: No. Unsaturated fat is a major source of energy for the body and aids in the absorption of vitamins A, D, E, and K, and carotenoids. Both animal- and plant-derived food products contain fat, and when eaten in moderation, unsaturated fat is important for proper growth, development, and maintenance of good health. As a food ingredient, unsaturated fat provides taste, consistency, and stability and helps you feel full. In addition, parents should be aware that unsaturated fats are an especially important source of calories and nutrients for infants and toddlers (up to 2 years of age), who have the highest energy needs per unit of body weight of any age group.

While unsaturated fats (monounsaturated and polyunsaturated) are beneficial when consumed in moderation, saturated and trans fats are not. Saturated fat and trans fat raise LDL cholesterol levels in the blood. Dietary cholesterol also raises LDL cholesterol and may contribute to heart disease even without raising LDL. Therefore, it is advisable to choose foods low in saturated fat, trans fat, and dietary cholesterol as part of a healthful diet.

< continued >

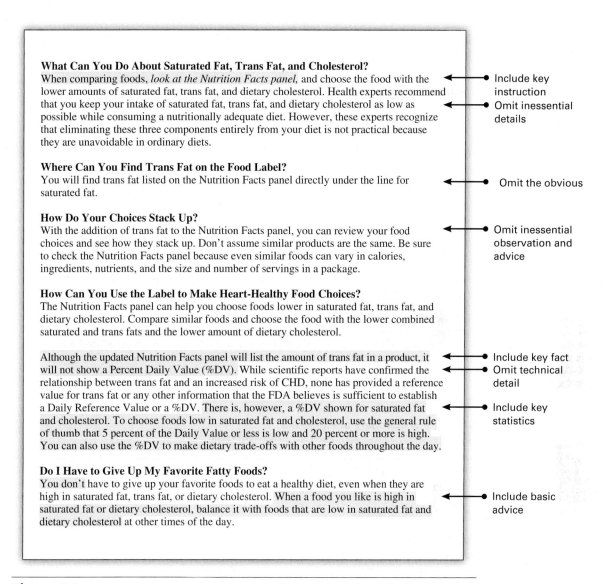

What Can You Do About Saturated Fat, Trans Fat, and Cholesterol?
When comparing foods, *look at the Nutrition Facts panel,* and choose the food with the lower amounts of saturated fat, trans fat, and dietary cholesterol. Health experts recommend that you keep your intake of saturated fat, trans fat, and dietary cholesterol as low as possible while consuming a nutritionally adequate diet. However, these experts recognize that eliminating these three components entirely from your diet is not practical because they are unavoidable in ordinary diets.

Include key instruction
Omit inessential details

Where Can You Find Trans Fat on the Food Label?
You will find trans fat listed on the Nutrition Facts panel directly under the line for saturated fat.

Omit the obvious

How Do Your Choices Stack Up?
With the addition of trans fat to the Nutrition Facts panel, you can review your food choices and see how they stack up. Don't assume similar products are the same. Be sure to check the Nutrition Facts panel because even similar foods can vary in calories, ingredients, nutrients, and the size and number of servings in a package.

Omit inessential observation and advice

How Can You Use the Label to Make Heart-Healthy Food Choices?
The Nutrition Facts panel can help you choose foods lower in saturated fat, trans fat, and dietary cholesterol. Compare similar foods and choose the food with the lower combined saturated and trans fats and the lower amount of dietary cholesterol.

Although the updated Nutrition Facts panel will list the amount of trans fat in a product, it will not show a Percent Daily Value (%DV). While scientific reports have confirmed the relationship between trans fat and an increased risk of CHD, none has provided a reference value for trans fat or any other information that the FDA believes is sufficient to establish a Daily Reference Value or a %DV. There is, however, a %DV shown for saturated fat and cholesterol. To choose foods low in saturated fat and cholesterol, use the general rule of thumb that 5 percent of the Daily Value or less is low and 20 percent or more is high. You can also use the %DV to make dietary trade-offs with other foods throughout the day.

Include key fact
Omit technical detail
Include key statistics

Do I Have to Give Up My Favorite Fatty Foods?
You don't have to give up your favorite foods to eat a healthy diet, even when they are high in saturated fat, trans fat, or dietary cholesterol. When a food you like is high in saturated fat or dietary cholesterol, balance it with foods that are low in saturated fat and dietary cholesterol at other times of the day.

Include basic advice

Figure 9.3 An Article to Be Summarized The writer has highlighted key phrases as part of the process of writing a summary, shown in Figure 9.4.

Source: Excerpt from *FDA Consumer* September–October 2003: 12–18.

Dietary Fats and Your Health (A Summary)

Saturated fat is a principal cause of bad cholesterol (LDL, or low density lipoprotein), but dietary cholesterol and trans fat also play a role. High LDL is a major risk factor for coronary heart disease, a leading killer in the United States. In addition to the listing of saturated fat and dietary cholesterol on food labels, the Food and Drug Administration (FDA) now requires that trans fat be listed.

Although small amounts of trans fat exist naturally in dairy products and other animal-based foods, most trans fat is a byproduct of hydrogenation, the addition of hydrogen to vegetable oil. This process increases shelf life and stabilizes the flavor of foods such as hard margarine, crackers, snack foods, and foods fried in hydrogenated oil.

Not all fats are bad. In fact, unsaturated fat serves as a major energy source, promotes vitamin absorption, enhances food taste, and helps us feel full. Consumed in moderation, unsaturated fat is essential to health and provides a vital calorie source for infants. However, saturated fats and trans fats, along with dietary cholesterol, should be avoided as much as possible.

To make wise choices about fats and cholesterol, read the Nutrition Facts label on packaged foods. Note the serving size, number of servings in the package, and the Percent Daily Value (%DV) of each ingredient. Although no Daily Value is yet established for trans fat, avoid foods with a high saturated fat or cholesterol DV (greater than 20 percent per serving), unless you can balance these with low DV foods during the day.

Figure 9.4 **A Summary of Figure 9.3**

Source: Adapted from *FDA Consumer* September–October 2003: 12–18.

Guidelines

for Summarizing Information

➤ **Fill out an audience and use profile.** First, determine exactly what your audience wants and why they want it.

➤ **Keep in mind what readers want.** Before you start your summary, remember that readers want a summary that is accurate, complete, readable, concise, and nontechnical in style.

➤ **Read the entire original.** When summarizing someone else's work, get a complete picture before writing a word.

➤ **Reread the original, highlighting essential material.** Focus on the essential message: thesis and topic sentences, findings, conclusions, and recommendations.

➤ **Edit the highlighted information.** Omit technical details, examples, explanations, or anything readers won't need for grasping the basics.

➤ **Rewrite in your own words.** Even if this first draft is too long, include everything essential for it to stand alone; you can trim later. In summarizing another's work, avoid direct quotations; if you must quote a crucial word or phrase, use quotation marks around the author's own words. Add no personal comments or other material, except for brief definitions, if needed.

➤ **Edit your own version.** When you have everything readers will need, edit for conciseness *(see Chapter 11, "Editing for Conciseness").*

    a. Cross out needless words—but keep sentences clear and grammatical:
       ~~As far as~~ Artificial intelligence ~~is concerned, the~~ technology is ~~only~~ in its infancy.

    b. Cross out needless prefaces such as
       The writer argues ...
       Also discussed is ...

    c. Use numerals for numbers, except to begin a sentence.

    d. Combine related ideas in order to emphasize relationships. *(See Chapter 11, "Combine Related Ideas.")*

➤ **Check your version against the original.** Verify that you have preserved the essential message and have added no comments—unless you are preparing an executive abstract (see "Executive Abstract" in this chapter).

➤ **Rewrite your edited version.** Add transitional expressions *(see Appendix B, "Transitions")* to emphasize the connections. Respect any stipulated word limit.

➤ **Document your source.** Cite the exact source below any summary that is not accompanied by its original. *(See Appendix A for documentation formats.)*

Special Types of Summaries

9.4 Differentiate among four special types of summaries

In preparing a report, proposal, or other document, you might summarize works of others as part of your presentation. But you will often summarize your own material as well. For instance, depending on its length, purpose, and audience, your document might include different forms of summarized information, in different locations, with different levels of detail (Vaughan). These four special types of summaries include the *closing summary, informative abstract, descriptive abstract,* and *executive abstract.* Figure 9.5 illustrates each of these types of summaries, their placement, and their purpose.

Closing Summary

A *closing summary* appears at the beginning of a long report's conclusion section. It helps readers review and remember the preceding major findings. This look back at "the big picture" also helps readers appreciate the conclusions and recommendations that typically follow the closing summary. (See Chapter 21, "Conclusion," and the Conclusion of Figure 21.3 for examples.)

Purpose and placement of closing summaries

Informative Abstract ("Summary")

Readers often appreciate condensed versions of long reports or proposals. Some readers like to see a capsule version before reading the complete document; others simply want to know the basics without having to read the whole document.

Purpose and placement of informative abstracts

    To meet reader needs, the *informative abstract* appears just after the title page. This type of summary encapsulates what the full version says: It identifies the need or issue that prompted the report; it describes the research methods used; it reviews the main facts and findings; and it condenses the conclusions and recommendations. (See the abstract for Figure 21.3 for an example.)

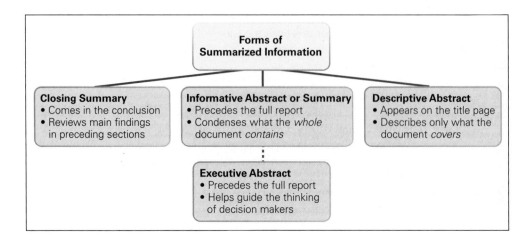

Figure 9.5 **Summarized Information Assumes Various Forms**

> NOTE *Actually, the title "Informative Abstract" is not used very often. You are more likely to encounter the title "Summary."*

The heading "Executive Summary" (or "Executive Abstract") is used for material summarized for managers who may not understand all the technical jargon a report might contain. (See below.) By contrast, a "Technical Summary" (or "Technical Abstract") is aimed at readers at the same technical level as the report's author. In short, you may need two or three levels of summary for report readers who have different levels of technical expertise. (See Chapter 21, "Front Matter," for more on the Summary section in a report.)

Descriptive Abstract ("Abstract")

Purpose and placement of descriptive abstracts

A *descriptive abstract* (usually one to three sentences on a report's title page) is another, more compressed form of summarized information. This type of abstract merely describes a report; it doesn't give the report's main points. Such an abstract helps people decide whether to read the report. Thus a descriptive abstract conveys only the nature and extent of a document. It presents the broadest view and offers no major facts from the original. Compare, for example, the abstract that follows with the article summary in Figure 9.4:

> This report explains different types of dietary fats and suggests ways consumers can make informed choices about healthy eating.

On the job, you might prepare informative abstracts for a boss who needs the information but who has no time to read the original. Or you might write descriptive abstracts to accompany a bibliography of works that you are recommending to colleagues or clients (an annotated bibliography).

Executive Abstract

A special type of informative abstract, the *executive abstract* (also called an *executive summary*) essentially falls at the beginning and "replaces" the entire report. Aimed at decision makers rather than technical audiences, an executive abstract generally has more of a persuasive emphasis: Its purpose is to motivate readers to act on the information. Executive abstracts are crucial in cases where readers have no time to read the entire original document and they expect the writer to help guide their thinking. ("Tell me how to think about this," instead of, "Help me understand this.") Unless the reader stipulates a specific format, organize your executive abstract to answer these questions:

- What is the issue?
- What was found?
- What does it mean?
- What should be done?

The executive summary in Figure 9.6 provides a concise yet full overview of a much longer, Web-based report from the Centers for Disease Control and Prevention (CDC). The full report is available via hyperlinks in the left margin; for busy readers, the

(margin note) Purpose and placement of executive abstracts

(margin note) Questions to answer in an executive abstract

CDC Centers for Disease Control and Prevention
CDC 24/7: Saving Lives, Protecting People™

SEARCH

CDC A-Z INDEX ⌄

Infection Control

Disinfection and Sterilization

Updates

Authors

Executive Summary

Introduction, Methods, Definition of Terms

A Rational Approach to Disinfection and Sterilization

Disinfection of Healthcare Equipment

Factors Affecting the Efficacy of Disinfection and Sterilization

Cleaning

Disinfection   +

Sterilization   +

Reuse of Single-Use Medical Devices

Conclusion & Web Resources

Recommendations for Disinfection and Sterilization in Healthcare Facilities

CDC > Infection Control > Disinfection and Sterilization > Executive Summary

Guideline for Disinfection and Sterilization in Healthcare Facilities (2008)

Executive Summary

The Guideline for Disinfection and Sterilization in Healthcare Facilities, 2008, presents evidence-based recommendations on the preferred methods for cleaning, disinfection and sterilization of patient-care medical devices and for cleaning and disinfecting the healthcare environment. This document supercedes the relevant sections contained in the 1985 Centers for Disease Control (CDC) Guideline for Handwashing and Environmental Control. [1] Because maximum effectiveness from disinfection and sterilization results from first cleaning and removing organic and inorganic materials, this document also reviews cleaning methods. The chemical disinfectants discussed for patient-care equipment include alcohols, glutaraldehyde, formaldehyde, hydrogen peroxide, iodophors, *ortho*-phthalaldehyde, peracetic acid, phenolics, quaternary ammonium compounds, and chlorine. The choice of disinfectant, concentration, and exposure time is based on the risk for infection associated with use of the equipment and other factors discussed in this guideline. The sterilization methods discussed include steam sterilization, ethylene oxide (ETO), hydrogen peroxide gas plasma, and liquid peracetic acid. When properly used, these cleaning, disinfection, and sterilization processes can reduce the risk for infection associated with use of invasive and noninvasive medical and surgical devices. However, for these processes to be effective, health-care workers should adhere strictly to the cleaning, disinfection, and sterilization recommendations in this document and to instructions on product labels.

In addition to updated recommendations, new topics addressed in this guideline include

1. inactivation of antibiotic-resistant bacteria, bioterrorist agents, emerging pathogens, and bloodborne pathogens;
2. toxicologic, environmental, and occupational concerns associated with disinfection and sterilization practices;
3. disinfection of patient-care equipment used in ambulatory settings and home care;
4. new sterilization processes, such as hydrogen peroxide gas plasma and liquid peracetic acid; and
5. disinfection of complex medical instruments (e.g., endoscopes).

⌃ Top of Page

Figure 9.6 An Executive Summary Also called an Executive Abstract, this summary of a Web-based report might be the only section that will be read by many in its intended audience.

Source: Centers for Disease Control and Prevention.

summary provides enough information. For other readers, such as those who need to implement the new guidelines, the summary offers a guide for where and how to look for information within the longer report.

Summarizing Information for Social Media

9.5 Exercise caution when summarizing for social media

Summaries in social media formats

Information in summary format is increasingly attractive to today's readers, who often feel bombarded by more content than they can handle. Consider, for example, the numerous tweets, Facebook posts, LinkedIn announcements, email digests, and other summary formats that people receive on a daily basis. These short formats allow people to read a summary and, if an item is of interest, click on a link to read the longer form article, report, or news story. In this way, such summaries help people explore a wide range of new information, including recent research, technical updates, news stories, and medical innovations.

Social media considerations when writing summaries

Yet, although summarized information on social media does provide advantages to busy readers, these short bursts maybe read so quickly that readers miss key points or ignore ideas that are not presented with the accuracy of the original published document. For instance, a Facebook post with a catchy headline may work to grab the attention of many readers, but if that headline (or the short summary, or the visuals used) is not accurate, the summary is doing a disservice to the original study and to your audience. The challenge for technical writers who create summaries for social media is to keep the information clear and accurate but at the same time condense the information into the right size for a post or tweet. Consider the following points:

- A summary that adds material not found in the original document or omits critical information from the original can be misleading. Likewise, misrepresenting the original document's point of view is inappropriate and will not give readers a clear understanding of the full story. Remember that social media posts travel widely and quickly, so any inaccuracies in your summary may have global implications.

- A summary written and designed to grab reader attention and get more clicks or likes may end up distorting the original idea if the post uses inappropriate images or exaggerated wording. A good social media summary will maintain the original meaning but make the information accessible to a wide audience. For instance, the sentence "Certain kinds of fat can raise your bad cholesterol" is more accurate than "Eating too many French fries will cause a heart attack."

Checklist

Summaries

Use the following Checklist when summarizing information.

❏ Does the summary contain only the essential message? (See "Considering Audience and Purpose" in this chapter.)

❏ Can the summary stand alone? (See "Considering Audience and Purpose" and "Guidelines for Summarizing Information" in this chapter.)

❏ Is the summary accurate when checked against the original? (See "What Readers Expect from a Summary" and "Guidelines for Summarizing Information" in this chapter.)

❏ Is the summary free of any additions to the original? (See "Guidelines for Summarizing Information" in this chapter.)

❏ Is the summary free of needless details? (See the introductory paragraph and "Guidelines for Summarizing Information" in this chapter.)

❏ Is the summary economical yet clear and comprehensive? (See "What Readers Expect from a Summary" and "Guidelines for Summarizing Information" in this chapter.)

❏ Is the summary's level of technicality appropriate for its audience? (See "What Readers Expect from a Summary" in this chapter.)

❏ Is the source of this summary documented? (See "Guidelines for Summarizing Information" in this chapter.)

❏ Does the descriptive abstract merely tell what the original is about? (See "Descriptive Abstract" in this chapter.)

❏ Does the executive abstract tell readers how to think about the information? (See "Executive Abstract" in this chapter.)

❏ If intended as a social media post, is the summary accurate and clear? (See "Summarizing Information for Social Media" in this chapter.)

Projects

For all projects, check with your instructor about whether to present your findings in class, bring drafts to class for discussion, upload your project to the class learning management system (LMS), and/or use the LMS forum or discussion boards to collaborate and review each activity below.

General

1. Find an article about your major field or area of interest and write both an informative abstract and a descriptive abstract for that article.

2. Find a long article (at least five pages) and summarize it, following the step-by-step process described in this chapter, and keeping accuracy, completeness, conciseness, and nontechnical language always in mind. Capture the essence and main points of the original article in no more than one-fourth the length. Use your own words, and do not distort the original. Submit a copy of the original along with your summary.

Team

In small groups, choose a topic for discussion: an employment problem, a campus problem, plans for an event, suggestions for energy conservation, or the like. (A possible topic: Should employers have the right to require lie detector tests, drug tests, or HIV tests for their employees?) Discuss the topic for one class period, taking notes on significant points and conclusions. Afterward, organize and edit your notes according to the directions for writing summaries. Write a summary of the group discussion in no more than 200 words. As a group, compare your individual summaries for accuracy, emphasis, conciseness, and clarity.

Digital and Social Media

In class, form teams whose members have similar majors or interests. Decide on a technical topic related to your major. Look for social media posts related to this topic. For example, for aerospace engineering, look at the NASA Facebook page and Twitter feed. For computer science, try the Association of Computing Machinery (ACM). For electrical engineering, look for the Institute of Electrical and Electronics Engineers (IEEE). Examine how these organizations summarize technical ideas and concepts when creating social media posts. Are the summaries intended for a broad audience or for people in that field? Do the summaries seem accurate? Create a short presentation of your findings for class; present this in class or make it available through the class learning management system.

Global

Find a long article (longer than five pages) from a technical, scientific, financial, or similar publication that deals with an issue that is global in scope (for instance, an article about the use of pesticides in different countries or one that discusses ways in which financial decisions have worldwide impacts). Summarize this article for an audience comprising two different groups of nonspecialists: one group of United States citizens and another group of citizens in a different country. Learn what you can about the other country by going online and determining key scientific, political, or financial issues. Or bring knowledge you have from your own travel (such as a study abroad experience) or background, or from a class that you took. What will you highlight in the summary for the U.S. readers, and what will you highlight in the article for the non-U.S. readers, and why? Bring both copies to class to discuss with others.

Part 3
Organization, Style, and Visual Design

Chapter 10
Organizing for Readers

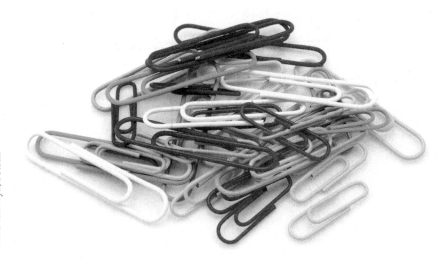

❝When I first start any writing project, I need to be sure that whatever I'm writing is clear *to me* first, and that the way I've organized it makes sense *to me*. But then, it's important for me to take that material and become more objective by trying to understand how my audience thinks. I ask myself, 'How can I make this logical to my audience? Will they understand what I want them to understand?'

Organizing is the key to helping your audience understand the material. Develop the type of outlining or listing or brainstorming tool that works best for you, but find one that works, and use it consistently. Then you'll be comfortable with that general strategy whenever you sit down to write, especially under a rigid deadline.❞

—Anne Brill, Environmental Engineer

Learning Objectives

10.1 Work from an introduction-body-conclusion structure

10.2 Create informal and formal outlines

10.3 Prepare a storyboard for a long document

10.4 Shape effective paragraphs

10.5 Chunk information into discrete units

10.6 Provide overviews of longer documents

10.7 Organize information for social media postings

In order to comprehend your thinking, readers need information organized in a way that makes sense to *them*. But data rarely materializes or thinking rarely occurs in neat, predictable sequences. Instead of forcing readers to make sense of unstructured information, we shape this material for their understanding. As we organize a document, we face questions such as these:

- What relationships do the collected data suggest?
- What should I emphasize?
- In which sequence will readers approach this material?
- What belongs where?
- What do I say first? Why?
- What comes next?
- How do I end the document?

Questions in organizing for readers

To answer these questions, we rely on a variety of organizing strategies.

The Typical Shape of Workplace Documents

10.1 Work from an introduction-body-conclusion structure

Standard introduction/ body/conclusion pattern

Organize your material to make the document logical from the reader's point of view. Begin with the basics: Useful documents of any length (memo, letter, long report, and so on) typically follow the pattern shown in Figure 10.1: introduction, body, and conclusion. The introduction attracts the reader's attention, announces the writer's viewpoint, and previews what will follow. The body delivers on the promise implied in the introduction. The body explains and supports the writer's viewpoint, achieving *unity* by remaining focused on that viewpoint and *coherence* by carrying a line of thought from sentence to sentence in a logical order. Finally, the conclusion has various purposes: it might reemphasize key points, take a position, predict an outcome, offer a solution, or suggest further study. Good conclusions give readers a clear perspective on what they have just read.

A nonstandard structure also can be effective in certain cases

There are many ways of adapting this standard structure. For example, Figure 10.2 provides visual features (columns, colors), headings, and an engaging layout. Although organized differently from the previous document, Figure 10.2 does provide an introduction, a body, and a conclusion. The heading "What is arsenic?" represents a form of introduction. The next several headings answer the question posed in the introduction, forming, in essence, the body of the document. The final heading, asking about EPA standards, represents a form of conclusion, moving beyond data and description to the topic of policy and use. In organizing any document, we typically begin with the time-tested strategy known as *outlining*.

Outlining

10.2 Create informal and formal outlines

Outlining is essential

Even basic documents require at least an introduction-body-conclusion outline to start with, or a few ideas jotted down in list form. Longer documents require a more detailed outline so that you can visualize your document overall and ensure that ideas flow logically from point to point.

An Outlining Strategy

Start by searching through the information you have gathered and creating a random list of key topics your document should include. For instance, in preparing the drinking water document in Figure 10.2, you might start by simply listing all the types of information you think readers need or expect:

Start by creating a list of essential information

- Explain what the EPA is doing about arsenic in drinking water
- Define what arsenic is
- Explain how arsenic gets into drinking water
- List some of the effects of arsenic (stomach, heart, cancer)
- Include specific data
- Mention/explain the Safe Water Drinking Act
- Refer to/define MCLGs

Powell   Rabkin

MEMORANDUM

To: Department Managers
From: Jill McCreary, General Manager
Date: December 8, 20XX
Subject: Diversity training initiative

As part of our ongoing efforts to highlight the company's commitment to diversity, we recently conducted two surveys: one directed to company employees and one to our retail buyers. We have just received the survey results from our outside analysts. The employee survey indicates that the members of all departments appreciate our efforts to create a diverse and comfortable work environment. The customer survey indicates that our company is well regarded for marketing products in ways that appeal to diverse buyers. However, both surveys also illuminate areas in which we could do even better. As a result, we will be initiating a new series of diversity training workshops early next year. Let me explain the survey findings that have led to this initiative.

● Introduction announces the topic and provides an overview of what will follow

First, the employee survey indicates that our workforce is rated "highly diverse" in terms of gender, with nearly equal representation of male and female employees in both managerial and nonmanagerial positions; however, we could do better in terms of minority representation at the managerial level. Meanwhile, the customer survey demonstrates that our customers are "very satisfied" with the diversity of our marketing materials, but that we fail to provide enough materials for our native Spanish-speaking buyers.

● Body provides the evidence and data to support the claims made in the introduction

Those are the survey highlights (see the attached analysis for a more detailed picture). Again, we are doing well, but could do better. We feel that the best solution to address our weaker areas is to conduct a second series of diversity training workshops in the upcoming 12 months. We hope that these workshops, which are often illuminating to both new employees and those who have attended diversity trainings earlier, will help keep the word "diversity" at the forefront of everyone's thoughts when hiring and mentoring employees and creating marketing materials. More information will follow, but for now please emphasize to your department employees the importance and value of these workshops.

● Conclusion summarizes by taking a position and making recommendations

Figure 10.1 **Document with a Standard Introduction-Body-Conclusion Structure**

Use of visuals, color, and columns makes the organization clear

"What is arsenic?" paragraph is placed above subsequent sections, indicating that it is the introduction

The body section is chunked into short paragraphs with headings in the form of reader questions

"What is EPA's standard?" section is placed at the bottom of the page, indicating that it is a conclusion

JUST THE FACTS FOR CONSUMERS

ARSENIC IN YOUR DRINKING WATER

What is arsenic?

Arsenic is a toxic chemical element that is unevenly distributed in the Earth's crust in soil, rocks, and minerals.

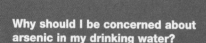

How does arsenic get into my drinking water?

Arsenic occurs naturally in the environment and as a by-product of some agricultural and industrial activities. It can enter drinking water through the ground or as runoff into surface water sources.

How is arsenic in drinking water regulated?

In 1974, Congress passed the Safe Drinking Water Act. This law directs EPA to issue non-enforceable health goals and enforceable drinking water regulations for contaminants that may cause health problems. The goals, which reflect the level at which no adverse health effects are expected, are called maximum contaminant level goals (MCLGs). The MCLG for arsenic is 0 parts per billion (ppb).

The enforceable standard for arsenic is a maximum contaminant level (MCL). MCLs are set as close to the health goals as possible, considering cost, benefits, and the ability of public water systems to detect and remove contaminants using suitable treatment technologies.

Why should I be concerned about arsenic in my drinking water?

Although short-term exposures to high doses (about a thousand times higher than the drinking water standard) cause adverse effects in people, such exposures do not occur from public water supplies in the U.S. that comply with the arsenic MCL.

Some people who drink water containing arsenic in excess of EPA's standard over many years could experience skin damage or problems with their circulatory system, and may have an increased risk of getting cancer. Health effects might include:

◆ Thickening and discoloration of the skin, stomach pain, nausea, vomiting, diarrhea, and liver effects;

◆ Cardiovascular, pulmonary, immunological, neurological (e.g., numbness and partial paralysis), reproductive, and endocrine (e.g., diabetes) effects;

◆ Cancer of the bladder, lungs, skin, kidney, nasal passages, liver, and prostate.

What is EPA's standard for arsenic in drinking water?

To protect consumers served by public water systems from the health risks of long-term (chronic) arsenic exposure, EPA recently lowered the arsenic MCL from 50 ppb to 10 ppb.

Figure 10.2 Document with a Nonstandard but Well-Organized Structure

Source: "Arsenic in Your Drinking Water," U.S. Environmental Protection Agency, March 2007.

Now you can reorganize this list, as shown below.

A simple list like the one above usually suffices for organizing a short document like the memo in Figure 10.1. However, for a more complex document, transform your list into a deliberate map that will guide readers from point to point. Create an introduction, body, and conclusion and then decide how you will divide each of these parts into subtopics. An outline for Figure 10.2 might look like this:

> I. Introduction—Define arsenic.
> II. Body
>   A. Explain how arsenic gets into drinking water.
>   B. Explain how it is regulated (Safe Drinking Water Act/MCLGs, Maximum Contaminant Level Goals).
>   C. List some of the health effects of arsenic (visible effects, diseases, cancers).
> III. Conclusion—Describe the EPA's standards.

Then organize the information into an outline

The Formal Outline

In planning a long document, an author or team rarely begins with a formal outline. But eventually in the writing process, a long or complex document calls for much more than a simple list. Figure 10.3 shows a formal outline for the report examining the health effects of electromagnetic fields (see Chapter 21, "An Outline and Model for Analytical Reports").

Long documents call for formal outlines

> NOTE *Long reports often begin directly with a statement of purpose. For the intended audience (i.e., generalists) of the report outlined in Figure 10.3, however, the technical topic must first be defined so that readers understand the context. Also, each level of division yields at least two items. If you cannot divide a major item into at least two subordinate items, retain only your major heading.*

Most word processing programs (e.g., Microsoft Word, Apple Pages) offer an outline tool (search for "outline" via the program's online help). A formal outline created in this way easily converts to a table of contents for the finished document, as shown in Chapter 21.

> NOTE *Because they serve mainly to guide the writer, minor outline headings (such as items [a] and [b] under II.A.2 in Figure 10.3) may be omitted from the table of contents or the report itself. Excessive headings make a document seem fragmented.*

In technical documents, alphanumeric notation often is replaced by decimal notation. Compare the following with part "A" of the DATA SECTION from Figure 10.3.

> 2.0 DATA SECTION
>   2.1 Sources of EMF Exposure
>   2.1.1 power lines
>   2.1.2 home and office
>   2.1.2.1 kitchen
>   2.1.2.2 workshop [and so on]
>   2.1.3 natural radiation
>   2.1.4 risk factors
>   2.1.4.1 current intensity
>   2.1.4.2 source proximity [and so on]

Part of a formal outline using decimal notation

Children Exposed to EMFs: A Risk Assessment

I. INTRODUCTION
    A. Definition of electromagnetic fields
    B. Background on the health issues
    C. Description of the local power line configuration
    D. Purpose of this report
    E. Brief description of data sources
    F. Scope of this inquiry

II. DATA SECTION [Body]
    A. Sources of EMF exposure
        1. power lines
        2. home and office
            a. kitchen
            b. workshop [and so on]
        3. natural radiation
        4. risk factors
            a. current intensity
            b. source proximity
            c. duration of exposure
    B. Studies of health effects
        1. population surveys
        2. laboratory measurements
        3. workplace links
    C. Conflicting views of studies
        1. criticism of methodology in population studies
        2. criticism of overgeneralized lab findings
    D. Power industry views
        1. uncertainty about risk
        2. confusion about risk avoidance
    E. Risk-avoidance measures
        1. nationally
        2. locally

III. CONCLUSION
    A. Summary and overall interpretation of findings
    B. Recommendations

Figure 10.3 A Formal Outline Using Alphanumeric Notation In an outline, alphanumeric notation refers to the use of letters and numbers.

The decimal outline makes it easier to refer readers to specifically numbered sections (e.g., "See Section 2.1.2"). If you are using the outline tool in your word processing program, you can choose the notation system you prefer.

You may wish to expand your *topic outline* into a *sentence outline*, in which each sentence serves as a topic sentence for a paragraph in the document:

> 2.0 DATA SECTION
>
>   2.1 Although the 2 million miles of power lines crisscrossing the United States have been the focus of the EMF controversy, potentially harmful waves also are emitted by household wiring, appliances, electric blankets, and computer terminals.

A sentence outline

Sentence outlines are used mainly in collaborative projects in which various team members prepare different sections of a long document.

> NOTE *The neat and ordered outlines in this book show the final* **products** *of writing and organizing, not the* **process,** *which is often initially messy and chaotic. Many writers don't start out with an outline at all! Instead, they scratch and scribble with pencil and paper or click away at the keyboard, making lots of false starts as they hammer out some kind of acceptable draft; only then do they outline to get their thinking straight.*

Not until you finish the final draft of a long document do you compose the finished outline. This outline serves as a model for your table of contents, as a check on your reasoning, and as a way of revealing to readers a clear line of thinking.

> NOTE *No single form of outline should be followed slavishly. The organization of any document ultimately is determined by the reader's needs and expectations. In many cases, specific requirements about a document's organization and style are spelled out in a company's style guide (see Chapter 13, "Using Style Guides and Style Sheets").*

Guidelines

for Outlining

- ➤ **List key topics and subtopics to be included in your document.** Determine what information is important to include.
- ➤ **Set up a standard outline.** Start with a typical introduction, body, and conclusion structure, even if you plan to vary the structure later.
- ➤ **Consider using your word processing program's outlining tool.**
- ➤ **Place key topics and subtopics where they fit within your standard outline.** Keep your introduction brief, setting the stage for the rest of your document. Include your specific data in the body section, to back up what you promised in your introduction. Do not introduce new data in the conclusion.
- ➤ **Use alphanumeric or decimal notation consistently throughout the outline.**
- ➤ **Avoid excessive subtopics.** If your outline is getting into multiple levels of detail too often, think of ways to combine information. Do not go to another level unless there are at least two distinct subtopics at that level.
- ➤ **Refine your outline as you write your document.** Continue revising the outline until you complete the document.

Storyboarding

10.3 Prepare a storyboard for a long document

Visualize each section of your outline

As you prepare a long document, one useful organizing tool is the *storyboard*, a sketch of the finished document.

Figure 10.4 displays one storyboard module based on Section II.A of the outline in Figure 10.3. Much more specific and visual than an outline, a storyboard maps out each section (or module) of your outline, topic by topic, to help you see the shape and appearance of the entire document in its final form. Working from a storyboard, you can rearrange, delete, and insert material as needed—without having to wrestle with a draft of the entire document.

To create a storyboard such as the example in Figure 10.4, you can use a word processing program or a collaborative shared writing app such as Google docs. If your project involves a team, shared documents can be helpful because everyone on the team has access to the same document, with no confusion about who has the most updated version. If you need to present your storyboard to a larger group, try converting the file into a presentation (PowerPoint, Google Slides, Prezi, or other).

NOTE *Storyboards can help you organize your document, but you might also want to try creating a storyboard after writing a full draft, for a bird's-eye view of the document's organization.*

Paragraphing

10.4 Shape effective paragraphs

Readers look for orientation, for shapes they can recognize. But a document's larger design (introduction, body, conclusion) depends on the smaller design of each paragraph.

Shape information into paragraphs

Paragraphs have various shapes and purposes (introduction, conclusion, or transition), but the focus here is on standard *support paragraphs*. Although part of the document's larger design, each support paragraph can usually stand alone in meaning.

The Support Paragraph

What support paragraphs do

All the sentences in a standard support paragraph relate to the main point, which is expressed as the *topic sentence*. Following are three examples of topic sentences:

Topic sentences

| As sea levels rise, New York City faces increasing risk of hurricane storm surge.

| A video display terminal can endanger the operator's health.

| Chemical pesticides and herbicides are both ineffective and hazardous.

Each topic sentence introduces an idea, judgment, or opinion. But in order to grasp the writer's exact meaning, people need explanation. Consider the third statement above:

| Chemical pesticides and herbicides are both ineffective and hazardous.

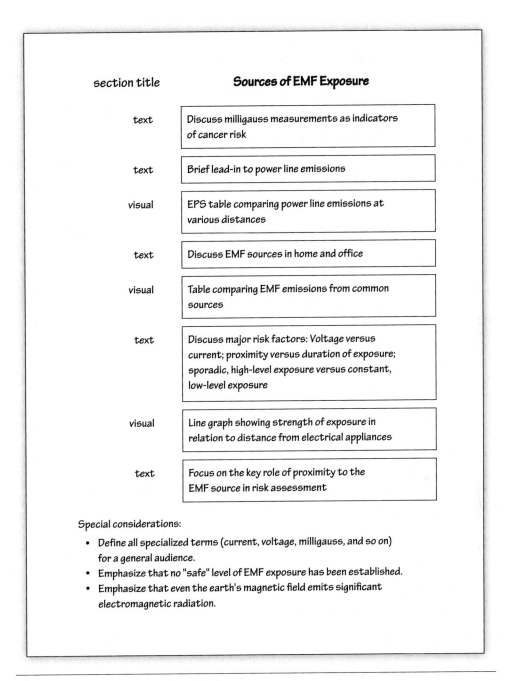

Figure 10.4 One Module from a Storyboard Notice how the module begins with the section title, describes each text block and each visual, and includes suggestions about special considerations. (To see how this storyboard turned into the body section of a report, see Chapter 21, "Body").

Imagine you are a researcher for the Epson Electric Light Company, assigned this question: Should the company (1) begin spraying pesticides and herbicides under its power lines, or (2) continue with its manual (and nonpolluting) ways of minimizing foliage and insect damage to lines and poles? If you simply responded with the preceding assertion, your employer would have further questions:

- Why, exactly, are these methods ineffective and hazardous?

- What are the problems? Can you explain?

To answer the previous questions and to support your assertion, you need a fully developed paragraph (note that sentences are numbered for later discussion; these numbers are not footnotes):

Introduction (1-topic sentence) Body (2–6)

Conclusion (7–8)

[1]**Chemical pesticides and herbicides are both ineffective and hazardous.** [2]Because none of these chemicals has permanent effects, pest populations invariably recover and need to be resprayed. [3]Repeated applications cause pests to develop immunity to the chemicals. [4]Furthermore, most of these products attack species other than the intended pest, killing off its natural predators, thus actually increasing the pest population. [5]Above all, chemical residues survive in the environment (and living tissue) for years, often carried hundreds of miles by wind and water. [6]This toxic legacy includes such biological effects as birth deformities, reproductive failures, brain damage, and cancer. [7]Although intended to control pest populations, these chemicals ironically threaten to make the human population their ultimate victims. [8]I therefore recommend continuing our manual control methods.

Most standard support paragraphs in technical writing have an introduction-body-conclusion structure. They begin with a clear topic (or orienting) sentence stating a generalization. Details in the body support the generalization.

The Topic Sentence

Why the topic sentence is essential

Readers look to a paragraph's opening sentences for the main idea. The topic sentence should appear *first* (or early) in the paragraph, unless you have good reason to place it elsewhere. Think of your topic sentence as "the one sentence you would keep if you could keep only one" (U.S. Air Force Academy 11). In some instances, a main idea may require a "topic statement" consisting of two or more sentences, as in this example:

A topic statement can have two or more sentences

The most common strip-mining methods are open-pit mining, contour mining, and auger mining. The specific method employed will depend on the type of terrain that covers the coal.

The topic sentence or topic statement should focus and forecast. Don't write *Some pesticides are less hazardous and often more effective than others* when you mean *Organic pesticides are less hazardous and often more effective than their chemical counterparts.* The first version is vague; the second helps us focus and tells us what to expect from the paragraph.

Paragraph Unity

A paragraph is unified when all its content belongs—when every word, phrase, and sentence directly expands on the topic sentence. Following is an example of a unified paragraph:

Characteristics of unified paragraphs

> **Solar power offers an efficient, economical, and safe solution to the Northeast's energy problems.** To begin with, solar power is highly efficient. Solar collectors installed on fewer than 30 percent of roofs in the Northeast would provide more than 70 percent of the area's heating and air-conditioning needs. Moreover, solar heat collectors are economical, operating for up to twenty years with little or no maintenance. These savings recoup the initial cost of installation within only 10 years. Most important, solar power is safe. It can be transformed into electricity through photovoltaic cells (a type of storage battery) in a noiseless process that produces no air pollution—unlike coal, oil, and wood combustion. In contrast to its nuclear counterpart, solar power produces no toxic waste and poses no catastrophic danger of meltdown. Thus, massive conversion to solar power would ensure abundant energy and a safe, clean environment for future generations.

A unified paragraph

One way to damage unity in the paragraph above would be to veer from the focus on *efficient*, *economical*, and *safe* toward material about the differences between active and passive solar heating or the advantages of solar power over wind power.

Every topic sentence has a key word or phrase that carries the meaning. In the pesticide-herbicide paragraph (see "The Support Paragraph" in this chapter), the key words are *ineffective* and *hazardous*. Anything that fails to advance their meaning throws the paragraph—and the readers—off track.

Paragraph Coherence

In a coherent paragraph, everything not only belongs, but also sticks together: Topic sentence and support form a *connected line of thought*, like links in a chain.

Characteristics of coherent paragraphs

Paragraph coherence can be damaged by (1) short, choppy sentences; (2) sentences in the wrong order; (3) insufficient transitions and connectors for linking related ideas; or (4) an inaccessible line of reasoning. Here is how the solar energy paragraph might become incoherent:

> Solar power offers an efficient, economical, and safe solution to the Northeast's energy problems. Unlike nuclear power, solar power produces no toxic waste and poses no danger of meltdown. Solar power is efficient. Solar collectors could be installed on fewer than 30 percent of roofs in the Northeast. These collectors would provide more than 70 percent of the area's heating and air-conditioning needs. Solar power is safe. It can be transformed into electricity. This transformation is made possible by photovoltaic cells (a type of storage battery). Solar heat collectors are economical. The photovoltaic process produces no air pollution.

An incoherent paragraph

In the above paragraph, the second sentence, about safety, belongs near the end. Also, because of short, choppy sentences and insufficient links between ideas, the paragraph reads more like a list than a flowing discussion. Finally, a concluding sentence is needed to complete the chain of reasoning and to give readers a clear perspective on what they've just read.

Here, in contrast, is the original, coherent paragraph with sentences numbered for later discussion (these numbers are not footnotes); transitions and connectors are shown in boldface. Notice how this version reveals a clear line of thought:

A coherent paragraph

> [1]Solar power offers an efficient, economical, and safe solution to the Northeast's energy problems. [2]**To begin with,** solar power is highly efficient. [3]Solar collectors installed on fewer than 30 percent of roofs in the Northeast would provide more than 70 percent of the area's heating and air-conditioning needs. [4]**Moreover,** solar heat collectors are economical, operating for up to twenty years with little or no maintenance. [5]**These savings** recoup the initial cost of installation within only ten years. [6]**Most important,** solar power is safe. [7]**It** can be transformed into electricity through photovoltaic cells (a type of storage battery) in a noiseless process that produces no air pollution—unlike coal, oil, and wood combustion. [8]**In contrast** to its nuclear counterpart, solar power produces no toxic waste and poses no danger of catastrophic meltdown. [9]**Thus,** massive conversion to solar power would ensure abundant energy and a safe, clean environment for future generations.

We can easily trace the sequence of thoughts in the previous paragraph:

1. The topic sentence establishes a clear direction.
2–3. The first reason is given and then explained.
4–5. The second reason is given and explained.
6–8. The third and major reason is given and explained.
9. The conclusion reemphasizes the main point.

To reinforce the logical sequence, related ideas are combined in individual sentences, and transitions and connectors signal clear relationships. The whole paragraph sticks together. For more on transitions and other connectors, see Appendix B, "Transitions."

Paragraph Length

Why paragraph length is important

Paragraph length depends on the writer's purpose and the reader's capacity for understanding. Writing that contains highly technical information or complex instructions may use short paragraphs or perhaps a list. In writing that explains concepts, attitudes, or viewpoints, support paragraphs generally run from 100 to 300 words. But word count really means very little. What matters is *how thoroughly the paragraph makes your point.*

Try to avoid too much of anything. A clump of short paragraphs can make some writing seem choppy and poorly organized, but a stretch of long paragraphs can be tiring. A well-placed short paragraph—sometimes just one sentence—can highlight an important idea.

NOTE Paragraphs that will be read on a computer, tablet, or phone need to be short and to the point. Short paragraphs, as well as lists, work well on small screens.

Chunking

10.5 Chunk information into discrete units

Each organizing technique discussed in this chapter is a way of *chunking* information: breaking it down into discrete, digestible units, based on the readers' needs and the document's purpose. Well-chunked material generally is easier to follow and is more visually appealing.

Break information down into smaller units

Chunking enables us to show which pieces of information belong together and how the various pieces are connected. For example, a discussion about research in technical communication might be divided into two chunks:

- Procedural Stages
- Inquiry Stages

A major topic chunked into subtopics

Each of these units then divides into smaller chunks:

- Procedural Stages
  - Searching for Information
  - Recording Your Findings
  - Documenting Your Sources
  - Writing the Document
- Inquiry Stages
  - Asking the Right Questions
  - Exploring a Balance of Views
  - Achieving Adequate Depth in Your Search
  - Evaluating Your Findings
  - Interpreting Your Findings

Subtopics chunked into smaller topics

Any of these segments that become too long might be subdivided again.

> NOTE Chunking requires careful decisions about exactly how much is enough and what constitutes sensible proportions among the parts. Don't overdo it by creating such tiny segments that your document ends up looking fragmented and disconnected.

In addition to chunking information verbally, we can chunk it visually. Notice how the visual display in Figures 7.1A and 7.1B makes relationships immediately apparent. (For more on visual design, see Chapter 12.)

Using visuals for chunking

Finally, you can chunk information visually by using white space, headings, lists, and other techniques of effective page and screen design. Whether the final document is designed for print or for digital delivery, a well-designed page or screen provides immediate clues about where readers should look and how they should proceed to read your document.

Using document design techniques to chunk information

Chunking on
the Web versus
chunking on a
printed page

When you work on a document to be delivered as hard copy (print), you can assume that readers will take longer to read the material than they would with, say, a tweet or a Facebook posting. In general, that's because the printed page is easier on the eye and causes readers to slow down. Research continues to show that when people read on screens, on the other hand, they do more skimming and have less patience for long blocks of text (Liu).

Chunking
information to fit
the medium

Yet most documents get converted to PDF and uploaded to an organization's Web site. You may also end up writing content that is solely intended for publication on a Web page or on social media. So, be sure you have a clear understanding of the final delivery medium for your document and organize your content in visual chunks that best fit the medium. See the section "Organizing for Social Media and Global Audiences" later in this chapter for more on organizing for social media. Also, see Chapter 13 for information on page design and Chapters 24 and 25 for details on writing for blogs, wikis, Web pages, and social media.

Providing an Overview

10.6 Provide overviews of longer documents

Show the big
picture

Once you've settled on a final organization for your document, give readers an immediate preview of its contents by answering their initial questions:

What readers
want to know
immediately

- What is the purpose of this document?
- Why should I read it?
- What information can I expect to find here?

Readers will also have additional, more specific questions, but first they want to know what the document is all about and how it relates to them.

An overview should be placed near the beginning of a document, but you may also want to provide section overviews at the beginning of each section in a long document. The following is an overview of a long report on groundwater contamination.

A report
overview

About This Report
This report contains five sections. Section One describes the scope and scale of groundwater contamination in Jackson County. Section Two offers background on previous legislation related to groundwater. Section Three shows the most recent data from the Jackson County Groundwater Project, and Section Four compares that data to national averages. Section Five offers recommendations and ideas for next steps.

Variety of
overview types

Overviews come in various shapes and sizes. The overview for this book, for example, appears under the heading "How This Book Is Organized" in this book's Preface. An informative abstract of a long document also provides an overview, as in the sample formal report in Chapter 21. An overview for an oral presentation appears as an introduction to that presentation (see Chapter 23, "Outline Your Presentation"). Whatever its shape or size, a good overview gives readers the "big picture" to help them navigate the document or presentation and understand its details.

Organizing for Social Media and Global Audiences

10.7 Organize information for social media postings

As noted earlier in this chapter, a 2005 study showed that when people read on screens they do more skimming and have less patience for long blocks of text (Liu). In the 10+ years that have followed this study, the number of social media platforms has increased dramatically. Today, most people get their news, updates, and technical information from Facebook, Twitter, Instagram, Reddit, and other social media platforms—not from print media. These readers are moving quickly across small chunks of content, making rapid-fire decisions about what information to delve into more deeply and what to skip or ignore. For technical writing, your best bet is to think about 1–2 sentences, where the first sentence is the topic sentence and the second sentence is a brief summary that includes a link to additional, more detailed information. Visuals, so long as they are accurate and appropriate, can help you organize a social media post into something that helps readers understand the facts and access additional content.

How to organize social media posts

Social media is viewed by a worldwide audience, and different cultures have varying expectations about how information should be organized. For instance, a paragraph in English typically begins with a topic sentence, followed by related supporting sentences; any digression from this main idea is considered harmful to the paragraph's unity. But some cultures consider digression a sign of intelligence or politeness. To native readers of English, the long introductions and digressions in certain Spanish or Russian documents might seem tedious and confusing, but a Spanish or Russian reader might view the more direct organization of English as abrupt and simplistic (Leki 151). Even same-language cultures might have different expectations (do a search on "British English for business" to discover some key differences in how U.S. versus British businesspeople approach the organization of content).

Special considerations when writing for global readers

Checklist

Organizing Information

Use the following Checklist as you organize information:

- ❏ Does the document employ a standard or varied introduction/body/conclusion structure? (See "The Typical Shape of Workplace Documents" in this chapter.)
- ❏ Will the outline allow me to include all the necessary data for the document? (See "Outlining" in this chapter.)
- ❏ Is this outline organized using alphanumeric or decimal notation? (See "The Formal Outline" in this chapter.)
- ❏ Have I created a storyboard to supplement my formal outline? (See "Storyboarding" in this chapter.)
- ❏ Is the information chunked into discrete, digestible units for the proper medium (print, PDF, Web, social media)? (See "Chunking" in this chapter.)

❏ Does each paragraph include these features? (See "Paragraphing" in this chapter.)
- Topic sentence (introduction)
- Unity (body that supports the topic sentence)
- Coherence (connected line of thought leading to a conclusion)

❏ If appropriate, does the document include an overview, offering a larger picture of what will follow? (See "Providing an Overview" in this chapter.)

❏ Have I considered how social media and my audience's specific cultural expectations should influence my choices about organization? (See "Organizing for Social Media and Global Audiences" in this chapter.)

Projects

For all projects, check with your instructor about whether to present your findings in class, bring drafts to class for discussion, upload your project to the class learning management system (LMS), and/or use the LMS forum or discussion boards to collaborate and review each activity below.

General

1. For each document below, use the outlining strategy described in the "An Outlining Strategy" section of this chapter to list the topics that need to be covered. Then, organize your list into a formal outline (see "The Formal Outline" in this chapter) most suited to readers of this particular type of document.

- Instructions for operating a power tool
- A report analyzing the desirability of a proposed trash incinerator or other environmentally sensitive project in your area
- A detailed breakdown of your monthly budget to trim excess spending
- A report investigating the success of a no-grade policy at other colleges

Team

Assume your group is preparing a report titled "The Negative Effects of Strip Mining on the Cumberland Plateau Region of Kentucky." After brainstorming and researching, you all settle on four major topics:

- Economic and social effects of strip mining
- Description of the strip-mining process
- Environmental effects of strip mining
- Description of the Cumberland Plateau

Conduct additional research to produce a list of subtopics, such as these:

- Strip mining method used in this region
- Location of the region
- Permanent land damage
- Water pollution
- Geological formation of the region
- Unemployment
- Increased erosion

- Natural resources of the region
- Increased flood hazards

Using a shared document (such as Google docs or a writing feature of your class learning management system), arrange these subtopics under appropriate topic headings. Then, use the outline feature from a word processing program to create the body of a formal outline. Appoint one group member to present the outline in class.

Digital and Social Media

Find a recent Facebook post, tweet, or other social media post from a research group or scientific/technical state or federal government agency. Analyze the posting in terms of the earlier section in this book ("organizing for social media and global audiences"). How long is the post? Does it contain a topic sentence and some form of a summary? Is this post organized in a format that makes you want to click to read more? Now, find the report or study referred to and compare the report's summary section to the content of the social media post. How accurate is the post relative to the overall study? What information was left out? Present your findings in class.

Global

Find a document that presents the same information in several languages (assembly instructions, for example). Even without being able to understand all of the languages used, see if you can spot any changes made in the use of headings, the length of paragraphs, or the extent to which information is organized and chunked. Interview a language professor or a professor in the business school to find out why these choices may have been made.

Chapter 11
Editing for a Professional Style and Tone

❝The amount of time I spend on revision depends on the document. Internal emails and memos get at least one careful review before being distributed. All letters and other documents to outside readers get detailed attention, to make sure that what is being said is actually what was intended. There's the issue of contractual obligations here—and also the issue of liability, if someone, say, were to misinterpret a set of instructions and were injured as a result. Also there's the issue of customer relations: Most people want to transact with businesses that display 'likability' on the interpersonal front. So, getting the style just right is always a priority.**❞**

—*Andi Wallin, Communications Manager for a power tool manufacturer*

Learning Objectives

11.1 Write clear sentences that can be understood in one reading

11.2 Write concise sentences that convey meaning in the fewest words

11.3 Write fluent sentences that provide clear connections, variety, and emphasis

11.4 Use precise language that conveys your exact meaning

11.5 Achieve a tone that connects with your audience and avoids bias

11.6 Understand style and tone in a global, legal, and ethical context

11.7 Consider style and tone in a digital context

No matter how technical your document, your audience will not understand the content unless the style is *readable,* with sentences easy to understand and words chosen precisely.

Every bit as important as *what* you have to say is *how* you decide to say it. Your particular writing style is a blend of these elements:

A definition of style

- The way in which you construct each sentence
- The length of your sentences
- The way in which you connect sentences
- The words and phrases you choose
- The tone you convey

What determines your style

Readable style, of course, requires correct grammar, punctuation, and spelling. But correctness alone is no guarantee of readability. For example, the following response to a job application is mechanically correct but hard to read:

Style is more than mechanical correctness

Inefficient style

> We are in receipt of your recent correspondence indicating your interest in securing the advertised position. Your correspondence has been duly forwarded for consideration by the personnel office, which has employment candidate selection responsibility. You may expect to hear from us relative to your application as the selection process progresses. Your interest in the position is appreciated.

Notice how hard you had to work to extract information from the previous paragraph when it could have been expressed this simply:

Readable style

> Your application for the advertised position has been forwarded to our human resources office. As the selection process moves forward, we will be in touch. Thank you for your interest.

Inefficient style makes readers work harder than they should.

Style can be inefficient for many reasons, but especially when it does the following:

Ways in which style goes wrong

- Makes the writing impossible to interpret
- Takes too long to make the point
- Reads like a story from primary school
- Uses imprecise or needlessly big words
- Sounds stuffy and impersonal

Regardless of the cause, inefficient style results in writing that is less informative and less persuasive than it should be. Also, inefficient style can be unethical when it confuses or misleads the audience, whether intentionally or unintentionally.

To help your audience spend less time reading, you must spend more time revising for a style that is *clear, concise, fluent, exact*, and *likable*.

Editing for Clarity

11.1 Write clear sentences that can be understood in one reading

Clear writing enables people to read each sentence only once in order to fully grasp its meaning. The following suggestions will help you edit for clarity.

Avoid Ambiguous Pronoun References

Pronouns (*he, she, it, their*, and so on) must clearly refer to the noun they replace. The following sentence, for example, contains an ambiguous referent (Are the patients or the warm days on their way out?):

Ambiguous referent

> Our patients enjoy the warm days while **they** last.

Depending on whether the referent (or antecedent) for *they* is *patients* or *warm days*, the sentence can be clarified in two ways, as in the following examples:

> While these warm days last, our patients enjoy them.

Clear referents

or

| Our terminal patients enjoy the warm days.

Similarly, the following sentence is ambiguous (Who is the competitive one—Jack or his assistant?):

| Jack resents his assistant because **he** is competitive.

Ambiguous
referent

This sentence can also be clarified in two ways, by providing clear referents:

| Because his assistant is competitive, Jack resents him.

or

Clear referents

| Because Jack is competitive, he resents his assistant.

(See Appendix B, "Faulty Agreement—Pronoun and Reference," for more on pronoun references, and "Guidelines for Avoiding Biased Language" in this chapter for avoiding sexist bias in pronoun use.)

Avoid Ambiguous Modifiers

A modifier is a word (usually an adjective or an adverb) or a group of words (usually a phrase or a clause) that provides information about other words or groups of words. If a modifier is too far from the words it modifies, the message can be ambiguous. Position modifiers to reflect your meaning. The following sentence is ambiguous, because we do not know if *only* modifies *press* or *emergency*:

| **Only** press the red button in an emergency.

Ambiguous
modifier

This ambiguity can be clarified in two ways by positioning the modifiers to reflect the intended meaning:

| Press **only** the red button in an emergency.

or

Clear modifiers

| Press the red button in an emergency **only.**

(See Appendix B, "Dangling and Misplaced Modifiers" for more on modifiers.)

Exercise 1

Edit each sentence below to eliminate ambiguities in pronoun reference or to clarify ambiguous modifiers.

   a. Janice dislikes working with Claire because she's impatient.
   b. Bill told Fred that he was mistaken.
   c. Only use this phone in a red alert.
   d. Just place the dishes back in the cabinets after 8 P.M.

Unstack Modifying Nouns

Too many nouns in a row can create confusion and reading difficulty. One noun can modify another (as in "software development"). But when two or more nouns modify a noun, the string of words becomes hard to read and ambiguous, as in the following example (Is the session being evaluated, or are the participants being evaluated?):

Ambiguous use | Be sure to leave enough time for a **training session participant** evaluation.
of stacked nouns

With no articles, prepositions, or verbs, readers cannot sort out the relationships among the nouns. The ambiguity can be corrected in two ways by providing those elements:

| Be sure to leave enough time **for** participants **to evaluate** the training session.

Clear sentences | or
with nouns
unstacked | Be sure to leave enough time **to evaluate** participants in **the** training session.

Arrange Word Order for Coherence and Emphasis

In coherent writing, everything sticks together; each sentence builds on the preceding sentence and looks ahead to the one that follows. In similar fashion, sentences generally work best when the beginning looks back at familiar information and the end provides the new (or unfamiliar) information:

Familiar		**Unfamiliar**
My dog	has	fleas.
Our boss	just won	the lottery.
This company	is planning	a merger.

The above pattern also emphasizes the new information. Every sentence has a key word or phrase that sums up the new information and that usually is emphasized best at the end of the sentence. The first example below places faulty emphasis on *shipment*, while the second places correct emphasis on *refund*:

Faulty emphasis | We expect a **refund** because of your error in our shipment.

Correct emphasis | Because of your error in our shipment, we expect a **refund.**

Similarly, the first example below places faulty emphasis on *element*, while the second places correct emphasis on *trust*:

Faulty emphasis | In a business relationship, **trust** is a vital element.

Correct emphasis | A business relationship depends on **trust.**

One exception to placing key words last occurs with an imperative statement (a command, an order, an instruction), with the subject [*you*] understood. For instance, each step in a list of instructions should begin with an action verb (*insert, open, close, turn, remove, press*), as in the following two examples:

| **Disable** the alarm before activating the system.   Correct emphasis

| **Remove** the protective seal.

With the opening key word, readers know immediately what action to take.

Exercise 2

Edit to unstack modifying nouns or to rearrange word order for coherence and emphasis.

a. Develop online editing system documentation.

b. I recommend these management performance improvement incentives.

c. Our profits have doubled since we automated our assembly line.

d. Education enables us to recognize excellence and to achieve it.

e. In all writing, revision is required.

f. Sarah's job involves fault analysis systems troubleshooting handbook preparation.

Use Active Voice Whenever Possible

In general, readers grasp the meaning more quickly and clearly when the writer uses the active voice ("I did it") rather than the passive voice ("It was done by me"). In active voice sentences, a clear agent performs a clear action on a recipient:

| ***Agent*** | ***Action*** | ***Recipient*** | Active voice
|---|---|---|
| Joe | lost | your report. |

Passive voice, in contrast, reverses this pattern, placing the recipient of the action in the subject slot:

| ***Recipient*** | ***Action*** | ***Agent*** | Passive voice
|---|---|---|
| Your report | was lost | by Joe. |

Sometimes the passive eliminates the agent altogether (in the following example, the reader doesn't know who lost the report):

| Your report was lost.   Passive voice

Passive voice is unethical if it obscures the person or other agent when that person or agent responsible should be identified.

 Some writers mistakenly rely on the passive voice because they think it sounds more objective and important—whereas it often makes writing wordy and evasive. The following example uses active voice and is concise and direct:

| **I underestimated** labor costs for this project. (7 words)   Concise and
                                                                  direct (active)

In contrast, the following example uses passive voice and is wordy and indirect:

Wordy and indirect (passive)

| Labor costs for this project **were underestimated by me.** (9 words)

In the following examples, passive voice leads to evasiveness (Who underestimated the labor costs? Who made the shipping mistake?):

Evasive (passive)

| Labor costs for this project **were underestimated**.

| A **mistake was made** in your shipment. (By whom?)

In reporting errors or bad news, use the active voice, for clarity and sincerity.

The passive voice creates a weak and impersonal tone, as in the following example:

Weak and impersonal

| An offer **will be made** by us next week.

In contrast, active voice creates a strong and personal tone:

Strong and personal

| **We will make** an offer next week.

Use the active voice when you want action. Otherwise, your statement will have no power, as in the following example:

Weak passive

| If my claim is not settled by May 15, the Better Business Bureau **will be contacted,** and their advice on legal action **will be taken.**

In contrast, active voice gives a statement power:

Strong active

| If you do not settle my claim by May 15, **I will contact** the Better Business Bureau for advice on legal action.

Notice above how the second, active version emphasizes the new and significant information by placing it at the end.

Ordinarily, use the active voice for giving instructions. Notice how the use of passive voice in the following two sentences makes the instructions sound like suggestions:

Passive

| The bid **should be sealed.**

| Care **should be taken** with the dynamite.

In contrast, the use of active voice in the following corrected sentences increases the likelihood that the instructions will be followed:

Active

| **Seal** the bid.

| **Be careful** with the dynamite.

Avoid shifts from active to passive voice in the same sentence, as in the following example:

Faulty shift

| During the meeting, project members **spoke** and **presentations were given.**

In contrast, the following corrected sentence retains active voice throughout:

Correct

| During the meeting, project members **spoke** and **gave** presentations.

Exercise 3

Convert these passive voice sentences to concise, forceful, and direct expressions in the active voice.

a. The evaluation was performed by us.

b. Unless you pay me within three days, my lawyer will be contacted.

c. Hard hats should be worn at all times.

d. It was decided to reject your offer.

e. Our test results will be sent to you as soon as verification is completed.

Use Passive Voice Selectively

Use the passive voice when your audience has no need to know the agent, as in the following examples:

| Mr. Jones **was brought** to the emergency room.   Correct passive
| The bank failure **was publicized** statewide.

Use the passive voice when the agent is not known or when the object is more important than the subject, as in the following examples:

| Fred's article **was published** last week.   Correct passive
| All policy claims **are kept** confidential.

Prefer the passive when you want to be indirect or inoffensive (as in requesting the customer's payment or the employee's cooperation, or to avoid blaming someone—such as your supervisor) (Ornatowski 94). The following examples represent blunt uses of active voice:

| **You have not paid** your bill.   Blunt use of
| **You need to overhaul** our filing system.   active voice

In contrast, the following examples represent appropriately indirect uses of passive voice:

| This bill **has not been paid.**   Appropriate
| Our filing system **needs to be overhauled.**   (indirect) use of
passive voice

Finally, use the passive voice if the person behind the action needs to be protected, as in the following examples:

| The criminal **was identified.**   Correct passive
| The embezzlement scheme **was exposed.**

Exercise 4

The sentences below lack proper emphasis because of inappropriate use of the active voice. Convert each to passive voice.

   a. Joe's company fired him.

   b. A power surge destroyed more than a dozen of our network routers.

   c. You are paying inadequate attention to worker safety.

   d. You are checking temperatures too infrequently.

   e. You did a poor job editing this report.

Avoid Overstuffed Sentences

Give no more information in one sentence than readers can retain and process. The following sentence is overstuffed:

Overstuffed
> Publicizing the records of a private meeting that took place three weeks ago to reveal the identity of a manager who criticized our company's promotion policy would be unethical.

Clear things up by sorting out the relationships, as in the following revision (other versions are possible, depending on the intended meaning):

Revised
> In a private meeting three weeks ago, a manager criticized our company's policy on promotion. It would be unethical to reveal the manager's identity by publicizing the records of that meeting.

Even short sentences can be hard to interpret if they have too many details, as in the following overstuffed sentence:

Overstuffed
> Send three copies of Form 17-e to all six departments, unless Departments A or B or both request Form 16-w instead.

Exercise 5

Unscramble this overstuffed sentence by making shorter, clearer sentences.

A smoke-filled room causes not only teary eyes and runny noses but also can alter people's hearing and vision, as well as creating dangerous levels of carbon monoxide, especially for people with heart and lung ailments, whose health is particularly threatened by secondhand smoke.

Editing for Conciseness

11.2 **Write concise sentences that convey meaning in the fewest words**

Concise writing conveys the most information in the fewest words. But it does not omit those details necessary for clarity. Use fewer words whenever fewer will do.

But remember the difference between *clear writing* and *compressed writing* that is impossible to decipher, as is the case with the following compressed sentence:

> Send new vehicle air conditioner compression cut-off system specifications to engineering manager advising immediate action.

Compressed

In contrast, the following correction is much clearer:

> The cut-off system for the air conditioner compressor on our new vehicles is faulty. Send the system specifications to our engineering manager so they can be modified.

Clear

First drafts rarely are concise. Trim the fat.

Avoid Wordy Phrases

Each phrase below can be reduced to one word:

due to the fact that	=	because
the majority of	=	most
readily apparent	=	obvious
a large number	=	many
aware of the fact that	=	know

Wordy phrases and their substitutes

Eliminate Redundancy

A redundant expression says the same thing twice, in different words, as the following examples:

completely eliminate	**end** result
enter **into**	consensus **of opinion**
mental awareness	**utter** devastation
mutual cooperation	**the month of** August

Redundant phrases

Avoid Needless Repetition

Unnecessary repetition clutters writing and dilutes meaning. The following sentence is needlessly repetitious:

> In trauma victims, breathing is restored by **artificial respiration.** Techniques of **artificial respiration** include mouth-to-mouth **respiration** and mouth-to-nose **respiration.**

Needlessly repetitious

Repetition in the above passage disappears when sentences are combined, as the following concise version demonstrates:

> In trauma victims, breathing is restored by artificial respiration, either mouth-to-mouth or mouth-to-nose.

Concise

> NOTE Don't hesitate to repeat, or at least rephrase, material (even whole paragraphs in a longer document) if you feel that readers need reminders. Effective repetition helps avoid cross-references like these: "See page 23" or "Review page 10."

Exercise 6

Make these sentences more concise by eliminating wordy phrases, redundancy, and needless repetition.

 a. I have admiration for Professor Jones.

 b. Due to the fact that we made the lowest bid, we won the contract.

 c. On previous occasions we have worked together.

 d. We have completely eliminated the bugs from this program.

 e. This report is the most informative report on the project.

 f. This offer is the most attractive offer I've received.

Avoid *There* Sentence Openers

Many *There is* or *There are* sentence openers can be eliminated. The following sentence contains a needless *there*:

Needless *There* | **There is** a danger of explosion in Number 2 mineshaft.

This revised version of the sentence shows how the sentence can be easily corrected:

Revised | Number 2 mineshaft is in danger of exploding.

Dropping such openers places the key words at the end of the sentence, where they are best emphasized.

Of course, in some contexts, proper emphasis would call for a *There* opener, as in the following example:

Appropriate *There* opener | People have often wondered about the rationale behind Boris's sudden decision. There were several good reasons for his dropping out of the program.

Avoid Some *It* Sentence Openers

Avoid beginning a sentence with *It*—unless the *It* clearly points to a specific referent in the preceding sentence: "This document is excellent. It deserves special recognition." The following sentence contains a needless *It* opener:

Needless *It* opener | **It** is necessary to complete both sides of the form.

This revised version shows how the *It* opener can be easily removed:

Revised | Please complete both sides of the form.

Delete Needless Prefaces

Instead of delaying the new information in your sentence, get right to the point. The following sentence contains a wordy preface:

Wordy preface | **I am writing this letter because** I wish to apply for the position of copy editor.

In contrast, the following concise correction removes the preface:

| Please consider me for the position of copy editor.   Concise

Similarly, the wordy preface in the following sentence is unnecessary:

| **As far as sustainable energy is concerned,** the technology is only in its infancy.   Wordy preface

Note how the preface can be easily removed to create a more concise sentence:

| Sustainable energy technology is only in its infancy.   Concise

---

Exercise 7

Make these sentences more concise by eliminating *There* **and** *It* **openers and needless prefaces.**

a. There was severe fire damage to the reactor.

b. There are several reasons why Jane left the company.

c. It is essential that we act immediately.

d. It has been reported by Bill that several safety violations have occurred.

e. This letter is to inform you that I am pleased to accept your job offer.

f. The purpose of this report is to update our research findings.

Avoid Weak Verbs

Prefer verbs that express a definite action: *open, close, move, continue, begin.* Avoid weak verbs that express no specific action: *is, was, are, has, give, make, come, take.* Although in some cases, such verbs are essential to your meaning, as in "Dr. Phillips is operating at 7 AM" or "Take me to the laboratory"), all forms of *to be* (*am, are, is, was, were, will, have been, might have been*) are weak. Substitute a strong verb for conciseness. The following sentence uses a weak form of *to be*:

| My recommendation **is** for a larger budget.   Weak verb

Note how the following corrected version of the sentence sounds stronger:

| I **recommend** a larger budget.   Strong verb

Don't disappear behind weak verbs and their baggage of needless nouns and prepositions. For example, take into consideration is a wordy verb phrase using a form of *take*:

| Please **take into consideration** my offer.   Wordy verb phrase

Notice how the following corrected version of the sentence uses a concise verb instead:

| Please **consider** my offer.   Concise verb

Strong verbs, or action verbs, suggest an assertive, positive, and confident writer. Here are examples of weak verbs converted to strong verbs:

Weak verbs and their replacements

has the ability to	=	Can
give a summary of	=	summarize
make an assumption	=	Assume
come to the conclusion	=	conclude
make a decision	=	decide

Exercise 8

Edit each of these wordy and vague sentences to eliminate weak verbs.

  a. Our disposal procedure is in conformity with federal standards.

  b. Please make a decision today.

  c. We need to have a discussion about the problem.

  d. I have just come to the realization that I was mistaken.

  e. Your conclusion is in agreement with mine.

Avoid Excessive Prepositions

Excessive prepositions needlessly clutter a sentence and make it difficult to read, as in the following example:

Excessive prepositions

The recommendation first appeared **in** the report written **by** the supervisor in January **about** that month's productivity.

Notice how the corrected version of the sentence below uses appropriate prepositions and is easy to read:

Appropriate prepositions

The recommendation first appeared in the supervisor's productivity report for January.

Each prepositional phrase below can be reduced to avoid the overuse of prepositions:

Prepositional phrases and their replacements

with the exception of	=	except for
in the near future	=	soon
at the present time	=	now
in the course of	=	during
in the process of	=	during (or while)

Avoid Nominalizations

Nouns manufactured from verbs (nominalizations) are harder to understand than the verbs themselves, as the following example demonstrates:

Nominalization

We ask for the **cooperation** of all employees.

Notice how simply using a clear verb form makes the sentence easy to understand:

| We ask that all employees **cooperate.**   Clear verb form

Similarly, the following sentence containing a nominalization is difficult to grasp on a single reading:

| Give **consideration** to the possibility of a career change.   Nominalization

Again, simply using a clear verb form makes the sentence easy to understand in one reading:

| **Consider** a career change.   Clear verb form

Besides causing wordiness, nominalizations can be vague—by hiding the agent of an action. Verbs are generally easier to read because they signal action. The following sentence is unclear, because the reader cannot tell who should take the action:

| A **valid requirement** for immediate action exists.   Nominalization

In contrast, the following sentence uses a precise verb form instead to clarify who should take the action:

| We **must act** immediately.   Precise verb form

Following are some nominalizations restored to their action verb forms:

conduct an investigation of	=	investigate
provide a description of	=	describe
conduct a test of	=	test

Nouns traded for verbs

Nominalizations drain the life from your style. In cheering for your favorite team, you wouldn't say "Blocking of that kick is a necessity!" instead of "Block that kick!"

> NOTE *Avoid excessive economy. For example, "Employees must cooperate" would not be an acceptable alternative to the first example in this section. But, for the final example, "Block that kick" would be acceptable.*

Exercise 9

Make these sentences more concise by eliminating needless prepositions, *to be* constructions, and nominalizations.

a. In the event of system failure, your sounding of the alarm is essential.

b. These are the recommendations of the chairperson of the committee.

c. Our acceptance of the offer is a necessity.

d. Please perform an analysis and make an evaluation of our new system.

e. A need for your caution exists.

f. Power surges are associated, in a causative way, with malfunctions of computers.

Make Negatives Positive

A positive expression is easier to understand than a negative one. For example, the following sentence's use of a negative phrase makes it sound indirect and wordy:

Indirect and
wordy

| Please do not be late in submitting your report.

In contrast, the following positive version of the sentence is direct and concise:

Direct and
concise

| Please submit your report on time.

Sentences with multiple negative expressions are even harder to translate, as in the following confusing and wordy sentence:

Confusing and
wordy

| Do **not** distribute this memo to employees who have **not** received a security clearance.

Notice how the following positive revision is clear and concise:

Clear and concise

| Distribute this memo only to employees who have received a security clearance.

Besides directly negative words (*no, not, never*), some indirectly negative words (*except, forget, mistake, lose, uncooperative*) also force readers to translate, as in the following confusing and wordy sentences:

Confusing and
wordy

| **Do not neglect** to activate the alarm system.
| My diagnosis was **not inaccurate.**

In contrast, the following positive revisions are clear and consise:

Clear and concise

| **Be sure** to activate the alarm system.
| My diagnosis was **accurate.**

Some negative expressions, of course, are perfectly correct, as when expressing disagreement:

Correct negatives

| This is **not** the best plan.
| Your offer is **unacceptable.**

Prefer positives to negatives, though, whenever your meaning allows. Following are wordy negative phrases replaced with concise positive verbs:

Trading negatives
for positives

did not succeed	=	Failed
does not have	=	Lacks
did not prevent	=	Allowed
not unless	=	only if

Clean Out Clutter Words

Clutter words stretch a message without adding meaning. Here are some of the most common: *very, definitely, quite, extremely, rather, somewhat, really, actually, currently, situation, aspect, factor*. Notice how the following sentence is cluttered with words that get in the way of grasping the meaning of the sentence:

> **Actually**, one **aspect** of a business **situation** that could **definitely** make me **quite** happy would be to have a **somewhat** adventurous parter who **really** shared my **extreme** attraction to risks.

Cluttered

In contrast, the following revision is concise and easy to understand:

> I seek an adventurous business partner who enjoys risks.

Concise

Delete Needless Qualifiers

Qualifiers such as *I feel, it seems, I believe, in my opinion,* and *I think* express uncertainty or soften the tone and force of a statement. The following sentence uses *I think* and *seems* appropriately:

> Despite Frank's poor grades last year he will, **I think,** do well in college.
> Your product **seems** to meet our needs.

Appropriate qualifiers

But when you are certain, eliminate the qualifier so as not to seem tentative or evasive. The following sentences contain the needless qualifiers *It seems that* and *appear to*:

> **It seems that** I've made an error.
> We **appear to** have exceeded our budget.

Needless qualifiers

> NOTE *In communicating across cultures, keep in mind that a direct, forceful style might be considered offensive.*

Exercise 10

Make these sentences more concise by changing negatives to positives and by clearing out clutter words and needless qualifiers.

a. Our design must avoid nonconformity with building codes.

b. Never fail to wear protective clothing.

c. We are currently in the situation of completing our investigation of all aspects of the accident.

d. I appear to have misplaced the contract.

e. Do not accept bids that are not signed.

f. It seems as if I have just wrecked a company car.

Editing for Fluency

11.3 Write fluent sentences that provide clear connections, variety, and emphasis

Fluent sentences are easy to read because they provide clear connections, variety, and emphasis. Their varied length and word order eliminate choppiness and monotony. Fluent sentences enhance *clarity*, emphasizing the most important ideas.

Fluent sentences also enhance *conciseness,* often replacing several short, repetitious sentences with one longer, more economical sentence. To write fluently, use the following strategies.

Combine Related Ideas

A series of short, disconnected sentences is not only choppy and wordy but also unclear; readers are forced to insert transitions between ideas and decide which points are most important. Note the disconnected series of ideas in the following sentence:

Disconnected series of ideas

> Jogging can be healthful. You need the right equipment. Most necessary are well-fitting shoes. Without this equipment you take the chance of injuring your legs. Your knees are especially prone to injury. (5 sentences)

In contrast, this revision provides a clear, concise, and fluent combination of ideas:

Clear, concise, and fluent combination of ideas

> Jogging can be healthful if you have the right equipment. Shoes that fit well are most necessary because they prevent injury to your legs, especially your knees. (2 sentences)

Most sets of information can be combined to form different relationships, depending on what you want to emphasize. Imagine that the following set of facts describes an applicant for a junior management position with your company.

- Roy James graduated from an excellent management school.
- He has no experience.
- He is highly recommended.

Assume that you are a personnel director, conveying your impression of this candidate to upper management. To convey a negative impression, you might combine the information in this way:

Strongly negative emphasis

> Although Roy James graduated from an excellent management school and is highly recommended, **he has no experience.**

The *independent clause* (in boldface) receives the emphasis. (See also Appendix B, "Faulty Subordination, for more about subordination.) But if you are undecided yet leaning in a negative direction, you might write:

Slightly negative emphasis

> Roy James graduated from an excellent management school and is highly recommended, **but** he has no experience.

In this sentence, the information both before and after *but* appears in independent clauses. Joining them with the coordinating word *but* suggests that both sides of the issue are equally important (or "coordinate"). Placing the negative idea last, however, gives it a slight emphasis. (See also Appendix B, "Faulty Coordination," for more about coordination.)

Finally, to emphasize strong support for the candidate, you could say this:

| Although Roy James has no experience, **he graduated from an excellent management school and is highly recommended.**

Positive emphasis

In the preceding example, the initial information is subordinated by *although*, giving the final information the weight of an independent clause.

> NOTE *Combine sentences only to simplify the reader's task. Overstuffed sentences with too much information and too many connections can be hard for readers to sort out. (See "Overstuffed Sentences" in this chapter.)*

Exercise 11

Combine each set of sentences below into one fluent sentence that provides the requested emphasis.

Examples:

SENTENCE SET   John is a loyal employee.

John is a motivated employee.

John is short-tempered with his colleagues.

COMBINED FOR   Even though John is short-tempered with his

POSITIVE EMPHASIS   colleagues, he is a loyal and motivated employee.

a. The job offers an attractive salary.
   It demands long work hours.
   Promotions are rapid.
   (*Combine for negative emphasis.*)

b. The job offers an attractive salary.
   It demands long work hours.
   Promotions are rapid.
   (*Combine for positive emphasis.*)

c. Company X gave us the lowest bid.
   Company Y has an excellent reputation.
   (*Combine to emphasize Company Y.*)

d. Superinsulated homes are energy efficient.
   Superinsulated homes create a danger of indoor air pollution.
   The toxic substances include radon gas and urea formaldehyde.
   (*Combine for a negative emphasis.*)

e. Computers cannot *think* for the writer.
   Computers eliminate many mechanical writing tasks.
   They speed the flow of information.
   (*Combine to emphasize the first assertion.*)

Vary Sentence Construction and Length

Related ideas often need to be linked in one sentence so that readers can grasp the connections. The following sentence contains disconnected ideas:

Disconnected ideas

| The nuclear core reached critical temperature. The loss-of-coolant alarm was triggered. The operator shut down the reactor.

In contrast, the following revision connects the ideas for readers:

Connected ideas

| As the nuclear core reached critical temperature, triggering the loss-of-coolant alarm, the operator shut down the reactor.

But an idea that should stand alone for emphasis needs a whole sentence of its own, as in the following sentence:

Correct

| Core meltdown seemed inevitable.

However, an unbroken string of long or short sentences can bore and confuse readers, as can a series with identical openings in the following boring and repetitive sentence:

Boring and repetitive

| There are a number of drawbacks about diesel engines. **They** are noisy. **They** are difficult to start in cold weather. **They** cause vibration. **They** also give off an unpleasant odor. **They** cause sulfur dioxide pollution.

This varied revision provides readers with a more interesting reading experience:

Varied

| Diesel engines have a number of drawbacks including noisiness, cold-weather starting difficulties, vibrations, and odor. Most seriously, they cause sulfur dioxide pollution.

Similarly, when you write in the first person, overusing *I* makes you appear self-centered. (Some organizations require use of the third person, avoiding the first person completely, for all manuals, lab reports, specifications, product descriptions, and so on.)

Do not, however, avoid personal pronouns if they make the writing more readable (say, by eliminating passive constructions).

Use Short Sentences for Special Emphasis

All this talk about combining ideas might suggest that short sentences have no place in good writing. Wrong. Short sentences (even one-word sentences) provide vivid emphasis. They stick in a reader's mind.

Finding the Exact Words

11.4 Use precise language that conveys your exact meaning

Too often, language can *camouflage* rather than communicate. People see many reasons to hide behind language, as when they do the following:

Situations in which people often hide behind language

- Speak for their company but not for themselves
- Fear the consequences of giving bad news
- Are afraid to disagree with company policy

Before using any jargon, think about your specific audience and ask yourself: "Can I find an easier way to say exactly what I mean?" Only use jargon that improves your communication.

Use Acronyms Selectively

Acronyms are words formed from the initial letter of each word in a phrase (as in *LOCA* from *loss of coolant accident*) or from a combination of initial letters and parts of words (as in *bit* from *binary digit* or *pixel* from *picture element*). Acronyms *can* communicate concisely—but only when the audience knows their meaning, and only when you use the term often in your document. The first time you use an acronym, spell out the words from which it is derived, as in the following example:

> **Modem** ("modulator + demodulator"): a device that converts, or "modulates," computer data in electronic form into a sound signal that can be transmitted and then reconverted, or "demodulated," into electronic form for the receiving computer.

An acronym defined

Avoid Triteness

Worn-out phrases (clichés) make writers seem too lazy or too careless to find exact, unique ways of saying what they mean. Following are examples of worn-our phrases:

make the grade	the chips are down
in the final analysis	not by a long shot
close the deal	last but not least
hard as a rock	welcome aboard
water under the bridge	over the hill

Worn-out phrases

Exercise 13

Edit these sentences to eliminate useless jargon and triteness.

a. To optimize your financial return, prioritize your investment goals.

b. The use of this product engenders a 50 percent repeat consumer encounter.

c. We'll have to swallow our pride and admit our mistake.

d. Managers who make the grade are those who can take daily pressures in stride.

Avoid Misleading Euphemisms

A form of understatement, a euphemism is an expression aimed at politeness or at making unpleasant subjects seem less offensive. Thus, *we powder our noses* or *use the boys' room* instead of *using the bathroom; we pass away* or *meet our Maker* instead of *dying*.

When a euphemism is appropriate	When euphemisms avoid offending or embarrassing people, they are perfectly legitimate. Instead of telling a job applicant he or she is *unqualified*, we might say, *Your background doesn't meet our needs*. In addition, there are times when friendliness and interoffice harmony are more likely to be preserved with writing that is not too abrupt, bold, blunt, or emphatic (MacKenzie 2).
When a euphemism is deceptive	Euphemisms, however, are unethical if they understate the truth when only the truth will serve. In the sugarcoated world of misleading euphemisms, bad news disappears:

Examples of euphemisms that mislead

- Instead of being *laid off* or *fired*, workers are *surplused* or *deselected*, or the company is *downsized*.
- Instead of *lying* to the public, the government *engages in a policy of disinformation*.
- Instead of *wars* and *civilian casualties*, we have *conflicts* and *collateral damage*.

Plain talk is always better than deception. If someone offers you a job *with limited opportunity for promotion*, expect a *dead-end job*.

Avoid Overstatement

Exaggeration sounds phony. Be cautious when using superlatives such as *best, biggest, brightest, most*, and *worst*. Recognize the differences among *always, usually, often, sometimes*, and *rarely*; among *all, most, many, some*, and *few*. The following sentences contain needless overstatements:

Overstatements

| You never listen to my ideas.

| This product will last forever.

| Assembly-line employees are doing shabby work.

Unless you mean *all employees*, qualify your generalization with *some,/* or *most*—or even better, specify *20 percent*.

Exercise 14

Edit these sentences to eliminate euphemism, overstatement, or unsupported generalizations.

a. I finally must admit that I am an abuser of intoxicating beverages.

b. I was less than candid.

c. This employee is poorly motivated.

d. Most entry-level jobs are boring and dehumanizing.

e. Clerical jobs offer no opportunity for advancement.

f. Because of your absence of candor, we can no longer offer you employment.

Avoid Imprecise Wording

Words listed as synonyms usually carry different shades of meaning. Do you mean to say *I'm slender, You're slim, She's lean,/* or *He's scrawny*? The wrong choice could be disastrous.

Imprecision can create ambiguity. For instance, is *send us more personal information* a request for more information that is personal or for information that is more personal? Does your client expect *fewer* or *less* technical details in your report? See Appendix B, "Usage" for a table of words that are commonly confused.

Be Specific and Concrete

General words name broad classes of things, such as *job, computer,* or *person.* Such terms usually need to be clarified by more specific ones. Following are two examples of more specific and concrete wordings:

job	=	senior accountant for Softbyte Press	General terms traded for specific terms
person	=	Sarah Jones, production manager	

The more specific your words, the more a reader can visualize your meaning.

Abstract words name qualities, concepts, or feelings (*beauty, luxury, depression*) whose exact meaning has to be nailed down by *concrete* words—words that name things we can visualize. Notice how much better you can visualize the following concrete substitutes for the abstract words *beautiful* and *depressed*:

a **beautiful** view	=	snowcapped mountains, a wilderness lake, pink ledge, ninety-foot birch trees	Abstract terms traded for concrete terms
a **depressed** worker	=	suicidal urge, insomnia, feelings of worthlessness, no hope for improvement	

Informative writing *tells* and *shows.* The lack of specific wording in the following sentence makes only a general statement:

One of our **workers** was **injured** by a **piece of equipment recently.**	General

In contrast the following revision provides specific details to show rather than merely tell:

Alan Hill suffered a **broken thumb** while working on a **lathe yesterday.**	Specific

Lastly, don't write *thing* when you mean *lever, switch, micrometer,* or *scalpel.*

> NOTE *In some instances, of course, you may wish to generalize for the sake of diplomacy. Instead of writing "Bill, Mary, and Sam have been posting to their personal Facebook pages instead of working," you might prefer to generalize: "Some employees...."*

Most good writing offers both general and specific information. The most general material appears in the topic statement and sometimes in the conclusion because these parts, respectively, set the paragraph's direction and summarize its content.

Exercise 15

Edit these sentences to make them more precise and informative.

a. Inflammatory airway disease is discussed in this report.

b. Your crew damaged a piece of office equipment.

c. His performance was admirable.

d. This thing bothers me.

Use Analogies to Sharpen the Image

Analogy versus comparison

Ordinary comparison shows similarities between two things of the same class (two different brand bicycles; two methods of cleaning dioxin-contaminated sites). Analogy, on the other hand, shows some essential similarity between two things of different classes (for example, the human brain and computer storage).

Analogies are good for emphasizing a point (*Some rain is now as acidic as vinegar*). They are especially useful in translating something abstract, complex, or unfamiliar, as long as the easier subject is broadly familiar to readers. Analogy therefore calls for particularly careful analyses of audience.

Analogies can save words and convey vivid images. *Collier's Encyclopedia* describes the tail of an eagle in flight as "spread like a fan." The following sentence from a description of a trout feeder mechanism uses an analogy to clarify the positional relationship between two working parts:

Analogy

> The metal rod is inserted (and centered, crosslike) between the inner and outer sections of the clip.

Without the analogy *crosslike*, we would need something like this to visualize the relationship:

Missing analogy

> The metal rod is inserted, perpendicular to the long plane and parallel to the flat plane, between the inner and outer sections of the clip.

Besides naming things, analogies help *explain* things. This next analogy helps clarify an unfamiliar concept (dangerous levels of a toxic chemical) by comparing it to something more familiar (human hair).

Analogy

> A dioxin concentration of 500 parts per trillion is lethal to guinea pigs. One part per trillion is roughly equal to the thickness of a human hair compared to the distance across the United States. (*Congressional Research Report 15*)

Adjusting Your Tone

11.5 Achieve a tone that connects with your audience and avoids bias

Your tone is your personal trademark—the personality that takes shape between the lines. The tone you create depends on (1) the distance you impose between yourself and the reader and (2) the attitude you show toward the subject.

How tone is created

Assume, for example, that a friend is going to take over a job you've held. You're writing your friend instructions for parts of the job. Here is your first sentence:

> Now that you've arrived in the glamorous world of office work, put on your running shoes; this is no ordinary manager-trainee job.

Informal tone

The example sentence imposes little distance between you and the reader (it uses the direct address, *you,* and the humorous suggestion to *put on your running shoes*). The ironic use of *glamorous* suggests just the opposite: that the job holds little glamor.

For a different reader (say, the recipient of a company training manual), you would choose some other opening:

> As a manager trainee at GlobalTech, you will work for many managers. In short, you will spend little of your day seated at your desk.

Semiformal tone

The tone now is serious, no longer intimate, and you express no distinct attitude toward the job. For yet another audience (clients or investors who will read an annual report), you might alter the tone again:

> Manager trainees at GlobalTech are responsible for duties that extend far beyond desk work.

Formal tone

Here the businesslike shift from second- to third-person address makes the tone too impersonal for any writing addressed to the trainees themselves.

We already know how tone works in speaking. When you meet someone new, for example, you respond in a tone that defines your relationship. Notice how the tone of the following series of introductions moves from most formal, then to semiformal, and finally to informal, while the distance between the speaker and the recipient shrinks:

> Honored to make your acquaintance.
> How do you do?
> Nice to meet you.
> Hello.
> Hi.
> What's happening?

Tone announces interpersonal distance

Each of these greetings is appropriate in some situations and inappropriate in others.

Whichever tone you decide on, be consistent throughout your document. Avoid shifting from an overly informal tone to an overly formal tone, as in the following inconsistent pairing of sentences:

> My office isn't fit for a pig. It is ungraciously unattractive.

Inconsistent tone

Instead, maintain a consistently informal or formal tone, as in the following revision:

Consistent tone

| My office is so shabby that it's an awful place to work.

In general, strive for a professional yet friendly tone. Look for good examples in the workplace from colleagues and managers whose email and other communication meet this standard.

Tone announces attitude

Besides setting the distance between writer and reader, your tone implies your attitude toward the subject and the reader, as in the following different ways of telling someone when dinner will be:

| We dine at seven.

| Dinner is at seven.

| Let's eat at seven.

| Let's chow down at seven.

| Let's strap on the feedbag at seven.

| Let's pig out at seven.

The words you choose tell readers a great deal about where you stand. For instance, in announcing a meeting to review your employee's job evaluation, would you invite this person to *discuss* the evaluation, *talk it over, have a chat*, or *chew the fat*? Decide how casual or serious your attitude should be.

Guidelines

for Deciding about Tone

➤ **Use a formal or semiformal tone** in writing for superiors, professionals, or academics (depending on what you think the reader expects).

➤ **Use a semiformal or informal tone** in writing for colleagues and subordinates (depending on how close you feel to your reader).

➤ **Use an informal tone** when you want your writing to be conversational, or when you want it to sound like a person talking. But always be professional.

➤ **Avoid a negative tone** when conveying unpleasant information.

➤ **Above all, find out what tone your particular readers prefer**. When in doubt, do not be too casual!

Consider Using an Occasional Contraction

Unless you have reason to be formal, use (but do not overuse) contractions. Balance an *I am* with an *I'm*, a *you are* with a *you're*, and an *it is* with an *it's*. Keep in mind that contractions rarely are acceptable in formal business writing.

*NOTE The contracted version often sounds less emphatic than the two-word version— for example, "**Don't** handle this material without protective clothing" versus "**Do not** handle this material without protective clothing." If your message requires emphasis, do not use a contraction.*

Address Readers Directly

Use the personal pronouns *you* and *your* to connect with readers. Readers often relate better to something addressed to them directly. The following sentence doesn't address readers directly and sounds impersonal:

| Students at this college will find the faculty always willing to help.   Impersonal tone

In contrast, the following revision using *you* sounds more personal:

| As a student at this college, **you** will find the faculty always willing to help.   Personal tone

*NOTE Use **you** and **your** only in letters, memos, instructions, and other documents intended to correspond directly with a reader. By using **you** and **your** in situations that call for first or third person, such as description or narration, you might end up writing something awkward like this: "When you are in northern Ontario, you can see wilderness and lakes everywhere around you."*

Exercise 16

The sentences below suffer from pretentious language, unclear expression of attitude, missing contractions, or indirect address. Adjust the tone.

a. Further interviews are a necessity to our ascertaining the most viable candidate.

b. All employees are hereby invited to the company picnic.

c. Employees must submit travel vouchers by May 1.

d. Persons taking this test should use the HELP option whenever they need it.

e. I am not unappreciative of your help.

f. My disapproval is far more than negligible.

Use *I* and *We* When Appropriate

Instead of disappearing behind your writing, use *I* or *We* when referring to yourself or your organization. The following sentence doesn't use *I* and as a result sounds distant:

| The writer of this letter would like a refund.   Distant tone

In contrast, the following revision uses *I* and establishes an appropriate distance:

| **I** would like a refund.   Appropriate distance

A message becomes doubly impersonal when both writer and reader disappear, as in the following example:

Impersonal tone ┃ The requested report will be sent next week.

The following revision brings both the writer and reader to the surface, establishing a personal tone:

Personal tone ┃ **We** will send the report **you** requested next week.

Prefer the Active Voice

Because the active voice is more direct and economical than the passive voice, it generally creates a less formal tone. (Review "Use Active Voice Whenever Possible" and "Use Passive Voice Selectively" in this chapter for use of active and passive voice.)

Exercise 17

These sentences have too few *I* or *We* constructions or too many passive constructions. Adjust the tone.

- **a.** Payment will be made as soon as an itemized bill is received.
- **b.** You will be notified.
- **c.** Your help is appreciated.
- **d.** Our reply to your bid will be sent next week.
- **e.** Your request will be given consideration.
- **f.** This writer would like to be considered for your opening.

Emphasize the Positive

Whenever you offer advice, suggestions, or recommendations, try to emphasize benefits rather than flaws. (For more on delivering bad news, see Chapter 16.) The following sentence emphasizes the negative and sounds highly critical:

Critical tone ┃ Because of your division's lagging productivity, a management review may be needed.

In contrast, this revision emphasizes the positive and establishes and encouraging tone:

Encouraging tone ┃ A management review might help boost productivity in your division.

Avoid an Overly Informal Tone

How tone can be too informal

Achieving a conversational tone does not mean writing in the same way we would speak to friends at a favorite hangout. *Substandard usage* ("He ain't got none," "I seen it today") is unacceptable in workplace writing; and so is *slang* ("hurling," "bogus," "bummed").

Profanity ("This idea sucks," "pissed off," "What the hell") not only conveys contempt for the audience but also triggers contempt for the person using it. *Colloquialisms* ("O.K.," "a lot," "snooze") tend to appear more in speaking than in writing.

Tone is offensive when it violates the reader's expectations: when it seems disrespectful, tasteless, distant and aloof, too "chummy," casual, or otherwise inappropriate for the topic, the reader, and the situation.

How tone can offend

A formal or academic tone is appropriate in countless writing situations: a research paper, a job application, a report for the company president. In a history essay, for example, you would not refer to George Washington and Abraham Lincoln as "those dudes, George and Abe." Whenever you begin with rough drafting or brainstorming, your initial tone might be overly informal and is likely to require some adjustment during subsequent drafts.

When to use an academic tone

Avoid Personal Bias

If people expect an impartial report, try to keep your own biases out of it. Imagine, for example, that you have been assigned to investigate the causes of an employee-management confrontation at your company's Omaha branch. Your initial report, written for the New York central office, is intended simply to describe what happened. Here is how an unbiased description might begin:

> At 9:00 a.m. on Tuesday, January 21, eighty female employees set up picket lines around the executive offices of our Omaha branch, bringing business to a halt. The group issued a formal protest, claiming that their working conditions were repressive, their salary scale unfair, and their promotional opportunities limited.

A factual account

Note the absence of implied judgments; the facts are presented objectively. A biased version of events, from a protestor's point of view, might read like this:

> Last Tuesday, sisters struck another blow against male supremacy when eighty women employees paralyzed the company's repressive and sexist administration for more than six hours. The timely and articulate protest was aimed against degrading working conditions, unfair salary scales, and lack of promotional opportunities for women.

A biased version

Judgmental words (*male supremacy, repressive, degrading, paralyzed, articulate*) inject the writer's attitude about events, even though it isn't called for. In contrast to this bias, the following version patronizingly defends the status quo:

> Our Omaha branch was the scene of an amusing battle of the sexes last Tuesday, when a group of irate feminists, eighty strong, set up picket lines for six hours at the company's executive offices. The protest was lodged against alleged inequities in hiring, wages, working conditions, and promotion for women in our company.

A biased version

(For more on how framing of the facts can influence reader judgments, see Chapter 8, "Evaluate the Evidence.")

> NOTE *Being unbiased doesn't mean remaining "neutral" about something you know to be wrong or dangerous (Kremers 59). If, for instance, you conclude, based on your careful research, that the Omaha protest was clearly justified, say so.*

Avoid Biased Language

Types of biased language

Language that is offensive or makes unwarranted assumptions is inappropriate for workplace communication and may harm the reputation of a company or organization. For example, letters addressed to "Dear Sir" may leave an unfavorable impression of the company on all readers and especially on women. Unnecessary mention of gender, marital status, physical appearances, and ethnicity may also alienate readers.

Mention difference only when relevant

The words you choose should demonstrate respect for all people. References to individuals and groups should include only as much detail as is necessary for the particular point you are trying to make. The *Publication Manual of the American Psychological Association* (APA), 6th edition, says it well when it states that "[p]art of writing without bias is recognizing that differences should be mentioned only when relevant. Marital status, sexual orientation, racial and ethnic identity, or the fact that a person has a disability should not be mentioned gratuitously" (71).

When such items are important to the document and need to be included, the APA recommends that writers pay special attention to how individuals and groups are labeled (72) and recognize that "[p]references for terms referring to racial and ethnic groups change often" (75). They recommend that, if uncertain, writers consider their "two basic guidelines of specificity and sensitivity"—in other words, how specific does the writer need to be and how can the language be as unbiased and sensitive as possible (75).

For example, the APA offers these specific suggestions:

- **Age**: When age is an important and relevant part of the discussion, use *girl* and *boy* to refer to people under age 12; use *young woman* and *young man* for people ages 13–17; use women and men for those ages 18 and over (76).

- **Disabilities**: In line with the recommendations of the University of Kansas's Research and Training Center on Independent Living, the APA suggests using "people-first language," such as *people with autism* rather than *autistics*.

Exercise 18

The sentences below suffer from negative emphasis, excessive informality, biased expressions, or offensive usage. Adjust the tone.

a. If you want your workers to like you, show sensitivity to their needs.

b. The union has won its struggle for a decent wage.

c. The group's spokesman demanded salary increases.

d. Each employee should submit his vacation preferences this week.

e. The explosion left me blind as a bat for nearly an hour.

f. This dude would be an excellent employee if only he could learn to chill out.

g. The Latino mechanic fixed the problem in under an hour.

NOTE *For a detailed and a thoughtful discussion on this topic, refer to the* Publication Manual of the American Psychological Association, *6th edition, pages 70–77, or visit the APA Web site. You can also visit your campus's writing center for information on how to avoid bias in your writing.*

Exercise 19

Find examples of overly euphemistic language (such as "chronologically challenged") or of insensitive language (such as "lame excuse"). Discuss examples in class.

Guidelines
for Avoiding Biased Language

➤ **Use neutral expressions**. Use "chair" or "chairperson" rather than "chairman" and "postal worker" rather than "postman."

➤ **Don't mention gender if it is not relevant**. Don't use "the female police officer" unless you are talking about police officers in the context of gender.

➤ **Avoid sexist pronouns.** You can avoid using a sexist pronoun by revising the sentence ("A writer who revises will succeed" rather than "A writer will succeed if he revises"), by using a plural form ("Writers will succeed if they revise"), or by occasionally using paired pronouns ("A writer will succeed if he or she revises").

➤ **Drop condescending diminutive endings.** Avoid words such as "poetess" and "majorette" and simply use "poet" and "major" instead.

➤ **Use "Ms."** Unless you know that a person prefers a traditional title, avoid "Mrs." and "Miss." Males are not identified by marital status, nor should females have to be.

➤ **Avoid potentially judgmental expressions.** For example, instead of "Third World" or "underdeveloped" use "newly industrialized" or "developing."

➤ **Never use demeaning expressions.** Avoid expressions such as "lame excuse," "the blind leading the blind," or "that's queer."

➤ **Use person-first language for people with disabilities or medical conditions.** Avoid terms that place the disability over the person or that suggest pity. Use "person with a disability" rather than "disabled person" and "person with AIDS" rather than "AIDS victim."

➤ **Use age-appropriate designations.** The APA recommends that when age is an important and relevant part of the discussion, use "girl" and "boy" to refer to people under age 12; use "young woman" and "young man" for people ages 13–17; use "women" and "men" for those ages 18 and over (76).

➤ **In quoting sources that ignore nonsexist or nonbiased standards, consider these options:**

   a. Insert [*sic*] ("thus" or "so") following the first instance of sexist usage.

   b. Use ellipses (see Appendix B, "Ellipses") to omit the biased phrasing.

   c. Paraphrase instead of quoting.

   d. Substitute or insert non-biased words between brackets.

Global, Legal, and Ethical Implications of Style and Tone

11.6 Understand style and tone in a global, legal, and ethical context

Style and tone have global implications

The style guidelines in this chapter apply to standard English in North America. But technical communication is a global process: Practices and preferences differ widely in various cultural contexts.

Cultures differ in their style preferences

For example, some cultures prefer long sentences and elaborate language to convey an idea's full complexity. Others value expressions of respect, politeness, praise, and gratitude more than clarity or directness (Hein 125–26; Mackin 349–50). Writing in non-English languages tends to be more formal than in English, and some languages rely heavily on passive voice (Weymouth 144). French readers, for example, may prefer an elaborate style that reflects sophisticated and complex modes of thinking. In contrast, our "plain English," conversational style might connote simplemindedness, disrespect, or incompetence (Thrush 277).

Challenges in writing for translation

Documents to be translated into other languages pose special challenges. In translation or in a different cultural context, some words have insulting or negative connotations; for example, in certain cultures, "male" and "female" refer only to animals (Coe, "Writing for Other Cultures" 17). Notable translation disasters include the Chevrolet *Nova*—*no va* means "doesn't go" in Spanish—and the Finnish beer *Koff* for an English-speaking market (Gesteland 20; Victor 44). Many U.S. idioms (*breaking the bank, cutthroat competition, sticking your neck out*) and cultural references (*the crash of '29, Beantown*) make no sense outside of U.S. culture (Coe, "Writing" 17–19). Slang (*bogus, fat city*) and colloquialisms (*You bet, Gotcha*) can seem too informal and crude. In short, offensive writing (including inappropriate humor) can alienate audiences—from both you *and* your culture (Sturges 32).

Style and tone also have legal and ethical implications

Chapter 4 discusses how workplace writing is regulated by laws against libel, deceptive advertising, and defective information. One common denominator among these violations is poor word choice. We are each accountable for the words we use—intentionally or not—in framing the audience's perception and understanding. Imprecise or inappropriate word choice can spell big trouble, as seen in the following examples.

- **Assessing risk.** Is the investment you are advocating "a sure thing" or merely "a good bet," or even "risky"? Are you announcing a "caution," a "warning," or a "danger"? Should methane levels in mineshaft #3 "be evaluated" or do "they pose a definite explosion risk"? Never downplay risks.

- **Offering a service or product.** Are you proposing to "study the problem," to "explore solutions to the problem," or to "eliminate the problem"? Do you "stand behind" your product or do you "guarantee" it? Never promise more than you can deliver.

- **Giving instructions.** Before inserting the widget between the grinder blades, should I "switch off the grinder" or "disconnect the grinder from its power source" or "trip the circuit breaker," or do all three? Always triple-check the clarity of your instructions.

- **Comparing your product with competing products.** Instead of referring to a competitor's product as "inferior," "second-rate," or "substandard," talk about your own "first-rate product" that "exceeds (or meets) standards." Never run down the competition.

- **Evaluating an employee** (T. Clark 75–76). In a personnel evaluation, do not refer to the employee as a "troublemaker" or "unprofessional," or as "too abrasive," "too uncooperative," "incompetent," or "too old" for the job. Focus on the specific requirements of this job; offer factual instances in which these requirements have been violated: "Our monitoring software recorded five visits by this employee to X-rated Web sites during working hours." Or "This employee arrives late for work on average twice weekly, has failed to complete assigned projects on three occasions, and has difficulty working with others." Instead of expressing personal judgments, offer the facts. Be sure everyone involved knows exactly what the standards are well beforehand. Otherwise, you risk violating federal laws against discrimination and libel (damaging someone's reputation) and you may face a lawsuit.

Situations in which word choice has ethical or legal consequences

Digital Writing and Editing

11.7 Consider style and tone in a digital context

Most of today's workplace writing takes place in digital forms, including Facebook, Instagram, LinkedIn and other social media posts; tweets; text messages; email; and more. Email is the most common form of daily communication at work, but increasingly, social media is being used almost exclusively when it comes to marketing, customer relations, and advertising.

Email and social media in workplace writing

Yet despite so much writing taking place in digital forms and on social media, writers often pay little attention to the style and tone of a message. Humor and sarcasm, which might be an appropriate tone when posting a reply to a friend, are typically not appropriate in the workplace. The speed of email messages and social media posts, combined with a lack of interpersonal cues (facial expressions, voice), can cause many a workplace writer to inadvertently create frustration and even legal problems (see previous section) due to an inappropriately worded message. Even with autocorrect and other tools, email can be fraught with spelling errors.

Cautions of writing in digital environments

In general, keep the tone and style of workplace email and social media posts brief, professional, and polite. See Chapter 14 for a thorough discussion of email in the workplace (pay special attention to Chapter 14, "Email Style and Tone" and to Figures 14.1, 14.2, and 14.3 for examples). See also Chapter 24 for related discussion on writing for blogs, wikis, and Web pages and Chapter 25 for information on writing for social media.

Professional style and tone are critical when using email and social media at work

The limits of spell check, autocorrect, and other editing apps

Many of the guidelines in this chapter are built-in features of today's digital writing applications. When you write an email message or text message, draft a document, or write a social media post, grammar and spell check programs (as well as word substitution programs like autocorrect) make suggestions or even change your content. Grammar checkers search for ambiguous pronoun references, overuse of passive voice, *to be* verbs, *There* and *It* sentence openers, negative constructions, clutter words, needless prefaces and qualifiers, overly technical language, jargon, sexist language, and so on. Autocorrect and spell check apps look for words judged to be incorrect and then suggest or insert replacements. But these digital editing tools can be extremely imprecise and should be used with caution.

Think critically when using digital editing tools

Autocorrect will not solve all grammar and spelling problems. For example, both *its* and *it's* are spelled correctly, but only one of them means "it is." The same is true for *their* and *there* (*their* is a possessive pronoun, as in "their books," while *there* is an adverb, as in "There is my dog"). Spell check is great for finding words that are spelled incorrectly, but don't count on it to find words that are *used* incorrectly or for typos that create the wrong word but are correctly spelled words on their own, such as *howl* instead of *how*. Autocorrect not only checks your spelling, it also recommends phrases or complete sentences, based on past experience. Grammar checkers work well to help you locate possible problems, but do not rely solely on what the program tells you. For example, not every sentence that the grammar checker flags as "long" should be shortened. Use these tools wisely and with common sense. In the end, style, tone, and other features should be your choice and should be based on your specific audience and purpose. Pay special attention to messages sent via email or text: slow down and be sure the suggested word is correct before pressing "send."

> NOTE *None of the rules offered in this chapter applies universally. Ultimately, your own sensitivity to meaning, emphasis, and tone—the human contact—will determine the effectiveness of your writing style.*

Checklist

Style

Use the following Checklist when editing your work or the work of others.

Clarity

❏ Does each pronoun clearly refer to the noun it replaces? (See "Avoid Ambiguous Pronoun References" in this chapter.)

❏ Is each modifier close enough to the word or words it defines or explains? (See "Avoid Ambiguous Modifiers" in this chapter.)

❏ Are modifying nouns unstacked? (See "Unstack Modifying Nouns" in this chapter.)

❏ Do most sentences begin with the familiar information and end with new information? (See "Arrange Word Order for Coherence and Emphasis" in this chapter.)

❏ Are sentences in active rather than passive voice, unless the agent is immaterial? (See "Use Active Voice Whenever Possible" and "Use Passive Voice Selectively" in this chapter.)

❏ Does each sentence provide only as much information as readers are able to process easily? (See "Avoid Overstuffed Sentences" in this chapter.)

Conciseness

❏ Is the piece free of wordiness, redundancy, or needless repetition? (See "Avoid Wordy Phrases," "Eliminate Redundancy" and "Avoid Needless Repetition" in this chapter.)

❏ Is it free of needless sentence openers and prefaces? (See "Avoid *There* Sentence Openers," "Avoid Some *It* Sentence Openers," and "Delete Needless Prefaces" in this chapter.)

❏ Have unnecessary weak verbs been converted to verbs that express a definite action? (See "Avoid Weak Verbs" in this chapter.)

❏ Have excessive prepositions been removed and nominalizations restored to their verb forms? (See "Avoid Excessive Prepositions" and "Avoid Nominalizations" in this chapter.)

❏ Have negative constructions been converted to positive ones, as needed? (See "Make Negatives Positive" in this chapter.)

❏ Is the piece free of clutter words and needless qualifiers? (See "Clean Out Clutter Words" and "Delete Needless Qualifiers" in this chapter.)

Fluency

❏ Are related ideas subordinated or coordinated and combined appropriately? (See "Combine Related Ideas" in this chapter.)

❏ Are sentences varied in construction and length? (See "Vary Sentence Construction and Length" in this chapter.)

❏ Are short sentences used for special emphasis? (See "Use Short Sentences for Special Emphasis" in this chapter.)

Word Choice

❏ Is the wording simple, familiar, unambiguous, and free of useless jargon? (See "Prefer Simple and Familiar Wording" and "Avoid Useless Jargon" in this chapter.)

❏ Is each acronym spelled out upon first use? (See "Use Acronyms Selectively" in this chapter.)

❏ Is the piece free of triteness, misleading euphemisms, and overstatement? (See "Avoid Triteness," "Avoid Misleading Euphemisms," and "Avoid Overstatement" in this chapter.)

❏ Does the wording precisely convey the intended meaning? (See "Avoid Imprecise Wording" in this chapter.)

❏ Are general or abstract terms clarified by more specific or concrete terms? (See "Be Specific and Concrete" in this chapter.)

❏ Are analogies used to clarify and explain? (See "Use Analogies to Sharpen the Image" in this chapter.)

❏ Have I reviewed the spelling, grammar, and word choice to be sure spell check or autocorrect didn't insert any errors? (See "Digital Writing and Editing" in this chapter.)

Tone

❏ Is the tone appropriate and consistent for the situation and audience? (See "Adjusting Your Tone" and "Guidelines for Deciding about Tone" in this chapter.)

❏ Is the level of formality what the intended audience would expect? (See "Adjusting Your Tone" in this chapter.)

❏ Is the piece free of implied bias, sexist language, or potentially offensive usage? (See "Avoid Personal Bias," "Avoid Biased Language," and "Guidelines for Avoiding Biased Language" in this chapter.)

❏ Does the piece display sensitivity to cultural differences? (See "Global, Legal, and Ethical Implications of Style and Tone" in this chapter.)

❏ Is the word choice ethically and legally acceptable? (See "Global, Legal, and Ethical Implications of Style and Tone" in this chapter.)

Projects

For all projects, check with your instructor about whether to present your findings in class, bring drafts to class for discussion, upload your project to the class learning management system (LMS), and/or use the LMS forum or discussion boards to collaborate and review each activity below.

General

Using "Checklist: Style", revise the following selections. (Hint: Use the brief example in this chapter's introduction as a model for revision.)

a. Letter to a local newspaper

In the absence of definitive studies regarding the optimum length of the school day, I can only state my personal opinion based upon observations made by me and upon teacher observations that have been conveyed to me. Considering the length of the present school day, it is my opinion that the day is excessive length-wise for most elementary pupils, certainly for almost all of the primary children.

To find the answer to the problem requires consideration of two ways in which the problem may be viewed. One way focuses upon the needs of the children, while the other focuses upon logistics, transportation, scheduling, and other limits imposed by the educational system. If it is necessary to prioritize these two ideas, it would seem most reasonable to give the first consideration to the primary reason for the very existence of the system, i.e., to meet the educational needs of the children the system is trying to serve.

b. Memo to employees

We are presently awaiting an on-site inspection of the designated professional library location by corporate representatives relative to electrical adaptations necessary for the computer installation. Meanwhile, all staff members are asked to respect the off-limits designation of the aforementioned location, as requested, due to the liability insurance provisions in regard to the computers.

Team

Use a document written for this course or a memo, letter, or email you received from work or school. In small groups, discuss areas where the style and tone could be improved based on concepts from this chapter. Each person should then draft a new version; share versions among the team and provide feedback to each other. Then, use this feedback to draft one new version that the team agrees is best. (Hint: Try using a collaborative writing app, such as Google docs or a writing app available in your class learning management system.)

Digital and Social Media

Go to a Web site, Facebook page, or the Twitter feed for an organization you have worked for or volunteered with. Or, find a site related to your major. Look for several recent posts and analyze the word choice, tone, and style based on items discussed in this chapter, especially passive and active voice, biased usage, and tone. If you were an employee of this organization, what tips would you offer to help improve the writing in a social media setting? Revise one of the posts to be clearer, more direct, and more useful for readers of social media.

Global

Search on the phrase "international business writing" to learn about the workplace tone and style preferences of one particular country. Then, in a one-page memo to your instructor and classmates, describe the style preferences of that country and give examples of how these preferences differ from the style guidelines presented (i.e., for North American English) in this chapter.

Chapter 12
Designing Visual Information

❝Every presentation, report, or specification I create is dependent on visuals. Among my fellow scientists, charts and graphs help condense large sets of data into patterns and trends that we can grasp quickly. And even when I write a document for lawyers or other non-scientists at the company, I use visuals to make complicated information easy for them to understand. Managers appreciate charts that illustrate the various pieces of the puzzle that go into researching a new product. Presentation and spreadsheet software make it easy to create professional looking visuals—but I always check the visual carefully before using it in my presentation or document because, as we all know, computers don't catch everything.**❞**

—Nanette Bauer, Research Scientist at a large biotechnology company

Learning Objectives

12.1 Explain why visuals are important in technical communication

12.2 Know when visuals are appropriate and how to select them

12.3 Create tables to display quantitative and/or qualitative information

12.4 Create graphs to translate numbers into shapes, shades, and patterns

12.5 Create charts to display quantitative and cause-and-effect relationships

12.6 Use graphic illustrations when pictures are more effective than words

12.7 Understand how to use color and present visuals appropriately

12.8 Identify ethical issues when using visuals

12.9 Understand how social media and culture affect your choice of visuals

Visuals make data easier to interpret and remember

Whether on their own or combined with text, visuals are powerful way to communicate complex topics to a wide range of readers. Because visuals focus and organize information, they make data easier to interpret and remember. By offering powerful new ways of looking at data, visuals also reveal meanings that might otherwise remain buried in lists of facts and figures. Visuals are especially powerful when used in social media settings, where people read quickly and may only remember the visual, not the detailed textual information.

Why Visuals Matter

12.1   Explain why visuals are important in technical communication

Readers want more than just raw information; they want this material shaped and enhanced so they can understand the message at a glance. Visuals help us answer questions posed by readers as they process information:

- Which information is most important?
- Where, exactly, should I focus?
- What do these numbers mean?
- What should I be thinking or doing?
- What should I remember about this?
- What does it look like?
- How is it organized?
- How is it done?
- How does it work?

Typical audience questions in processing information

When people look at a visual pattern, such as a graph, they see it as one large pattern—the Big Picture that conveys information quickly and efficiently. For instance, the following line graph has no verbal information. The axes are not labeled, nor is the topic identified. But one quick glance, without the help of any words or numbers, tells you that the trend, after a period of gradual rise, has risen sharply. The graph conveys information in a way plain text never could.

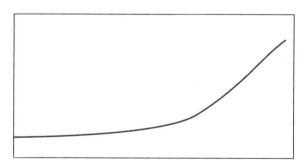

The trend depicted in the above graph would be hard for readers to visualize by just reading the long list of numbers in the following passage:

> The time required for global population to grow from 5 to 6 billion was shorter than the interval between any of the previous billions. It took just 12 years for this to occur, just slightly less than the 13 years between the fourth and fifth billion, but much less time than the 118 years between the first and second billion...

Technical data in prose form can be hard to interpret

When all this information is added to the original graph, as in Figure 12.1, the numbers become much easier to comprehend and compare.

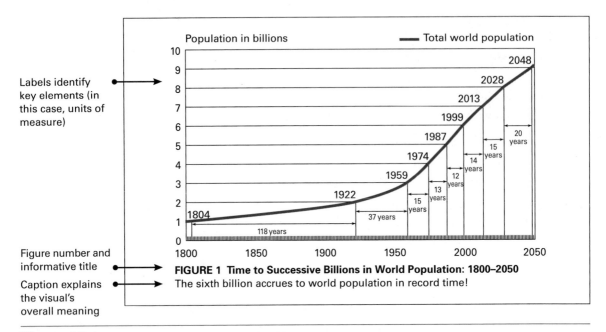

Labels identify key elements (in this case, units of measure)

Figure number and informative title

Caption explains the visual's overall meaning

FIGURE 1 Time to Successive Billions in World Population: 1800–2050
The sixth billion accrues to world population in record time!

Figure 12.1 **A Graph that Conveys the Big Picture**

Source: United Nations (1995b); U.S. Census Bureau, International Programs Center, International Database and Unpublished Tables.

> NOTE *Visuals enhance—but should not replace—essential discussion in your written text. In your document refer to the visual by number ("see Figure 1") and explain what to look for and what it means. For more on introducing and interpreting visuals in a document, see "Guidelines for Presenting Visuals" in this chapter.*

When to Use Visuals and How to Choose the Right Ones

Get to know then when and how of visuals first

12.2 Know when visuals are appropriate and how to select them

Deciding on whether to use a visual and then choosing and creating the right type of visual for the situation requires careful thought about the variety of visuals available and what's best for your audience and purpose. For this reason, you need to spend time considering when to use visuals and what types of visuals are best suited to your situation.

When to Use Visuals

Use visuals in situations like these

In general, you should use visuals whenever they can make your point more clearly than text or when they can enhance your text. Use visuals to clarify and support your discussion, not just to decorate your document. Use visuals to direct the audience's focus or help them remember something. There may be organizational reasons for using visuals; for example, some companies may always expect a chart or graph as part of their annual report. Certain industries, such as the financial sector, routinely use graphs and charts (such as the graph of the daily Dow Jones Industrial Average).

Types of Visuals to Consider

This section offers an overview of the most common types of visual displays, sorted into four categories: tables, graphs, charts, and graphical illustrations. The rest of this chapter provides more details and examples of each. Note how each type of visual offers a different perspective for helping readers understand and process information.

TABLES display data across columns and rows for easy comparison.

Numeric tables
Compare
exact values

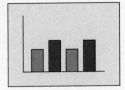

Prose tables
Organize textual
information

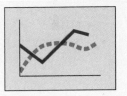

GRAPHS translate numbers into shapes, shades, and patterns.

Bar graphs
Show comparisons

Line graphs
Show trends
over time

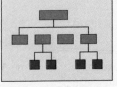

CHARTS depict relationships via arrows, lines, and other design elements.

Pie charts
Relate parts or
percentages to
the whole

**Organizational
charts**
Show the hierarchy
and structure of an
organization

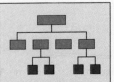

Flowcharts
Trace the steps or
decisions in a
procedure or
process

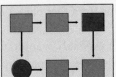

Tree charts
Show how parts of an
idea or concept are
related

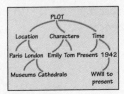

**Gantt and PERT
charts**
Depict how the
phases of a project
relate to each other

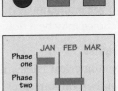

Pictograms
Use icons and symbols
to represent the
displayed items

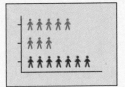

continued

GRAPHIC ILLUSTRATIONS rely mainly on pictures and images.

Illustrations
Present realistic but simplified views

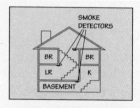

Diagrams
Show what's inside a device, how an item is assembled, or the conceptual elements of a process or system

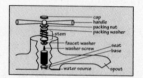

Maps
Visualize position, location, and other geographic features

Photographs
Show exactly what items look like

Symbols and icons
Represent concepts in forms that resemble an item or idea

Infographics
Tell a story or explain an idea visually through icons, graphics, and text

U.S. Dept. of Health & Human Services, Centers for Disease Control and Prevention

YuryZap/Shutterstock

Choosing the Right Visual

To select the most effective display, answer these questions:

Questions about a visual's purpose and audience

- **What is the purpose for using this visual?**
  - To convey facts and figures alone, a table may be the best choice. But if I want my audience to draw conclusions from that data, I may use a graph or chart to show comparisons.
  - To show parts of a mechanism, I probably want to use an exploded or cutaway diagram, perhaps together with a labeled photograph.
  - To give directions, I may want to use a diagram.
  - To show relationships, my best choice may be a flowchart or graph.

- **Who is my audience for these visuals?**
  - Expert audiences tend to prefer numerical tables, flowcharts, schematics, and complex graphs or diagrams that they can interpret for themselves.
  - General audiences tend to prefer basic tables, graphs, diagrams, and other visuals that direct their focus and interpret key points extracted from the data.
  - Cultural differences might come into play in the selection of appropriate visuals.

NOTE *Although visual communication has global appeal, certain displays might be inappropriate in certain cultures. For more on cultural considerations in selecting visuals, see "Social Media and Cultural Considerations" in this chapter.*

- **What form of information will best achieve my purpose for this audience?**
  - Is my message best conveyed by numbers, shapes, words, pictures, or symbols?
  - Will my audience most readily understand a particular type of display?
  - Where will this visual be displayed (on social media; on a Web site; in print; in a combination of these formats)?

Although several alternatives might work, one particular type of visual (or a combination) usually is superior. The best option, however, may not be available. Your audience or organization may express its own preferences, or choices may be limited by lack of equipment (software, scanners, digitizers), insufficient personnel (graphic designers, technical illustrators), or budget. Or, the visual you have in mind may not be suitable for the format (for example, a visual with lots of small fonts and detail may work in print but not as a social media post). Regardless of the limitations, your basic task is to enable the audience to interpret the visual correctly. *[margin: Not all visual options may be available to you]*

The many kinds of visuals you can use in your documents are described throughout this chapter. Regardless of type, certain requirements apply to all visuals. These requirements include the following:

- Using a title and number for each visual *[margin: Features required in any visual]*
- Keeping the design of the visual clean and easy to read
- Labeling all parts of the visual and providing legends as needed
- Placing the visual near the text it is helping to describe
- Citing the sources of your visual material (both the source of the data and, when appropriate, the source of the actual visual—for instance, the creator of the bar chart or the person who took the photograph)

See the Guidelines boxes throughout this chapter for more information about using specific types of visuals, and use the Planning Sheet for Preparing Visuals (Figure 12.2) as you consider each visual.

Using Software to Create Visuals

Whichever visuals you choose, some of the visuals discussed in this chapter can be created using software you probably already use quite frequently. For instance, spreadsheet software, (Microsoft Excel; Numbers for Mac; Google Sheets) can generate a variety of tables, graphs, and charts based on the data in the spreadsheet. Presentation and word processing apps (Microsoft PowerPoint or Word; Apple Keynote or Pages; Google Docs or Slides) contain basic drawing tools that allow you to annotate or draw simple figures. *[margin: Basic applications for creating visuals]*

Focusing on Your Purpose

- What is this visual's purpose (to instruct, persuade, create interest)? _____

- What forms of information (numbers, shapes, words, pictures, symbols) will this visual depict? _____

- What kind of relationship(s) will the visual depict (comparison, cause-effect, connected parts, sequence of steps)? _____

- What judgment, conclusion, or interpretation is being emphasized (that profits have increased, that toxic levels are rising, that X is better than Y)? _____

- Is a visual needed at all? _____

Focusing on Your Audience

- Is this audience accustomed to interpreting visuals? _____

- Is the audience interested in specific numbers or an overall view? _____

- Should the audience focus on one exact value, compare two or more values, or synthesize a range of approximate values? _____

- Which type of visual will be most accurate, representative, accessible, and compatible with the type of judgment, action, or understanding expected from the audience? _____

- In place of one complicated visual, would two or more straightforward ones be preferable? _____

- Are there any specific cultural considerations? _____

Focusing on Your Presentation

- What enhancements, if any, will increase audience interest (colors, patterns, legends, labels, varied typefaces, shadowing, enlargement or reduction of some features)? _____

- How will you present this material: as hard copy? In PDF? As a poster? As part of an instruction manual, proposal, or report? As an infographic or social media post? _____

- For greatest utility and effect, where in the presentation does this visual belong? _____

Figure 12.2   **Planning Sheet for Preparing Visuals**

For more sophisticated visuals, apps like Adobe Illustrator or CorelDRAW allow you to sketch, edit, and refine diagrams and drawings. High-end drawings are usually produced using computer-aided design (CAD) tools. Photos can be highlighted and refined using Adobe Photoshop or other photo editing software. But unlike the spreadsheet or word processing programs that you use every day, these tools require a good deal of practice and skill and can't be learned overnight. You might need to work with the graphic design department at your organization. Or, if you want to learn more, there are plenty of online tutorials available, including at the Lynda.com site.

Advanced applications for creating visuals

NOTE See if your college or university provides students with access to online tutorials or workshops on some of the programs mentioned above.

Tables

12.3 Create tables to display quantitative and/or qualitative information

A table is a powerful way to display dense textual information such as specifications or comparisons. Numerical tables such as Table 12.1 present *quantitative information* (data that can be measured). Prose tables present *qualitative information* (prose descriptions, explanations, or instructions). Table 12.2 combines numerical data, probability estimates, comparisons, and instructions.

Purpose of tables

NOTE Including a caption with your visual enables you to analyze or interpret the trends or key points you want readers to recognize (as in Table 12.1).

Years of Potential Life Lost Before Age 75				
(Per 100,000 population under age 75)				
	Heart Disease		Unintentional Injuries	
Year	Male	Female	Male	Female
1990	2356.0	948.5	1162.1	607.4
2000	1766.0	774.6	1026.5	573.2
2005	1559.0	680.2	1137.2	647.9
2009	1399.2	606.5	1028.2	604.6
2010	1370.8	593.6	1025.2	616.4
% change, 1990–2010	−41.8	−37.4	−11.8	+1.5

Title explains the table's purpose

Each column has a clear heading

Numbers are aligned properly for ease of reading

Where helpful, data are tallied

This table compares years of potential life lost before age 75 due to heart disease and unintentional injury in males and females. In both males and females, years lost due to heart disease decreased between 1990 and 2010. Years lost due to unintentional injury decreased only slightly in males and increased very slightly in females.

A caption explains the numerical relationships

TABLE 12.1   A Numerical Table
Source: Health, United States 2012, from the U.S. National Center for Health Statistics.

Column headings lead into the information

Phrases are brief and aligned for ease of reading

Numbers enhance the verbal information

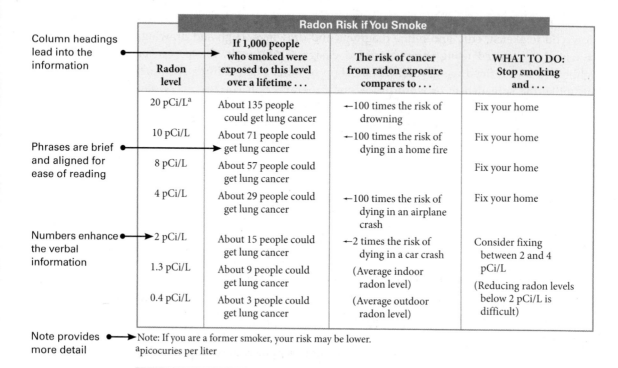

Radon Risk if You Smoke			
Radon level	**If 1,000 people who smoked were exposed to this level over a lifetime . . .**	**The risk of cancer from radon exposure compares to . . .**	**WHAT TO DO: Stop smoking and . . .**
20 pCi/L[a]	About 135 people could get lung cancer	←100 times the risk of drowning	Fix your home
10 pCi/L	About 71 people could get lung cancer	←100 times the risk of dying in a home fire	Fix your home
8 pCi/L	About 57 people could get lung cancer		Fix your home
4 pCi/L	About 29 people could get lung cancer	←100 times the risk of dying in an airplane crash	Fix your home
2 pCi/L	About 15 people could get lung cancer	←2 times the risk of dying in a car crash	Consider fixing between 2 and 4 pCi/L
1.3 pCi/L	About 9 people could get lung cancer	(Average indoor radon level)	(Reducing radon levels below 2 pCi/L is difficult)
0.4 pCi/L	About 3 people could get lung cancer	(Average outdoor radon level)	

Note provides more detail

Note: If you are a former smoker, your risk may be lower.
[a]picocuries per liter

TABLE 12.2 A Prose Table Displays numerical and verbal information.
Source: Home Buyer's and Seller's Guide to Radon, by the U.S. Environmental Protection Agency.

Audience and purpose of tables

    No table should be overly complex for its audience. Table 12.3, designed for expert readers, is hard to interpret for people who aren't specialists in this area because it presents too much information at once. For laypersons, use fewer tables and keep them as simple as possible. All visuals should be created with audience and purpose in mind. For instance, an accountant doing an audit might need a table listing exact amounts, whereas the average public stockholder reading an annual report would prefer the "big picture" in an easily grasped bar graph or pie chart (Van Pelt 1). Similarly, scientists might find the complexity of data shown in Table 12.3 perfectly appropriate, but a nonexpert audience (say, environmental groups) might prefer the clarity and simplicity of a chart.

    Tables work well for displaying exact values, but often graphs or charts are easier to interpret. Geometric shapes (bars, curves, circles) are generally easier to remember than lists of numbers (Cochran et al. 25).

NOTE Any visual other than a table is usually categorized as a figure, and so titled ("Figure 1 Aerial View of the Panhandle Mine Site").

Toxic Chemical Releases by Industry: 2010						
[In millions of pounds (4,438.7 represents 4,438,700,000), except as indicated.]						
Industry	2010 SIC[1] code	Total on- and off-site releases	On-site release			Off-site releases/ transfers to disposal
			Total[2]	Point source air emissions	Surface water discharges	
Total[3]	(X)	4,438.7	3,920.7	1,381.3	222.6	518.0
Metal mining	10	1,245.7	1,244.7	1.8	0.7	1.0
Coal mining	12	12.9	12.9	0.1	0.2	-
Food and kindred products	20	153.2	145.8	35.1	83.1	7.3
Tobacco products	21	3.2	2.8	2.4	0.1	0.4
Textile mill products	22	7.4	6.5	4.8	0.3	0.9
Apparel and other textile products	23	0.7	0.5	0.4	-	0.2
Lumber and wood products	24	33.0	31.0	27.0	0.1	2.0
Furniture and fixtures	25	6.2	6.1	5.4	0.0	0.1
Paper and allied products	26	215.0	209.6	146.2	18.7	5.3
Printing and publishing	27	15.0	14.7	7.4	-	0.3
Chemical and allied products	28	544.7	500.3	168.6	44.5	44.4
Petroleum and coal products	29	75.0	71.9	34.6	17.1	3.1
Rubber and misc. plastic products	30	75.3	65.8	51.3	0.1	9.5
Leather and leather products	31	2.1	1.0	0.7	0.0	1.1
Stone, clay, glass products	32	51.2	45.8	38.1	2.1	5.5
Primary metal industries	33	477.5	198.1	35.9	39.4	279.4
Fabricated metals products	34	58.6	38.8	23.7	2.3	19.8
Industrial machinery and equipment	35	14.3	10.7	4.1	0.2	3.6
Electronic, electric equipment	36	20.3	13.8	6.5	3.6	6.4
Transportation equipment	37	74.8	63.5	51.1	0.2	11.2
Instruments and related products	38	8.7	7.9	5.1	1.0	0.8
Miscellaneous	39	7.1	4.9	3.9	0.1	2.2

Can cause information overload for nontechnical audiences

X Not applicable. [1]Standard Industrial Classification, see text, Section 12. Labor Force. [2]Includes on-site disposal to underground injection for Class I wells, Class II to V wells, other surface impoundments, land releases, and other releases, not shown separately. [3]Includes industries with no specific industry identified, not shown separately.

TABLE 12.3 A Complex Table Causes Information Overload This table is too complex for nonexpert readers but suitable for those who have training and expertise in chemical and environmental engineering or related fields.
Source: Annual Toxics Release Inventory, Environmental Protection Agency 2011.

How to Construct a Table

1. Number the table in its order of appearance and provide a title that describes exactly what is being measured.
2. Label stub, column, and row heads (*Number of Awards; 2005; Pell Grant*) to orient readers.
3. Specify units of measurement or use familiar symbols and abbreviations (*$, hr.*). Define specialized symbols or abbreviations ($\mathring{A}$ = *angstrom, db = decibel*) in a footnote.

TABLE 14.4 ■ Federal Student Financial Assistance: 2005–2011

STUB HEAD Number of Awards (1000)[a]	2005[b]	2008	2009	2010	2011[b]
Total	19,898	23,759	29,867	31,982	34,391
Pell Grant	5,167	6,157	8,094	8,873	9,413
TEACH Grant	(X)	8	31	36	44
Work-Study	710	678	733	713	713
Perkins Loan	727	488	441	493	493
Direct Student Loan	2,971	3,730	6,109	16,647	23,728
Family Educ. Loan	10,323	12,698	14,459	5,220	(X)

[a]As of June 30. [b]Estimate. (X) Not available.

Source: U.S. Census Bureau. *Statistical Abstract of the United States: 2012*. 131st ed.
Washington, DC: U.S. Census Bureau, 2011: 186. <http://www.census.gov/compendia/statab>.

4. Compare data vertically (in columns) instead of horizontally (rows). Columns are easier to compare. Try to include row or column averages or totals, as reference points for comparing individual values.

5. Use horizontal rules to separate headings from data. In a complex table, use vertical rules to separate columns. In a simple table, use as few rules as clarity allows.

6. List items in a logical order (alphabetical, chronological, decreasing cost). Space listed items for easy comparison. Keep prose entries as brief as clarity allows.

7. Convert fractions to decimals. Align decimals and all numbers vertically. Keep decimal places for all numbers equal. Round insignificant decimals to whole numbers.

8. Use *x*, *NA*, or a dash to signify any omitted entry, and explain the omission in a footnote (*Not available, Not applicable*).

9. Use footnotes to explain entries, abbreviations, or omissions. Label footnotes with lowercase letters so readers do not confuse the notation with the numerical data.

10. Cite data sources beneath any footnotes. When adapting or reproducing a copyrighted table for a work to be published, obtain written permission.

11. If the table is too wide for the page, turn it 90 degrees with the left side facing page bottom. Or use two tables.

12. If the table exceeds one page, write "continues" at the bottom and begin the next page with the full title, "continued," and the original column headings.

Graphs

12.4 Create graphs to translate numbers into shapes, shades, and patterns

Purpose of graphs

Graphs translate numbers into shapes, shades, and patterns. Graphs display, at a glance, the approximate values, the point being made about those values, and the relationship being emphasized. Graphs are especially useful for depicting comparisons, changes over time, patterns, or trends.

A graph's horizontal axis shows categories (the independent variables) to be compared, such as years within a period (1990, 2000, 2010). The vertical axis shows the range of values (the dependent variables) for comparing the categories, such as the number of deaths from heart failure in a given year. A dependent variable changes according to activity in the independent variable (say, a decrease in quantity over a set time, as in Figure 12.3).

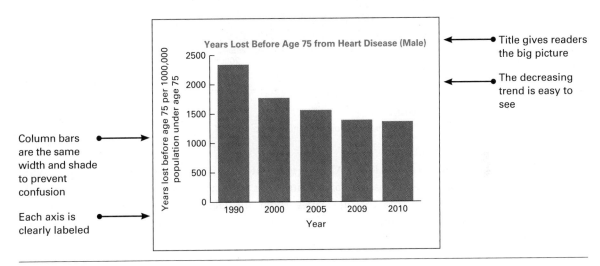

Figure 12.3 **A Simple Bar Graph**
Source: Health, United States 2012, from the U.S. National Center for Health Statistics.

Bar Graphs

Generally easy to understand, bar graphs show discrete comparisons, such as year-by-year or month-by-month. Each bar represents a specific quantity. You can use bar graphs to focus on one value or to compare values over time.

Bar graphs show comparisons

SIMPLE BAR GRAPH.   A simple bar graph displays one trend or theme. The graph in Figure 12.3 shows one trend extracted from Table 12.1, years lost before age 75 from heart disease (males). If the audience needs exact numbers, you can record exact values above each bar.

MULTIPLE-BAR GRAPH.   Bar graphs can display several relationships at the same time. Figure 12.4 contrasts two data sets to show comparative trends. Use a different pattern or color for each data set, and include a key so readers will know which color or pattern goes with which set. In general, don't include more than three data sets on one graph.

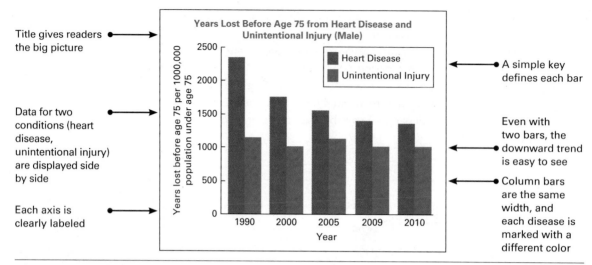

Title gives readers the big picture

Data for two conditions (heart disease, unintentional injury) are displayed side by side

Each axis is clearly labeled

A simple key defines each bar

Even with two bars, the downward trend is easy to see

Column bars are the same width, and each disease is marked with a different color

Figure 12.4 **A Multiple-Bar Graph**

Source: *Health, United States 2012*, from the U.S. National Center for Health Statistics.

HORIZONTAL-BAR GRAPH. Horizontal-bar graphs are good for displaying a large series of bars arranged in order of increasing or decreasing value, as in Figure 12.5. This format leaves room for labeling the categories horizontally (*Doctorate,* and so on).

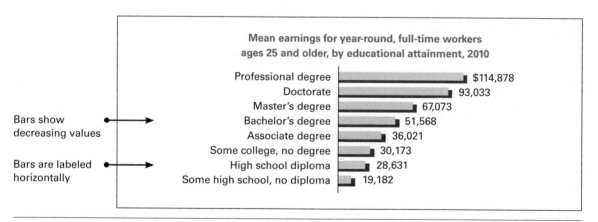

Bars show decreasing values

Bars are labeled horizontally

Figure 12.5 **A Horizontal-Bar Graph** Accommodates lengthy labels.

Source: Bureau of Labor Statistics.

STACKED-BAR GRAPH. Instead of displaying bars side-by-side, you can stack them. Stacked-bar graphs show how much each data set contributes to the whole. Figure 12.6 displays other comparisons from Table 12.1. To avoid confusion, don't display more than four or five sets of data in a single bar.

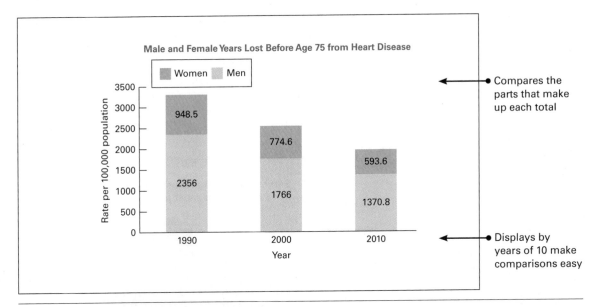

Figure 12.6   **A Stacked-Bar Graph** Displays of 10 make comparisons easy.

100 PERCENT BAR GRAPH.   This type of bar graph shows the value of each part that makes up 100 percent (see Figure 12.7). The more data, the harder such graphs are to interpret; consider using a pie chart (see "Pie Charts" in this chapter) instead.

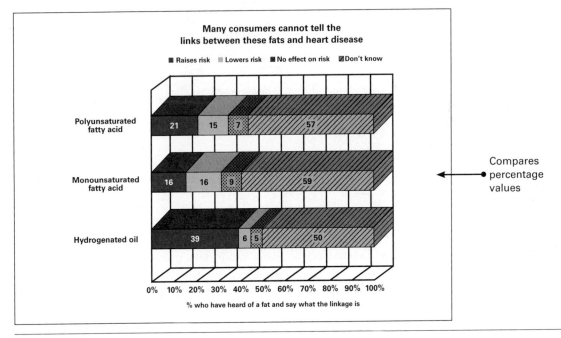

Figure 12.7   **A 100 Percent Bar Graph**
Source: Center for Food Safety and Applied Nutrition, U.S. Food and Drug Administration.

3-D BAR GRAPH. Graphics software makes it easy to shade and rotate images for a three-dimensional view. The 3-D perspective in Figure 12.8 engages our attention and visually emphasizes the data.

NOTE Although 3-D graphs can enhance and dramatize a presentation, an overly complex graph can be misleading or hard to interpret. Use 3-D only when a two-dimensional version will not serve as well. Never sacrifice clarity and simplicity for the sake of visual effect.

3-D approach allows data to be displayed by different types (service-providing and goods-providing)

Units are placed on top of each bar for easy comparison

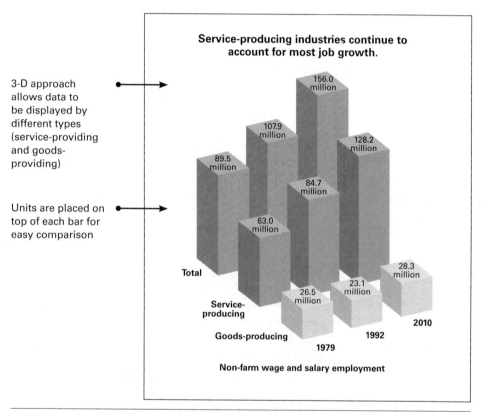

Figure 12.8 **3-D Bar Graph** Adding a third axis creates the appearance of depth.
Source: Bureau of Labor Statistics.

Line Graphs

Line graphs show trends over time

A line graph can accommodate many more data points than a bar graph (for example, a twelve-month trend, measured monthly). Line graphs help readers synthesize large bodies of information in which exact quantities don't need to be emphasized.

SIMPLE LINE GRAPH. A simple line graph, as in Figure 12.9, plots time intervals (or categories) on the horizontal scale and values on the vertical scale.

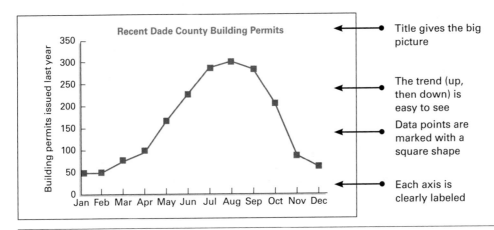

Figure 12.9   **A Simple Line Graph** Displays one relationship.

MULTILINE GRAPH.   A multiline graph displays several relationships simultaneously, as in Figure 12.10. Include a caption to explain the relationships readers are supposed to see and the interpretations they are supposed to make.

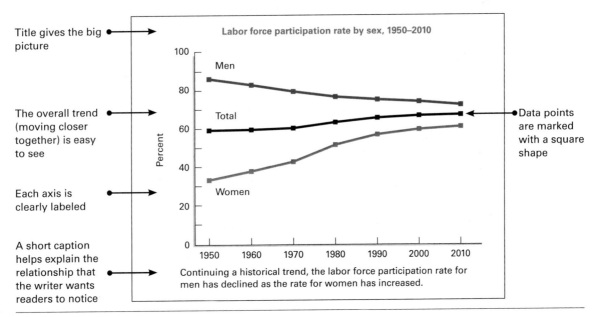

Figure 12.10   **A Multiline Graph** Displays multiple relationships.
Source: Bureau of Labor Statistics.

DEVIATION LINE GRAPH.   Extend your vertical scale below the zero baseline to display positive and negative values in one graph, as in Figure 12.11. Mark values below the baseline in intervals parallel to those above it.

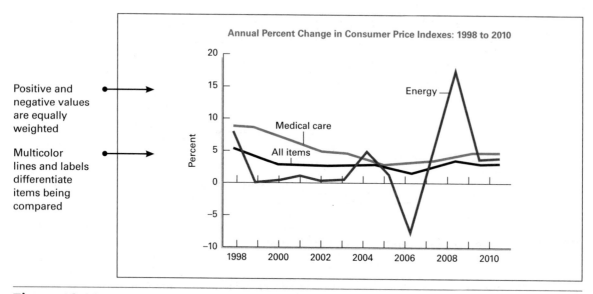

Positive and negative values are equally weighted

Multicolor lines and labels differentiate items being compared

Figure 12.11 **A Deviation Line Graph** Displays negative and positive values.

Source: Chart prepared by U.S. Bureau of the Census.

BAND OR AREA GRAPH. By shading in the area beneath the main plot lines, you can highlight specific information. Figure 12.12 is another version of the Figure 12.9 line graph.

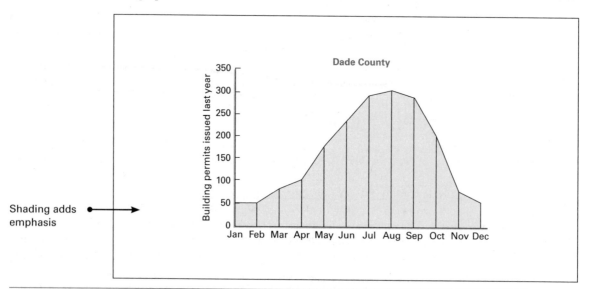

Shading adds emphasis

Figure 12.12 **A Simple Band Graph** Uses shading to highlight information.

MULTIPLE-BAND GRAPH. The multiple bands in Figure 12.13 depict relationships among sums instead of the direct comparisons depicted in the Figure 12.10 multiline graph. Despite their visual appeal, multiple-band graphs are easy to misinterpret: In a multiline graph, each line depicts its own distance from the zero baseline. But in a multiple-band graph, the very top line depicts the *total* distance from the zero baseline, with each band below it being a part of that total. Always clarify these relationships for your audience.

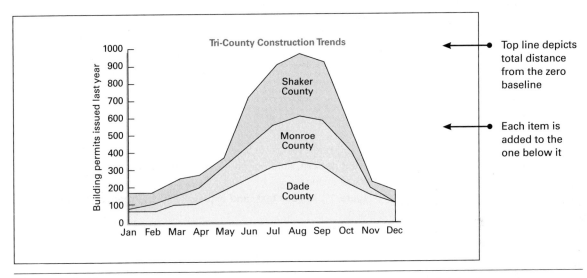

Figure 12.13   A Multiple-Band Graph Depicts relationships among sums instead of direct comparisons.

Guidelines
for Creating Tables and Graphs

For all types of tables and graphs, provide a clear title and credit your sources.

Tables
> **Don't include too much information in a single table**. Overly complex tables are confusing. Limit your table to two or three areas of comparison. Or use multiple tables.
> **Provide a brief but descriptive title**. Announce exactly what is being compared.
> **Label the rows and columns**.
> **Line up data and information clearly**. Use neat columns and rows and plenty of white space between items.
> **Keep qualitative information and quantitative data brief**. When including high numbers (more than three digits), abbreviate the numbers and indicate "in thousands," "in millions," and so on. When using text in a table, limit the number of words.
> **Provide additional information, if necessary**. Add footnotes or a caption at the bottom of the table to explain anything readers may not understand at first glance.

Bar Graphs
> **Use a bar graph only to compare values that are noticeably different**. Small value differences will yield bars that look too similar to compare.

➤ **Keep the graph simple and easy-to-read**. Don't plot more than three types of bars in each cluster. Avoid needless visual details.

➤ **Number your scales in units familiar to the audience**. Units of 1 or multiples of 2, 5, or 10 are best.

➤ **Label both scales to show what is being measured or compared**. If space allows, keep all labels horizontal for easier reading.

➤ **Use tick marks to show the points of division on your scale**. If the graph has many bars, extend the tick marks into *grid lines* to help readers relate bars to values.

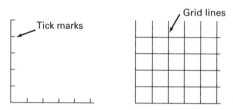

➤ **Make all bars the same width**. (unless you are overlapping them).

➤ **In a multiple-bar graph, use a different pattern, color, or shade for each bar in a cluster**. Provide a key, or legend, identifying each pattern, color, or shade.

➤ **Refer to the graph by number ("Figure 1") in your text, and explain what the reader should look for.** Or include a prose caption with the graph.

Line Graphs

Follow the guidelines above for bar graphs, with these additions:

➤ **Display no more than three or four lines on one graph.**

➤ **Mark each individual data point used in plotting each line.**

➤ **Make each line visually distinct (using color, symbols, and so on).**

➤ **Label each line so readers know what the given line represents.**

➤ **Avoid grid lines that readers could mistake for plotted lines.**

Charts

12.5 Create charts to display quantitative and cause-and-effect relationships

Purpose of charts The terms *chart* and *graph* often are used interchangeably. Technically, a chart displays relationships (quantitative or cause-and-effect) that are *not* plotted on a coordinate system (*x* and *y* axes).

Pie Charts

Pie charts relate parts or percentages to the whole Easy for most people to understand, a pie chart displays the relationship of parts or percentages to the whole. Readers can compare the parts to each other as well as to the whole (to show how much was spent on what, how much income comes from which sources, and so on). Figure 12.14 shows a simple pie chart. Figure 12.15 is an exploded pie chart. Exploded pie charts highlight various pieces of the pie.

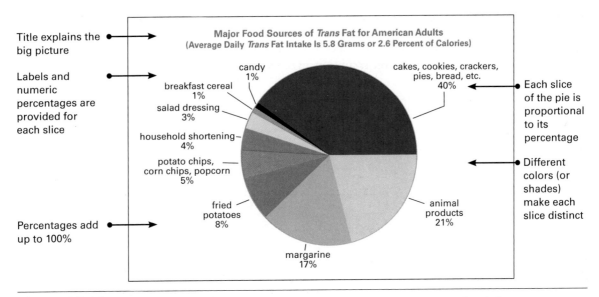

Figure 12.14   A Simple Pie Chart Shows the relationships of parts or percentages to the whole.
Source: U.S. Food and Drug Administration.

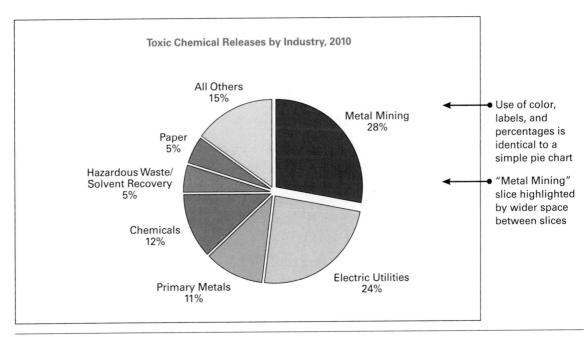

Figure 12.15   An Exploded Pie Chart Highlights various slices.
Source: U.S. Environmental Protection Agency. (See Table 12.3 for data.)

Organization Charts

Organization charts show the hierarchy and structure of an organization

An organization chart shows the hierarchy and relationships between different departments and other units in an organization, as in Figure 12.16.

Upper box represents the top of the hierarchy

Arrows show top-down relationships

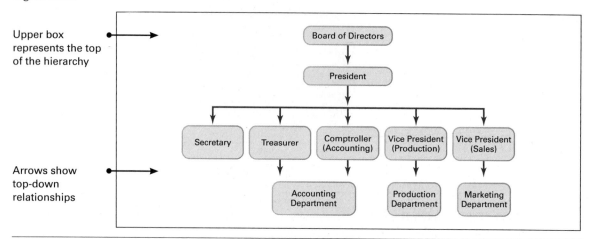

Figure 12.16   An Organization Chart Shows how different people or departments are ranked and related.

Flowcharts

Flowcharts trace the steps or decisions in a procedure or process

A flowchart traces a procedure or process from beginning to end. Figure 12.17 illustrates the procedure for helping an adult choking victim.

Tree Charts

Tree charts show how parts of an idea or concept are related

Whereas flowcharts display the steps in a process, tree charts show how the parts of an idea or concept are related. Figure 12.18 displays part of an outline for this chapter so that readers can better visualize relationships. The tree chart seems clearer and more interesting than the prose listing.

Gantt and PERT Charts

Gantt and PERT charts depict how the phases of a project relate to each other

Named for engineer H. L. Gantt (1861–1919), Gantt charts depict how the parts of a project relate. A series of bars or lines (time lines) indicates start-up and completion dates for each phase or task in a project. Gantt charts are useful for planning and tracking a project. The Gantt chart in Figure 12.19 illustrates the schedule for a manufacturing project. A PERT (Program Evaluation and Review Technique) chart uses shapes and arrows to outline a project's main activities and events (Figure 12.20). Both types of charts can be created with project management software such as Microsoft Project.

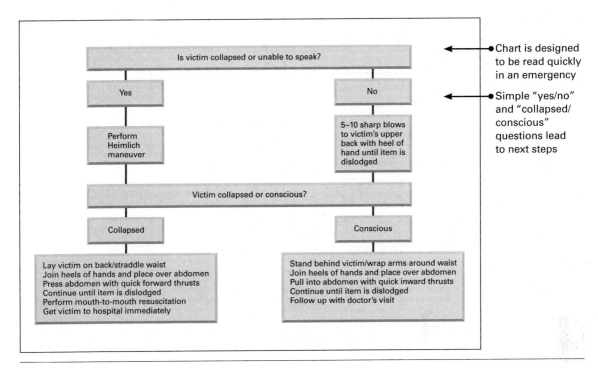

Figure 12.17 A Flowchart Depicts a sequence of events, activities, steps, or decisions.

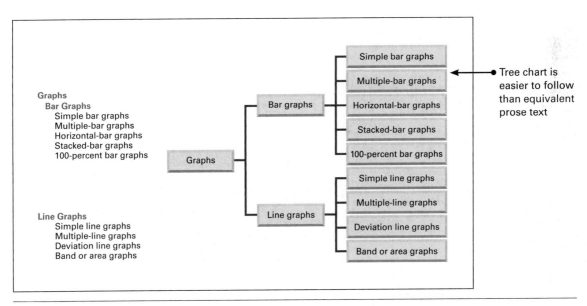

Figure 12.18 An Outline Converted to a Tree Chart Shows which items belong together and how they are connected.

Bars indicate
event dates and
their overlaps

Activity stages
are roughly
chronological

Numerical
data provide
further specific
information

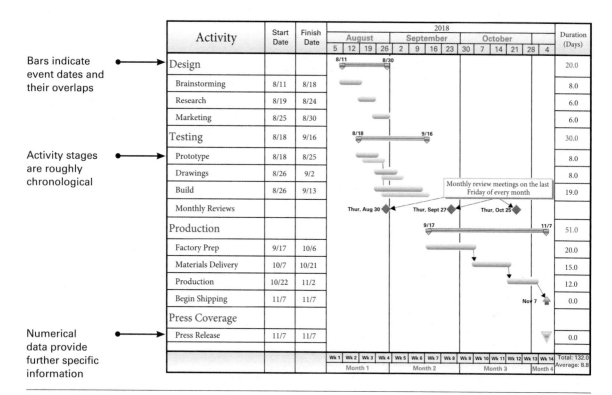

Figure 12.19 **A Gantt Chart** Depicts how the phases of a project interrelate.

Source: Chart created in *FastTrack Schedule*.™ Reprinted by permission from AEC Software.

Rectangles
indicate key
activities while
ovals represent
milestones

Heavy arrows
indicate the
critical path
(milestones to be
achieved) through
the project

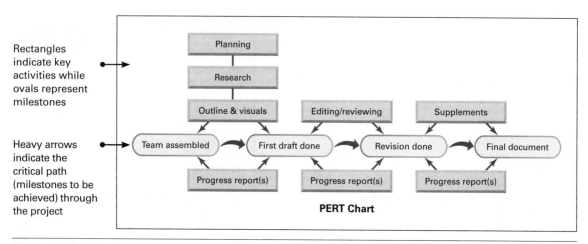

Figure 12.20 **A PERT Chart** This chart maps out the key activities and milestones ("Team assembled," "First draft done," and so on) for a major technical report to be produced by a collaborative team.

Pictograms

Pictograms are something of a cross between a line graph and a chart. Like line graphs, pictograms display numerical data, often by plotting it across x and y axes. But like a chart, pictograms use icons, symbols, or other graphic devices rather than simple lines or bars. In Figure 12.21, stick figures illustrate population changes during a given period. Pictograms are visually appealing and can be especially useful for nontechnical or multicultural audiences.

Pictograms use icons and symbols to represent the displayed items

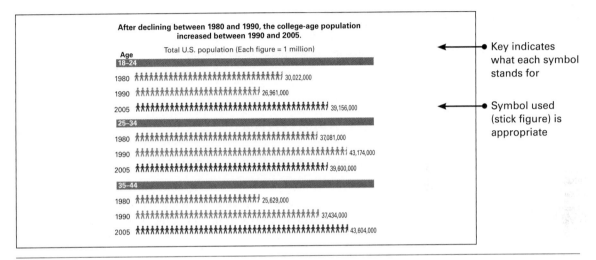

Figure 12.21   A Pictogram In place of lines and bars, icons and symbols lend appeal and clarity.
Source: U.S. Bureau of the Census.

Guidelines

for Creating Charts

Pie Charts

➤ **Make sure the parts of the pie add up to 100 percent.**

➤ **Differentiate and label each slice clearly.** Use different colors or shades for each slice, and label the category and percentage of each slice.

➤ **Keep all labels horizontal.** Make the chart easy to read.

➤ **Combine very small pie slices.** Group categories with very small percentages under "other."

Organization Charts

➤ **Move from top to bottom or left to right.** Place the highest level of hierarchy at the top (top-to-bottom chart) or at the left (left-to-right chart).

➤ **Use downward- or rightward-pointing arrows.** Arrows show the flow of hierarchy from highest to lowest.

➤ **Keep boxes uniform and text brief.** Shape may vary slightly according to how much text is in each box. Maintain a uniform look. Avoid too much text in any box.

Flowcharts, Tree Charts, and Gantt/PERT Charts

➤ **Move from top to bottom or left to right.** The process must start at the top (top-to-bottom chart) or left (left-to-right chart).

➤ **Use connector lines.** Show relationships between the parts.

➤ **Keep boxes uniform and text brief.** See the tips for organization charts above.

Pictograms

➤ **Follow the guidelines for bar graphs.** (See "Guidelines for Tables and Charts" in this chapter.)

➤ **Use symbols that are universally recognized.**

➤ **Keep the pictogram clean and simple (avoid too much visual clutter).**

Graphic Illustrations

12.6 Use graphic illustrations when pictures are more effective than words

Purpose of graphic illustrations

Illustrations can range from simple drawings to diagrams, photographs, maps, icons and symbols, infographics, or any other visual that relies mainly on pictures rather than on data or words. For example, the diagram of a safety-belt locking mechanism in Figure 12.22 accomplishes what the verbal text alone cannot: it portrays the mechanism in operation.

Verbal text that requires a visual supplement

The safety-belt apparatus includes a tiny pendulum attached to a lever, or locking mechanism. Upon sudden deceleration, the pendulum swings forward, activating the locking device to keep passengers from pitching into the dashboard.

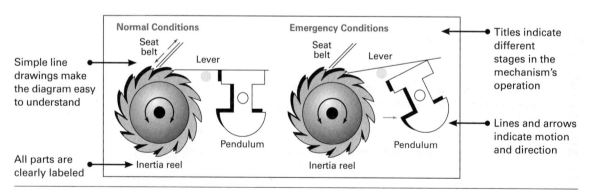

Figure 12.22 A Diagram of a Safety-Belt Locking Mechanism Shows how the basic parts work together.
Source: U.S. Department of Transportation.

Illustrations are invaluable when you need to convey spatial relationships or help your audience see what something actually looks like. Drawings can often illustrate more effectively than photographs because a drawing can simplify the view, omit unnecessary features, and focus on what is important.

Diagrams

Diagrams are especially effective for presenting views that could not be captured by photographing or actually observing the item.

Diagrams show how items function or are assembled

EXPLODED DIAGRAMS. Exploded diagrams show how parts of an item are assembled, as in Figure 12.23. These diagrams often appear in technical repair or maintenance manuals. This diagram could be annotated with numbers, or a caption, or other information to explain the relationship of parts to the whole.

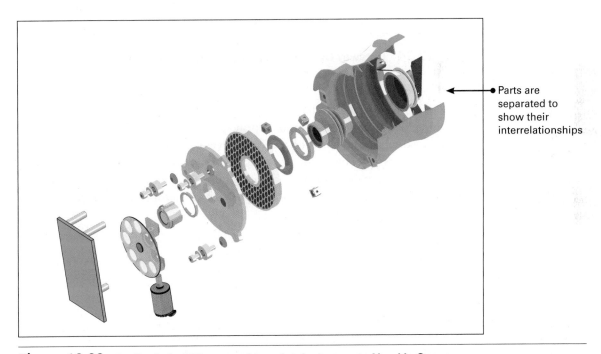

Parts are separated to show their interrelationships

Figure 12.23 **An Exploded Diagram of Imaging Instruments Used in Space**
Source: NASA.

CUTAWAY DIAGRAMS. Cutaway diagrams show the item with its exterior layers removed to reveal interior sections, as in Figure 12.24. Unless the specific viewing perspective is immediately recognizable (as in Figure 12.24), name the angle of vision: "top view," "side view," and so on.

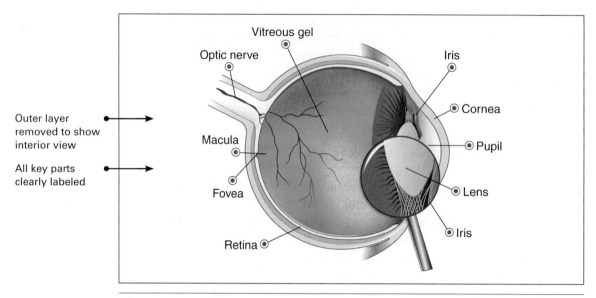

Outer layer
removed to show
interior view

All key parts
clearly labeled

Figure 12.24 **Cutaway Diagram of an Eye** Shows what is inside.
Source: Courtesy of National Eye Institute, National Institutes of Health (NEI/NIH).

BLOCK DIAGRAMS. Block diagrams are simplified sketches that represent the relationship between the parts of an item, principle, system, or process. Because block diagrams are designed to illustrate *concepts* (such as current flow in a circuit), the parts are represented as symbols or shapes. The block diagram in Figure 12.25 illustrates how any process can be controlled automatically through a feedback mechanism. Figure 12.26 shows the feedback concept applied as the cruise-control mechanism on a motor vehicle.

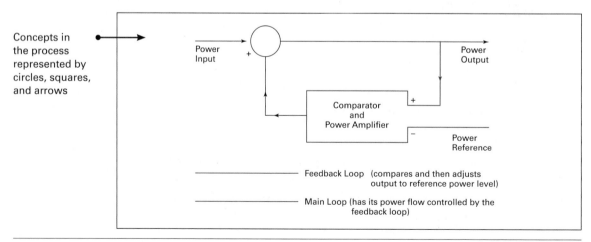

Concepts in
the process
represented by
circles, squares,
and arrows

Figure 12.25 **A Block Diagram Illustrating the Concept of Feedback**

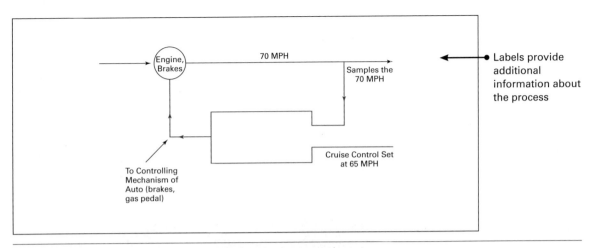

Figure 12.26   A Block Diagram Illustrating a Cruise-Control Mechanism Depicts a specific application of the feedback concept.

Specialized diagrams generally require the services of graphic artists or technical illustrators. The client requesting or commissioning the visual provides the art professional with an *art brief* (often prepared by writers and editors) that spells out the visual's purpose and specifications. The art brief is usually reinforced by a *thumbnail sketch,* a small, simple sketch of the visual being requested. For example, part of the brief addressed to the medical illustrator for Figure 12.24 might read as follows:

- **Purpose:** to illustrate for laypersons major parts of internal anatomy of the eye
- **View:** full cutaway, sagittal
- **Range:** entire eyeball at roughly 500 percent scale
- **Depth:** medial cross-section
- **Structures omitted:** retinal blood vessels, sclera, ciliary and lateral rectus muscles, and other accessory structures
- **Structures included:** gross anatomy of eyeball—delineated by color, shape, shading, and texture, each connected with peripheral labels by roughly 1.5-point leader lines
- **Structures highlighted:** iris, cornea, pupil, lens, vitreous gel, retina, fovea, macula, and posterior junction with the optic nerve

An art brief for Figure 12.24

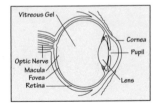

A thumbnail sketch of Figure 12.24

Photographs

Photographs are especially useful for showing, in a realistic manner, exactly how something looks. Unlike a diagram or line drawing, which highlights certain parts of an item, photographs can show the entire device or even parts of it but with much more detail. Photographs are useful, but if the photo is not taken with planning and precision, it can provide too much detail or fail to emphasize the important parts of the device.

Photos taken from your cell phone can be useful for capturing the moment or helping remind you of important details. Cell phone cameras are increasing in their

Photographs show exactly what things look like

capabilities, and for some documents, you might be able to take a very good photo just using your phone. But for the most effective workplace photographs, use a professional photographer who knows all about angles, lighting, lenses, and special film or digital editing options.

Deciding between a photo or a diagram

Figures 12.27 and 12.28 show the difference between a photo and a diagram. The photo helps your audience see the entire mechanism in a realistic manner. Although you can also insert labels into a photo (using Adobe Photoshop or other apps), a diagram is more effective in terms of directing reader attention to specific parts. For a user manual about lasers that you might be asked to create, you might consider using the photograph on the cover (to show what the device looks like) and the diagram in the first section (to illustrate the major parts and functioning of the laser).

Figure 12.27 **Shows a Complex Mechanism** Free tunable laser.
Source: YuryZap/Shutterstock.

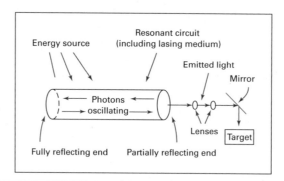

Figure 12.28 **A Simplified Diagram of Figure 12.30** Major parts of the laser.

Maps

Maps visualize position, location, and other geographic features

Besides being visually engaging, maps are especially useful for showing comparisons and for helping readers *visualize* position, location, and relationships among various data. Figure 12.29 synthesizes statistical information in a format that is accessible and understandable. Color enhances the comparisons.

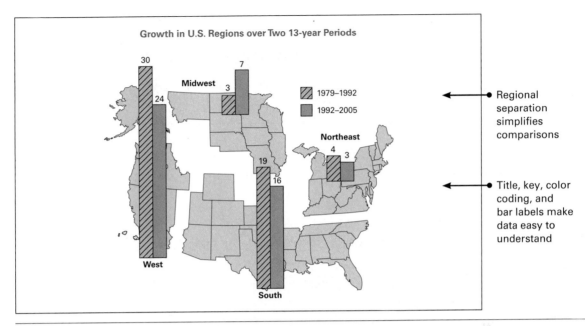

Figure 12.29 **A Map Rich in Statistical Significance** Shows the geographic distribution of data.
Source: U.S. Bureau of the Census.

Symbols and Icons

Symbols and icons can convey information visually to a wide range of audiences. Because such visuals do not rely on text, they are often more easily understood by international audiences, children, or people who may have difficulty reading. Symbols and icons are used in airports and other public places as well as in documentation, manuals, or training material. Some of these images are developed and approved by the International Organization for Standardization (ISO). The ISO makes sure the images have universal appeal and conform to a single standard, whether used in a printed document or on an elevator wall.

> Symbols and icons represent concepts in forms that resemble and item or idea

The words *symbol* and *icon* are often used interchangeably. Technically, icons tend to resemble the item they represent: An icon of a file folder on your computer, for example, looks like a real file folder. Symbols can be more abstract. Symbols still get the meaning across but may not resemble, precisely, what they represent, for example a biohazard symbol. Figure 12.30 shows some familiar icons and symbols.

> How symbols and icons differ

Ready-to-use icons and symbols can be found in clip-art collections, from which you can import and customize images by using a drawing program. Because of its generally unpolished appearance, consider using clip art only for in-house documents

> Limitations of clip art icons and symbols

Simple drawings reduce chances for confusion

The first three images are icons (representative); the last two are symbols (abstract)

Figure 12.30 **Internationally Recognized Icons and Symbols**

or for situations in which your schedule or budget preclude obtaining original artwork.

> NOTE Be sure the image you choose is "intuitively recognizable" to multicultural readers ("Using Icons" 3).

Infographics

Infographics tell a story or explain an idea visually through icons, graphics, and text

Infographics help you tell a story or explain an idea visually through a combination of graphics and text. The graphics are typically designed to look like icons or symbols. Each section of the infographic provides a visual snapshot of a larger idea, story, or activity. Although infographics typically combine text and images, the infographic is a visual document, almost like a small flyer or poster, where complex ideas are explained in a manner accessible to a broad audience, from experts to everyday readers.

In the Centers for Disease Control and Prevention (CDC) infographic in Figure 12.31, a heading that rhymes (burn, learn) makes the message easy to remember. The heading is followed by a series of icons that represent an idea: if young people do 60 minutes of soccer, basketball, biking, or swimming a day, they will do better at school. This simplified message represents a research study that correlates student grades with exercise, summarized later in the infographic. Notice the use of color (three colors of high contrast, applied consistently) and a question that draws readers into the document.

How to create infographics

You can create infographics using apps such as Piktochart or Canva. Tools such as these allow you to import data, text, and images and create visuals from this content. Free apps usually offer a limited range of templates, but these apps do allow you to create simple infographics that can be effective. For more complex infographics, you can work with a graphic designer or try your hand at Adobe Illustrator or similar graphics programs. Infographics often accompany social media posts; be sure to see how your infographic looks at that size before posting.

Figure 12.31 **An Infographic**
Source: U.S. Department of Health & Human Services, Centers for Disease Control and Prevention.

Guidelines

for Creating Graphic Illustrations

Simple Drawings and Diagrams

➤ **Let simple drawings stand on their own.** As long as a drawing is placed near to where it is discussed in a document, viewers will understand its purpose.

➤ **Call out diagrams.** Because a diagram is more complex than a simple drawing, mention why it is being used.

➤ **Explain how diagram parts fit together or operate.** Viewers will need clear explanations.

➤ **Use lines and arrows to indicate direction and motion in diagrams.** For diagrams that show action, directional markers help viewers understand the action.

➤ **Keep diagram illustrations simple.** Only show viewers what they need to see.

➤ **Label each important part of a diagram.** Help viewers get a full understanding of what they are looking at.

Photographs

➤ **Simulate the readers' angle of vision.** Consider how they would view the item or perform the procedure.

➤ **Trim (crop) the photograph.** Eliminate needless detail the viewer does not need to see.

➤ **Provide a sense of scale for an object unfamiliar to readers.** Include a person, a ruler, or a familiar object (such as a hand) in the photo.

➤ **Supplement the photograph with diagrams.** This way, you can emphasize selected features.

➤ **Obtain permission.** If your document will be published, get a signed release from any person in your photograph.

➤ **Explain what readers should look for in the photo.** Do this in your discussion or use a caption. If necessary, label all the parts readers need to identify.

➤ **If you are not the photographer, consider quality, copyright, and ethical use of any photograph.** Online, you can find a wealth of quality photography sites, such as Creative Commons, Shutterstock, or the Flickr Commons. If a photograph is not copyright free, obtain written permission from the copyright holder (and pay a reprint fee, if required) before republishing it. For ethical and legal reasons, do not alter someone else's photograph in any way, although you may add an annotation, caption, or citation information.

Maps

➤ **Use maps from credible sources.** Reliable, copyright free sources include the U.S. Census Bureau or other government agencies.

➤ **Keep colors to a minimum.** Make sure any maps you use are easy to read on a computer or in print.

➤ **Avoid overly complex maps.** Use maps that show only what your readers need to see.

Symbols and Icons

➤ **Use internationally recognized symbols and icons.** Check out the International Standards Organization (ISO) Web site for more information.

➤ **Make sure you use any symbols and icons according to their copyright status.** Check the Web site where you found the symbols or icons to see if these items require permission or if they have some other kind of licensing statement (such as Creative Commons) that allows limited use with permission or attribution.

Infographics

➤ **Choose a topic that has a story to tell.** Good examples might include how to get more exercise; how climate change is changing a particular part of the economy; how effective the flu vaccine has been this season.

➤ **Strive for a balanced combination of visuals and text.** However, let the visuals be the main attraction.

➤ **Write and design the infographic for readers with a broad range of technical backgrounds.**

➤ **Keep your design simple.** This is especially important if the infographic will be posted to social media.

Using Color and Presenting Visuals

12.7 Understand how to use color and present visuals appropriately

Color often makes a presentation more interesting, focusing viewers' attention and helping them identify various elements. In Figure 12.19, for example, color helps viewers sort out the key schedule elements of a Gantt chart for a major project: activities, time lines, durations, and meetings. Color can also make complex data and relationships easier to understand. In Figure 12.32, a time-series map uses a color range to demonstrate changes to the earth's temperature from 1884 to the present. The original is actually in video format, making the temperature change data and its implications that much more visible over time.

Benefits of using color within visuals

Even the simplest use of color, such as the two-color PDF document shown in Figure 1.2 (Chapter 1), can help organize the reader's understanding, provide orientation, and emphasize important material. For PDF documents, which most people read on the screen but some may choose to print, a simple color scheme may be best. For Web pages, the use of color can be more complex. See Chapter 13 for more on page design and Chapter 24 for more on Web pages.

Benefits of using color as a visual

Other ways color can be useful and important for your audience include using color to organize, orient, and emphasize.

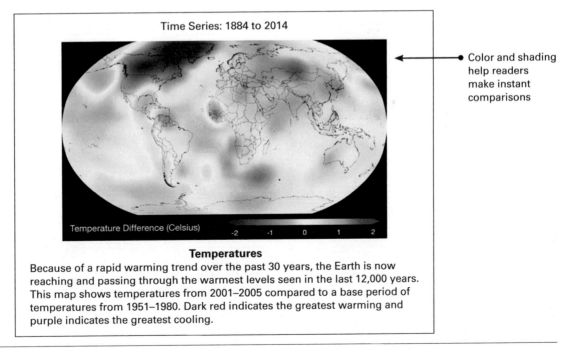

Time Series: 1884 to 2014

Color and shading help readers make instant comparisons

Temperature Difference (Celsius)   -2   -1   0   1   2

Temperatures
Because of a rapid warming trend over the past 30 years, the Earth is now reaching and passing through the warmest levels seen in the last 12,000 years. This map shows temperatures from 2001–2005 compared to a base period of temperatures from 1951–1980. Dark red indicates the greatest warming and purple indicates the greatest cooling.

Figure 12.32 **Colors Used to Show Relationships**
Source: NASA.

Use Color to Organize

How color reveals organization

Readers look for ways of organizing their understanding of a document. Color can reveal structure and break material up into discrete blocks that are easier to locate, process, and digest, as shown in Figure 12.33.

- A color background screen can set off like elements such as checklists, instructions, or examples.
- Horizontal colored rules can separate blocks of text, such as sections of a document or areas of a page.
- Vertical rules can set off examples, quotations, captions, and so on.

Use Color to Orient

Readers look for signposts that help them find their place or locate what they need, as shown in Figure 12.34.

Color used to organize →

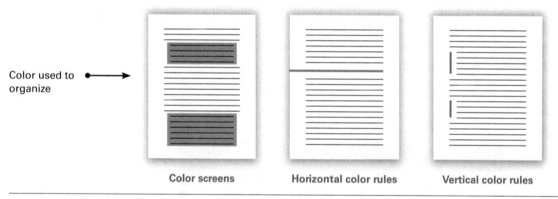

Color screens   Horizontal color rules   Vertical color rules

Figure 12.33 **Color Used to Organize**

Color used to orient →

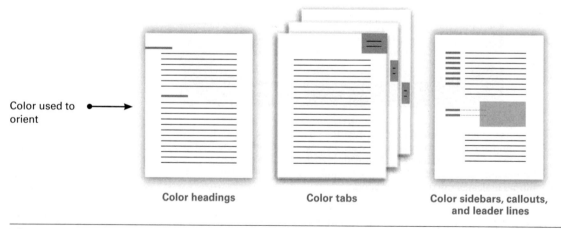

Color headings   Color tabs   Color sidebars, callouts, and leader lines

Figure 12.34 **Color Used to Orient**

- Color can help headings stand out from the text and differentiate major headings from minor ones.
- Color tabs and boxes can serve as location markers.
- Color sidebars (for marginal comments), callouts (for labels), and leader lines (dotted lines for connecting a label to its referent) can guide the eyes.

How color provides orientation

Use Color to Emphasize

Readers look for places to focus their attention in the document, as shown in Figure 12.35.

How color emphasizes

- Color type can highlight key words or ideas.
- Color can call attention to cross-references or to links on a Web page.
- A color or a ruled box can frame a warning, caution, note, hint, or any other item that needs to stick in people's minds.

For more on using color within visuals, as well as using color as a visual within a document, see the following Guidelines.

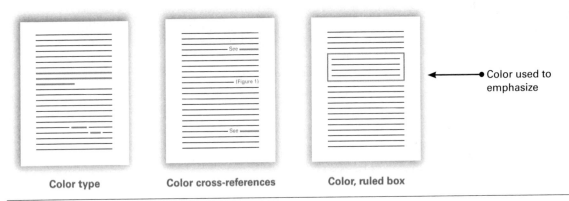

Color type   Color cross-references   Color, ruled box

Figure 12.35 **Color Used to Emphasize**

Guidelines

for Incorporating Color

Using Color within Visuals

➤ **Use color sparingly in tables, graphs, and charts.** Except in a pie chart, which may use as many colors as there are segments of the pie, avoid using more than 4–5 colors within a table, graph, or chart. The simpler the visual, the fewer colors you should use.

➤ **Use appropriate colors in all visuals you create yourself.** Stick to basic, primary colors. Viewers do not want to look at a multiple bar chart that uses overly bright, fluorescent, or pastel colors. Likewise, viewers don't want to have a difficult time differentiating between overly muted colors in a visual.

➤ **When choosing graphic illustrations, avoid overly flashy images.** Stick to colorful but professional looking images. Avoid any diagrams, photographs, maps, symbols, icons, or infographics that have too much "visual noise" and are difficult to read.

Using Color as a Visual

➤ **Use color sparingly.** Color gains impact when used selectively. It loses impact when overused (*Aldus Guide* 39). Use no more than three or four distinct colors—including black and white (White, *Great Pages* 76).

➤ **Apply color consistently to like elements throughout your document.** (Wickens 117).

➤ **Make color redundant.** Be sure all elements are first differentiated in black and white: by shape, location, texture, type style, or type size. Different readers perceive colors differently or, in some cases, not at all (White, *Great Pages* 76).

➤ **Use a darker color to make a stronger statement.** The darker the color, the more important the material. Darker items can seem larger and closer than lighter objects of identical size.

➤ **Make color type larger or bolder than text type.** For text type, use a high-contrast color (dark against a light background). Color is less visible on the page than black ink on a white background. The smaller the image or the thinner the ruled line, the stronger or brighter the color needed (White, *Editing* 229, 237).

➤ **Create contrast.** For contrast in a color screen, use a very dark type against a very light background, say a 10 to 20 percent screen (Gribbons 70). The larger the screen area, the lighter the background color needed.

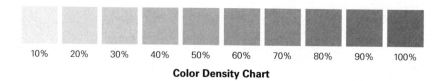

10%   20%   30%   40%   50%   60%   70%   80%   90%   100%

Color Density Chart

Presenting Visuals

Placing visuals

Once you've chosen the appropriate type of visual, then created it (including adding color if appropriate), or chosen a visual from another source, you need to fit the visual within your document. Place your visuals where they best serve the needs of your readers: as close as possible to the related discussion if they are central to that discussion or in an appendix if they are peripheral to your discussion.

Introducing visuals

Introduce your visuals within the text by referring to them by number (Figure 1, Table 4, etc.) and explaining what they mean (e.g., "As Figure 3.4 shows ..."), making sure the figure and table numbers match the cross-references in the text.

Framing and sizing visuals

Make your visuals user-friendly in their presentation by framing them with plenty of white space, eliminating visual "noise" (excessive lines, bars, numbers, and inessential information) and sizing each visual for the right proportion and emphasis on the page.

For more on how to present visuals, see the following Guidelines.

Guidelines

for Presenting Visuals

➤ **Place the visual where it will best serve your readers.** If it is central to your discussion, place the visual as close as possible to the material it clarifies. If the visual is peripheral to your discussion or of interest to only a few readers, place it in an appendix. Tell readers when to consult the visual and where to find it.

➤ **Never refer to a visual that readers cannot easily locate.** In a long document, don't be afraid to repeat a visual if you discuss it a second time.

➤ **Never crowd a visual into a cramped space.** Frame the visual with plenty of white space, and position it on the page for balance. To achieve proportion with the surrounding text, consider the size of each visual and the amount of space it will occupy.

➤ **Number the visual and give it a clear title and labels.** Your title should tell readers what they are seeing. Label all the important material and cite the source of data or of graphics.

➤ **Match the visual to your audience.** Don't make it too elementary for specialists or too complex for nonspecialists.

➤ **Introduce the visual.** In your introduction, tell readers what to expect:

> As Table 2 shows, operating costs have increased 7 percent annually since 1990. (Informative.)
> See Table 2. (Uninformative.)

➤ **Interpret the visual.** Visuals alone make ambiguous statements (Girill, "Technical Communication and Art" 35); pictures need to be interpreted. Instead of leaving readers to struggle with a page of raw data, explain the relationships displayed. Follow the visual with a discussion of its important features:

> This cost increase means that . . .

➤ **Explain the visual.** Always tell readers what to look for and what it means.

➤ **Use prose captions to explain important points made by the visual.** Use a smaller type size so that captions don't compete with text type (*Aldus Guide* 35).

➤ **Eliminate "visual noise."** Excessive lines, bars, numbers, colors, or patterns will overwhelm readers. In place of one complicated visual, use two or more straightforward ones.

➤ **Be sure the visual can stand alone.** Even though it repeats or augments information already in the text, the visual should contain everything readers will need to interpret it correctly.

Ethical Considerations

12.8 Identify ethical issues when using visuals

Although you may be perfectly justified in presenting data in its best light, you are ethically responsible for avoiding misrepresentation. Any one set of data can support contradictory conclusions. Even though your numbers may be accurate, your visual display could be misleading. You are also ethically responsible for obtaining permission for and citing the sources of visuals you obtain from other sources.

The ethical use of visuals is critical

Present the Real Picture

Visual relationships in a graph should accurately portray the numerical relationships they represent. Begin the vertical scale at zero. Never compress the scales to reinforce your point.

Notice how visual relationships in Figure 12.36 become distorted when the value scale is compressed or fails to begin at zero. In version A, the bars accurately depict the numerical relationships measured from the value scale. In version B, item Z (400) is depicted as three times X (200). In version C, the scale is overly compressed, causing the shortened bars to understate the quantitative differences.

Deliberate distortions are unethical because they imply conclusions that are contradicted by the actual data.

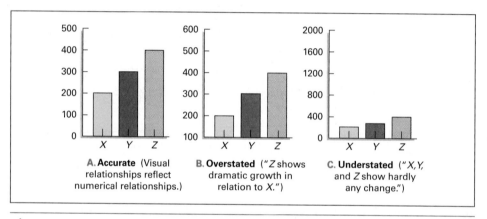

Figure 12.36 **An Accurate Bar Graph and Two Distorted Versions** Absence of a zero baseline in B shrinks the vertical axis and exaggerates differences among the data. In C, the excessive value range of the vertical axis dwarfs differences among the data.

Present the Complete Picture

An accurate visual should include all essential data, without getting bogged down in needless detail. Figure 12.37 shows how distortion occurs when data that would provide a complete picture are selectively omitted. Version A accurately depicts the numerical relationships measured from the value scale. In version B, too few points are plotted. Always decide carefully what to include and what to leave out.

Don't Mistake Distortion for Emphasis

When you want to emphasize a point (a sales increase, a safety record, etc.), be sure your data support the conclusion implied by your visual. For instance, don't use inordinately large visuals to emphasize good news or small ones to downplay bad news. When using clip art, pictograms, or drawn images to dramatize a comparison, be sure the relative size of the images or icons reflects the quantities being compared.

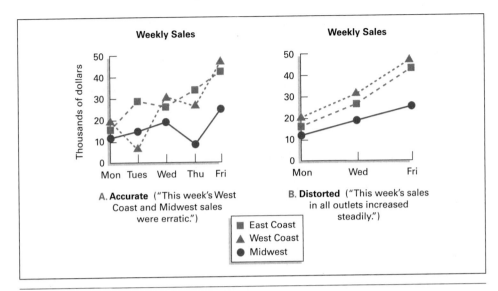

Figure 12.37 An Accurate Line Graph and a Distorted Version Selective omission of data points in B causes the lines to flatten, implying a steady increase rather than an erratic pattern of sales, as more accurately shown in A.

A visual accurately depicting a 100 percent increase in phone sales at your company might look like version A in Figure 12.38. Version B overstates the good news by depicting the larger image four times the size, instead of twice the size, of the smaller image. Although the larger image is twice the height, it is also twice the *width,* so the total area conveys the visual impression that sales have *quadrupled.*

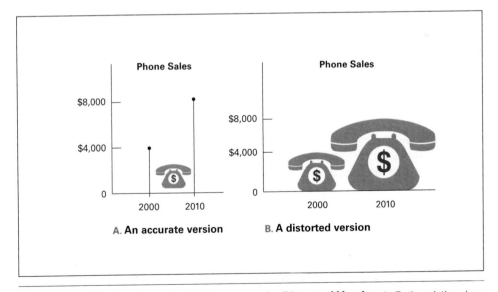

Figure 12.38 An Accurate Pictogram and a Distorted Version In B, the relative sizes of the images are not equivalent to the quantities they represent.

Visuals have their own rhetorical and persuasive force, which you can use to advantage—for positive or negative purposes, for the reader's benefit or detriment (Van Pelt 2). Avoiding visual distortion is ultimately a matter of ethics.

Use Copyright Free Visuals or Credit the Work of Others

Presenting the real picture, presenting the complete picture, and avoiding distortion are not the only ethical issues related to using visuals in professional documents. Writers must also be keenly aware of whether the visuals they use from other sources are copyright free or that have a licensing statement (such as those listed with Creative Commons) stating the terms by which you can use the visual. Also, always cite the visual's source, both when the source is picked up directly from someone else or based on someone else's data. See the following Guidelines.

Guidelines

for Obtaining and Citing Visual Material

Copyright
The Internet is a rich source for visuals. It's tempting to cut and paste a visual you find online directly into your document. But you need to be very careful about checking the copyright status of the particular image you wish to use. To avoid problems, follow these guidelines:

➤ **Look for visuals that are either copyright-free or cleared for purposes that match your desired use.** Start by checking Creative Commons, where you can learn more about the different licenses and can search for images. Also note that Flickr has an entire section devoted to images with Creative Commons licenses.

➤ **Use public domain sources.** Many government sites, such as the U.S. Census Bureau and NASA, are a good source of photographs, charts, graphs, and other visuals that don't require copyright permission.

➤ **Follow fair use guidelines** (see Chapter 7, "Consider This: Frequently Asked Questions about Copyright"), which allow for limited use of copyrighted material without permission (for example, for a class project that will not have widespread distribution).

Citing visuals created by someone else
➤ **Cite the source of the visual.** Even if you are following copyright guidelines, you should still properly cite the source of the visuals you use. *Failing to cite the source of a visual could constitute plagiarism.*

➤ **If the visual is available on the Internet, provide the Web address or other information so your reader can locate it.**

Attributing the source of your original visual
➤ **Cite the source of the data you used to create your visual.** Even if you create a visual yourself, the data for that visual may come from another source. For example, in Figure 12.10, the graph was created by a writer using Excel, but she used data downloaded from the U.S. Bureau of Labor Statistics.

➤ **If the data is available on the Internet, provide the Web address or other information so your reader can check the original source.**

Social Media and Cultural Considerations

12.9 Understand how social media and culture affect your choice of visuals

Social media posts rely heavily on visual information. People post photos, emojis, and memes in order to make a point with few to no words. While these postings can be effective (a photo of everyone celebrating a new product release at work, for example), purely visual information read by so many people can be misunderstood. Humor, sarcasm, and images that might make sense to you and your immediate friends could potentially be meaningless to others. So, while visual communication can serve as a universal language, visuals only work well as long as the graphic or image is not misinterpreted.

Visuals and social media posts

For example, not all cultures read left to right, so a chart designed to be read left to right that is read in the opposite direction could be misunderstood. Color is also a cultural consideration: U.S. audiences associate red with danger and green with safety. But in Ireland, green or orange carry may strong political connotations. In Muslim cultures, green is a holy color (Cotton 169). Icons and symbols as well can have offensive connotations. Hand gestures (including emojis) are especially problematic: some Arab cultures consider the left hand unclean; a pointing index finger—on either hand—may signify rudeness in Venezuela or Sri Lanka (Bosley 5–6).

Meaning is in the eye of the beholder

Checklist

Visuals

Use the following Checklist when creating visuals or using visuals from another source.

Content

❑ Does the visual serve a valid purpose (clarification, not mere ornamentation)? (See "When to Use Visuals" in this chapter.)

❑ Is the level of complexity appropriate for the audience? (See "Choosing the Right Visual" in this chapter.)

❑ Is the visual titled and numbered? (See "Choosing the Right Visual" and "Guidelines for Presenting Visuals" in this chapter.)

❑ Is the visual clean and easy to read? (See "Choosing the Right Visual" and "Guidelines for Presenting Visuals" in this chapter.)

❑ Are all patterns identified by label or legend? (See "Choosing the Right Visual" and "Guidelines for Presenting Visuals" in this chapter.)

❑ Are all values or units of measurement specified (grams per ounce, millions of dollars)? (See "Ethical Considerations" in this chapter.)

❑ Do the visual relationships represent the numeric relationships accurately? (See "Ethical Considerations" in this chapter.)

❑ Are captions and explanatory notes provided as needed? (See "Guidelines for Presenting Visuals" in this chapter.)

❑ Are all data sources cited? (See "Choosing the Right Visual" and "Guidelines for Obtaining and Citing Visual Material" in this chapter.)

❏ Is the visual introduced, discussed, interpreted, integrated with the text, and referred to by number? (See "Guidelines for Presenting Visuals" in this chapter.)

❏ Can the visual itself (along with any captions and labels) stand alone in terms of meaning? (See "Guidelines for Presenting Visuals" in this chapter.)

Style

❏ Is this the best type of visual for my purpose and audience? (See "Choosing the Right Visual" in this chapter.)

❏ Is the visual uncrowded, uncluttered, and free of "visual noise"? (See "Guidelines for Presenting Visuals" in this chapter.)

❏ Is color used tastefully and appropriately? (See "Using Color and Presenting Visuals" in this chapter.)

❏ Is the visual ethically acceptable? (See "Ethical Considerations" in this chapter.)

❏ Does the visual respect readers' cultural values? (See "Social Media and Cultural Considerations" in this chapter.)

Placement

❏ Is the visual easy to locate? (See "Guidelines for Presenting Visuals" in this chapter.)

❏ Do all design elements (title, line thickness, legends, notes, borders, white space) achieve balance? (See "Presenting Visuals" and "Guidelines for Presenting Visuals" in this chapter.)

❏ Is the visual positioned on the page to achieve balance? (See "Guidelines for Presenting Visuals" in this chapter.)

❏ Is the visual set off by adequate white space or borders? (See "Guidelines for Presenting Visuals" in this chapter.)

❏ Is the visual placed near the text it is helping to describe? (See "Choosing the Right Visual" and "Guidelines for Presenting Visuals" in this chapter.)

❏ If posted to the Web or on social media, will the visual be readable and easy to access? (See "Social Media and Cultural Considerations" in this chapter.)

Projects

For all projects, check with your instructor about whether to present your findings in class, bring drafts to class for discussion, upload your project to the class learning management system (LMS), and/or use the LMS forum or discussion boards to collaborate and review each activity below.

General

1. The following statistics are based on data from three colleges in a large western city. They compare the number of applicants to each college over six years.

    • In 2017, X college received 2,714 applications for admission, Y college received 2,840, and Z college 1,992.
    • In 2018, X college received 2,872 applications for admission, Y college received 2,615, and Z college 2,112.
    • In 2019, X college received 2,868 applications for admission, Y college received 2,421, and Z college 2,267.

Display these data in a line graph, a bar graph, and a table. Which version seems most effective for someone who (a) wants exact figures, (b) wonders how overall enrollments are changing, or (c) wants to compare enrollments at each college in a certain year? Include a caption interpreting each version.

2. Devise an organization chart showing the lines of responsibility and authority in an organization where you work.

3. Devise a pie chart to depict your yearly expenses. Title and discuss the chart.

4. In textbooks or professional journal articles, locate each of these visuals: a table, a multiple-bar graph, a multiline graph, a diagram, and a photograph. Evaluate each according to the revision checklist, and discuss the most effective visual in class.

5. Choose the most appropriate visual for illustrating each of these relationships.

   a. A comparison of three top brands of skis, according to cost, weight, durability, and edge control.

   b. A breakdown of your monthly budget.

   c. The percentage of college graduates finding desirable jobs within three months after graduation, over the last 10 years.

   d. A breakdown of the process of corn-based ethanol production.

   e. A comparison of five cereals on the basis of cost and nutritional value.

6. Display each of these sets of information in the visual format most appropriate for the stipulated audience. Complete the planning sheet in Figure 12.37 for each visual. Explain why you selected the type of visual as most effective for that audience. Include with each visual a caption that interprets and explains the data.

   a. (For general readers.) Assume that the Department of Energy breaks down energy consumption in the United States (by source) into these percentages: In 1980, coal, 18.5; natural gas, 32.8; hydro and geothermal, 3.1; nuclear, 1.2; oil, 44.4. In 1990, coal, 20.3; natural gas, 26.9; hydro and geothermal, 3.8; nuclear, 4.0; oil, 45.0. In 2000, coal, 23.5; natural gas, 23.8; hydro and geothermal, 7.3; nuclear, 4.1; oil, 41.3. In 2010, coal, 20.3; natural gas, 25.2; hydro and geothermal, 9.6; nuclear, 6.3; oil, 38.6.

   b. (For experienced investors in rental property.) As an aid in estimating annual heating and air-conditioning costs, here are annual maximum and minimum temperature averages from 1975 to 2010 for five Sunbelt cities (in Fahrenheit degrees): In Jacksonville, the average maximum was 78.4; the minimum was 57.6. In Miami, the maximum was 84.2; the minimum was 69.1. In Atlanta, the maximum was 72.0; the minimum was 52.3. In Dallas, the maximum was 75.8; the minimum was 55.1. In Houston, the maximum was 79.4; the minimum was 58.2. (From U.S. National Oceanic and Atmospheric Administration.)

   c. (For the student senate.) Among the students who entered our school four years ago, here are the percentages of those who graduated, withdrew, or are still enrolled: In Nursing, 71 percent graduated; 27.9 percent withdrew; 1.1 percent are still enrolled. In Engineering, 62 percent graduated; 29.2 percent withdrew; 8.8 percent are still enrolled. In Business, 53.6 percent graduated; 43 percent withdrew; 3.4 percent are still enrolled. In Arts and Sciences, 27.5 percent graduated; 68 percent withdrew; 4.5 percent are still enrolled.

7. Anywhere on campus or at work, locate a visual that needs revision for accuracy, clarity, appearance, or appropriateness. Look at lab manuals, newsletters, financial aid or admissions brochures, student or faculty handbooks, newspapers, textbooks, Web sites, social media pages. Using the planning sheet in Figure 12.37 and "Checklist: Visuals" as guides, revise the visual. Submit a copy of the original, along with a memo explaining your improvements. Be prepared to discuss your revision in class.

Team

Assume your instructor has been asked to test a few graphic and document design apps for students. Your group's task is to test one app and to make a recommendation. In small groups, check with your school's learning technologies department (or your school's computer lab) and ask for a listing of available graphic design apps. Select one and learn how to use it. Design at least four representative visuals. In a memo or presentation to your instructor and classmates, describe the app briefly and tell what it can do. Would you recommend that the school purchase a site license for this product? Explain. Submit your report, along with the sample graphics you have composed.

Digital and Social Media

Search for a free app to create infographics, or ask your instructor if your school provides you with access to any similar apps. Look for data that would be suitable for creating an infographic. (Hint: Try the Pew Research Center, then click on "social trends" or "Internet & tech" for recent surveys and other data about people and technology.)

Sketch out your idea for the infographic, then try using a template in the app to create your infographic. Keep in mind that people want to know the story behind the data or research. Think about how well your infographic would display if it were posted to social media.

Global

The International Organization for Standardization (ISO) is a group devoted to standardizing a range of material, including technical specifications and visual information. If you've ever been in an airport and seen the many international signs directing travelers to the restroom or informing them not to smoke, you have seen ISO signs. Go to the ISO Web site to learn about ISO icons and symbols. Show some of the icons and symbols, and explain why these work for international audiences.

Chapter 13
Designing Pages and Documents

"Every project I'm involved with requires careful planning about document design. If a book only needs a light revision, I often end up revising the design features, too. I do have a coworker who designs documents professionally, though, so if I need something fancy, he designs it and I follow the specs. But I also think about page design with routine workplace communication; even a simple memo or email needs to be carefully formatted and structured so people will actually read what's written."

—*Lorraine Patsco, Director of Prepress and Multimedia Production*

Learning Objectives

13.1 Explain the importance of print and digital document design in the workplace

13.2 Discuss the everyday design skills technical communicators need to have

13.3 Consider a variety of techniques for designing a reader-friendly document

13.4 Develop an audience and use profile to guide your design

13.5 Explain how digital documents have special design requirements

What page design does

Page design, the layout of words and graphics, determines the look of a document. Well-designed pages invite readers into the document, guide them through the material, and help them understand and remember the information.

Page Design in Print and Digital Workplace Documents

13.1 **Explain the importance of print and digital document design in the workplace**

Technical documents rarely get undivided attention

People read work-related documents only because they have to. If there are easier methods of getting the information, people will use these methods first. In fact, busy readers often only skim a document, or they refer only to certain sections during a meeting or presentation. Amid frequent distractions, readers want to be able to leave the document and then return and locate what they need easily.

Before actually reading a document, people usually scan it for a sense of what it's about and how it's organized. An audience's first impression tends to involve a purely visual, aesthetic judgment: "Does this look like something I want to read, or like too much work?" Instead of an unbroken sequence of paragraphs, readers look for charts, diagrams, lists, various type sizes and fonts, different levels of headings, and other aids to navigation. Having decided at a glance whether your document is visually appealing, logically organized, and easy to navigate, readers will draw conclusions about the value of your information, the quality of your work, and your overall credibility.

Readers are attracted by documents that appear inviting and accessible

To appreciate the impact of page design, consider Figures 13.1 and 13.2. Notice how the information in Figure 13.1 resists interpretation. Without design cues, we have no way of chunking that information into organized units of meaning. Figure 13.2 shows the same information after a design overhaul.

How page design transforms a document

Today, pages come in all forms, including hard copy (printed), PDF, Web-based, e-reader formats, social media posts, and more. Despite this proliferation of formats, the most common technical and workplace documents continue to take the shape of a printed page, designed to be read in portrait mode, similar to the pages of a book. Résumés, letters, memos, formal reports, proposals, journal articles, white papers, legal briefs, and most other workplace documents still follow this general shape and format. (Notice, for example, how Microsoft Word and other word processing programs default to the standard U.S. letter size of 8 1/2 × 11 inches.)

Shape and format of printed documents

Many of these documents are designed to be read both in hard copy and on the screen as a PDF. This chapter focuses primarily on pages for such document types. Later in this chapter, a section on digital documents offers more information about PDF as well as other digital formats. Also, later chapters discuss specific digital documents such as email (Chapter 14); résumés (Chapter 16); blogs, wikis, and Web pages (Chapter 24); and social media posts (Chapter 25).

PDF and other digital formats

Design Skills Needed by Technical Communicators

13.2 **Discuss the everyday design skills technical communicators need to have**

In large organizations, page design may be the domain of specialists in the graphics department. But in smaller organizations, technical writers are often responsible for writing *and* designing the document. Even when a graphic design department exists, these groups do not typically get involved with smaller, everyday documents. So, you need to be familiar with page design software and some of the standard features that are available to you.

Most technical communicators are writers and designers

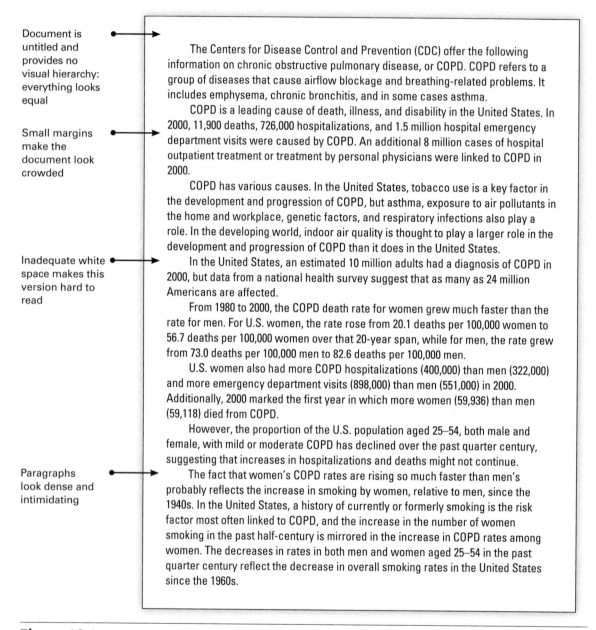

Document is untitled and provides no visual hierarchy: everything looks equal

Small margins make the document look crowded

Inadequate white space makes this version hard to read

Paragraphs look dense and intimidating

The Centers for Disease Control and Prevention (CDC) offer the following information on chronic obstructive pulmonary disease, or COPD. COPD refers to a group of diseases that cause airflow blockage and breathing-related problems. It includes emphysema, chronic bronchitis, and in some cases asthma.

COPD is a leading cause of death, illness, and disability in the United States. In 2000, 11,900 deaths, 726,000 hospitalizations, and 1.5 million hospital emergency department visits were caused by COPD. An additional 8 million cases of hospital outpatient treatment or treatment by personal physicians were linked to COPD in 2000.

COPD has various causes. In the United States, tobacco use is a key factor in the development and progression of COPD, but asthma, exposure to air pollutants in the home and workplace, genetic factors, and respiratory infections also play a role. In the developing world, indoor air quality is thought to play a larger role in the development and progression of COPD than it does in the United States.

In the United States, an estimated 10 million adults had a diagnosis of COPD in 2000, but data from a national health survey suggest that as many as 24 million Americans are affected.

From 1980 to 2000, the COPD death rate for women grew much faster than the rate for men. For U.S. women, the rate rose from 20.1 deaths per 100,000 women to 56.7 deaths per 100,000 women over that 20-year span, while for men, the rate grew from 73.0 deaths per 100,000 men to 82.6 deaths per 100,000 men.

U.S. women also had more COPD hospitalizations (400,000) than men (322,000) and more emergency department visits (898,000) than men (551,000) in 2000. Additionally, 2000 marked the first year in which more women (59,936) than men (59,118) died from COPD.

However, the proportion of the U.S. population aged 25–54, both male and female, with mild or moderate COPD has declined over the past quarter century, suggesting that increases in hospitalizations and deaths might not continue.

The fact that women's COPD rates are rising so much faster than men's probably reflects the increase in smoking by women, relative to men, since the 1940s. In the United States, a history of currently or formerly smoking is the risk factor most often linked to COPD, and the increase in the number of women smoking in the past half-century is mirrored in the increase in COPD rates among women. The decreases in rates in both men and women aged 25–54 in the past quarter century reflect the decrease in overall smoking rates in the United States since the 1960s.

Figure 13.1 Ineffective Page Design This design provides no visual cues to indicate how the information is structured, what main ideas are being conveyed, or where readers should focus.

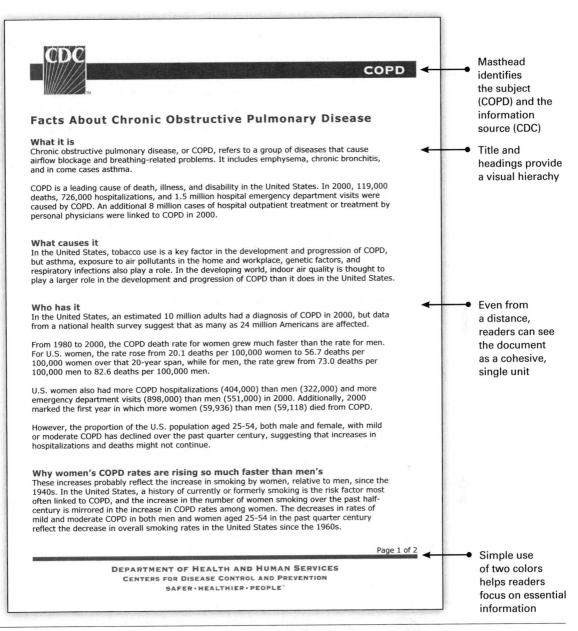

Masthead identifies the subject (COPD) and the information source (CDC)

Title and headings provide a visual hierachy

Even from a distance, readers can see the document as a cohesive, single unit

Simple use of two colors helps readers focus on essential information

Figure 13.2 Effective Page Design Notice how this revision of Figure 13.1 uses a title, white space, headings, and color to help readers. This document is suitable for online viewing or for printing hard copies.

Source: Centers for Disease Control and Prevention, http://www.cdc.gov/nceh/airpollution/copd/pdfs/copdfaq.pdf.

Word Processing and Desktop Publishing

Designing basic documents

Most word processing programs offer a wide range of page design features, including column settings, font choices, headers and footers, paragraph styles, lists, options for inserting visuals, automatic table of contents, outlining, and more. Style settings allow you to maintain consistency across common features like headings. These programs also come loaded with a number of templates, which are pre-designed formats that can be used or modified for your specific document. (More on style sheets and templates in the next section.) Since your document is probably written with Microsoft Word or a similar program, using the same program to format the document can often be the easiest approach.

Designing complex documents

But for complex documents, particularly longer ones that contain more visuals and require sophisticated layout, page flows, and formatting, you may need to work with desktop publishing and page layout programs (DTP). In fact, most companies create their technical marketing and other visual-heavy documents this way. Popular apps include Adobe InDesign and FrameMaker as well as Microsoft Publisher. These programs take training and time to master because they offer many more features, including options for version control and collaborative writing and commenting.

> NOTE Your college or university may provide students with access to workshops as well as online tutorials (at Lynda.com or other sites) for learning how to use desktop publishing programs. If you take and complete a workshop or tutorial, be sure to add this item to your résumé.

Using Styles and Templates

Benefits of using styles

"Styles" are pre-formatted options in word processing and DTP programs that make it easy for you to maintain a consistent look and feel across particular features of your document. For example, in Figure 13.2, the headings starting with "What it is" are all formatted in the same font, size, and layout (set flush to the left margin). You could accomplish this same look by applying each individual characteristic, but then, if you decide to change one of these headings, you will need to go back and change all the others. By using a style, any changes you make will automatically apply to all other items with the same style. Some styles come with the program (for instance, Microsoft Word has styles called "Heading 1, Heading 2, Normal"). You can also create your own style based on custom formatting. Some organizations create custom styles and require everyone to use these.

Benefits of using templates

Styles are for individual parts of the page, such as headings or paragraphs. Templates, on the other hand, apply to the entire document. All word processing and DTP programs come with dozens of templates for résumés, brochures, letters, sales proposals, and other document types. Templates can be a good start, because they look nice and may save you on design time, especially for a small project. But you need to ensure that the template you select is consistent with the organization's guidelines. More importantly, templates may look good at first but may not be suited to your specific audience and purpose (see next section on Creating a Design that Works for Your Readers).

Using Style Guides and Style Sheets

Style guides are documents that describe an organization's rules for document design and language use. These guides typically provide details about the use of fonts, logos, trademarks, and other visual features, as well as rules for written elements such as use of punctuation, capitalization, and preferred spellings of product names. Some organizations have detailed style guides; technical writers often contribute to the development of these guides. Style guides help ensure a consistent look across a company's various documents and publications.

The difference between style guides and style sheets

Sometimes, even with an organizational style guide in place, you may still need to create a style sheet. Style sheets specify the specific design elements of a particular document. Especially if you are working as part of a team of writers and designers, everyone needs to be using the same typefaces, headings, and other elements. Here are two examples of what you might find in a style sheet:

- The first time you use or define a specialized term, highlight it with *italics* or **boldface.**
- In headings, capitalize prepositions of five or more letters ("Between," "Versus").

Possible style sheet entries

The more complex the document, the more specific the style sheet should be. All writers and editors should have a copy. Consider keeping the style sheet on a shared document (Google Drive or your organization's internal file server) for easy access and efficient updating.

Creating a Design That Works for Your Readers

13.3 Consider a variety of techniques for designing a reader-friendly document

Approach your design decisions to achieve a consistent look, to highlight certain material, and to aid navigation. First, consider the overall look of your pages, and then consider the following three design categories: styling the words and letters, adding emphasis, and using headings for access and orientation.

How to approach document design

> NOTE *All design considerations are influenced by the budget for a publication. For print documents, adding a single color (say, to major headings) can double the printing cost.*

If your organization prescribes no specific guidelines, the general design principles that follow should serve in most situations.

Shaping the Page

In shaping a page, consider its look, feel, and overall layout. The following suggestions will help you shape appealing and usable pages.

Start by shaping the page

PROVIDE PAGE NUMBERS, HEADERS, AND FOOTERS. For a long document, count your title page as page i, without numbering it, and number all front matter pages, including the table of contents and abstract, with lowercase Roman numerals (ii, iii, iv). Number the first text page and subsequent pages with arabic numerals (1, 2, 3). Along with page numbers, *headers* or *footers* (or *running heads* and *feet*) appear in the top or bottom page margins, respectively. These provide chapter or article titles, authors' names, dates, or other publication information. (For more on running heads and feet, see "Guidelines for Using Headings" in this chapter.)

USE A GRID. Readers make sense of a page by looking for a consistent underlying structure, with the various elements located where they expect them. With a view of a page's Big Picture, you can plan the size and placement of your visuals and calculate the number of lines available for written text. Most important, you can rearrange text and visuals repeatedly to achieve a balanced and consistent design (White, *Editing* 58). Figure 13.3 shows a sampling of grid patterns. Brochures and newsletters typically employ a two- or three-column grid. Web pages often use a combined vertical/horizontal grid. Figure 13.2 uses a single-column grid, as do most memos, letters, and reports. (Grids are also used in storyboarding; see Chapter 10, "Storyboarding.")

USE WHITE SPACE TO CREATE AREAS OF EMPHASIS. Sometimes, what is *not* on the page can make a big difference. Areas of text surrounded by white space draw the reader's eye to those areas.

Well-designed white space imparts a shape to the whole document, a shape that orients readers and lends a distinctive visual form to the printed matter by keeping related elements together, by isolating and emphasizing important elements, and by providing breathing room between blocks of information.

In the examples in Figure 13.4, notice how the white space pulls your eye toward the pages in different ways. Each example causes the reader to look at a different

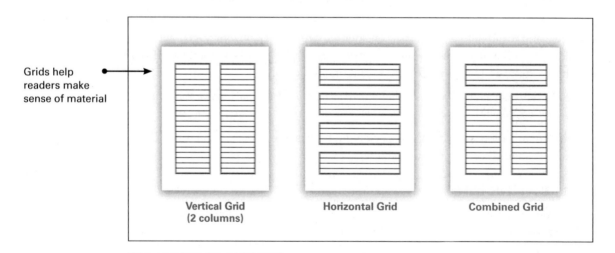

Grids help readers make sense of material

Vertical Grid (2 columns)   Horizontal Grid   Combined Grid

Figure 13.3 **Grid Patterns** By subdividing a page into modules, grids provide a blueprint for your page design as well as a coherent visual theme for the document's audience.

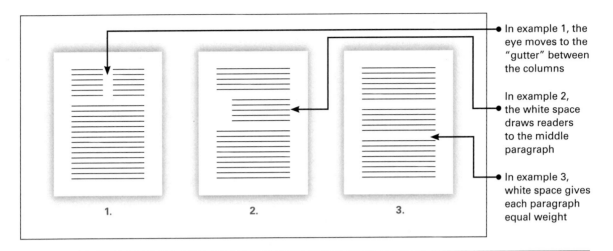

In example 1, the eye moves to the "gutter" between the columns

In example 2, the white space draws readers to the middle paragraph

In example 3, white space gives each paragraph equal weight

Figure 13.4 White Space White space creates areas of emphasis.

place on the page first. White space can keep a page from seeming too cluttered, and pages that look uncluttered, inviting, and easy to follow convey an immediate sense of reader-friendliness.

PROVIDE AMPLE MARGINS.   Small margins crowd the page and make the material look difficult. On your 8½-by-11-inch page, leave margins of at least 1 or 1½ inches. If the manuscript is to be bound in some kind of cover, widen the inside margin to two inches.

Headings, lines of text, or visuals that abut the right or left margin, without indentation, are designated as *flush right* or *flush left.*

Choose between *unjustified* text (uneven or "ragged" right margins) and *justified* text (even right margins). Each arrangement creates its own "feel."

> To make the right margin even in justified text, the spaces vary between words and letters on a line, sometimes creating channels or rivers of white space. The eyes are then forced to adjust continually to these space variations within a line or paragraph. Because each line ends at an identical vertical space, the eyes must work hard to differentiate one line from another (Felker 85). Moreover, in order to preserve the even margin, words at line's end are often hyphenated, and frequently hyphenated line endings can be distracting.

Justified lines are set flush left and right

> Unjustified text, on the other hand, uses equal spacing between letters and words on a line, and an uneven right margin (as was traditionally produced by a typewriter). For some readers, a ragged right margin makes reading easier. These differing line lengths can prompt the eye to move from one line to another (Pinelli et al. 77). In contrast to justified text, an unjustified page looks less formal, less distant, and less official.

Unjustified lines are set flush left only

Justified text is preferable for books, annual reports, and other formal materials. Unjustified text is preferable for more personal forms of communication such as letters, memos, and in-house reports.

KEEP LINE LENGTH REASONABLE. Long lines tire the eyes, especially on the screen. The longer the line, the harder it is for the reader to return to the left margin and locate the beginning of the next line (White, *Visual Design* 25).

> Notice how your eye labors to follow this apparently endless message that seems to stretch in lines that continue long after your eye was prepared to move down to the next line. After reading more than a few of these lines, you begin to feel tired and bored and annoyed, without hope of ever reaching the end.

Short lines force the eyes back and forth (Felker 79). "Too-short lines disrupt the normal horizontal rhythm of reading" (White, *Visual Design* 25).

> Lines that are too
>
> short cause your eye
>
> to stumble from one
>
> fragment to another.

The number of characters per line for an 8½-by-11-inch single-column page will vary depending on the type of document and the font size. Books tend to have more characters per line, whereas reports, proposals, and other workplace documents tend to use 11 or 12 point fonts for body copy, thus yielding a relatively consistent number of characters and words per line. Longer lines call for larger type and wider spacing between lines (White, *Great Pages* 70).

Line length is also affected by the number of columns (vertical blocks of print) on your page. Two-column pages often appear in newsletters and brochures, but most formal technical documents (reports, proposals) use single column pages.

> NOTE: *Line length is a special consideration if the page will be viewed both in print and PDF. Even though PDF pages are meant to resemble hard copy, digital text can be hard on the eyes. Aim for good use of white space in all documents, but especially for PDF.*

KEEP LINE SPACING CONSISTENT. For any document likely to be read completely (letters, memos, instructions), single-space within paragraphs and double-space between paragraphs. Instead of indenting the first line of single-spaced paragraphs, separate them with one line of space. For longer documents likely to be read selectively (proposals, formal reports), increase line spacing within paragraphs by one-half space. Indent these paragraphs or separate them with one extra line of space.

> NOTE *Although academic papers generally call for double spacing, most workplace documents do not.*

TAILOR EACH PARAGRAPH TO ITS PURPOSE. Readers often skim a long document to find what they want. Most paragraphs, therefore, begin with a topic sentence forecasting the content. As you shape each paragraph, follow these suggestions:

- Use a long paragraph (no more than fifteen lines) for clustering material that is closely related (such as history and background, or any information best understood in one block).

- Use short paragraphs for making complex material more digestible, for giving step-by-step instructions, or for emphasizing vital information.

- Avoid "orphans," leaving a paragraph's opening line at the bottom of a page, and "widows," leaving a paragraph's closing line at the top of the page.

Shape each paragraph

MAKE LISTS FOR EASY READING. Whenever you find yourself writing a series of related items within a paragraph, consider using a list instead, especially if you are describing a series of tasks or trying to make certain items easy to locate. Types of items you might list: advice or examples, conclusions and recommendations, criteria for evaluation, errors to avoid, materials and equipment for a procedure, parts of a mechanism, or steps or events in a sequence. Notice how the items just mentioned, integrated into the previous sentence as an *embedded list*, become easier to grasp and remember when displayed below as a *vertical list*. Types of items you might display in a vertical list:

An embedded list is part of the running text

- Advice or examples

- Conclusions and recommendations

- Criteria for evaluation

- Errors to avoid

- Materials and equipment for a procedure

- Parts of a mechanism

- Steps or events in a sequence

A vertical list draws readers' attention to the content of the list

A list of brief items usually needs no punctuation at the end of each line. A list of full sentences or questions requires appropriate punctuation after each item. For more on punctuating embedded and vertical lists, see Appendix B, "Lists," or consult your organization's style guide.

Depending on the list's contents, set off each item with some kind of visual or verbal signal, as in Figure 13.5. If the items follow a strict sequence or chronology (say, parts of a mechanism or a set of steps), use Arabic numbers (*1, 2, 3*) or the words *First, Second, Third*. If the items require no strict sequence (as in the sample vertical list above), use dashes, asterisks, or bullets. For a checklist, use open boxes.

Introduce your list with a forecasting phrase ("Topics to review for the exam:") or with a sentence ("To prepare for the exam, review the following topics:"). For more on introducing a list, see Appendix B, "Lists."

Phrase all listed items in parallel grammatical form (see Appendix B, "Faulty Parallelism"). When items suggest no strict sequence, try to impose some logical ranking (most to least important, alphabetical, or some such). Set off the list with extra white space above and below.

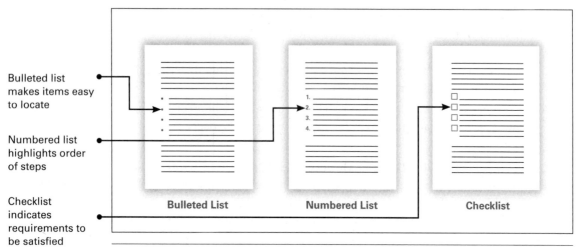

Bulleted list makes items easy to locate

Numbered list highlights order of steps

Checklist indicates requirements to be satisfied

Bulleted List   **Numbered List**   **Checklist**

Figure 13.5 Vertical Lists Lists help organize material for easy reading and comprehension.

NOTE *A document with too many vertical lists appears busy, disconnected, and splintered (Felker 55).*

Guidelines

for Shaping the Page

➤ **Picture the document's overall look and feel when you make design choices about pages.** If your company has other documents that resemble the one you are designing, use these to guide your choices.

➤ **Select an appropriate grid pattern.** Use a single-column grid for basic documents such as letters or reports; use a two-column grid for manuals, and either a two- or a three-column grid for brochures and newsletters.

➤ **Use white space to make pages easier to navigate.**

➤ **Use adequate margins.** On standard-size paper (8½ × 11 inches), use 1-inch or 1.5-inch margins. For bound documents use a 2-inch margin on the bound side.

➤ **Keep line lengths easy on the eye.** Adequate margins help keep line length reasonable.

➤ **For PDF documents, use white space to break up text and make it easier for people to read on a screen.**

Styling the Words and Letters

The significance of type choices

After shaping the page, decide on the appropriate typefaces (fonts), type sizes, and capitalization. Typography, the art of type styling, consists of choices among various typefaces. *Typeface*, or *font*, refers to all the letters and characters in one particular family such as Times, Helvetica, or New York. Each typeface has its own personality: Some convey seriousness; others convey humor; still others convey a technical or businesslike quality. Choice of typeface can influence reading speed by as much as 30 percent (Chauncey 36).

All typefaces divide into two broad categories: *serif* and *sans serif* (Figure 13.6). Serifs are the fine lines that extend horizontally from the main strokes of a letter:

SELECT AN APPROPRIATE TYPEFACE. In selecting a typeface, consider the document's purpose. If the purpose is to help patients relax, choose a combination that conveys ease; fonts that imitate handwriting are often a good choice, but they can be hard to read if used in lengthy passages. If the purpose is to help engineers find technical data in a table or chart, use Helvetica or some other sans serif typeface—not only because numbers in sans serif type are easy to see but also because engineers will be more comfortable with fonts that look precise. Figure 13.7 offers a sampling of typeface choices.

For visual unity, use different sizes and versions (**bold,** *italic,* SMALL CAP) of the same typeface throughout your document. For example, you might decide on Times for an audience of financial planners, investors, and others who expect a traditional font. In this case, use Times 14 point bold for the headings, 12 point regular (Roman) for the body copy, and 12 point italic, sparingly, for emphasis.

If the document contains illustrations, charts, or numbers, use Helvetica 10 point for these; use a smaller size for captions (brief explanation of a visual) or sidebars (marginal comments). You can also use one typeface (say, Helvetica) for headings and another (say, Times) for body copy. In any case, use no more than two different typeface families in a single document—and use them consistently.

Serif type makes printed body copy more readable because the horizontal lines "bind the individual letters" and thereby guide the reader's eyes from letter to letter (White, *Visual Design* 14). Serif fonts look traditional, the sort you see in newspapers and formal reports.   ← ● Serif type

Sans serif type is purely vertical. Clean looking and "businesslike," sans (French for "without") serif is ideal for technical material (numbers, equations, etc.), marginal comments, headings, examples, tables, and captions, and any other material set off from the body copy (White, *Visual Design* 16). Sans serif is also more readable in *projected* environments such as overhead transparencies and PowerPoint slides.   ← ● Sans serif type

Figure 13.6 **Serif Versus Sans Serif Typefaces** Each version makes its own visual statement.

Times New Roman is a standard serif typeface.

Palatino is a slightly less formal serif alternative.

Helvetica is a standard sans serif typeface.

Arial seems a bit more readable than Helvetica.

Chicago makes a bold statement.

A font that imitates handwriting can be hard to read in long passages.

Ornate or whimsical fonts generally should be avoided.

Figure 13.7   Sample Typefaces Except for special emphasis, choose traditional typefaces (Times Roman, Helvetica). Decorative typefaces are hard to read and inappropriate for most workplace documents.

USE TYPE SIZES THAT ARE EASY TO READ.   To map out a page, designers measure the size of type and other page elements (such as visuals and line length) in picas and points (Figure 13.8).

The height of a typeface, the distance from the top of the *ascender* to the base of the *descender*, is measured in points.

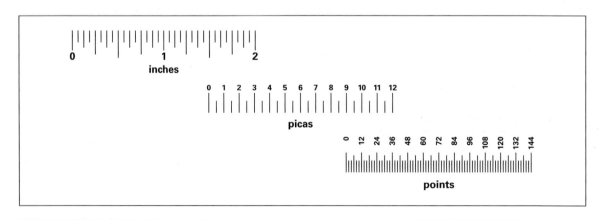

Figure 13.8 Sizing the Page Elements One pica equals roughly 1/6 of an inch and one point equals 1/12 of a pica (or 1/72 of an inch).

Standard type sizes for body copy run from 10 to 12 point, depending on the typeface. Use different sizes for other elements: headings, titles, captions, sidebars, or special emphasis. Whatever the element, use a consistent type size throughout your document. For overhead transparencies or computer projection in oral presentations, use 18 or 20 point type for body text and 20 or greater for headings.

USE FULL CAPS SPARINGLY.   Long passages in full capitals (uppercase letters) are hard to recognize and remember because uppercase letters lack ascenders and descenders, and so all words in uppercase have the same visual outline (Felker 87). The longer the passage, the harder readers work to grasp your emphasis.

Use full caps as section headings (INTRODUCTION) or to highlight a word or phrase (WARNING: NEVER TEASE THE ALLIGATOR). As with other highlighting options discussed below, use them sparingly.

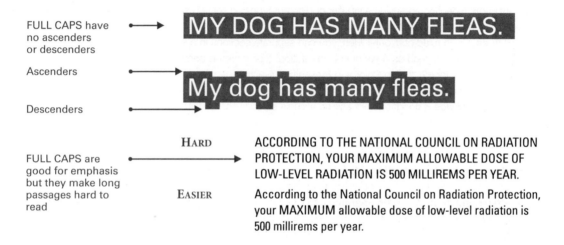

FULL CAPS have no ascenders or descenders

Ascenders

Descenders

MY DOG HAS MANY FLEAS.

My dog has many fleas.

FULL CAPS are good for emphasis but they make long passages hard to read

HARD

ACCORDING TO THE NATIONAL COUNCIL ON RADIATION PROTECTION, YOUR MAXIMUM ALLOWABLE DOSE OF LOW-LEVEL RADIATION IS 500 MILLIREMS PER YEAR.

EASIER

According to the National Council on Radiation Protection, your MAXIMUM allowable dose of low-level radiation is 500 millirems per year.

Guidelines
for Styling the Words and Letters

➤ **Use a serif font (such as Times New Roman) for formal documents such as reports, legal communication, and letters.** Also use serif fonts for newspapers, magazines, and other documents where readers' eyes will need to move across long lines of text.

➤ **Use a sans serif font (such as Helvetica) for captions, most visuals (charts, graphs, and tables), and engineering specifications.**

➤ **Create visual unity by using the same typeface throughout.** Different sizes and versions (bold, italic) are fine within the same typeface.

➤ **Keep fonts at sizes that people can read.** Standard body copy should be between 10 and 12 points. Use different sizes for headings, titles, captions, and so on.

Adding Emphasis

Ways to add
emphasis

Once you have selected the appropriate font, you can use different features, such as boldface or italics, to highlight important elements such as headings, special terms, key points, or warnings. The following Guidelines offer some basic highlighting options.

Guidelines
for Adding Emphasis

You can use page layout and typographic devices to highlight and add emphasis to certain items or sections of your document. But be careful: excessive highlights make a document look busy and make it difficult to read.

➤ **Use indentation.** Indenting and using a smaller or different type can set off examples, explanations, or any material that should be differentiated from body copy.

➤ **Use ruled lines.** Using ruled horizontal lines, you can separate sections in a long document or you can box or set off crucial information such as a warning or a caution. (For more on background screens, ruled lines, and ruled boxes, see Chapter 12, "Using Color and Presenting Visuals.")

When using typographic devices for highlighting, keep in mind that some options are better than others:

➤ **Use boldface. Boldface** is good for emphasizing a single sentence or brief statement and is seen by readers as "authoritative" (Aldus Guide 42).

➤ **Use italics.** More subtle than boldface, *italics* can highlight words, phrases, book titles, or anything else one might otherwise underline. But long passages of italic type can be hard to read.

➤ **Use small type in various situations.** Small type sizes (usually sans serif) work well for captions and credit lines and as labels for visuals or to set off other material from the body copy.

➤ **Use large type and dramatic fonts sparingly.** Avoid large type sizes and dramatic typefaces—unless you really need to convey forcefulness.

➤ **Use color sparingly.** Color is appropriate in some documents, but only when used sparingly. (Chapter 12, "Using Color and Presenting Visuals" discusses how color can influence audience perception and interpretation of a message.) Whichever options you select, be consistent: Highlight all headings at one given level identically; set off all warnings and cautions identically. And never mix too many highlights.

Using Headings for Access and Orientation

Benefits of using
headings

Readers of a long document often look back or jump ahead to sections that interest them most. Headings announce how a document is organized, point readers to what they need, and divide the document into accessible blocks or "chunks." An informative heading can help a person decide whether a section is worth reading. Besides cutting down on reading and retrieval time, headings help readers remember information.

LAY OUT HEADINGS BY LEVEL. Like a good road map, your headings should clearly announce the large and small segments in your document. When you write your material, think of it in chunks and subchunks. In preparing any long document, you most likely have developed a formal outline (see Chapter 10, "The Formal Outline"). Use the logical divisions from your outline as a model for laying out the headings in your final draft.

Figure 13.9 shows how headings vary in their position and highlighting, depending on their rank. However, because of space considerations, Figure 13.9 does not show that each higher-level heading yields at least two lower-level headings.

Many variations of the heading format in Figure 13.9 are possible. For example, some heading formats use decimal notation. (For more on decimal notation, see Chapter 10, "The Formal Outline.")

DECIDE HOW TO PHRASE YOUR HEADINGS. Depending on your purpose, you can phrase your headings in various ways (*Add Useful Headings* 17):

HEADING TYPE	EXAMPLE	WHEN TO USE	
Topic headings use a word or short phrase.	**Usable Page Design**	When you have lots of headings and want to keep them brief. Or to sound somewhat formal. Frequent drawback: too vague.	When to use which type of heading
Statement headings use a sentence or explicit phrase.	**How to Create a Usable Page Design**	To assert something specific about the topic. Occasional drawback: wordy and cumbersome.	
Question headings pose the questions in the same way readers are likely to ask them.	**How Do I Create a Usable Page Design?**	To invite readers in and to personalize the message, making people feel directly involved. Occasional drawbacks: too "chatty" for formal reports or proposals; overuse can be annoying.	

To avoid verbal clutter, brief *topic headings* can be useful in documents that have numerous subheads (as in a textbook or complex report)—as long as readers understand the context for each brief heading. *Statement headings* work well for explaining how something happens or operates (say, "How the Fulbright Scholarship Program Works"). *Question headings* are most useful for explaining how to do something because they address the actual questions readers will have (say, "How Do I Apply for a Fulbright Scholarship?").

Phrase your headings to summarize the content as concisely as possible. But remember that a vague or overly general heading can be more misleading or confusing than no heading at all (Redish et al. 144). Compare, for example, a heading titled "Evaluation" versus "How the Fulbright Commission Evaluates a Scholarship Application"; the second version announces exactly what to expect.

SECTION HEADING

In a formal report, center each section heading on the page. Use full caps and a type size roughly 4 points larger than body copy (say, 16 point section heads for 12 point body copy), in boldface. Avoid overly large heads, and use no other highlights, except possibly a second color.

Major Topic Heading

Place major topic headings at the left margin (flush left), and begin each important word with an uppercase letter. Use a type size roughly 2 points larger than body copy, in boldface. Start the copy immediately below the heading, or leave one space below the heading.

Minor Topic Heading

Set minor topic headings flush left. Use boldface, italics (optional), and a slightly larger type size than in the body copy. Begin each important word with an uppercase letter. Start the copy immediately below the heading, or leave one space below the heading.

Subtopic Heading. Incorporate subtopic headings into the body copy they head. Place them flush left and set them off with a period. Use boldface and roughly the same type size as in the body copy.

Figure 13.9 **One Recommended Format for Headings** Note that each head is set one extra line space below any preceding text. Also, different type sizes reflect different levels of heads.

MAKE HEADINGS VISUALLY CONSISTENT AND GRAMMATICALLY PARALLEL.
Feel free to vary the format shown in Figure 13.9—as long as you are consistent. When drafting your document, you can use the marks *h1, h2, h3,* and *h4* to indicate heading levels. All *h1* headings would then be set identically, as would each lower level of heading. For example, on a word-processed page, level one headings might use 14 point, bold upper case type, and be centered on the page; level two headings would then be 12 point, bold in upper and lower case and set flush left with the margin (or extended into the margin); level three headings would be 11 point bold, set flush left; level four would be 10 point bold, flush left, with the text run in. If your organization or instructor prefers a particular style guide (such as MLA or APA, that guide may have their own rules for headings.

Along with being visually consistent, headings of the same level should also be grammatically parallel. For example, if you phrase headings in the form of reader questions, make sure all are phrased in this way at that level. Or if you are providing instructions, begin each heading with the verb (shown in italics) that names the required action: "To avoid damaging your DVDs: (1) *Clean* the DVD drive heads. (2) *Store* DVDs in appropriate containers—and so on.

Guidelines

for Using Headings

➤ **Ordinarily, use no more than four levels of headings (section, major topic, minor topic, subtopic).** Excessive heads and subheads make a document seem cluttered or fragmented.

➤ **Divide logically.** Be sure that beneath each higher-level heading, you have at least two headings at the next-lower level.

➤ **Insert one additional line of space above each heading.** For double-spaced text, triple-space before the heading and double-space after; for single-spaced text, double-space before the heading and single-space after.

➤ **Never begin the sentence right after the heading with "this," "it," or some other pronoun referring to the heading.** Make the sentence's meaning independent of the heading.

➤ **Never leave a heading floating as the final line of a page.** If at least two lines of text cannot fit below the heading, carry it over to the top of the next page.

➤ **Use running heads (headers) or feet (footers) in long documents.** Include a chapter or section heading across the top or bottom of each page (see Figure 13.10). In a document with single-sided pages, running heads or feet should always be placed consistently, typically flush right. In a document with double-sided pages, such as a book, the running heads or feet should appear flush left on left-hand pages and flush right on right-hand pages.

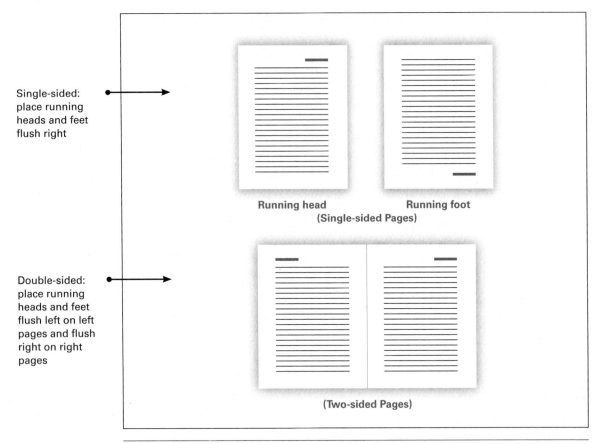

Single-sided: place running heads and feet flush right

Double-sided: place running heads and feet flush left on left pages and flush right on right pages

Running head **Running foot**
(Single-sided Pages)

(Two-sided Pages)

Figure 13.10 Running Heads and Feet Running heads and feet help readers find material and stay oriented.

Audience Considerations in Page Design

13.4 Develop an audience and use profile to guide your design

In deciding on a format, work from a detailed audience and use profile. Know your audience and their intended use of your information. Create a design to meet their particular needs and expectations (Wight 11):

How readers' needs determine page design

- If people will use your document for reference only (as in a repair manual), make sure you have plenty of headings.
- If readers will follow a sequence of steps, show that sequence in a numbered list.
- If readers need to evaluate something, give them a checklist of criteria (as in this book, at the end of most chapters).

- If readers need a warning, highlight the warning so that it cannot possibly be overlooked.

- If readers have asked for a one-page report or résumé, save space by using a 10-point type size.

- If readers will be encountering complex information or difficult steps, or if readers will be using the document mainly in PDF (on a screen), widen the margins, increase all white space, and shorten the paragraphs.

Consider also your audience's cultural expectations. For instance, Arabic and Persian text is written from right to left instead of left to right (Leki 149). In other cultures, readers move up and down the page, instead of across. A particular culture might be offended by certain colors or by a typeface that seems too plain or too fancy (Weymouth 144). Ignoring a culture's design conventions can be interpreted as disrespect.

> NOTE *Even the most brilliant page design cannot redeem a document with worthless content, chaotic organization, or unreadable style. The value of any document ultimately depends on elements beneath the visual surface.*

Designing Digital Documents

13.5 Explain how digital documents have special design requirements

As discussed in this chapter, many workplace documents are designed to be read in both print and PDF. But for some documents, additional digital formats may also be appropriate. For instance, a lawn and garden equipment manufacturer creates a User Guide for its lawnmowers; the guides are available in hard copy with the purchase of the mower and can also be viewed or downloaded in PDF from the company Web site. But the company may also decide to create a separate Web page with updated information about best practices for environmentally friendly lawn care (including keeping your mower blade set properly) as well as links to other information about watering and fertilizing your lawn. The company may also create a phone (and tablet) app that provides basic user information about the mower and also checks the local forecast and provides tips for lawn care given rain and weather conditions. Although the guidelines in this chapter apply most closely to print and PDF pages, many of these ideas (about white space, line length, grid patterns) are adaptable to other digital document types. In addition, pay special attention to the features discussed below.

Types of digital documents

Adobe Acrobat™ and PDF files

PDF stands for "portable document format." PDF documents retain their formatting and appear exactly the same in print as they do on the screen. Typically, PDF files are much more difficult for readers to alter than, say, a word processing document, thus protecting the integrity of your document's layout and design. You can create PDF

Characteristics of PDF files

files in most word processing or desktop publishing programs by choosing "PDF" as the file type when you click on "Save as."

Advantages of using PDFs

Companies are all too familiar with how often people lose or misplace user manuals and instructions. Putting these documents on a Web site or company social media feed, for easy access and downloading, saves these organizations countless hours of time and expense. Customers are also happy when they can find the manual they've been looking for and download it quickly.

PDF files can also be sent as email attachments, which is common in today's workplace as a way to distribute a report, handbook, or other document. Keep in mind, though, that PDF files, especially ones with lots of images and color, can be large in size. Some people's email accounts may limit the size of incoming attachments. (For more on PDFs and email, see Chapter 15.)

PDF documents can be viewed with Adobe Acrobat Reader or other PDF readers. To work on PDF files, you may need to upgrade to Adobe Acrobat Professional (part of the Adobe Creative Cloud). Acrobat Professional offers a number of writing, editing, and design features including allowing you to do the following:

- Combine individual PDF files (handy for taking several PDF files and creating one single document)
- Edit the document, insert text, or extract pages using a series of markup and editing tools
- Comment on sections of the document using the comment tools
- Circulate the document for review and digital signatures

Another advantage of PDF is that although they resemble print documents, PDFs are digital and, thus, searchable. Say, for example, you are reading a PDF report on changes in rainwater acidity over the past ten years. If you are interested in the results for your city or state, you can simply search, using the "Find" command (usually "control-F"), for the word or phrase you are looking for. PDFs also let you create "active" tables of contents and tabs, as well as embedded hyperlinks (links to Web sites). Both features allow readers to click on the topic or link they are interested in.

Web Pages

Characteristics of Web pages

Many of the guidelines discussed in this chapter also apply to Web pages. If you look at a well-designed Web page, you will notice features such as grid structures; use of white space for emphasis; attention to line length and margins; consistent spacing; strategic uses of color; and lists for easy reading. Several of these features are especially important on Web pages; for instance, content must be written in discrete chunks, so line length and white space are especially critical.

How Web pages differ from PDFs

Unlike typical print and PDF pages, however, Web pages must be designed to accommodate the shape of a screen. Most computer screens are more "landscape" than "portrait"—wider than they are high. On phones and tablets, readers can rotate the device and read a Web page in either format. So, pages must provide for plenty of marginal width, and lines of text can't be too long.

Most word processing programs offer features that let you save documents as Web pages. This approach works well for very simple Web pages, but for most situations, Web designers use sophisticated tools, such as Adobe Dreamweaver. Like desktop publishing, these apps offer many features and templates but have a steep learning curve. Even with these tools, designers often need to tinker with margins, headings, fonts, and other layout features in order to achieve the design most suited to their audience and purpose.

Tools for designing Web pages

For more on Web page design, as well as a discussion about the related topics of blogs and wikis, see Chapter 24. If this topic interests you, you may want to take a course, workshop, or online tutorial.

Tablets, Smartphones, and E-reader Pages

Today's workplace documents might be read on a large computer screen, a smaller laptop, a tablet, a phone, or an e-reader. Tablets are getting smaller, and phones are getting bigger; e-readers (such as the Kindle and others) come in a variety of sizes. As noted in this chapter, many organizations use PDFs as a way to ensure that document formatting does not change no matter where the document is viewed. This approach works well for the sort of documents referred to in this chapter (résumés, letters, memos, reports, and so forth). But there may be times when you will be part of a team writing and designing documents that are meant to be viewed on a specific device.

Special design considerations for small screen devices

For example, emergency room physicians might not have time to locate and consult the hospital's large, print copy of the *Physician's Desk Reference*; instead, they may prefer a more compact version for their phone or Kindle. The same may be true for engineers who need technical manuals that can be accessed outdoors, say, on a job site where a new bridge is being built.

If you know in advance that your document will be viewed on a small screen, you'll need to work with a team that includes specially trained designers who will use tools such as the Adobe Creative Suite to create a document suitable for that device. Just as with Web page design, it may still be necessary to come back to the key guidelines in this chapter to ensure the use of the most appropriate margins, headings, fonts, white space, and other layout features for your audience.

Social Media Posts

Social media posts are not considered documents per se, and the shape of most posts is not that of a traditional document page. But social media posts do offer small amounts of information that then typically will include a link so readers can access a longer document. Twitter provides numerous examples of short posts (tweets) that invite readers to learn more. Facebook, Instagram, LinkedIn, and other social media platforms often include both text and visuals, sometimes as "memes," that, like tweets, encourage readers

Special design considerations for social media

to view more via a link to a more formal document. For example, when NASA makes a new discovery, a social media post may lead to a white paper or report with detailed scientific findings. For these social media posts, your use of visuals and your choice of fonts is important. Allow plenty of white space so the information is not crammed into a small area. See Chapter 12, "Social Media and Cultural Considerations" for a discussion about how people read social media posts; see also Chapter 25 for information on how to write and design information for social media.

Checklist

Page Design

Use the following Checklist when designing print or digital documents.

Shape of the Page

❑ Are page numbers, headers, and footers used consistently? (See "Shaping the Page" and "Guidelines for Shaping the Page" in this chapter.)

❑ Does the grid structure provide a consistent visual theme? (See "Shaping the Page" and "Guidelines for Shaping the Page" in this chapter.)

❑ Does the white space create areas of emphasis? (See "Shaping the Page" and "Guidelines for Shaping the Page" in this chapter.)

❑ Are the margins ample? (See "Shaping the Page" and "Guidelines for Shaping the Page" in this chapter.)

❑ Is line length reasonable? (See "Shaping the Page" and "Guidelines for Shaping the Page" in this chapter.)

❑ Is the right margin unjustified? (See "Shaping the Page" and "Guidelines for Shaping the Page" in this chapter.)

❑ Is line spacing appropriate and consistent? (See "Shaping the Page" and "Guidelines for Shaping the Page" in this chapter.)

❑ Is each paragraph tailored to suit its purpose? (See "Shaping the Page" and "Guidelines for Shaping the Page" in this chapter.)

❑ Are paragraphs free of "orphan" lines or "widows"? (See "Shaping the Page" and "Guidelines for Shaping the Page" in this chapter.)

❑ Is a series of parallel items within a paragraph formatted as a list (numbered or bulleted, as appropriate)? (See "Shaping the Page" and "Guidelines for Shaping the Page" in this chapter.)

Style of Words and Letters

❑ In general, are versions of a single typeface used throughout the document? (See "Styling the Words and Letters" and "Guidelines for Styling the Words and Letters" in this chapter.)

❑ If different typefaces are used, are they used consistently? (See "Styling the Words and Letters" and "Guidelines for Styling the Words and Letters" in this chapter.)

❑ Are typefaces and type sizes chosen for readability? (See "Styling the Words and Letters" and "Guidelines for Styling the Words and Letters" in this chapter.)

❑ Do full caps highlight only single words or short phrases? (See "Styling the Words and Letters" and "Guidelines for Styling the Words and Letters" in this chapter.)

Emphasis, Access, and Orientation

❑ Are boldface and italics used appropriately? (See "Guidelines for Adding Emphasis" in this chapter.)

❑ Is the highlighting consistent and tasteful? (See "Guidelines for Adding Emphasis" in this chapter.)

❏ Do headings clearly announce the large and small segments in the document? (See "Using Headings for Access and Orientation" in this chapter.)

❏ Are headings formatted to reflect their specific level in the document? (See "Using Headings for Access and Orientation" in this chapter.)

❏ Is the phrasing of headings consistent with the document's purpose? (See "Using Headings for Access and Orientation" in this chapter.)

❏ Are headings visually consistent and grammatically parallel? (See "Using Headings for Access and Orientation" in this chapter.)

Audience Considerations

❏ Does this design meet the audience's needs and expectations? (See "Audience Considerations in Page Design" in this chapter.)

❏ Does this design respect the cultural conventions of the audience? (See "Audience Considerations in Page Design" in this chapter.)

❏ Does the design take social media into consideration, if applicable? (See "Social Media Posts" in this chapter.)

Projects

For all projects, check with your instructor about whether to present your findings in class, bring drafts to class for discussion, upload your project to the class learning management system (LMS), and/or use the LMS forum or discussion boards to collaborate and review each activity below.

General

1. Find an example of effective page design based on the guidelines and discussion in this chapter. Photocopy or download a selection (two or three pages), and attach a memo explaining to your instructor and classmates why this design is effective. Be specific in your evaluation. Now do the same for an example of ineffective page design, making specific suggestions for improvement. Bring or post your examples and explanations to class, and be prepared to discuss them.

As an alternative assignment, imagine you are a technical communication consultant and address each memo to the manager of the organization that produced each document.

2. The following are headings from a set of instructions for listening. Rewrite the headings to make them parallel.

- You Must Focus on the Message
- Paying Attention to Nonverbal Communication
- Your Biases Should Be Suppressed
- Listen for Main Ideas
- Distractions Should Be Avoided
- Provide Verbal and Nonverbal Feedback
- Making Use of Silent Periods
- Keeping an Open Mind Is Important

3. Using "Checklist: Page Design", redesign an earlier assignment or a document you've prepared on the job. Submit to your instructor the revision and the original, along with a memo explaining your improvements. Discuss your design in class.

4. On campus or at work, locate a document (PDF or print) with a design that needs revision. Candidates include career counseling handbooks, financial aid handbooks, student or faculty handbooks, software or computer manuals, medical information, newsletters, or registration procedures. Redesign the whole document or a two- to five-page selection from it. Submit to your instructor a copy of the original, along with a memo explaining your improvements. Be prepared to discuss your revision in class.

Team

Working in small groups, redesign a document you select or one that your instructor provides. Use a collaborative writing app (Google docs or your class LMS) to discuss the document's features and determine how you will proceed. Prepare a detailed explanation of your group's revision. Appoint a group member to present your revision to the class.

Digital and Social Media

Technical documents are often created using preformatted templates, which are available in Microsoft Word, Apple Pages, Google Docs, and similar apps. Templates allow you to create sophisticated, attractive documents quickly and easily. But this ease of use can restrict your options to change or customize a design for a specific audience and purpose. In groups of two, consider the heart surgery brochure described on the first page of this chapter. Find a few brochure templates and compare your findings to the design principles presented in the Guidelines boxes in this chapter. What are the pros and cons of using a template for this situation? How much are you able to manipulate the template's settings? For example, can you change an all caps heading to upper- and lowercase? Can you change the colors?

Global

Find a document that presents the same information in several languages (assembly instructions for various products are often written in two or three languages, for example). Evaluate the design decisions made in these documents. For example, are the different languages presented side by side or in different sections? Write a memo to your instructor evaluating the document and making recommendations for improvement.

Part 4
Specific Documents and Applications

Chapter 14
Email

nicoelnino/123RF

"I work for a bank, and our environment is regulated by many government organizations, both national and international. I write and read dozens of emails every day and am well aware that every email has potential legal implications. When I hire an intern or new employees, I always remind them to write emails that are professional and factual and to avoid wisecracks and rude remarks. Also, at our company, we don't use email to send confidential information, and we don't use our work email for anything that is not work related.**"**

—*Aneta Dorrigan, Financial Executive*

Learning Objectives

14.1 Identify the workplace uses of email
   and the three types of workplace
   emails

14.2 Think carefully about audience and
   purpose when writing workplace
   emails

14.3 Identify the components of a
   workplace email message

14.4 Write an email using a professional
   style and tone

14.5 Recognize ethical, legal, and global
   issues affecting email use

Email Basics and Types

14.1 Identify the workplace uses of email and the three types of workplace emails

In today's workplace and professional settings, the primary way people communicate is
with email. According to one survey, people spend more than five hours per workday
checking their email (Naragon). Email is easy to use and offers both rapid speed and wide
reach: with one keystroke, an email message can immediately reach the inboxes of thousands of people and can easily be forwarded to others. Email also provides organizations
with written documentation—an electronic trail—that helps track a project or conversation
and that may prove critical for project management and legal purposes down the road.

Workplace uses
of email

Email is especially useful when people are in different time zones or have different work schedules. If for example you are based in the United States and members
of your team are in China, a 12-hour time difference means that email is much better than using a chat session or real-time video conference. Or, if you work late at
night and need to communicate with colleagues who start their workday early in the
morning, you can send your email at 2 A.M. without disturbing anyone, and your colleagues will wake up to the information they need.

Time zone and
scheduling
benefits of using
email

At work, email is used for everything from routine correspondence ("Received
the file, thanks") to formal messages from management ("Please see the information
below regarding our newest product line"). In general, it can be helpful to think of
most workplace email as fitting into one of these three categories:

Definition of
primary email

A *primary email* is one in which the body of the email contains all of the content. In other words, there is no separate attachment; the email message itself contains all the information the writer wishes to convey. For example, in Figure 14.1, Rachel Byerson sends a short message to one of her staff. (Notice that the content is brief. For longer, more complicated content, you would use a transmittal email with an attachment.)

Definition of
transmittal email

A *transmittal email* is similar to a transmittal memo (see Chapter 15). In these cases, email is used to describe, in brief, the background and context for a longer, more formal attachment. For example, in Figure 14.2, Harvey Keck uses email to send a contract (in PDF) to a consultant he wishes to hire.

Definition of
formatted email

A *formatted email* is similar to primary email in that the body of the email contains most of the content. But formatted email also uses links to provide readers with access to more detailed information. Formatted email is typically more formal, often using the company logo. In Figure 14.3, Lamdet Publishers uses a formal email to communicate to the entire organization information about a new workplace health and wellness initiative.

Email is short and
respectful

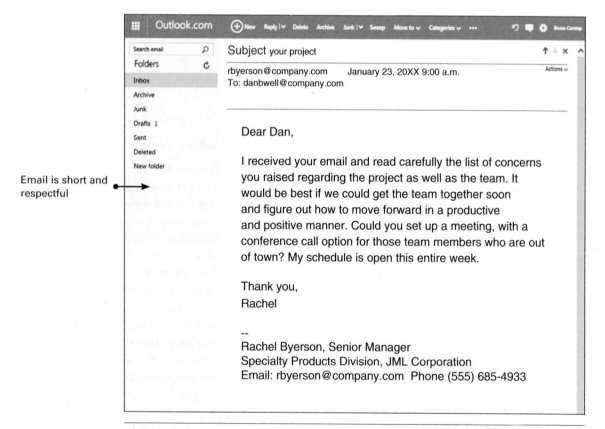

Figure 14.1 A Primary Email In this email, the body of the message contains all of the content (there are no other attachments).
Source: Microsoft Outlook 2013.

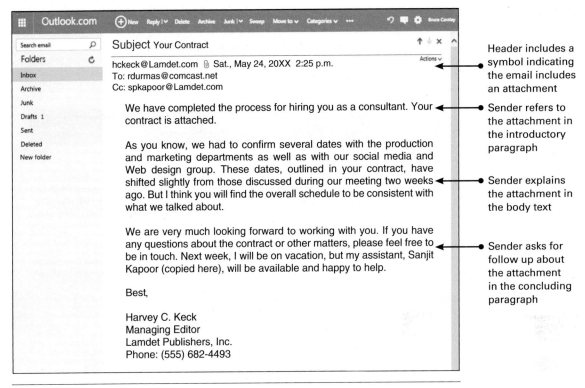

Header includes a symbol indicating the email includes an attachment

Sender refers to the attachment in the introductory paragraph

Sender explains the attachment in the body text

Sender asks for follow up about the attachment in the concluding paragraph

Figure 14.2 A Transmittal Email This email includes an attachment notification and refers to the attached contract in all three paragraphs.

Source: Microsoft Outlook 2013.

Considering Audience and Purpose

14.2 Think carefully about audience and purpose when writing workplace emails

No matter how careful you are about sending email only to one particular person or group, in the end, you have little control over who the final readers will be. For instance, you might intend to address only your project team, with a cc to your manager, but because email can be forwarded so easily, your message could end up being read by a company vice president. As people work together and become more familiar and comfortable with each other, email exchanges often shift from a professional tone to a more casual, off-the-cuff tone, which may work within the team but could be completely misunderstood if that email ends up being read by others. Because email can travel so widely with so little effort (just press "send" and there it goes, from your in box to half the company), audience considerations are crucial.

Suppose, for example, that after a long week on a difficult assignment, you email a colleague with a lengthy list of complaints about one of the engineers not holding up his end of the project. You quickly press "Send" and head out for the weekend. At the

Audience considerations

Internal and external audiences for email

Company letterhead signals the importance of the email

Fonts, hyperlinks, and bulleted list encourage ease of reading

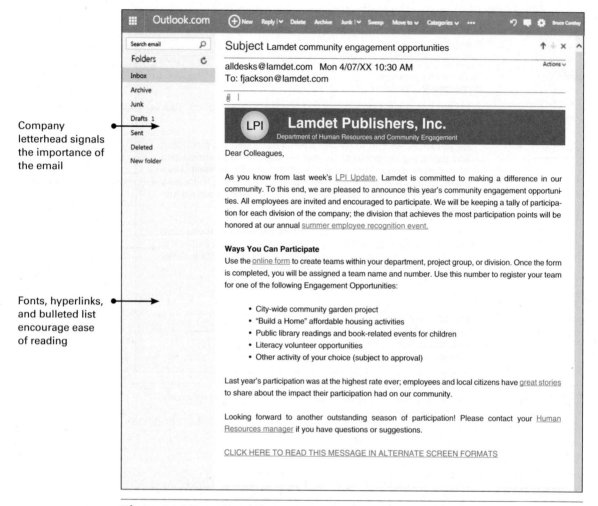

Figure 14.3 A Formatted Email These formal emails are typically sent from an organization to all employees. Compared with email between individuals, this type of email uses formatting for the company heading, fonts, hyperlinks, and a bulleted list to encourage ease of reading. Emails of this type, whether internal or external, are typically written collaboratively by the marketing, human resources, or similar units within the organization.
Source: Microsoft Outlook 2013.

regular Monday morning status meeting, you are surprised to find that your message was forwarded—inadvertently or deliberately—to other colleagues, including the engineer you wrote about. As this example demonstrates, when writing email, always assume the message will travel far beyond its intended recipient. This way of thinking could save you from sending emails that you may later come to regret. Your audience might be *internal* (most email is sent within the company or organization), but you might also be writing emails to send to *external* readers, such as potential or existing customers or clients (either as individual correspondence or as regular updates that people signed up for). In both cases, internal or external, you still need to consider audience carefully.

Purpose considerations

While the audience for an email might be vast, the purpose of email should be specific: to schedule meetings, update or brainstorm with team members on a project, contact colleagues for answers you need, provide clients with important updates, and send other forms of workplace communication. When using email at work, whether internally or externally, remember that people are busy. Make the purpose clear and the message brief; use a transmittal email with attachments if the topic is long and complicated. Be sure your purpose is strictly work related.

Email Parts and Format

14.3 Identify the components of a workplace email message

Almost all email programs contain a standard heading section, with fields for "To," "From," and "Subject" (or, instead of Subject, "Re," which is short for "in reference to"). You can also include people other than the recipient by using the "cc" field (short for "courtesy copy"). A final field, "bcc," stands for "blind carbon copy" and is used when you want to hide the email addresses in that field. (See "Ethical, Legal, and Global Issues When Using Email" later in this chapter for more on using the bcc function.)

Email parts

Like many workplace documents, effective email messages should have a brief introduction that gets right to the point, a clear body sentence or section, and a brief conclusion (which, if appropriate, often requests action, such as "please get back to me"). Figure 14.2 illustrates a workplace email with three paragraphs (introduction, body, conclusion), while Figure 14.1, although only one paragraph, still starts with an introductory sentence, makes a suggestion, then concludes by requesting a meeting.

Standard email format

Depending on the level of formality, the style of a particular workplace, and the specific situation, it may be appropriate for an email to start off with a salutation ("Dear Dan") and/or end with a closing ("Best, Harvey"). Most workplace emails also end with a signature block, containing the sender's name and contact information. (Typically you can create a signature block in the preferences settings of most email apps.)

Optional format elements

Everyday email correspondence, especially between individuals (such as Rachel Byerson's email in Figure 14.1), need not use any special fonts or visuals. But more formal email messages sent from an organization (such as Lamdet Publishers' announcement in Figure 14.3 or the email to its customers in Figure 14.5) may be formatted with the company logo and appropriate fonts and layout, including links and alternative ways to view the information.

Using special formats and visuals

Email Style and Tone

14.4 Write an email using a professional style and tone

Workplace email differs from personal email

The word "netiquette" (a word created by combing "Internet" and "etiquette") is used to suggest that online, people need to pay special attention to style and tone. In the workplace, appropriate netiquette means being professional, respectful, and polite at all times. Workplace email should be professional not only because of the potential for critical misunderstandings and errors but also, and importantly, out of respect for your colleagues and coworkers. Also, keep in mind that any email message written at work may be used in legal proceedings. So a professional, factual style and tone are critical.

Workplace email needs to be professional

Yet in an attempt to keep up with the sheer volume of email at work, even extremely careful writers often become careless as they dash off dozens or more emails daily. The kind of writing appropriate for email or text exchanges with your family or friends—full of humor, casual phrasing ("uh huh," "cool"), an informal tone (Hey!), or shouting (by using ALL CAPS)—is not suitable for workplace email. Also, avoid angry, sarcastic, or insulting messages that don't serve any real purpose other than to start a long, out of control email chain. Use language that is respectful in tone and does not blame. Be polite, professional, and thoughtful. If the situation is complicated and seems highly charged, suggest a phone call or meeting instead and keep your reply brief (Figure 14.1 offers a good example of a diplomatic, respectful email that is kept brief and clear and does not engage). When corresponding with customers or clients, remember that your words represent the entire organization, not just your individual point of view.

Proofread and check spelling on email

At work, spelling and accuracy matter. If you are part of a team involved in a complicated technical project, and you don't proofread your email carefully, one seemingly small mistake in a number, formula, weight, measure, date, or other detail could send your project into a death spiral. Because we tend to type emails very quickly and send them off almost just as quickly, we often don't apply the same careful scrutiny to email content that we might for a formal report or proposal. Especially if your email contains key technical information, check it over carefully. Have a colleague proofread it. Write your first draft using Word (or similar), then when you are sure the information is accurate and the tone is appropriate, paste the content into the email window. Use spell check and autocorrect, but as discussed in Chapter 11, these apps are not foolproof, so check everything carefully.

Interpersonal Issues and Email

Email omits social and physical cues

As described in Chapter 5 (See "Interpersonal Issues in Global Teams"), email omits important social and physical cues such as facial expressions, tone of voice, eye contact, and immediate feedback. An awkward situation, therefore, makes it tempting to send an email rather than use the phone or meet personally. In Figure 14.4, Frank is upset about his annual performance review. Rather than wait, organize his thoughts, and request a meeting with his supervisor Mitch, Frank hurriedly writes an inappropriate email. The speed that makes email so efficient for many workplace discussions can hamper good judgement when a message is written quickly and the writer presses "send" before reconsidering.

Like many volatile situations, the one described in Figure 14.4 would be easier to resolve in a face-to-face meeting. If distance is an issue, try a video conference or phone call, not more email. Before the meeting, the writer could take time to list his concerns, ask a colleague or friend for advice, review his list again, and come up with a skillful, strategic approach that he could practice in advance. Using a logical, professional style is always more appropriate at work than writing an email that may feel good when you first send it but will probably not have the effect you are hoping for.

Don't use email to avoid essential face-to-face contact

In general, don't use email for complex discussions. For example, in Figure 14.1, Rachel, a manager at a large manufacturer of vitamin products, received a long, rambling email from her employee Dan, complaining about problems and specific people on a team project. Rachel realizes that these problems are too complicated to solve via email—they involve not just Dan but an entire team. If she were to respond by email, she would lose the opportunity to hear Dan's points in person and to convey to Dan (using words as well as physical cues) that she takes his concerns seriously. Because Dan's email mentions team members by name, Rachel wisely chooses to set up a

Avoid using email for complex discussions

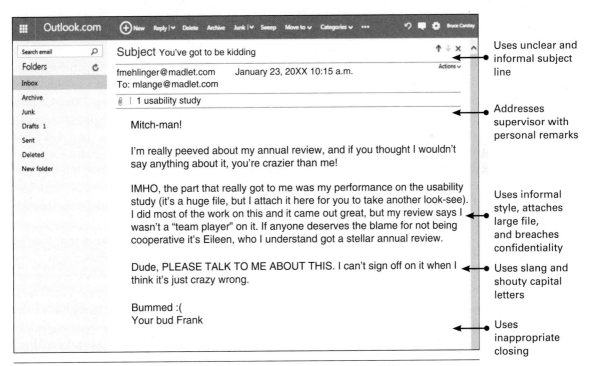

Figure 14.4 A Workplace Email Lacking Professional Style and Tone The style and tone of this email are inappropriate for the workplace.
Source: Microsoft Outlook 2013.

meeting with the entire team, because the issues need to be discussed by everyone involved, not just as a back and forth email exchange between Rachel and Dan.

Rachel therefore responds by asking Dan to set up a conference call. Notice that Rachel does not "take the bait"—she doesn't respond to Dan's specific complaints about colleagues, which could easily begin a long chain of replies that might get off topic or cause hurt feelings. Instead, Rachel uses a considerate, polite tone to diffuse the situation.

Meetings or phone calls can be better than a long email chain

In the case of Rachel and Dan, they do know each other and can base their communication on past interactions. But when you are emailing external readers, such as customers, you can only go by the information you have on hand (such as an online complaint form, a Yelp review, a social media post, or an email sent to the company). Many companies have created guidelines for how to respond to customers. Everything you put in writing provides a trail, so be sure that your words are honest and clear and that they represent the company policy. If the discussion gets complicated, you might offer to call the customer, if that approach is allowed by your organization.

Choose the Right Approach for the Situation

Consider alternatives to email in difficult situations

As discussed in the case of Dan and Rachel, if a project or situation becomes too complicated, sensitive, or emotionally charged, consider other options such as a face-to-face meeting, a phone call, a video conference, or another kind of online meeting that allows for voice and images. After the meeting, use email to summarize the meeting outcomes in a factual, objective manner. You might also set up an online forum or wiki where team members can continue the discussion, encouraging open collaboration over rapid-fire email discussions. Or, you might encourage short weekly team meetings that allow everyone to connect and share ideas and concerns. Doing so on a weekly basis might help diffuse any concerns before they spiral out of control on email.

Consider an in-person approach in positive situations

Also, keep in mind that for positive situations, such as when a coworker wins an award or completes a big project, email is a nice way to offer congratulations and showcase accomplishments to that individual and the entire company, but a phone call or in-person congratulations is also greatly appreciated (and can be a pleasant surprise that helps boost morale and makes a person feel valued). You might also use the weekly meeting to make these announcements in person.

Avoid huge attachments and provide opt-in and opt-out options for customers

Email also has technical limitations that can add to the unprofessional tone of the message. For instance, large email messages with big attachments might cause an email app to load very slowly or even crash. Recipients are justifiably annoyed by the amount of time required for large files to download. In Figure 14.4, Frank attaches a large PDF usability study—he calls it a "huge file"—even though it's clear from the email that Mitch already has a copy of this file. Frank's message is already inappropriate for a workplace setting, and this big attachment will just make matters worse. If you need to include attachments (photographs, long reports, word processing files, spreadsheets, or similar), ask the recipient in advance if he or she would prefer the attachment via email or placed on a shared server (such as the secure server at work, Google drive, Dropbox, or similar). For customers, be sure to take an "opt in" approach: give customers the option to say that they want to be on an email list (to receive updates, coupons, or the like)—and, provide a clear and easy way for them to get off the list (opt out).

Inappropriate formatting can also make an email appear unprofessional in tone and style. As described earlier in this chapter, internal email correspondence between individuals typically does not require fancy formatting. Use the default font (usually Times Roman or Helvetica) of your email application for primary and transmittal emails. For formatted emails from the company or organization that are designed more like a brief newsletter or announcement (internal or external), you can use visuals, fonts, hyperlinks, and other page design features similar to Figure 14.4. Follow the page design guidelines in Chapter 13. Keep in mind that for these types of emails, the use of bullets, white space, and headings will be helpful to readers who are checking email on smaller screens (phone or tablet). Your company may have a template for use when creating and sending official formatted emails.

Consider when to use non-standard email formatting

Email and External Audiences

Email intended for external audiences requires writers to consider a few more items. Existing clients are probably people you have worked with in other capacities. For example, if you are a lawyer, you might meet with clients a few times (say, for a real estate transaction) and then use email to continue the process, send forms to sign, and so forth. Or, if you are a contractor, you might meet with potential clients then follow up with an email that summarizes the meeting and offers a proposal. In these cases, even as you get to know your clients, remember that you are representing yourself and your organization and need to maintain a professional approach, no matter how friendly the transaction may be.

Using email with your clients

When using email with customers, as discussed previously, always provide opt-in and opt-out options. If the email is intended to provide regular updates (about sales, coupons, new features, technical updates, and the like), a formatted email creates a sense of professionalism. But remember that scammers also use emails that look credible and professional, so make sure the "From" address reflects your company name so readers know the email came from the company and is not a scam. To be extra-safe, create a statement that says something like "We will never solicit financial or personal information over email. Please call us if you have any questions." Make the opt out link easy to find. And even if customers opt out, make sure any automated reply thanks them for being customers.

Using email with your customers

Remember that while email to external audiences needs to be professional, it doesn't need to be stuffy or boring. It's fine to be friendly and engaging. The email in Figure 14.5 is a good example of how to be personal in tone and style, brief, direct, and engaging. In many organizations, these type emails are created by people with job titles such as *social media manager* or *social media marketing specialist*. See Chapter 25 for more on writing for social media.

External email doesn't need to be stuffy

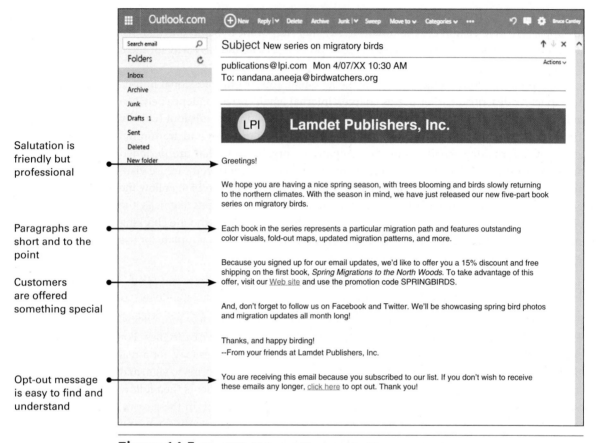

Salutation is friendly but professional

Paragraphs are short and to the point

Customers are offered something special

Opt-out message is easy to find and understand

Figure 14.5 Email To External Customers When customers sign up to receive email notifications, use a professional but friendly tone. Always provide a way for readers to opt out.
Source: Microsoft Outlook 2013.

Ethical, Legal, and Global Issues When Using Email

14.5 Recognize ethical, legal, and global issues affecting email use

Email is permanent

Because email is so widely used in workplace settings, you need to consider carefully certain ethical and legal issues. Email messages are archived and can be retrieved years later in the case of lawsuits, patent disputes, employee complaints, safety hearings, and many other potential situations. Messages intended only for inside a company or government agency or emails written only with certain readers in mind have and will continue to end up on the front page of national newspapers, on Twitter and other forms of social media, in government hearings, and in the courtroom. As we rely more and more on digital communication at work, especially email, the potential for ethical and legal issues looms large. What you write is permanent, despite the feeling that nobody will ever see that late-night email you sent at the end of a long day.

If you are an employee, what you write at work is typically owned by the company (see Chapter 7, "Consider This: Frequently Asked Questions about Copyright" for more on the works-for-hire doctrine, part of copyright law). Unless stated in writing, therefore, employers own the contents of your email; court decisions to date agree that under most situations, employers also have a right to monitor and read workplace employee email.

Workplace email is not private

Another ethical concern is when your reply is intended only for one person but you accidentally press "reply all" and send the email to everyone on the list, causing embarrassment to the original writer or worse, sharing information that was not ready for the entire team. Or, without asking, your colleague decides to forward your email, written just to a few people, to a company vice president. You are summoned to a meeting to explain your idea, without any time to prepare.

Workplace email can travel beyond its original destination

A related concern is with the "bcc" (blind courtesy copy) function. Unlike the cc function, where email addresses are visible, the bcc function is used to include recipients but not have their email address visible to others. This feature can be efficient if you want to send an email to lots of people and don't want to bog down their email header with dozens or even hundreds of addresses. (For example, some people use bcc to send out invitations.) But the bcc function can be questionable if you use it hide other recipients for less than honorable reasons. In Figure 14.1, Rachel, a senior manager, writes back to Dan, a team member. From Dan's perspective, the email is written to him and no one else (Rachel does not "cc" anyone.) But what if Rachel had bcc-ed the head of human resources or a company vice president, as a way to signal that Dan was a troublemaker? This use of bcc might be considered unethical.

Use the bcc function ethically

Almost every major news story—about political campaigns, airline or building catastrophes, corporate ethics cases, government agencies—contains some reference to email. Email messages can easily become the centerpiece of legal battles, often revealing more information than anyone could imagine and doing so in very public ways. In the now famous case of one U.S. corporation, those emails are still accessible even though the legal case took place over ten years ago (Silverstein; Leber). Many state and federal laws include penalties for destroying or altering information (including email) necessary for investigations. Even wisecracks and rude remarks can be used as evidence in harassment or discrimination cases. For email sent to external audiences (clients or customers), careless and inaccurate wording could mean a liability or other lawsuit; poor writing can drive away potential customers if the content seems unprofessional or unclear.

Legal issues and workplace email

In addition, email travels quickly. Just one click on the "Send" command, and your message can be routed to countless recipients. If these readers press "Forward," the message may travel to even more people. In any setting, but especially in companies that have international offices and clients, email can be read by dozens if not more people across the globe.

Email can quickly travel across countries

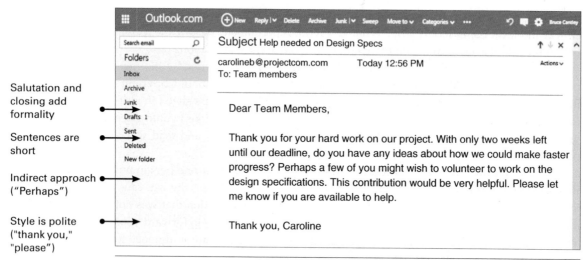

Salutation and closing add formality

Sentences are short

Indirect approach ("Perhaps")

Style is polite ("thank you," "please")

Figure 14.6 **Email Message Written to a Global Audience**
Source: Microsoft Outlook 2013.

All email is potentially global

Consider that any email could potentially be read by someone in another country; especially when you know that your audience contains people from many different cultures and countries, always write with a global audience in mind. Chapter 5 (Figure 5.3) illustrates an email message that is inappropriate for a global audience. Figure 14.6 below shows a more appropriate message. The "Guidelines for Writing and Using Email" in this chapter also lists items to consider when you email a global audience.

Guidelines

for Writing and Using Email

Audience and Purpose

➤ **Consider your audience.** If you are writing to a customer, client, stranger, or someone in authority, use a more formal tone than for a coworker or immediate supervisor. In those situations, use a formal salutation and closing (see "Email Parts and Format" in this chapter).

➤ **Consider your purpose.** If the situation is complicated or requires discussion and back-and-forth exchange, don't use email. Consider setting up a meeting, conference call, or video conference.

➤ **If your message is an official company communication, announcing a new policy or procedure or the like, consider writing a formal memo and sending as an attachment.** (See Chapter 15 for more memos and letters.)

➤ **Check and answer your email daily.** If you're really busy, at least acknowledge receipt and respond later.

➤ **Check your distribution list before each mailing.** Deliver your message to all intended recipients (but not to unintended ones).

➤ **Spell each recipient's name correctly.** There is no bigger turn-off for readers than seeing their name spelled incorrectly.

Formatting

- ➤ **For very brief email, stick with just one paragraph.** When more detail is required, follow the structure shown in Figure 14.2.
- ➤ **Don't indent paragraphs.** Instead, double-space between paragraphs.
- ➤ **End with a signature block.** Include your contact information.
- ➤ **Don't send huge or specially formatted attachments without first checking with the recipient.** Typically, files over 10 megabytes can cause download problems. Use a standard format, such as PDF.
- ➤ **Use formatting sparingly.** Headings, bullets, and font changes such as italics and bold are most appropriate for group announcements, such as Figure 14.5.
- ➤ **Use your email application's default font for everyday messages.** A decorative font is not a good choice for a workplace memo. If you do choose a font, stick with Times Roman or Helvetica.

Style, Tone, and Interpersonal Issues

- ➤ **Write a clear subject line.** Instead of "Test Data" or "Data Request," for example, be specific: "Request for Beta Test Data for Project 18."
- ➤ **Keep it short.** Readers are impatient and don't want to scroll through long screens of information. Use an attachment or direct readers to an online source for more information.
- ➤ **Be polite and professional.** When in doubt, use a formal tone. Avoid angry, personal attacks on other people (known as "flaming," this style has no place in workplace communication).
- ➤ **Use emoticons and abbreviations sparingly.** Use smiley faces and other emoticons strictly in informal messages to people you know well. Avoid these symbols when writing to international readers. The same goes for common email abbreviations such as BTW ("by the way"). Avoid ALL CAPS, which means that you are screaming.
- ➤ **Proofread and run the spell check before pressing "Send."** Misspellings not only affect your credibility and image, but they also can confuse your readers.
- ➤ **Don't use email when a more personal approach is called for or for complicated issues.** Although it may be uncomfortable, meeting in person is often the best way to resolve a complex situation. For group situations, even if everyone can't attend the meeting, use a conference call or video session. For external audiences, responses to customer complaints should include other ways to help the customer, including social media sites and an 800 number.
- ➤ **Focus on the message.** The survey cited at the start of this chapter notes that 69% of respondents said that they check email while watching TV or a movie (Naragon). While it might be tempting to do so, watching TV while reading workplace email could cause you to make mistakes—from spelling errors to unprofessional tone to incorrect data–that you might come to regret later.

Ethical and Legal Issues

- ➤ **Be careful about using "Reply All" if you only intend to reply to the original writer.**
- ➤ **Assume that your email is permanent and readable by anyone at any time.** "Forensic software" can find and revive deleted files.
- ➤ **Avoid wisecracks and rude remarks.** Any email judged harassing or discriminatory can have dire legal consequences.
- ➤ **Don't use email to send confidential information.** Avoid complaining, evaluating, or criticizing, and handle anything that should be kept private (say, an employee reprimand) in some other way.
- ➤ **Don't use your employer's email network for messages that are not work related.**
- ➤ **Before you forward a message, obtain permission from the sender.**

➤ **Use the "bcc" feature if you have a long list of email addresses and don't want to burden readers with scrolling through these.** But don't use "bcc" to hide from your recipient the other people who will be reading the message.

➤ **For external audiences, be professional but don't be afraid to be friendly.** Keep messages short. Always allow readers to opt out.

Global Issues

➤ **Avoid humor, slang, and idioms.** These items may be offensive in different cultures; also, this kind of language does not translate easily.

➤ **Write simple, short sentences that are easy to translate.**

➤ **Convey respect for your recipient.** A respectful tone and style is appreciated in any culture.

➤ **Don't be too direct or blunt.** Some cultures find directness offensive.

➤ **Be an active listener.** Don't respond immediately; read email carefully to get a sense of the cultural norms of the writer and the other readers.

Checklist

Email

Use the following Checklist when writing email.

❑ Is the message short and to the point, with a clear subject line? (See "Email Parts and Format" in this chapter.)

❑ For a longer email, do I use introduction/body/conclusion format? (See "Email Parts and Format" in this chapter.)

❑ Is email the best medium in this situation, versus a call or visit? (See "Interpersonal Issues and Email" in this chapter.)

❑ If a situation is complicated, have I used email to acknowledge the issue, but suggested other avenues for solving the problem? (See "Choose the Right Approach for the Situation" in this chapter.)

❑ Is the tone professional and courteous, avoiding insults and accusations? (See "Email Style and Tone" in this chapter.)

❑ Have I avoided emoticons, ALL CAPS, and abbreviations? (See "Email Style and Tone" in this chapter.)

❑ Have I been careful not to send overly large attachments? (See "Choose the Right Approach for the Situation" in this chapter.)

❑ Have I avoided using excessive fonts, colors, and backgrounds? (See "Choose the Right Approach for the Situation" in this chapter.)

❑ Have I maintained confidentiality and privacy? (See "Ethical, Legal, and Global Issues When Using Email" in this chapter.)

❑ Will my email be easy to understand and inoffensive to global readers? (See "Ethical, Legal, and Global Issues When Using Email" in this chapter.)

❑ Have I proofread, spell checked, and verified my distribution list? (See "Interpersonal Issues and Email" in this chapter.)

Projects

For all projects, check with your instructor about whether to present your findings in class, bring drafts to class for discussion, upload your project to the class learning management system (LMS), and/or use the LMS forum or discussion boards to collaborate and review each activity below.

General

Individually or in small groups, consider whether each of the following situations would be suitable as an email message. If not, why (and what other approach would you take)? If so, which type of email would you use: a primary email (Figure 14.1), a transmittal email (Figure 14.2), or a formatted email (Figure 14.3)? In which cases would you add an attachment, and why? Consider all items in the Guidelines box, especially interpersonal and legal issues.

- Sarah Burnes's memo about benzene levels (Chapter 1, "Case: Considering the Ethical Issues")
- The "Rational Connection" memo (Chapter 3, "Case: Connecting with the Audience")
- The "better" memo to the maintenance director (Chapter 3, "Organizational Constraints")
- Rosemary Garrido's letter to a potential customer (Chapter 3, Figure 3.4)
- The medical report written for expert readers (Chapter 2, Figure 2.3)
- A memo reporting illegal or unethical activity in your company
- A personal note to a colleague
- A request for a raise or promotion
- Minutes of a meeting
- Announcement of a no-smoking policy
- An evaluation or performance review of an employee
- A reprimand to an employee
- A notice of a meeting
- Criticism of an employee or employer
- A request for volunteers
- A suggestion for change or improvement in company policy or practice
- A gripe
- A note of praise or thanks
- A message you have received and have decided to forward to other recipients

Team

Working with a team of 2–3 other students, discuss how you would write, format, and distribute an email announcing a new policy at work. You can base this activity on a job or internship you've had, or you can select from the list in General Projects. Create your email message and present it to class.

Digital and Social Media

Imagine you work for a company that publishes digital books and podcasts related to your field (ecology; engineering; nursing; science; other). You have been asked to write monthly email updates to customers who have opted into the mailing list. What kind of updates would your customers care about, and how would you create each email so that it is friendly yet professional, short and to the point, and interesting but not too technical? (Hint: review how to assess your audience's technical background in Chapter 2.) Write and design a formatted email similar to the one shown in Figure 14.3. Now, write another email to members of your workplace team, explaining this new monthly update and inviting them to contribute content.

Global

You've been appointed by your manager to be team leader for an important new project. The team involves people from the following countries: China, Germany, India, Ireland, and the United States. Your first assignment is to come up with a communication plan. You decide that given the time differences, the team should use email as its primary means of communication. But you know that when people write using email, there can be misunderstandings due to tone, style, and levels of directness. These issues might be amplified by the differences in cultures between team members. Use material in this chapter, from Chapter 4, and on the Internet to research these issues. Write an email to the team outlining guidelines for communication for this project. Remember that the email you send needs to reflect the guidelines you are proposing.

Chapter 15
Workplace Memos and Letters

Shapercharge/iStock/Getty Images

❝At work, everyone is busy, and you need to be clear about why people should read your message and, if appropriate, act on it. You're competing with a constant stream of information to get the attention of your readers. To separate important information from routine emails, I use memos or letters. Even though these items are sent via email, usually as PDF attachments, memos and letters are less likely to be overlooked or deleted. Memos signal an important announcement or decision; letters convey an even more formal, official mode of communication. Effective memos and letters take time to research, write, and proofread. It's important to take deliberate steps in writing and revising for content, format, and tone. These official communications usually require a collaborative approach, involving several managers as well as the human resources or legal department. Sometimes, to slow the process down and give myself time to reflect, I print out the final version and proofread it carefully before sending.❞

—*Pat Hoffmann, Marketing Communication Manager for a software development company*

Learning Objectives

15.1 Understand the basics of memos, including their audience and purpose

15.1 Identify the parts and format of a standqward memo

15.1 Use proper tone in all memos

15.1 Write transmittal, summary, and routine miscellaneous memos

15.1 Understand the basics of letters, including their audience and purpose

15.6 Identify the parts, formats, and design features of workplace letters

15.7 Use proper tone in any letter

15.8 Explain the global and ethical implications of workplace letters

15.9 Effectively deliver bad or unwelcome news in a letter

15.10 Write inquiry, claim, sales, and adjustment letters

Memo Basics, Audience, and Purpose

15.1 Understand the basics of memos, including their audience and purpose

Definition of memos

"Memo" (short for "memorandum") is derived from the same Latin roots as the words "memorize," "remember," and "remind." Memos remind readers about important events, give directives, provide instructions and information, and make requests. Memos are important in workplace communication because unlike the daily, routine emails sent between individuals and teams, memos signal a more official communication, usually a message from the company, a manager or director, or another person or group acting in an official capacity.

Most memos are sent via email, either as the body of the email message or as a PDF attachment. Chapter 14 provides examples of email memos, which can be formatted (Figure 14.4) or unformatted (Figure 14.1), depending on the level of formality appropriate for your audience and purpose. PDF memos, usually sent as attachments, require a short cover email stating that readers should be sure to open and review the attachment. In some cases, organizations may choose to also send the memo on paper or ask readers to print the PDF copy (for example, if the legal department determines that both an electronic and a hard copy are required to comply with certain regulations). With the amount of email received at work, readers need to see a clear and compelling subject line to understand that the memo is important. *Memos and email*

Whether digital or paper, memos provide formal, written documentation about an event or issue. Organizations rely on memos to trace decisions and responsibilities, track progress, and check agreements and commitments made. Make sure the memo's content is specific, unambiguous, and accurate. Because memos are official documents, they can have far reaching ethical and legal implications. You should do as much research as needed to ensure that you have all your facts straight. This research might include consulting with other groups such as accounting, marketing, human resources, and legal. *Memos have ethical and legal implications*

To determine your approach to any particular memo, identify the various audience members who will receive it. Some companies use standard memo distribution lists, via email, for various audiences: a list for managers, a list for software developers, a list for the legal department, and so on. Keep in mind that digital distribution makes it easy for a memo intended for one audience to quickly be forwarded to others. *Audience considerations*

The purpose of your memo should also be clear: Is it to inform your audience? To persuade people to support a new plan? To motivate them to take action? To announce bad news? In addition to memos that take on these questions, memo format might also be used to write a short or informal report (see Chapter 20). If you understand the memo's primary purpose, you will be able to create an effective, clear subject line and keep this purpose at the forefront of your drafting and final revising process. *Purpose considerations*

Memo Parts and Format

15.2 Identify the parts and format of a standard memo

As noted above, memos are written in a variety of shapes and forms, depending on whether they are sent as the body of an email or as a more formal PDF attachment. This section describes the parts and format of a typical workplace memo. The following parts can be modified to fit your audience and purpose:

- The word **"Memo" or "Memorandum"** either centered or flush left at the top of the page, either in all capital letters or upper and lower case. You can use bold face for more visibility. *Standard parts of a memo*

- **A header that includes fields for the recipient ("To"), the sender ("From"), the date the memo was circulated, and the subject.** The subject field may be listed as "Re:," which stands for "in reference to" or "regarding."

- **The body of the memo,** which typically should include an introductory paragraph or sentence, a middle portion, and a conclusion.

- **A distribution notation,** including any people who were not listed in the "To" field but are copied on the memo. This notation is indicated by "cc" (courtesy copy) followed by the names of those individuals. Note that with email, there is also a "bcc" (blind copy) field that can be used to send copies but not allow other recipients to see who received a "bcc." See Chapter 14 for more about the proper and ethical use of bcc with email memos.

- **An enclosure notation,** used to indicate any other documents that accompany the memo.

Format of the memo body

The body copy (main text portion) of a memo should focus on one topic. Content should be complete yet compact, providing all the information readers need but not going into unnecessary detail. Organize the body of your memo by starting with a short introduction and then writing a paragraph or two to address the main issue. Conclude by suggesting a course of action or asking your readers to follow up.

Format for a print or PDF memo

In a print or PDF memo, typically you should double space between the header and the opening paragraph as well as between the paragraphs. Top and bottom margins are best set to 1 inch; left and right margins should be set to anywhere from 1 inch to 1.25 inches.

Format for an email memo

For memos sent as email messages, the "To," "From," "Subject," and "Date" fields in the email header take the place of these lines on the PDF or print memo. But, if you want to emphasize that your email is in fact a memo, you can repeat the word "Memo" or "Memorandum" as well as these fields within the body of the email. Format the body of your message using the same introduction, middle portion, and conclusion structure you would use in a print or PDF memo.

Using memo templates

Word processing programs provide a wide range of memo formats, called templates. These templates are pre-formatted for the required parts of a memo, including the header fields. For most apps, you simply select the memo template when starting a new file. In terms of design, templates can range from simple to complex. As tempting as it may be to simply choose a template, make sure the one you use is appropriate for your audience and purpose. Unless the situation specifically calls for a decorative format, strive for a tasteful, conservative look. When in doubt, ignore the templates and work from a blank document, based on the guidelines in this chapter and the sample memos illustrated here. Or, modify the template to fit your audience, purpose, and these guidelines.

See Figure 15.1 for more detail on the parts and format of memos and Figure 15.2 for an email with all of the required parts, correct formatting, and well-written body text.

NAME OF ORGANIZATION

MEMORANDUM   Center this label on the page or set it flush left (as shown)

To: Name and title of recipient
From: Your name and title
Date: (also serves as a chronological record for future reference)
Subject: Elements of a Usable Memo (or, replace Subject with Re for
             in reference to)

Subject Line
Be sure that the subject line clearly announces your purpose: "Recommendations for Software Security Upgrades" instead of "Software Security Upgrades." Capitalize the first letter of all major words. (Some organizations also use boldface for the subject line. Follow the guidelines for your workplace.)

Memo Text
Unless you have reason for being indirect (see "Decide on a Direct or Indirect Organizing Pattern" in this chapter), state your main point in the opening paragraph. Provide a context the recipient can recognize. (*As you requested in our January meeting, I am forwarding the results of our software security audit.*) For recipients unfamiliar with the topic, begin with a brief background paragraph.

Headings
When the memo covers multiple subtopics, include headings (as shown here). Headings (see Chapter 13, "Using Headings for Access and Orientation") help you organize, and they help readers locate information quickly.

Graphic Highlights
To improve readability you might organize facts and figures in a table (see Chapter 12, "Tables") or in bulleted or numbered lists (see Chapter 13, "Shaping the Page").

Paragraph and Line Spacing
Do not indent a paragraph's first line. Single-space within paragraphs and double-space between.

Subsequent Page Header
Be as brief as possible. If you must exceed one page, include a running head on each subsequent page, naming the recipient and date (*J. Baxter, 6/12/11, page 2*).

Distribution and Enclosure Notations
Use "cc" for courtesy copies. Use "Enclosure" or "Encl." to indicate additional documents.

Figure 15.1 Standard Parts of a Memo These elements can differ across organizations and professions, but most memos, especially longer ones, contain most of these parts. Because memos are read rapidly by busy people, formatting and layout are key.

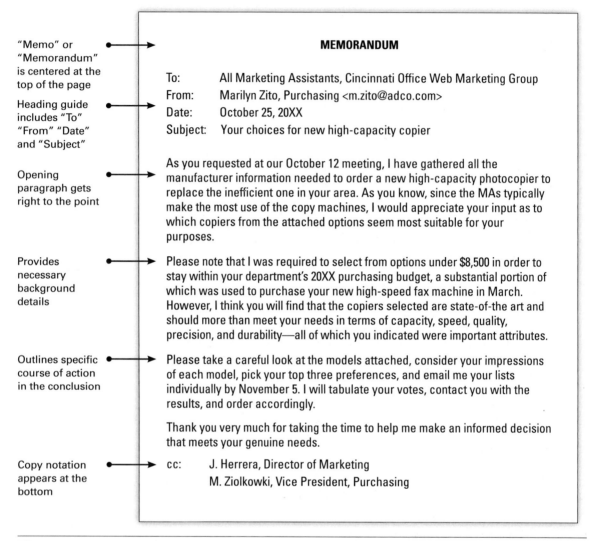

"Memo" or "Memorandum" is centered at the top of the page

Heading guide includes "To" "From" "Date" and "Subject"

Opening paragraph gets right to the point

Provides necessary background details

Outlines specific course of action in the conclusion

Copy notation appears at the bottom

MEMORANDUM

To: All Marketing Assistants, Cincinnati Office Web Marketing Group
From: Marilyn Zito, Purchasing <m.zito@adco.com>
Date: October 25, 20XX
Subject: Your choices for new high-capacity copier

As you requested at our October 12 meeting, I have gathered all the manufacturer information needed to order a new high-capacity photocopier to replace the inefficient one in your area. As you know, since the MAs typically make the most use of the copy machines, I would appreciate your input as to which copiers from the attached options seem most suitable for your purposes.

Please note that I was required to select from options under $8,500 in order to stay within your department's 20XX purchasing budget, a substantial portion of which was used to purchase your new high-speed fax machine in March. However, I think you will find that the copiers selected are state-of-the art and should more than meet your needs in terms of capacity, speed, quality, precision, and durability—all of which you indicated were important attributes.

Please take a careful look at the models attached, consider your impressions of each model, pick your top three preferences, and email me your lists individually by November 5. I will tabulate your votes, contact you with the results, and order accordingly.

Thank you very much for taking the time to help me make an informed decision that meets your genuine needs.

cc: J. Herrera, Director of Marketing
         M. Ziolkowski, Vice President, Purchasing

Figure 15.2 A Typical Memo The writer has provided a copy to each appropriate recipient—no one appreciates being left "out of the loop." Because of this memo's length, it is more suitable as a PDF attachment than as the body of an email message.

Memo Tone

15.3 Use proper tone in all memos

Memos are most often distributed internally within an organization and written for employees at different levels and functions. Memos are written to address questions such as the following:

- Who's doing what, and when, and where?

- How will any changes to existing policies or procedures affect employee workplace activities or conditions?

- What is the status of a project or recent meeting or discussion?

Memo topics often involve evaluations or recommendations about policies, procedures, and, ultimately, the *people with whom we work.*

Because people are sensitive to criticism (even when it is merely implied) and often resistant to change, an ill-conceived or aggressive tone can spell disaster for the memo's author. So, be especially careful about your tone. Consider, for instance, this memo—an evaluation of one company's training program for new employees:

> No one tells new employees what it's *really* like to work here—how to survive politically: For example, never tell anyone what you *really* think; never observe how few women are in management positions, or how disorganized things seem to be. New employees shouldn't have to learn these things the hard way. We need to demand clearer behavioral objectives.

A hostile tone

Instead of sounding angry and demanding, the following version comes across as more thoughtful and respectful:

> New employees would benefit from a concrete guide to the personal and professional traits expected in our company. Training sessions could focus on appropriate attitudes, manners, and behavior in business settings.

A more reasonable tone

> NOTE *A professional, reasonable tone is always important in all workplace communication but is especially critical when communicating via email. See Chapter 14 for more on tone and style in email.*

Achieving the right tone in your memos involves using some common sense. Put yourself in the shoes of your recipients and write accordingly. Be polite and avoid sounding bossy, condescending, and aggressive, or deferential and passive. Don't criticize, judge, or blame any individual or department. Don't resort to griping, complaining, and other negative commentary. Try to emphasize the positive. Finally, approach difficult situations reasonably. Instead of taking an extreme stance, or suggesting ideas that will never work, be practical and realistic.

Using common sense

The tone of a memo also comes across in the sequence in which you deliver the information. Depending on the sensitivity of your memo's subject matter, you may want to take a direct or an indirect approach. A direct approach begins with the "bottom line" in the first sentence (as well as in the subject line) and then presents the details or analysis to support your case. An indirect approach lays out the details of the case over several sentences (and leaves the subject line vague) before delivering the bottom line later in the paragraph.

Being direct or indirect

Readers generally prefer the direct approach because they want to know the bottom line without being told in advance how to feel about it. Assume, for example, that a company payroll manager has to announce to employees that their paychecks will be delayed by two days: This manager should take a direct approach, announcing the troubling news in both the subject line and the opening sentence and then explaining the causes of the problem:

Direct approach: Subject line states main point

Opening paragraph starts with bottom line

> MEMO
>
> To: All employees
>
> From: Meredith Rocteau, Payroll Manager
>
> Date: May 19, 20XX
>
> Subject: Delay in Paychecks
>
> I regret to inform you that those employees paid by direct deposit will experience a two-day delay in receiving their paychecks.
>
> This delay is due to a virus that infiltrated the primary computer server for our payroll system. Although we hired consultants to identify the virus and clean out the server, the process took nearly 48 hours.
>
> We apologize for the inconvenience.

However, when you need to convey exceedingly bad news or make an unpopular request or recommendation (as in announcing a strict new policy or employee layoffs), you might consider an indirect approach; this way you can present your case and encourage readers to understand your position before announcing the unpopular bottom line. The danger of the indirect approach, though, is that you may come across as evasive.

Indirect approach: Subject line is not specific about the main point

Offers an explanation before delivering the bottom line

The bottom line

> MEMO
>
> To: All employees
>
> From: J. Travis Southfield, Director of Human Resources
>
> Date: September 19, 20XX
>
> Subject: Difficult Economic Times
>
> Each employee of the AutoWorld family is a valued member, and each of you has played an important role in our company's expansion over the past 10 years.
>
> Yet as you all know, times are difficult right now for the automobile industry. Sales are down, financing is hard to obtain, and consumers are holding back on major purchases.
>
> In order to keep the company solvent, we must consider all options. Therefore, I have been informed by our company president, John Creaswell, that we must downsize. We will begin with options for retirement packages, but please be prepared for the possibility that layoffs may follow.
>
> We will have more information for you at an all-hands meeting tomorrow.

(For more on direct versus indirect organizing patterns, see "Decide on a Direct or Indirect Organizing Pattern" in this chapter.)

Consider whether to use email or a PDF attachment

Finally, a memo's tone comes across not only in the words you choose but also in the way you handle its distribution. If your topic is very short, not overly formal, and needs to reach everyone quickly, consider sending the memo as the body of an email. But if your topic is more formal and more detailed, send out a brief cover email

("Please see attached memo for information about this year's raises") with the memo as a PDF attachment. Also, be careful about who receives copies. Don't copy everyone at work when the content is only appropriate for a few, and don't leave vital people off your distribution (cc) list.

Common Types of Memos

15.4 Write transmittal, summary, and routine miscellaneous memos

Memo format can also be used for distributing short, informal reports, discussed in Chapter 21. However, for the purposes of this chapter, consider the following most common types of memos.

Transmittal Memo

A transmittal memo accompanies a package of materials such as a long report, a manuscript, or a proposal. Its purpose is to signal that the information is being sent from one place to another (providing a paper trail), to introduce the material, and to describe what is enclosed. A transmittal memo may be as simple as a sentence or a paragraph with a bulleted list describing the contents of the package, as in Figure 15.3. Because this transmittal memo is part of a longer, more formal document (a report), it is written on company letterhead and sent as a PDF attachment.

Summary or Follow-up Memo

A summary or follow-up memo provides a written record of a meeting or conversation or offers a recap of a topic discussed that was not resolved at the time. In addition to providing evidence that the meeting or conversation took place, summary and follow-up memos also ensure that each recipient has the same understanding of what was decided. Figure 15.4 shows a memo sent as an email that performs both a summary and follow-up function.

Routine Miscellaneous Memo

Routine miscellaneous memos cover a virtually infinite variety of topics. Such a memo, for example, may contain some type of announcement or update, such as announcing the closure of a parking ramp over the holidays for repair or an upcoming awards ceremony. Other such memos may request information or action, reply to an inquiry, or describe a procedure. Shorter, less formal memos of this type are usually sent as email. But if the memo has a formal purpose, especially if it comes from a group such as the Human Resources department to all employees, a more formal version may be preferable. Figure 15.5 shows one such example, formatted on company letterhead and sent as a PDF attachment. Chapter 14, Figure 14.4, shows another way to format a memo and send the formatted version as the body of an email message.

Use "Checklist: Memos" later in this section as you write a memo.

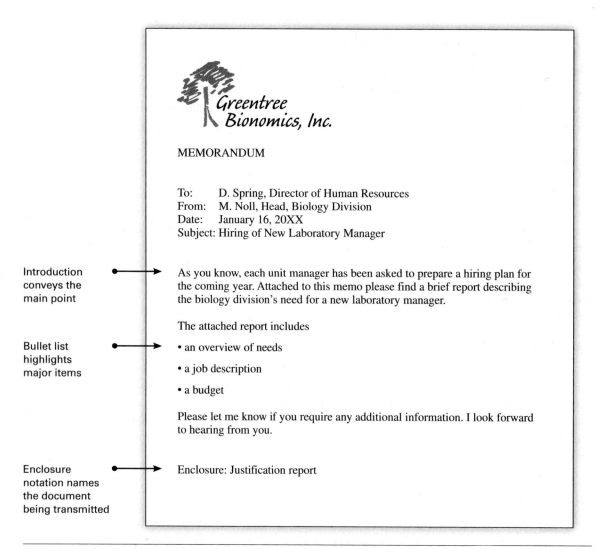

Figure 15.3 A Transmittal Memo A memo like this would be the first page of a longer document. If the memo was sent as the body of an email message, the enclosure notation would be removed because the email attachment would serve the same purpose.

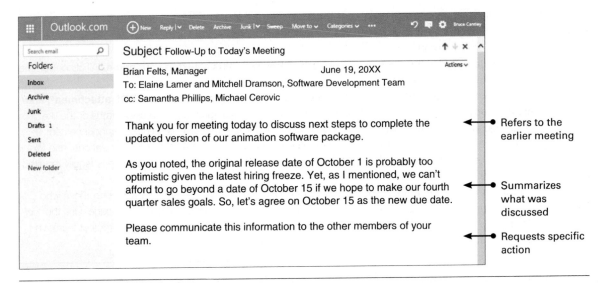

Figure 15.4 A Summary or Follow-Up Memo This type of memo provides a written record. Since the content is short and not overly formal, it takes the form of an email message.
Source: Microsoft Outlook 2013.

Guidelines

for Memos

> **Do not overuse or misuse memos.** Use email or the telephone when you need to ask a quick question or resolve a simple issue. For a sensitive topic, prefer a face-to-face conversation whenever possible.

> **Focus on one topic.** If you need to address more than one topic, consider a format other than a memo (for instance, a report).

> **Be brief but sufficiently informative.** Recipients expect memos that are short and to the point but not at the expense of clarity.

> **Be sure the tone of your memo is polite and respectful.** Use the rational connection (Figure 3.3) to show that you respect your readers, no matter how difficult the topic may be.

> **Avoid sounding too formal or too informal for the topic or audience.** A memo to the person in the next cubicle to ask for help on a project, for example, would be more informal than a memo to a company executive.

> **Use the appropriate organizational sequence (direct or indirect).** Prefer the direct approach when you need people to get the point quickly and the indirect approach when you have something difficult to say that needs to be softened.

> **Follow the standard format illustrated throughout this chapter.** Keep in mind that some organizations may have their own formatting requirements for various documents.

> **Use white space, headings, and bullets, as needed.** These features provide visible structure to your memo, as well as "chunking" all elements into easily digestible parts.

> **Use tables, charts, and other visuals to display quantitative information and to achieve emphasis, as needed.** A simple visual as part of a memo can be an effective way to make your point. See Chapter 12 for more on visual communication.

➤ **Check spelling, grammar, and style.** Autocorrect and spellcheck are great, but there is no single app that catches everything. Be sure to proofread and, if the memo is long or highly technical, ask a colleague to proofread before you press send.

➤ **On memos delivered in hard copy only, you can place your initials next to your name.** Initials beside the typed name provide more formality.

➤ **Determine whether to send your memo as the body of an email message, as a PDF attachment, and/ or in hard copy.** Use email for shorter memos, particularly ones that summarize a project status or the like (such as Figure 15.4). But use PDF attachments on letterhead if your message is an official announcement, such as Figure 15.5. Or, use a formatted email message, such as Figure 14.4. In some organizations, and for some purposes (such as tax notices, human resources announcements, and other corporate or legally required situations), hard copy may also be required.

➤ **Distribute to the right people.** Do not "spam" people with your memo. Be sure it reaches only those who need the information. At the same time, don't leave out anyone who needs to read your message. Use the "cc" field to include readers outside your primary audience (for instance, if you are writing to your project team, you may need to cc your manager to be sure she or he is included in the update).

Checklist

Memos

Use the following Checklist as you write a memo.

Content
❏ Is the information based on careful research? (See Chapter 7.)
❏ Is the message brief and to the point? (See "Memo Basics, Audience, and Purpose" in this chapter.)
❏ Are tables, charts, and other graphics used as needed? (See Chapter 12.)
❏ Are recipients given enough information to make an informed decision? (See "Memo Basics, Audience and Purpose" in this chapter.)
❏ Are the conclusions and recommendations clear? (See "Memo Parts and Format" in this chapter.)

Organization
❏ Is the direct or indirect pattern used appropriately to present the memo's bottom line? (See "Memo Tone" in this chapter.)
❏ Is the material "chunked" into easily digestible parts? (See Chapter 10, "Chunking.")

Style
❏ Is the writing clear, concise, and fluent? (See Chapter 11.)
❏ Is the tone appropriate? (See "Memo Tone" in this chapter.)
❏ Has the memo been carefully proofread? (See Chapter 6, "Make Proofreading Your Final Step.")

Format
❏ Does the memo have a complete heading? (See "Memo Parts and Format" in this chapter.)
❏ Does the subject line announce the memo's content and purpose? (See "Memo Parts and Format" in this chapter.)
❏ Do headings announce subtopics, as needed? (See "Memo Parts and Format" in this chapter.)
❏ If more than one reader is receiving a copy, does the memo include a distribution notation (cc:) to identify other recipients? (See "Memo Parts and Format" in this chapter.)
❏ Does the memo fit the mode of delivery (email, PDF, print copy)? (See "Memo Basics, Audience, and Purpose" in this chapter.)

(Continued)

Ethical, Legal, and Interpersonal Considerations
❑ Is the information specific, accurate, and unambiguous? (See "Memo Tone" in this chapter.)
❑ Is the message inoffensive to all parties? (See "Memo Tone" in this chapter.)
❑ Are all appropriate parties receiving a copy? (See "Memo Tone" in this chapter.)

Morris and Sutton, LLC

MEMORANDUM

To: All employees
From: Jorge Gonsalves, Human Resources
Date: January 12, 20XX
Subject: 401K matching policy

As the new year begins, we in Human Resources would like to remind you   ◄—● Describes the
about the company's generous 401K matching policy. We will match your   benefit plan
401K contributions 100% when you roll up to 10% of your salary into
your 401K.

Many companies will match only up to 5% of an employee's salary and
usually not at a 100% rate, so please take advantage of this program by   ◄—● Encourages
enrolling now. Enrollment is only open until March 1 and will not be open   participation
again until next January.

Please drop by the Human Resources office on the 6th floor to get a
handout that provides more detailed information or to speak with an HR   ◄—● Describes the
representative in person.   procedure

Thanks.

cc: Alison Sheffield, Manager, Human Resources

Figure 15.5 A Routine Miscellaneous Memo This type of memo can cover a wide variety of topics. The memo is sent as a PDF email attachment. Because this company's policy is to provide employees with all Human Resources notices in hard copy as well, print copies are also sent via regular mail.

Letter Basics, Audience, and Purpose

15.5 Understand the basics of letters, including their audience and purpose

At work, people often need to send messages that are more formal than an email or a memo. In these situations, the appropriate format is a letter. Workplace letters are written and sent for some of the following reasons:

When to send a letter instead of a memo or email

- To personalize the correspondence, conveying the sense that the message is prepared exclusively for the recipient

- To convey a dignified, professional impression

- To represent a company or organization

- To present a reasoned, carefully constructed case

- To respond to clients, customers, and others outside your organization in a formal, businesslike manner

- To provide an official notice or record (letters are often the required format for legal notifications)

Because your signature certifies your approval—and your responsibility—for the information being sent (as in Figure 15.6), precision is crucial in letters. Letters can be sent as PDF email attachments, in hard copy, or both. Some organizations require certain letters (such as changes to salaries or pension plans, or legal notifications) be sent in hard copy, sometimes by certified mail.

The rest of this chapter covers four common types of letters written at work: inquiry letters, claim letters, sales letters, and adjustment letters. (Job application letters and letters of transmittal are discussed in Chapters 16 and 21, respectively.)

Audience considerations

Your overall approach to a workplace letter is determined by the letter's audience and purpose. Begin by focusing on your audience: Who will be the recipient of this letter? (When possible, write to a named person, not the title of a position.) What is your relationship to this person? Is this a potential employer, a client, an associate, a stranger? Exactly what information and level of formality does this person expect? How might this person react to the contents of your letter? Answering these questions in advance will help you craft a letter that connects with its recipient.

Purpose considerations

Next, focus on your purpose: What do you want the recipient to do after reading your letter—offer a job, provide advice, grant a favor, accept bad news? Do you have multiple purposes in mind as in, say, obtaining a refund for a faulty product while also preserving your business relationship with that supplier? Answering these questions in advance will help you craft a letter that achieves the outcome you seek. A letter often has a *persuasive* purpose (see Chapter 3); therefore, proper tone is essential for connecting with the recipient.

Letter Parts, Formats, and Design Elements

15.6 Identify the parts, formats, and design features of workplace letters

Most workplace letters have the same basic components. This conventional and predictable arrangement enables recipients to locate what they need immediately, as in Figure 15.6.

Standard Parts

Many organizations have their own formats for letters. Depending on where you work, some of these parts may appear at different locations on the page. But in general, a letter contains the elements listed here.

HEADING AND DATE. If your stationery has a company letterhead, simply include the date a few lines below the letterhead, flush against the left margin. If you are not using letterhead, type the company name and address. If you are using your personal address, omit your name, because that will appear below your signature at the letter's end.

> 154 Sea Lane
> Harwich, MA 02163
>
>
> July 15, 20XX

Use the Postal Service's two-letter state abbreviations (e.g., MA for Massachusetts, WY for Wyoming) in your heading, in the inside address, and on the envelope.

INSIDE ADDRESS. Two to six line spaces below the heading, flush against the left margin, is the inside address (the address of the recipient).

> Dr. Ann Mello, Dean of Students
> Western University
> 30 Mogul Hill Road
> Stowe, VT 51350

Whenever possible, address a specifically named recipient, and include the person's title. Using "Mr." or "Ms." before the name is optional. (See Chapter 11, "Adjusting Your Tone," for avoiding sexist usage in titles and salutations.)

> NOTE *Depending on the letter's length, adjust the vertical placement of your return address and inside address to achieve a balanced page.*

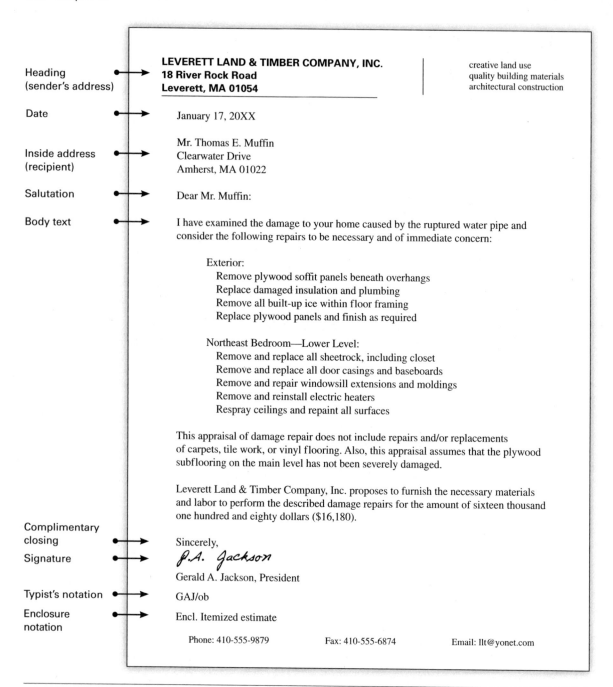

Heading (sender's address) →

LEVERETT LAND & TIMBER COMPANY, INC.
18 River Rock Road
Leverett, MA 01054

creative land use
quality building materials
architectural construction

Date →

January 17, 20XX

Inside address (recipient) →

Mr. Thomas E. Muffin
Clearwater Drive
Amherst, MA 01022

Salutation →

Dear Mr. Muffin:

Body text →

I have examined the damage to your home caused by the ruptured water pipe and consider the following repairs to be necessary and of immediate concern:

    Exterior:
      Remove plywood soffit panels beneath overhangs
      Replace damaged insulation and plumbing
      Remove all built-up ice within floor framing
      Replace plywood panels and finish as required

    Northeast Bedroom—Lower Level:
      Remove and replace all sheetrock, including closet
      Remove and replace all door casings and baseboards
      Remove and repair windowsill extensions and moldings
      Remove and reinstall electric heaters
      Respray ceilings and repaint all surfaces

This appraisal of damage repair does not include repairs and/or replacements of carpets, tile work, or vinyl flooring. Also, this appraisal assumes that the plywood subflooring on the main level has not been severely damaged.

Leverett Land & Timber Company, Inc. proposes to furnish the necessary materials and labor to perform the described damage repairs for the amount of sixteen thousand one hundred and eighty dollars ($16,180).

Complimentary closing →

Sincerely,

Signature →

G.A. Jackson

Gerald A. Jackson, President

Typist's notation →

GAJ/ob

Enclosure notation →

Encl. Itemized estimate

Phone: 410-555-9879   Fax: 410-555-6874   Email: llt@yonet.com

Figure 15.6 Standard Parts of a Workplace Letter in Block Format This writer is careful to stipulate not only the exact repairs and costs, but also those items excluded from his estimate. In the event of legal proceedings, a formal letter signifies a contractual obligation on the sender's part.

SALUTATION. The salutation, set two line spaces below the inside address, begins with *Dear* and ends with a colon (*Dear Ms. Smith:*). If you don't know the recipient's name, use the position title (*Dear Manager:*) or an attention line (see "Optional Parts" in this chapter). Only address the recipient by first name if that is the way you would address that individual in person.

| Dear Ms. Smith:

| Dear Managing Editor:

| Dear Professor Trudeau:

Typical
salutations

No satisfactory guidelines exist for addressing several people within an organization. *Dear Sir or Madam* sounds old-fashioned (although is still considered acceptable in certain settings and cultures; see discussion later in this chapter under "Global and Ethical Considerations"). *To Whom It May Concern* is vague and impersonal but may be all you can use if you don't have a name. Another option in this situation is to eliminate the salutation by using an attention line (see "Optional Parts" later in this section).

TEXT. Begin your letter text two line spaces below the salutation or subject line. Workplace letters typically include (1) a brief introductory paragraph (five or fewer lines) that identifies your purpose and connects with the recipient's interest, (2) one or more discussion paragraphs that present details of your message, and (3) a concluding paragraph that sums up and encourages action.

The shape of
workplace letters

Keep the paragraphs short, usually fewer than eight lines. If a paragraph goes well beyond eight lines, or if the paragraph contains detailed supporting facts or examples, as in Figure 15.6, consider using a vertical list.

COMPLIMENTARY CLOSING. The closing, two line spaces (returns) below the last line of text, should parallel the level of formality used in the salutation and should reflect your relationship to the recipient (polite but not overly intimate). *Yours truly* and *Sincerely* are the most common. Others, in order of decreasing formality, include

| Respectfully,

| Cordially,

| Best wishes,

| Regards,

| Best,

Align the closing with the letter's heading.

Complimentary
closings

SIGNATURE. Type your name and title on the fourth and fifth lines below and aligned with the closing. Sign in the space between the complimentary closing and typed name.

The signature block

> Sincerely yours,
>
> *Martha S. Jones*
>
> Martha S. Jones
> Personnel Manager

If you are representing your company or a group that bears legal responsibility for the correspondence, type the company's name in full caps two line spaces below your complimentary closing; place your typed name and title four line spaces below the company name and sign in the triple space between.

Signature block representing the company

> Yours truly,
> HASBROUCK LABORATORIES
>
> *Lester Fong*
>
> L. H. Fong
> Research Associate

For PDF, you can sign a hard copy and scan, or use one of the PDF tools (digital pencil or digital signature).

Optional Parts

Some letters have one or more of the following specialized parts. (Examples appear in the sample letters in this chapter.)

ATTENTION LINE.   Use an attention line when you write to an organization and do not know your recipient's name but are directing the letter to a specific department or position.

An attention line can replace your salutation

> Glaxol Industries, Inc.
> 232 Rogaline Circle
> Missoula, MT 61347
> Attention: Director of Research and Development

Use two line spaces below the inside address and place the attention line either flush with the left margin or centered on the page.

SUBJECT LINE.   Typically, subject lines are not used with letters, only with memos, but if your recipient is not expecting your letter, a subject line is a good way of catching a busy reader's attention.

A subject line can attract attention

> Subject: *Placement of the Subject Line*

Place the subject line below the inside address or attention line. You can italicize the subject to make it prominent.

TYPIST'S NOTATION. If someone else types your letter for you (common in the days of typewriters but rare today), your initials (in CAPS), a slash, and your typist's initials (in lower case) appear below the typed signature, flush with the left margin.

| JJ/pl

ENCLOSURE NOTATION. If you enclose other documents in the same envelope, indicate this one line space below the typist's notation (or writer's name and position), flush against the left margin. State the number of enclosures.

| Enclosure
| Enclosures 2
| Encl. 3

If the enclosures are important documents such as legal certificates, checks, or specifications, name them in the notation.

| Enclosures: 2 certified checks, 1 set of KBX plans

COPY (OR DISTRIBUTION) NOTATION. If you distribute copies of your letter to other recipients, indicate this by inserting the notation "copy" or "cc," followed by a colon, one line below the previous line (such as an enclosure line). The "cc" notation once stood for "carbon copy," but no one uses carbon paper any longer, so now it is said to stand for "courtesy copy."

| cc: office file
| Melvin Blount
|
| copy: S. Furlow
| B. Smith

Most copies are distributed on an FYI (For Your Information) basis, but writers sometimes use the copy notation to maintain a paper trail or to signal the primary recipient that this information is being shared with others (for example, other managers, different project team leaders, legal authorities).

Multiple notations would appear in this order: typist, enclosures, and then copy.

POSTSCRIPT. A postscript draws attention to a point you wish to emphasize that may not be directly related to the main point of the letter itself. Do not use a postscript if you forget to mention a key point in the body of the letter. Rewrite the body section instead. In the example below, the main point of the letter is to let the customer know about key dates when their copier/printer will be serviced during the coming year. The "p.s." is not directly related to this point but does provide some useful additional information.

| P.S.   We also wanted to let you know that during the week of January 15, all printer cartridges
| will be on sale at up to 25% off.

Place the postscript two lines after the signature, before any "cc" or "enclosure" notations.

Formats and Design Features

The following elements help make workplace letters look inviting, accessible, and professional.

LETTER FORMAT.   Although several formats are acceptable, and your company may have its own, the most popular format for workplace letters is called *block format*. In this format, every line begins at the left margin. Unlike books, for letters, do not right-justify the body text. Similar to memo templates, discussed previously, word processing programs provide a wide range of letter templates, too. These templates are pre-formatted for the required parts of a letter and have the margins and other layout features already set. As with memo templates, you simply select the letter template when starting a new file. Letter templates offer a variety of professional looking styles and designs. If you wish to use a template, choose one that is appropriate for your audience and purpose. Strive for a clean, professional look. When in doubt, ignore the templates and work from a blank document, based on the guidelines in this chapter and the sample memos illustrated here. Or, modify the template to fit your audience, purpose, and these guidelines.

QUALITY STATIONERY.   When printing hard copies for an official workplace letter, use the organization's formal letterhead. Otherwise, use high-quality, 20-pound bond, 8½-by-11-inch stationery with a minimum fiber content of 25% to convey a professional look.

UNIFORM MARGINS AND SPACING.   Workplace letters look best and are easiest to read if top and side margins are set between 1 inch and 1-½ inches. Use single spacing within paragraphs and double-spacing between. Vary these guidelines based on the amount of space required by the letter's text, but strive for a balanced look. When using letterhead, the top margin will already be set for you.

HEADERS FOR SUBSEQUENT PAGES.   If your letter is longer than one page, use a running header for each additional page with a notation identifying the recipient, date, and page number.

| Adrianna Fonseca, June 25, 20XX, page 2

Align your header with the right-hand margin. See Chapter 3, Figure 3.4 (second page) for an example.

> NOTE *Never use an additional page solely for the closing section. Instead, reformat or revise the content of the letter so that the closing appears on the first page, or so that at least two lines of text appear above the closing on the subsequent page.*

THE ENVELOPE.   For hard copy, use the organization's formal envelope. Otherwise, use a standard-size business envelope (usually #10) that matches your stationery. Place your own name and address in the upper-left corner. Single-space these elements. Most word-processing programs and printing apps have envelope printing options that automatically place these elements.

Letter Tone

15.7 Use proper tone in any letter

When you speak with someone face to face, you unconsciously modify your state-ments and facial expressions as you read and listen to the listener's signals: a smile, a frown, a raised eyebrow, a nod, a short vocal expression of agreement or disagree-ment. In a phone conversation, the person's voice can signal approval, dismay, anger, or confusion. Those cues allow you to modify your comments and vocal tone.

The importance of a letter's tone

With any form of written communication—memos, letters, email, texts—we often get so caught up in what we want to say and the points we want to make that we often forget that a flesh-and-blood person will be reading and reacting to what you say—or seem to say. This idea is especially important when it comes to letter writing, because, as discussed previously, letters are typically addressed to in-dividuals. Therefore, the tone of a letter is especially important to get right, not only to achieve your goals but also to show respect for the person at the other end of the message.

To achieve an appropriate tone, consider the factors discussed below, factors that affect the relationship between you, the sender, and your reader, the recipient. As you read through this section, refer to Figure 15.6, which maintains an appropri-ate tone in conveying bad news. Also, since letters are often sent electronically (as PDF attachments), keep in mind the items discussed in Chapter 14 under "Email Style and Tone."

Establish and Maintain a "You" Perspective

A letter displaying a "you" perspective puts the reader's interest and feelings first. To convey a "you" perspective, put yourself in the place of the person who will read your correspondence, and ask yourself how this recipient will react to what you have writ-ten. Even a single word or sentence, carelessly chosen or phrased, can offend. Consider the following sentence in a letter to a customer:

Prioritize the reader's needs, wants, and feelings

| Our record keeping is very efficient and we have looked into it, so this is obviously your error.

Offensive

This self-centered tone might be appropriate after numerous investigations into the customer's complaint and failed attempts to communicate your company's perspec-tive to the customer, but in your initial correspondence it would be offensive. Here is a more considerate version. Instead of expressing only the writer's point of view, this second version conveys respect for the reader's viewpoint.

Considerate and respectful	Although my paperwork shows that you were charged correctly, I will investigate this matter immediately by checking my files against our computer records.

Do not sign and mail the letter until you are certain that the needs and feelings of your reader consistently get top billing, even when you simultaneously must assert your own perspective.

Be Polite and Tactful

If you must express criticism, do so in a way that conveys good will and trust in the recipient. Avoid the following type of expression:

Tactless	I am shocked that your company lacks the standards to design and manufacture an alarm clock that actually works.

Although a company representative would be required to write a polite and thoughtful response to the above complaint, he or she might be inclined to look closely at the clock's warranty and offer only the most basic reimbursement.

In contrast, a polite and thoughtful letter might yield a full refund or a brand new replacement:

Polite	Although your clock worked reliably for several months, one of the internal mechanisms recently malfunctioned. I would appreciate your contacting me about an exchange or refund.

Use Plain English

Avoid the stuffy, puffed-up phrases some writers use to make their communication sound important. Even though a letter is more formal than a memo or an email, plain English still can get your point across. For example, consider the following closing section to an inquiry letter asking for help:

Stuffy phrasing	Humbly thanking you in anticipation of your kind assistance, I remain Faithfully yours,

The reader of this letter might feel spoken down to and therefore decide not to respond. However, in this next revised version, the reader would likely perceive the writer as a straight-talking equal and be more inclined to follow up:

Clear and direct	I would greatly appreciate any help you could offer. Best wishes,

Here are a few stuffy phrasings, with clearer, more direct translations:

Stuffy and unclear	**Clear and direct**
As per your request	As you requested
Contingent upon receipt of	As soon as we receive
Due to the fact that	Because

Be natural. Write as you would speak in a classroom or office: professionally and respectfully but clearly and directly.

> NOTE *In the legal profession (and others), phrases such as those shown above are known as "terms of art" and connote a specific meaning. In these cases, you may not be able to avoid such elaborate phrases.*

Decide on a Direct or Indirect Organizing Pattern

The reaction you anticipate should determine the organizational plan of your letter: either *direct* or *indirect.* (Figure 15.7 illustrates the choices.)

- Will the recipient feel pleased, angry, or neutral?
- Will the message cause resistance, resentment, or disappointment?

Questions for organizing your message

The direct pattern puts the main point in the first paragraph, followed by the explanation. Be direct when you expect the recipient to react with approval or when you want to convey immediately the point of your letter (e.g., in good news, inquiry, or application letters—or other routine correspondence).

When to be direct

If you expect the reader to resist or to need persuading, or if this person is from a different culture, consider an indirect plan. Give the explanation *before* the main point (as in requesting a pay raise or refusing a request).

When to be indirect

Research indicates that "readers will always look for the bottom line" (*Writing Reader-Friendly Documents* 14). Therefore, a direct pattern, even for certain types of bad news, may be preferable—as in complaining about a faulty product. For more on conveying bad news, see the section later in this chapter.

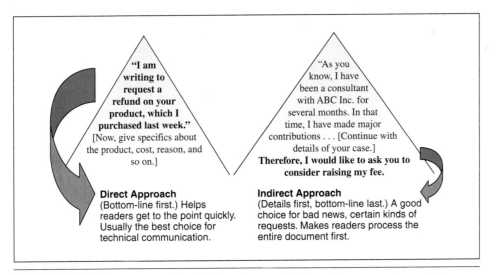

"I am writing to request a refund on your product, which I purchased last week." [Now, give specifics about the product, cost, reason, and so on.]

Direct Approach
(Bottom-line first.) Helps readers get to the point quickly. Usually the best choice for technical communication.

"As you know, I have been a consultant with ABC Inc. for several months. In that time, I have made major contributions . . . [Continue with details of your case.] **Therefore, I would like to ask you to consider raising my fee.**

Indirect Approach
(Details first, bottom-line last.) A good choice for bad news, certain kinds of requests. Makes readers process the entire document first.

Figure 15.7 Deciding on Your Writing Approach Use a direct approach most of the time. But when you need to convey difficult or negative information, use an indirect approach. Be as brief as possible.

NOTE Whenever you consider using an indirect pattern, think carefully about its ethical implications. Never try to deceive the recipient—and never create an impression that you have something to hide.

For more on direct versus indirect organizing patterns, see "Memo Tone" in this chapter.

Global and Ethical Considerations When Writing Letters

15.8 Explain the global and ethical implications of workplace letters

Know your audience

In today's global marketplace, you can expect to communicate with people from numerous countries and cultures. Many such people are nonnative speakers of English and/or have a range of cultural backgrounds. When writing a more formal document such as a letter, you need to learn all you can about the letter recipient's culture and preferences. Even something as standard as the salutation requires you to understand the customs and conventions appropriate for your recipient. For example, the salutation "Dear Madam or Sir," which earlier in this chapter was mentioned to be too formal in most U.S. settings, is considered quite appropriate in the U.K. (University of Nottingham).

Consider the direct vs. indirect approach

Many cultures value the indirect approach, versus the direct approach that is generally preferred in North American business communication. When a writer "gets right down to business" without focusing first on the relationship, this approach may be considered rude or inconsiderate. Some international audiences may instead expect a personalized introduction that compliments the recipient, inquires about the family, and dwells on other personal details before discussing the topic at hand. Also, in U.S. business communication, writers tend to value correspondence that is extremely clear and direct, to ensure one interpretation only. In contrast, readers from certain other cultures may prefer ambiguity in their correspondence, thereby allowing the recipient to infer his/her own meaning: In short, countless people across the globe may be insulted by a message that seems to be telling them what to think.

Discussing any controversial topic with international readers can be especially hazardous. Consider, for example, the following response to a customer's inquiry about a possible billing error (repeated here from "Establish and Maintain a 'You' Perspective" in this chapter): This example shows respect for the recipient's viewpoint.

An acceptable version for an audience that values directness

Although my paperwork shows that you were charged correctly, I will investigate this matter immediately by checking my files against our computer records.

However, while the previous version may be perfectly acceptable in the United States and with other cultures that value directness, someone from a different culture might prefer a version like this one:

A preferable version for an audience that values indirectness

Thank you for bringing your question about the possible billing error to my attention. I personally will investigate this matter immediately and do everything possible to answer your question to your full satisfaction.

Take special care in expressing disagreement. For example, instead of writing "I'm not so sure that's the best approach," try "Are there any other approaches?" or "Do you think that is the best approach?" Or, instead of writing "I disagree" or "We need to discuss this," prefer "That viewpoint is interesting" or "I had not thought of that."

Consider how to handle disagreement

Finally, the letter's closing requires similar attention and care. For example, informal complimentary closings such as "Cheers" or "Best" often are considered offensive; instead, choose a formal closing, such as "Respectfully," which seems to be a universally acceptable choice. Avoid excessive informality throughout your letter.

Consider how to close

Even the best intentions can violate ethical standards. For example, while trying to be polite and respectful, the writer might end up being evasive and misleading instead, thereby inadvertently deceiving the reader. In short, almost any type of international correspondence poses this dilemma: how to be clear and straightforward without appearing rude and insensitive. Do not allow your concern for diplomacy to overshadow the need for recipients to receive the information—as well as the understanding—they require to make sound decisions. Regardless of cultural differences, an ethical message ensures that the reader understands and interprets the information just as clearly and accurately as the writer does.

How good intentions can go wrong

You can learn more about intercultural communication by looking online at credible sources such as *Forbes*, the *Harvard Business Review*, the state and/or cultural departments of different countries, and the international student travel office at your college or university. Also, workplace colleagues and faculty members who have traveled or who are from different countries may be good sources of advice. Never send off any global correspondence until you have done diligent research. (For more on this topic, see Chapters 1, 3, and 5.)

Guidelines
for Letters in General

- ➤ **Determine whether the situation calls for a letter, memo, or email.** Use a letter to communicate formally with a client or customer or as required by company or legal policy.
- ➤ **Determine if the letter needs to be sent as an email attachment, in hard copy, or both.** If using a PDF only, you can sign your name using the pen tool, or you can use the digital signature option.
- ➤ **Use proper letter format and include all the required parts.** Unless your organization has its own guidelines, use block format and the parts discussed earlier.
- ➤ **Place the reader's needs first.** Always write from the "you" perspective, putting yourself in your reader's place.
- ➤ **Decide on the direct or indirect approach.** Generally speaking, take the direct approach for good news and the indirect approach for bad news.
- ➤ **Maintain a courteous, professional tone.** A professional tone creates goodwill and is more effective in the long run.
- ➤ **Avoid stuffy language.** Use clear wording, no matter how formal or important the letter. Stuffy language only comes across as phony.
- ➤ **Keep international readers in mind.** Don't assume that every letter you write is directed at a recipient whose first language is English or whose cultural values match your own.

Conveying Bad or Unwelcome News in Letters

15.9 Effectively deliver bad or unwelcome news in a letter

Bad news is a fact of life in the workplace

During your career, you may have to say no to customers, employees, and job applicants. You may have to make difficult requests, such as asking employees to accept higher medical insurance premiums or seeking an interview with a beleaguered official. You may have to notify consumers or shareholders about accidents or product recalls. You may need to apologize for errors—the list of possibilities goes on. In conveying bad news, you face a *persuasive* challenge (see Chapter 3): You must convince people to accept your message. As the bearer of unwelcome news and requests, you will need to offer reasonable explanations, incentives, or justifications—and your tone will need to be diplomatic, as in Figure 15.8.

Decide if a direct or indirect approach is best

In each instance, you will have to decide whether to build your case first or get right to the main point. Your choice will depend on the situation. If you are requesting a refund for a faulty printer, for example, you will probably want a direct approach because the customer service person, who could easily receive hundreds of letters each day, will get to your point quickly. But if you are announcing a 15% increase in your client service fees, you might want readers to process your justification first.

The following general guidelines apply to many situations you will face; these guidelines also complement the guidelines for each specific type of letter covered in this chapter.

Common Types of Letters

15.10 Write inquiry, claim, sales, and adjustment letters

Among the many types of business letters you may write on the job, the most common types are inquiry letters, claim letters, sales letters, and adjustment letters.

Inquiry Letters

Solicited and unsolicited inquiry letters

Inquiry letters ask questions and request a reply. They may be solicited (in response to an advertisement or announcement) or unsolicited (spontaneously written to request some type of information you need). For example, a Web developer might write a solicited inquiry to a Web design company offering consulting services based on the advertisement. If there has been no such advertised offer, the developer might write an unsolicited inquiry to the same company to ask if any Web development jobs might be available.

In a solicited inquiry, be brief and to the point, and be sure to reference the advertisement or announcement that prompted you to write. In an unsolicited inquiry,

LEVERETT LAND & TIMBER COMPANY, INC.
18 River Rock Road
Leverett, MA 01054

creative land use
quality building materials
architectural construction

January 17, 20XX

Mr. Thomas E. Shaler
19 Clearwater Drive
Amherst, MA 01022

Dear Mr. Shaler:

Thank you for bringing the matter of the ruptured water pipe to my attention. I was pleased to hear from you again these months after our firm completed construction of your living room addition, though I was of course sorry to hear about the water damage not only to the new construction but to the living room as a whole.

← Establishes "you" perspective immediately

Naturally, I understand your desire to receive compensation for your home's damage, especially taking into account how recently the extension was completed. In reviewing the blueprints for the extension, however, I find that the pipes were state-of-the-art and were fully insulated. In fact, it is the practice of Leverett Land & Timber not only to use the best materials available but also to exceed piping insulation requirements by as much as 50 percent. For this reason, we cannot fulfill your request to replace the piping at no cost and repair the water-damaged areas.

← Uses indirect approach by easing into bad news at end of second paragraph

Undoubtedly, your insurance will cover the damage. I suspect that the rupture was caused by insufficient heating of the living room area during this unusually cold winter, but homeowner's insurance will cover damages resulting from cold-ruptured pipes 95 percent of the time.

← Is Honest and clear without blaming the reader

Our policy is to make repairs at a 20 percent discount in situations like this. Though the pipe rupture was not our fault, we feel personally close to every project we do and to every client we serve. Please get in touch if you would like to discuss this matter further. I would also be happy to speak with your insurance company if you wish.

Remains polite and tactful, despite refusing the request

← Maintains the "you" perspective throughout

Sincerely,

P.A. Jackson

Gerald A. Jackson

Figure 15.8 Bad News Letter Note the reader-friendly tone throughout.

Guidelines

for Conveying Bad News

➤ **Don't procrastinate.** As much as people may dislike the news, they will feel doubly offended after being kept in the dark.

➤ **Never just blurt it out.** Set a considerate tone by prefacing your bad news with considerate terms such as *I regret, We're sorry,* or *Unfortunately.* Instead of flatly proclaiming *Your application has been denied,* give recipients information they can use: *Unfortunately, we are unable to offer you admission to this year's Program. This letter will explain why we made this decision and how you can reapply.* Provide a context that leads into your explanation (Dumont and Lannon 206–21).

➤ **Give a clear and honest explanation.** Don't make things worse by fogging or dodging the issue. (See Chapter 3, "Guidelines for Persuasion.")

➤ **When you need to apologize, do so immediately.** Place your apology right up front. Don't say *An error was made in calculating your construction bill.* Do say *We are sorry we made a mistake in calculating your construction bill.* Don't attempt to camouflage the error. Don't offer excuses or try to shift the blame.

➤ **Use the passive voice to avoid accusations but not to dodge responsibility.** Instead of *You used the wrong bolts,* say *The wrong bolts were used* ("Plain Language").

➤ **Do not use "you" to blame the reader.** Instead of *You did not send a deposit,* say *We have not received your deposit.*

➤ **Keep the tone friendly and personal.** Avoid patronizing or impersonal jargon such as *company policy or circumstances beyond our control.*

➤ **Consider the format.** Take plenty of time to write and revise the letter, even by hand, if a personal note is warranted. For exceedingly bad news—say, denial of a promotion—consider sending the letter and following up with a meeting. Never use form letters for important matters, and don't use a formal letter for a relatively minor issue; for example, to notify employees that a company softball game has been cancelled, an email would be sufficient.

➤ **Consider how to deliver your message.** A letter attached to an email might be efficient, but this lack of any personal connection could cause more harm than good (Timmerman and Harrison 380). If possible, consider making a phone call first, or stopping by the person's office, before sending the letter.

you are asking a busy person to spend the time to read your letter, consider your request, collect the information, and write a response. Therefore, keep your request reasonable and state the purpose clearly and concisely. Apologize for any imposition and express your appreciation. Avoid long, involved inquiries that are unlikely to be answered.

Figure 15.9 illustrates an unsolicited letter requesting information. Research consultant Alan Greene is preparing a report on the feasibility of solar energy for home heating in Alaska. After learning that a nonprofit research group has been experimenting with new and improved solar applications, Alan decides to write for details. Notice how he tries to make the respondent's task as easy as possible.

If your questions are too numerous or complex, you might alternately request a phone or Skype interview, to save the person the time involved in writing a complicated reply. The writer of the letter in Figure 15.10 takes this approach.

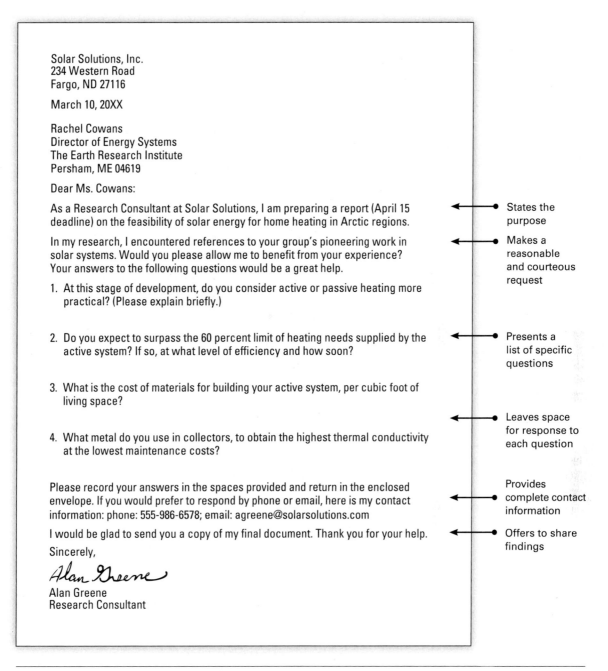

Solar Solutions, Inc.
234 Western Road
Fargo, ND 27116

March 10, 20XX

Rachel Cowans
Director of Energy Systems
The Earth Research Institute
Persham, ME 04619

Dear Ms. Cowans:

As a Research Consultant at Solar Solutions, I am preparing a report (April 15 deadline) on the feasibility of solar energy for home heating in Arctic regions. ◄——● States the purpose

In my research, I encountered references to your group's pioneering work in solar systems. Would you please allow me to benefit from your experience? Your answers to the following questions would be a great help. ◄——● Makes a reasonable and courteous request

1. At this stage of development, do you consider active or passive heating more practical? (Please explain briefly.)

2. Do you expect to surpass the 60 percent limit of heating needs supplied by the active system? If so, at what level of efficiency and how soon? ◄——● Presents a list of specific questions

3. What is the cost of materials for building your active system, per cubic foot of living space?

4. What metal do you use in collectors, to obtain the highest thermal conductivity at the lowest maintenance costs? ◄——● Leaves space for response to each question

Please record your answers in the spaces provided and return in the enclosed envelope. If you would prefer to respond by phone or email, here is my contact information: phone: 555-986-6578; email: agreene@solarsolutions.com ◄——● Provides complete contact information

I would be glad to send you a copy of my final document. Thank you for your help. ◄——● Offers to share findings

Sincerely,

Alan Greene

Alan Greene
Research Consultant

Figure 15.9 An Unsolicited Inquiry Letter This type of letter must be reader-friendly to increase the chance of getting a reply.

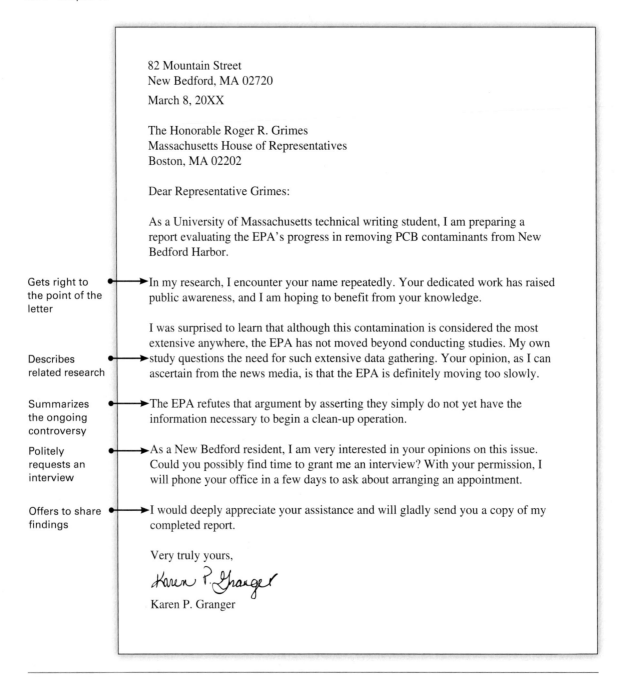

82 Mountain Street
New Bedford, MA 02720
March 8, 20XX

The Honorable Roger R. Grimes
Massachusetts House of Representatives
Boston, MA 02202

Dear Representative Grimes:

As a University of Massachusetts technical writing student, I am preparing a report evaluating the EPA's progress in removing PCB contaminants from New Bedford Harbor.

Gets right to the point of the letter → In my research, I encounter your name repeatedly. Your dedicated work has raised public awareness, and I am hoping to benefit from your knowledge.

Describes related research → I was surprised to learn that although this contamination is considered the most extensive anywhere, the EPA has not moved beyond conducting studies. My own study questions the need for such extensive data gathering. Your opinion, as I can ascertain from the news media, is that the EPA is definitely moving too slowly.

Summarizes the ongoing controversy → The EPA refutes that argument by asserting they simply do not yet have the information necessary to begin a clean-up operation.

Politely requests an interview → As a New Bedford resident, I am very interested in your opinions on this issue. Could you possibly find time to grant me an interview? With your permission, I will phone your office in a few days to ask about arranging an appointment.

Offers to share findings → I would deeply appreciate your assistance and will gladly send you a copy of my completed report.

Very truly yours,

Karen P. Granger
Karen P. Granger

Figure 15.10 Request for an Informative Interview Be as straightforward and polite as possible to get the reader interested in interviewing you.

Guidelines

for Inquiry Letters

➤ **Don't wait until the last minute.** Provide ample time for a response.

➤ **Whenever possible, write to a specific person.** If you need the name, call the organization and ask to whom you should address your inquiry.

➤ **Do your homework to ask the right questions.** A vague request such as "Please send me your data on ..." is likely to be ignored. Don't ask questions for which the answers are readily available elsewhere.

➤ **Explain who you are and how the information will be used.** If you appear to be from a competing company, your request will likely be ignored. But even in other situations, you will need to explain how you plan to use the requested data.

➤ **Write specific questions that are easy to understand and answer.** If you have multiple questions, put them in a numbered list to increase your chances of getting all the information you want. Consider leaving space for responses below each question.

➤ **Provide contact information.** If you can be reached via phone, email, and fax, provide all your numbers/addresses.

➤ **When possible, send your letter as a PDF attachment to an email.** The body of your email should take the form of a transmittal memo, asking your reader to please review the attachment.

➤ **If your letter is sent on paper, include a stamped, self-addressed envelope.** This courteous gesture will increase the likelihood of a response.

➤ **Say thank you and offer to follow up.** Offer to send a copy of the document in which you plan to use the information, if appropriate.

Claim Letters

Routine and arguable claim letters

In the workplace, things do not always run smoothly. Sometimes people make mistakes, systems break down, or companies make promises that can't be kept. Claim (or complaint) letters request adjustments for defective goods or poor services, or they complain about unfair treatment or something similar. Such letters fall into two categories: *routine claims* and *arguable claims.* Each situation calls for a different approach. Routine claims typically take a direct approach because the customer's claim is not debatable. Arguable claims present more of a persuasive challenge because they convey unwelcome news and are open to interpretation; arguable claims, therefore, typically take an indirect approach.

Figure 15.11 shows a routine claim letter. Jeffrey Ryder assumes the firm will honor the lifetime warranty, so he ends the letter by stating his claim clearly and with confidence, indicating exactly what he is requesting. Notice that an attention line directs the claim to the appropriate department, while a subject line makes clear the nature of the claim.

Figure 15.12 shows an arguable claim letter. Because the reply may not necessarily be in her favor, Sandra Alvarez uses a tactful and reasonable tone and an indirect approach to present her argument. Although she is courteous, she is also somewhat forceful, to reflect her insistence on an acceptable adjustment.

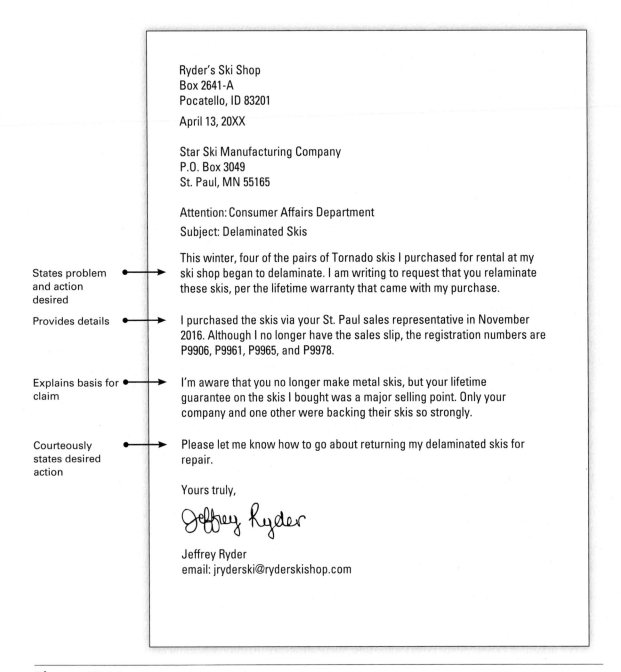

Ryder's Ski Shop
Box 2641-A
Pocatello, ID 83201

April 13, 20XX

Star Ski Manufacturing Company
P.O. Box 3049
St. Paul, MN 55165

Attention: Consumer Affairs Department

Subject: Delaminated Skis

States problem and action desired → This winter, four of the pairs of Tornado skis I purchased for rental at my ski shop began to delaminate. I am writing to request that you relaminate these skis, per the lifetime warranty that came with my purchase.

Provides details → I purchased the skis via your St. Paul sales representative in November 2016. Although I no longer have the sales slip, the registration numbers are P9906, P9961, P9965, and P9978.

Explains basis for claim → I'm aware that you no longer make metal skis, but your lifetime guarantee on the skis I bought was a major selling point. Only your company and one other were backing their skis so strongly.

Courteously states desired action → Please let me know how to go about returning my delaminated skis for repair.

Yours truly,

Jeffrey Ryder

Jeffrey Ryder
email: jryderski@ryderskishop.com

Figure 15.11 A Routine Claim Letter This type of claim letter is not debatable, but it still maintains a courteous tone.

OFFICE SYSTEMS, INC. 657 High Street Tulsa, OK 74120
(302) 655-5550 (v) (302) 655-5551 (fax) osys@sys.com (email)

January 23, 20XX

Consumer Affairs Department ETC.
Hightone Office Supplies
93 Cattle Drive
Houston, Texas 77028

Attention: Ms. Dionne Dubree

Dear Ms. Dubree:

Your company has an established reputation as a reliable wholesaler of office supplies. For eight years we have counted on that reliability, but a recent episode has left us annoyed and disappointed. ← Establishes early agreement

On January 29, we ordered 5 cartons of 700 MB HP CDs (#A74-866) and 13 cartons of Epson MX 70/80 black cartridges (#A19-556).

On February 5, the order arrived. But instead of the 700 MB HP CDs ordered, we received 650 MB Everlast CDs. And the Epson cartridges were blue, not the black we had ordered. We returned the order the same day. ← Presents facts to support claim

Also on the 5th, we called John Fitsimmons at your company to explain our problem. He promised delivery of a corrected order by the 12th. Finally, on the 22nd, we did receive an order—the original incorrect one—with a note claiming that the packages had been water damaged while in our possession. ← Offers more support

Our warehouse manager insists the packages were in perfect condition when he released them to the shipper. Because we had the packages only five hours and had no rain on the 5th, we are certain the damage did not occur here. ← Includes all relevant information

Responsibility for damages therefore rests with either the shipper or your warehouse staff. What bothers us is our outstanding bill from Hightone ($2,049.50) for the faulty shipment. We insist that the bill be canceled and that we receive a corrected statement. Until this misunderstanding, our transactions with your company were excellent. We hope they can be again. ← Requests a specific adjustment

We would appreciate having this matter resolved before the end of this month. ← Stipulates a reasonable response time

Yours truly,

Sandra Alvarez

Sandra Alvarez
Manager, Accounting

Figure 15.12 An Arguable Claim Letter This claim is debatable—in such situations, be sure to state your claim thoroughly and professionally.

Guidelines

for Claim Letters

Routine claim letters

➤ **Use a direct approach.** Describe the request or problem; explain the problem; and close courteously, restating the action you request.

➤ **Be polite and reasonable.** Your goal is not to sound off but to achieve results: a refund, a replacement, or an apology. Press your claim objectively yet firmly by explaining it clearly and by stipulating the reasonable action that will satisfy you. Do not insult the reader or revile the company.

➤ **Provide enough detail to clarify the basis for your claim.** Explain the specific defect. Identify the faulty item precisely, giving serial and model numbers, and date and place of purchase.

➤ **Conclude by expressing goodwill and confidence in the company's integrity.** Do not make threats or create animosity.

Arguable claim letters

➤ **Use an indirect approach.** People are more likely to respond favorably *after* reading your explanation. Begin with a neutral statement both parties can agree to—but that also serves as the basis for your request.

➤ **Once you've established agreement, explain and support your claim.** Include enough information for a fair evaluation: date and place of purchase, order number, dates of previous letters or calls, and background.

➤ **Conclude by requesting a specific action.** Be polite but assertive in phrasing your request.

Sales Letters

Purpose and tone of sales letters

Sales letters are written to persuade a current or potential customer to buy a company's product or try its services. Because people are bombarded by sales messages—in magazines, on billboards, on television, on the Internet—your letter must be genuinely persuasive and must get to the point quickly. Engage the reader immediately with an attention-grabbing statement or an intriguing question. Describe the product or service you offer, and explain its appeal. Conclude by requesting immediate action.

In the letter in Figure 15.13, restaurant owner Jimmy Lekkas opens with an attention-grabbing question that is hard to ignore and has universal appeal: good food, for free, right in the neighborhood. He then makes his case by explaining the history of his restaurant (which provides immediate credibility) and offering vivid descriptions of the food. He closes by asking readers to take action by a specific date.

Jimmy's Greek Kitchen
24-52 28th Street, Astoria, NY 11102
Phone: (555) 274-5672
Facebook: JimmysGreekKitchen
www.jimmysgreekkitchen.com

July 16, 20XX

Adriana Nikolaidis
26-22 30th Street #5
Astoria, NY 11102

Dear Ms. Nikolaidis:

Are you in the mood to sample the best Greek food in the neighborhood absolutely free of charge? We at the newly opened Jimmy's Greek Kitchen would like to say "Thank you for having us in your neighborhood" by inviting you to sample a variety of our authentic Greek specialties. ← Opens with an attention-grabbing question

If you've heard of or visited the famous Jimmy's in Chicago, you know that our fare has been pleasing Chicago diners for over 40 years. At long last, we have opened a companion restaurant in Astoria, and we are proud to offer you the same high-quality appetizers, entrees, and desserts, prepared to perfection. In fact, I trained our Astoria chef myself. ← Describes the long history and appeal of the restaurant

Ranging from charbroiled meats and grilled seafoods to vegetarian specialties and Greek favorites like tzatziki, pastitsio, and moussaka, Jimmy's is truly the best in town. Please have a look at the enclosed menu to see the full range of tasty foods we offer. ← Maintains appeal by describing the menu

Please take advantage of this special offer while it lasts. From now until August 31, just bring this letter to Jimmy's and lunch or dinner is on the house. You may choose any appetizer, entree, side order, beverage, and dessert on the menu—all free of charge. We hope that you will not only enjoy the dining experience but will tell others and come back to see us frequently. ← Ends by asking the reader to take action

Thank you,

Jimmy Lekkas

Jimmy Lekkas

Figure 15.13 **A Sales Letter** Sales letters must grab immediate attention, maintain interest, and evoke reader action.

Guidelines
for Sales Letters

➤ **Begin with a question or other attention-grabbing statement.** Induce the recipient to take notice.

➤ **Get to the point.** People resist reading long opening passages, especially if the message is unsolicited.

➤ **Spell out the benefits for the recipients.** Answer this implied question from the reader: "What do I stand to gain from this?"

➤ **Persuade with facts and with appeals to the senses.** Facts (such as the history of your company) appeal to logic. Graphic descriptions (such as the colors of your new cars or the types of food you offer) appeal to a different part of the brain—the emotions. Use both.

➤ **Tell the truth.** Despite your desire to sell something, it is unethical to lie, distort, exaggerate, or underestimate to make the sale.

➤ **Close by asking readers to take action.** Either ask for some reasonable action (such as "go to our Web site"), or offer an incentive (such as a free sample) to encourage follow-up.

Adjustment Letters

Positive adjustment letters

Adjustment letters are written in response to a claim letter from a customer. Even though most people never make formal complaints or follow up on warranties or product guarantees, companies generally will make a requested adjustment that seems reasonable.

Rather than quibbling over questionable claims, companies usually honor the request and show how much they appreciate the customer, as in Figure 15.14. In that example, writer Jane Duval apologizes graciously for a mistake. She omits an explanation because the error is obvious: Someone sent the wrong software. Once the reader has the information and apology, Duval shifts attention to a positive feature: the gift certificate. Note the "you" perspective, the friendly tone, and the incentive for further business.

Negative adjustment letters

Of course, if a claim is unreasonable or unjustified, the recipient usually will refuse the request. In refusing to grant a refund for a ten-speed bicycle, company representative Anna Jenkins needs to maintain a delicate balance (Figure 15.15). On the one hand, she must explain why she cannot grant the customer's request; on the other hand, she must be diplomatic in how she asserts that the customer is mistaken. Although Mrs. Gower may not be pleased by the explanation, it is thorough, reasonable, and courteous.

Use "Checklist: Letters" later in this section as you write a letter.

Software Unlimited
421 Fairview Road, Tulsa, OK 74321

May 2, 20XX

Mr. James Morris
P.O. Box 176
Little Rock, AR 54701

Dear Mr. Morris:

Your software should arrive by May 15. Sorry for the mixup. We don't make a
practice of sending Apple software to PC owners, but we do slip up once in a while. ◄────● Apologizes immediately

In appreciation for your patience and understanding, I've enclosed a $50 gift ◄────● Offers compensation
certificate. You can give it to a friend or apply it toward your next order. If you order
by phone, just give the certificate number, and the operator will credit your account.

Keep your certificate handy because you will be getting our new catalog soon. It ◄────● Looks toward the future

features 15 new business and utility programs that you might find useful.

Sincerely,

Jane Duval

Jane Duval
Sales Manager

Encl. Gift Certificate

Figure 15.14 A Positive Adjustment Letter Positive responses to claims ensure customer loyalty.

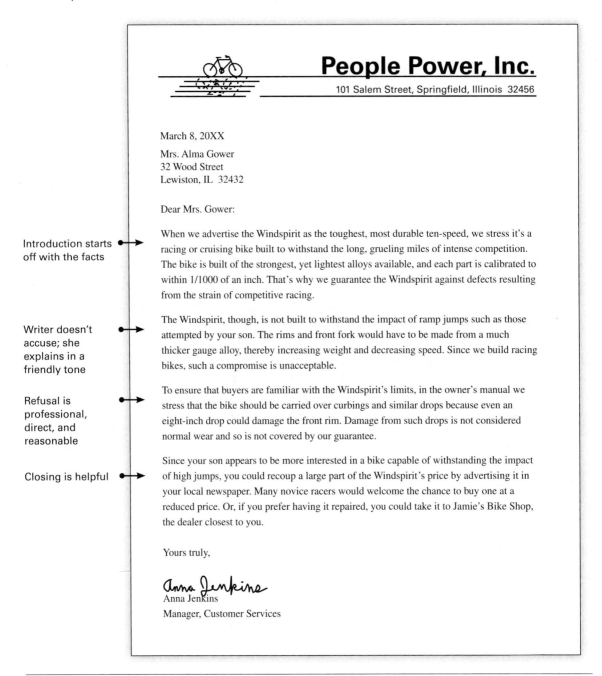

Introduction starts off with the facts

Writer doesn't accuse; she explains in a friendly tone

Refusal is professional, direct, and reasonable

Closing is helpful

People Power, Inc.

101 Salem Street, Springfield, Illinois 32456

March 8, 20XX

Mrs. Alma Gower
32 Wood Street
Lewiston, IL 32432

Dear Mrs. Gower:

When we advertise the Windspirit as the toughest, most durable ten-speed, we stress it's a racing or cruising bike built to withstand the long, grueling miles of intense competition. The bike is built of the strongest, yet lightest alloys available, and each part is calibrated to within 1/1000 of an inch. That's why we guarantee the Windspirit against defects resulting from the strain of competitive racing.

The Windspirit, though, is not built to withstand the impact of ramp jumps such as those attempted by your son. The rims and front fork would have to be made from a much thicker gauge alloy, thereby increasing weight and decreasing speed. Since we build racing bikes, such a compromise is unacceptable.

To ensure that buyers are familiar with the Windspirit's limits, in the owner's manual we stress that the bike should be carried over curbings and similar drops because even an eight-inch drop could damage the front rim. Damage from such drops is not considered normal wear and so is not covered by our guarantee.

Since your son appears to be more interested in a bike capable of withstanding the impact of high jumps, you could recoup a large part of the Windspirit's price by advertising it in your local newspaper. Many novice racers would welcome the chance to buy one at a reduced price. Or, if you prefer having it repaired, you could take it to Jamie's Bike Shop, the dealer closest to you.

Yours truly,

Anna Jenkins
Anna Jenkins
Manager, Customer Services

Figure 15.15 A Negative Adjustment Letter Negative responses to claims say "no" diplomatically but emphatically.

Guidelines

for Adjustment Letters

Granting Adjustments

➤ **Begin with the good news.** A sincere apology helps rebuild customers' confidence.

➤ **Explain what went wrong and how the problem will be corrected.** Without an honest explanation, you leave the impression that such problems are common or beyond your control.

➤ **Never blame employees as scapegoats.** To blame someone in the firm reflects poorly on the firm itself.

➤ **Do not promise that the problem will never recur.** Mishaps are inevitable.

➤ **End on a positive note.** Focus on the solution, not the problem.

Refusing Adjustments

➤ **Use an indirect organizational plan.** Explain diplomatically and clearly why you are refusing the request. Your goal is to convince the reader that your refusal results from a thorough analysis of the situation.

➤ **Be sure the refusal is unambiguous.** Don't create unrealistic expectations by using evasive language.

➤ **Avoid a patronizing or accusing tone.** Use the passive voice so as not to accuse the claimant, but do not hide behind the passive voice (see Chapter 11, "Use Passive Voice Selectively").

➤ **Close courteously and positively.** Offer an alternative or compromise, when it is feasible to do so.

Checklist

Letters

Use the following Checklist as you write a letter.

Content

❏ Does the situation call for a formal letter rather than a memo or email? (See "Letter Basics, Audience, and Purpose" in this chapter.)

❏ Is the letter addressed to the correct and specifically named person? (See "Letter Basics, Audience, and Purpose" in this chapter.)

❏ Have you determined the position or title of your recipient? (See "Standard Parts" in this chapter.)

❏ Does the letter contain all the standard parts? (See "Standard Parts" in this chapter.)

❏ Does the letter have all needed specialized parts? (See "Optional Parts" in this chapter.)

❏ Is the letter's main point clearly stated? (See "Standard Parts" in this chapter.)

❏ Is all the necessary information included? (See "Standard Parts" in this chapter.)

Arrangement

❏ Does the introduction engage the reader and preview the body section? (See "Standard Parts" in this chapter.)

❏ Is the direct or indirect approach used appropriately? (See "Decide on a Direct or Indirect Organizing Pattern" in this chapter.)

❏ Does the conclusion encourage the reader to act? (See "Standard Parts" in this chapter.)

❏ Is the format block? (See "Letter Parts, Formats, and Design Features" in this chapter.)

Style

❏ Does the letter convey a "you" perspective throughout? (See "Establish and Maintain a 'You' Perspective" in this chapter.)

❏ Is the letter in plain English (free of stuffy language)? (See "Use Plain English" in this chapter.)

❏ Is the tone professional, polite, and appropriately formal? (See "Be Polite and Tactful" in this chapter.)

❏ Is the letter designed for a tasteful, conservative look? (See "Letter Parts, Formats, and Design Features" in this chapter.)

❏ Is the style clear, concise, and fluent? (See Chapter 11.)

❏ Have you proofread with extreme care? (See Chapter 6, "Make Proofreading Your Final Step.")

Projects

For all projects, check with your instructor about whether to present your findings in class, bring drafts to class for discussion, upload your project to the class learning management system (LMS), and/or use the LMS forum or discussion boards to collaborate and review each activity below.

General

1. Think of an idea you would like to see implemented in your job (e.g., a way to increase productivity, improve service, increase business, or improve working conditions). Write a routine miscellaneous memo requesting action and persuading your audience that your idea is worthwhile.

2. Write and send (as a paper letter or a PDF attachment to an email) an unsolicited letter of inquiry about the topic you are investigating for an analytical report or research assignment. In your letter, you might request brochures, pamphlets, or other informative literature, or you might ask specific questions. Submit a copy of your letter, and the response, to your instructor.

3. Politicians and other officials receive dozens if not hundreds of emails daily; research has shown that a hard-copy letter, sent in the mail, is a more effective approach. Write a claim letter to a politician about an issue affecting your school or community. Or, write a claim letter to an appropriate school official to recommend an action on a campus problem.

4. Write a claim letter about a problem you've had with goods or services. State your case clearly and objectively, and request a specific adjustment.

5. The following sentences need to be overhauled before being included in a memo or letter. Identify the weakness in each statement, and revise as needed. For example, you would revise the accusatory *You were not very clear* to *We did not understand your message.*

   a. I need all the information you have about methane-powered engines.

   b. You people have sent me the wrong software!

   c. It is imperative that you let me know of your decision by January 15.

   d. I have become cognizant of your experiments and wish to ask your advice about the following procedure.

   e. You will find the following instructions easy enough for an ape to follow.

   f. As per your request, I am sending the country map.

   g. I am in hopes that you will call soon.

   h. We beg to differ with your interpretation of this leasing clause.

Team

1. Divide into teams and assume you own a company (create a name for it). Revise the following message so that it gets the intended results. Pay close attention to *tone.* Use a memo format. Make sure the subject line clearly forecasts your topic. Appoint one team member to present the revised memo in class.

> Too many employees are parking in front of the store rather than behind it. As a result of this infraction, customers have been complaining that they cannot find parking spaces. If employees don't start parking where they're supposed to, we'll lose customers. Lost customers means lost jobs. Since I've worked hard to build this company, I don't want to lose it because employees are too lazy to walk from the back lot. Let's keep our customers by keeping them happy.

2. Working in groups, respond to the following scenario. Appoint one group member to present the letter in class.

> As director of consumer affairs, you've received an adjustment request from Brian Maxwell. Two years ago, he bought a pair of top-of-the-line Gannon speakers. Both speakers, he claims, are badly distorting bass sounds, and he states that his local dealer refuses to honor the three-year warranty. After checking, you find that the dealer refused because someone had obviously tampered with the speakers. Two lead wires had been respliced; one of the booster magnets was missing; and the top insulation also was missing from one of the speaker cabinets. Your warranty specifically states that if speakers are removed from the cabinet or subjected to tampering in any way, the warranty is void. You must refuse the adjustment; however, because Maxwell bought the speakers from a factory-authorized dealer, he is entitled to a 30% discount on repairs. Write the refusal, offering this alternative. His address: 691 Concord Street, Biloxi, MS 71690.

Digital and Social Media

As described earlier in this chapter, word-processing programs offer templates for writing letters. (Templates are preformatted layouts that can be used to create résumés, memos, letters, and other documents.) Templates can help you get started with the writing. Templates can also be a problem, however, because instead of thinking for yourself about the audience, purpose, and appropriate organizational pattern and language usage, you may end up letting the template do the thinking. Look at the various templates available in your word-processing program or other business writing apps and make a list of ways a template may or may not work for your purposes. Revise the template to be more suitable for the audience and purpose in General Project 2, 3, or 4 (above).

Global

Interview a person whose work takes him or her to one or more countries outside the United States. Ask that person to describe the way letters or memos are used for international communication and whether any special issues involving grammar, forms of address, direct or indirect organizational patterns, or other features make letter writing different when addressing international audiences. Ask about the use of email instead of memos or letters (for example, ask about what situations call for a memo or letter versus an email, in this international setting). Next, write a memo to your instructor describing what you learned. Send your memo to your instructor as a PDF attachment to an email, or upload the memo to your learning management system.

Chapter 16
Résumés and Other Job-Search Materials

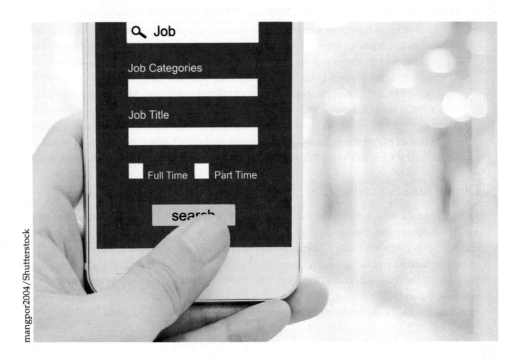

mangpor2004/Shutterstock

❝My company's recruiting and hiring team reviews dozens of entry-level job applications weekly. One main quality we look for in any candidate—regardless of technical qualifications—is that person's attention to detail. The first indication of this, of course, appears in the résumé and application letter. We expect these materials to be nothing less than professional in their content and presentation. Being a newly minted graduate is no excuse for a sloppy set of application materials.**❞**

—*Carol Jiminez, Personnel Manager, large civil engineering firm*

Learning Objectives

16.1 Identify your personal assets as a job applicant

16.2 Search for a job systematically by planning, focusing, exploring, and networking

16.3 Create an effective reverse chronological or functional résumé

16.4 Write an effective solicited or unsolicited application letter

16.5 Submit appropriately formatted digital application materials

16.6 Prepare an impressive dossier, portfolio, or e-portfolio

16.7 Succeed in a job interview and write professional follow-up correspondence

In today's job market, many applicants compete for few openings. Whether you are applying for your first professional job or changing careers, you need to market your skills effectively. At each stage of the application process, you must stand out among the competition.

Assessing Your Skills and Aptitudes

16.1 Identify your personal assets as a job applicant

Identify your assets

Begin your job search by assessing those qualities and skills you can offer a potential employer:

- Do I communicate well, and am I also a good listener?
- Do I work well in groups and with people from different backgrounds?
- Do I have experience or aptitude for a leadership role?
- Can I solve problems and get things done?
- Can I perform well under pressure?
- Can I work independently, with minimal supervision?
- Do I have any special skills (public speaking; working with people, software or other technical skills; aptitude with words, analytical skills, second or third languages; artistic/musical talent; mathematical aptitude)?
- Do I have any hobbies that could improve my job prospects?
- Would I prefer to work at a large company or a small one, or at a for-profit or a nonprofit organization?
- Do I like to travel, or would I prefer working in a single location?

Besides helping you focus your job search, your answers to these questions will come in handy when you write your résumé and prepare for job interviews.

Researching the Job Market

16.2 Search for a job systematically by planning, focusing, exploring, and networking

Search within a reasonable range, focusing on those fields that interest you most and fit you best.

Plan Your Strategy

Don't just dive in; work step by step

Begin your research well in advance of the time you need to have a job lined up. The question "Where do I start?" can be daunting: "Do I go to a career counselor first?" "Should I talk with friends and family members?" "Do I go straight to the Help Wanted section—in print or online?" "Which Web sites are the best?" Seemingly endless sources of information are available to job seekers. Proceed in a step-by-step, logical way—rather than going straight to the Internet and trying to navigate random Web sites.

Focus Your Search

Consult industry specific resources

Before you apply for specific jobs, learn about the industry: Consult relevant books, magazines, journals, and Web sites. Join a professional group related to your industry and either attend meetings or interact online. Identify key companies and organizations and research them. Try to arrange an informational interview with someone in your field. Even busy professionals are often willing to speak with interested job seekers who are not applying for a specific job. These people can offer general advice about the industry as a whole as well as specific information about their own company. Alumni from your program may be especially willing to meet with you or talk by phone.

Explore Employment Resources

Online, you can use nationwide portals such as Careerbuilder, Indeed, Monster, and SnagAJob to find job postings across the country or narrow your search to specific locations. Your county and/or state may also have Web sites and social media feeds specifically dedicated to promoting jobs in your area. Use Google to search on "Jobs in Minneapolis" or "Jobs in Boston" (or whatever city you wish) to see what might be available for your area.

Use online career sites

You can also check out the online Help Wanted sections of your local newspaper sites. Similar to these Help Wanted sections, Craigslist is also a popular way to review job openings of a general sort. But for jobs that are more technical and industry specific, look at Web sites or social media feeds for professional societies in your field. For example, electrical engineers can look through job postings on the Institute of Electrical and Electronics Engineers (IEEE) Web site. Almost all career areas have similar professional societies, some with local chapters.

Explore online help wanted and industry-specific sites

Social media is a valuable resource for any job search. Start with LinkedIn, the most popular site of its kind. By connecting you with others in your field and with companies as well as recruiters, LinkedIn can help you stay informed about job postings, sometimes before the postings are distributed more widely. If you are interested in a particular company or organization, check out their Twitter feed and Facebook pages. Many companies use social media feeds to post short job announcements, directing readers to a link for more information. Figure 16.1 shows a sample posting that might appear on Twitter or another social media site.

Look for jobs via LinkedIn

Figure 16.1   Electronic Job Posting Available on Twitter and LinkedIn Employers know they can reach thousands of people this way.

Look for specific
job postings

You can also go directly to the "employment" or "jobs" tab on a company or organization's Web site to see what jobs are currently available and to get more details on qualifications, due dates, and so forth. These Web sites will provide information about the organization's mission and vision, locations, and product lines or services.

Learn to Network

Talk with helpful
people

An important but often overlooked route to a good job is the *human connection*. Once you've focused in on a few potential career choices, go to your campus job placement office to meet with a career counselor and learn more about upcoming career fairs, recruiter visits to campus, and networking events. Speak with faculty, alumni, and others in the field. Network with acquaintances and family friends who may have other contacts.

Find an
internship
or volunteer
opportunity

When it comes to landing a job, experience can count more than grades. Find a summer job or internship in your field or do related volunteer work. Consider registering with agencies that provide temporary staffing. Even the most humble and temporary job offers the chance to make contacts and discover opportunities.

Use online social
networking

As mentioned above, social media can help you discover job openings. But more than that, LinkedIn and similar sites provide powerful ways for you to network with hundreds of others and market yourself. You can easily keep your profile—including recent work experience, résumé, and references—updated and ready to share with potential employers. Be sure to connect with former colleagues and classmates; these people are often your best bet for hearing about the most recent openings. Don't be shy about contacting people you don't know first-hand but to whom you are connected through a former coworker or classmate. Keep in mind that these are professional sites; make sure that anything you post is something you would want an employer to see. Be sure that *all* of your social media presence is professional and makes the best case for hiring you.

> NOTE *Although online job listings and résumé postings have provided new tools for job seekers, today's job searches require the same basic approach and communication skills that people have relied on for decades: well written and professionally formatted application materials.*

Résumés

16.3 Create an effective reverse chronological or functional résumé

What a résumé
does

Essentially, an applicant's personal advertisement for employment, a résumé gives an employer an instant overview. In fact, employers initially spend only 15 to 45 seconds looking at a résumé; during this scan, they are looking for a persuasive answer to the essential question: "What can you do for us?"

What employers
expect in a
résumé

Employers are impressed by a résumé that looks good, reads easily, appears honest, and provides only the relevant information an employer needs to determine whether the applicant should be interviewed. Résumés that are mechanically flawed, cluttered, sketchy, hard to follow, or seemingly dishonest simply get discarded.

Parts of a Résumé

Résumés contain these standard parts: contact information, career objectives, education, work experience, personal data and interests, and references. A résumé is not the place for such items as your desired salary and benefits or your requirements for time off. Omit your photograph as well as information that employers are not allowed to legally request (such as race, age, or marital status).

What to include—and not include—in a résumé

As you read through this section, refer to Figure 16.2 later in this section, which includes all the required parts of a résumé.

CONTACT INFORMATION. Make sure prospective employers know how to reach you. If you are between addresses, provide both addresses and check each contact point regularly. Be sure your email address and phone number are accurate. If you use an answering machine or voice mail, record an outgoing message that sounds friendly and professional. If you have your own Web site (professional, not personal), include the Web address. Remember that many employers will look at your Facebook page and often perform a Google search, too. Be sure your public profile is professional in tone and content.

CAREER OBJECTIVES. Spell out the kind of job you want. Avoid vague statements such as "A position in which I can apply my education and experience." Be specific: "An intensive-care nursing position in a teaching hospital, with the eventual goal of supervising and instructing." Tailor your career objective statement as you apply for different jobs, in order to match yourself with each position. State your immediate and long-range goals, including any plans to continue your education. If the company has branches, include *Willing to relocate.*

One hiring officer for a major computer firm offers this advice: "A statement should show that you know the type of work the company does and the type of position it needs to fill" (Beamon, qtd. in Crosby 3).

> NOTE *Below career objectives, you might insert a summary of qualifications. This section is vital in a computer-scannable résumé (Figure 16.6), but even in a conventional résumé, a "Qualifications" section can highlight your strengths. Make the summary specific and concrete: replace "proven leadership" with "team and project management," "special-event planning," or "instructor-led training"; replace "persuasive communicator" with "fundraising," "publicity campaigns," "environmental/public-interest advocacy," or "door-to-door canvassing." In short, allow the reader to visualize your activities.*

EDUCATION. Begin with your most recent schooling and work backward. Include the name of the school, degree completed, year completed, and your major and minor. Omit high school, unless the high school's prestige or your achievements there warrant its inclusion. List courses that have directly prepared you for the job you seek. If your class rank or grade point average is favorable, list it. Include specialized training during military service. If you finance your education by working, say so, indicating the percentage of your contribution.

Contact information is complete, including email address

Objective fits job criteria

Education section lists all relevant information

Work experience section lists most recent jobs first and includes relevant skills applied on the job

Leadership section combines awards, skills, and activities that may be relevant to a job

When no references are included, an "available on request" statement substitutes

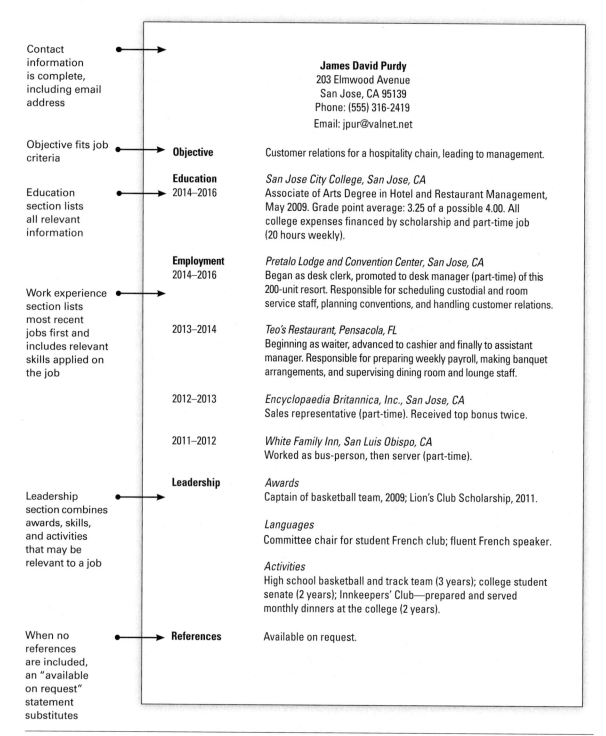

James David Purdy
203 Elmwood Avenue
San Jose, CA 95139
Phone: (555) 316-2419
Email: jpur@valnet.net

Objective   Customer relations for a hospitality chain, leading to management.

Education
2014–2016   *San Jose City College, San Jose, CA*
Associate of Arts Degree in Hotel and Restaurant Management, May 2009. Grade point average: 3.25 of a possible 4.00. All college expenses financed by scholarship and part-time job (20 hours weekly).

Employment
2014–2016   *Pretalo Lodge and Convention Center, San Jose, CA*
Began as desk clerk, promoted to desk manager (part-time) of this 200-unit resort. Responsible for scheduling custodial and room service staff, planning conventions, and handling customer relations.

2013–2014   *Teo's Restaurant, Pensacola, FL*
Beginning as waiter, advanced to cashier and finally to assistant manager. Responsible for preparing weekly payroll, making banquet arrangements, and supervising dining room and lounge staff.

2012–2013   *Encyclopaedia Britannica, Inc., San Jose, CA*
Sales representative (part-time). Received top bonus twice.

2011–2012   *White Family Inn, San Luis Obispo, CA*
Worked as bus-person, then server (part-time).

Leadership   *Awards*
Captain of basketball team, 2009; Lion's Club Scholarship, 2011.

Languages
Committee chair for student French club; fluent French speaker.

Activities
High school basketball and track team (3 years); college student senate (2 years); Innkeepers' Club—prepared and served monthly dinners at the college (2 years).

References   Available on request.

Figure 16.2   A Reverse Chronological Résumé Use this format to show a clear pattern of job experience.

WORK EXPERIENCE. If your experience relates to the job, list it before your education. List your most recent job and then earlier jobs. Include employers' names and dates of employment. Indicate whether a job was full-time, part-time (hours weekly), or seasonal. Describe your exact duties for each job, indicating promotions. If it is to your advantage, state why you left each job. Include military experience and relevant volunteer work. If you lack paid experience, emphasize your education, including internships and special projects.

PERSONAL DATA AND INTERESTS. List any awards, skills, activities, and interests that are *relevant* to the given position, such as memberships in professional organizations, demonstrations of leadership, languages, special skills, and hobbies that may be of interest to the employer.

REFERENCES. List three to five people who have agreed to provide strong assessments of your qualifications and who can speak on your behalf. Never list as references people who haven't first given you express permission. Your references should not be family members or non-work-related friends; instead, list former employers, professors, and community figures who know you well. If saving space is important, simply state at the end of your résumé, "References available upon request," to help keep the résumé to one page. But if the résumé already takes up more than one page, you probably should include your references. If you don't list references, prepare a separate reference sheet that you can provide on request. Include each person's job title, company address, and contact information.

> NOTE *Under some circumstances, you may—if you wish—waive the right to examine your references. Some applicants, especially those applying to professional schools, as in medicine and law, waive this right in concession to a general feeling that a letter writer who is assured of confidentiality is more likely to provide a balanced, objective, and reliable assessment of a candidate. Before you decide, seek the advice of your major adviser or a career counselor.*

PORTFOLIOS. To illustrate your skills and experience in areas such as marketing, engineering, or other fields that generate actual documents or visual designs, assemble a portfolio showing samples of your work. If you do have a portfolio, indicate this on your résumé, followed by "Available on request," as in Figure 16.6. (See "Guidelines for Dossiers, Portfolios, and E-Portfolios" in this chapter.)

Using Templates

Most word-processing programs, such as Microsoft Word or Apple Pages, provide templates that can be used as a basis for your résumé. Similarly, major online job sites, such as Monster.com, offer dozens of sample résumés and downloadable templates organized by the industry or type of job you are applying for. College career centers also have sample résumés for you to review. These templates, along with the examples provided in this chapter, will help you get a feel for the range of styles and approaches

Benefits of using templates

used by professionals in different fields and at different stages of their careers (entry level versus, say, middle management).

Cautions with using templates

Templates and examples can help you get started (and can save you a lot of time with formatting), but you still need to pay attention to key features that make your résumé suitable for your audience and purpose. For instance, be aware of which organizational pattern (for example, chronological versus functional) is appropriate.

Organizing Your Résumé

Reverse chronological and functional résumés

Organize your résumé to convey the strongest impression of your qualifications, skills, and experience. A résumé like the one in Figure 16.2 is known as a *reverse chronological résumé*, listing the most recent school and job first. If you have limited experience or education, gaps in your work history (e.g., due to illness, raising children), or if you have frequently switched career paths, create a *functional résumé* (Figure 16.3) to highlight skills relevant to a particular job.

Guidelines

for Writing and Designing Your Résumé

➤ **Begin drafting your résumé well before your job search.**

➤ **Tailor your résumé for each job.** Read the advertised job requirements and adjust your career objective accordingly—but realistically. Tailor your work experience, personal data, and personal interests to emphasize certain areas for certain jobs—but do not distort the facts.

➤ **Try to limit the résumé to a single page but keep it uncluttered and tasteful.** If the résumé looks cramped, you might need to go to a second page—in which case you could have room to list your references, but only if those references have given you permission in advance.

➤ **Stick to experience relevant to the job.** Don't list everything you've ever done.

➤ **Use action verbs and key words.** Action verbs (*supervised, developed, built, taught, installed, managed, trained, solved, planned, directed*) stress your ability to produce results. If your résumé is likely to be scanned electronically or if you post it online, list keywords as nouns (*leadership skills, software development, data processing, editing*) below your contact information and your statement of objective.

➤ **Use bold, italic, underlining, colors, fonts, bullets, and punctuation thoughtfully, for emphasis.** Do not use highlighting or punctuation to be artsy. Keep punctuation consistent and as simple as possible.

➤ **Never invent or distort credentials.** Make yourself look as good as the *facts* allow. Companies routinely investigate claims made in résumés, and people who lie will certainly not be hired.

➤ **Use templates and examples to get started.** But check with a career counselor and with others in the field (alumni, mentors, or trusted colleagues) to be sure your final résumé is organized appropriately for the particular career and job you are interested in.

➤ **As a rule, do not include hyperlinks on your résumé.** Instead, use an e-portfolio or dossier to provide access to additional information.

➤ **Proofread, proofread, proofread.** Don't rely on autocorrect, spell check, or grammar checkers. Famous résumé mistakes include winning a "bogus award" instead of a "bonus award" and "ruining" rather than "running" a business.

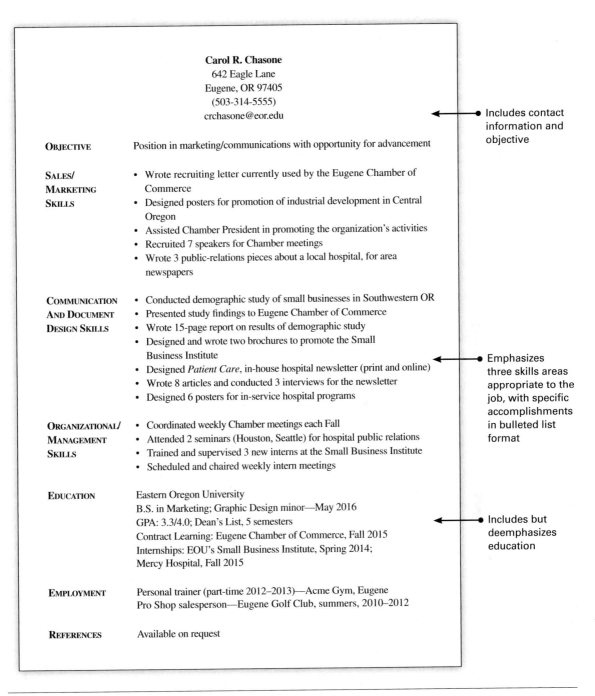

Carol R. Chasone
642 Eagle Lane
Eugene, OR 97405
(503-314-5555)
crchasone@eor.edu

→ Includes contact information and objective

OBJECTIVE   Position in marketing/communications with opportunity for advancement

SALES/ MARKETING SKILLS
- Wrote recruiting letter currently used by the Eugene Chamber of Commerce
- Designed posters for promotion of industrial development in Central Oregon
- Assisted Chamber President in promoting the organization's activities
- Recruited 7 speakers for Chamber meetings
- Wrote 3 public-relations pieces about a local hospital, for area newspapers

COMMUNICATION AND DOCUMENT DESIGN SKILLS
- Conducted demographic study of small businesses in Southwestern OR
- Presented study findings to Eugene Chamber of Commerce
- Wrote 15-page report on results of demographic study
- Designed and wrote two brochures to promote the Small Business Institute
- Designed *Patient Care*, in-house hospital newsletter (print and online)
- Wrote 8 articles and conducted 3 interviews for the newsletter
- Designed 6 posters for in-service hospital programs

→ Emphasizes three skills areas appropriate to the job, with specific accomplishments in bulleted list format

ORGANIZATIONAL/ MANAGEMENT SKILLS
- Coordinated weekly Chamber meetings each Fall
- Attended 2 seminars (Houston, Seattle) for hospital public relations
- Trained and supervised 3 new interns at the Small Business Institute
- Scheduled and chaired weekly intern meetings

EDUCATION   Eastern Oregon University
B.S. in Marketing; Graphic Design minor—May 2016
GPA: 3.3/4.0; Dean's List, 5 semesters
Contract Learning: Eugene Chamber of Commerce, Fall 2015
Internships: EOU's Small Business Institute, Spring 2014;
Mercy Hospital, Fall 2015

→ Includes but deemphasizes education

EMPLOYMENT   Personal trainer (part-time 2012–2013)—Acme Gym, Eugene
Pro Shop salesperson—Eugene Golf Club, summers, 2010–2012

REFERENCES   Available on request

Figure 16.3 A Functional Résumé Use this format to focus on skills and potential instead of employment chronology. (Note that certain items in the above skills categories overlap.)

Application Letters

16.4 Write an effective solicited or unsolicited application letter

What an application letter does

An *application letter*, also known as a *cover letter*, complements your résumé. The letter's main purpose is to explain how your credentials fit the particular job and to convey a sufficiently informed, professional, and likable persona for the prospective employer to decide that you should be interviewed. Another purpose of the letter is to highlight specific qualifications or skills; for example, you might have listed "Java programming" on your résumé, but for one particular job application, you may wish to call attention to this item in your cover letter:

> My résumé notes that I am experienced with Java programming. I also tutor Java programming students in our school's learning center.

Solicited and unsolicited application letters

Sometimes you will apply for positions that are advertised (*solicited applications*). At other times, you will write prospecting letters to organizations that have not advertised an opening but that might need someone like you (*unsolicited applications*). In either case, tailor your letter to the situation.

Solicited Application Letters

An application letter—whether solicited or unsolicited—consists of an introduction, body (one or a few short paragraphs), and conclusion as in Figure 16.4. Imagine you are James Purdy (see his résumé in Figure 16.2). On one popular on-line job site, you see the advertisement in Figure 16.1 (see "Explore Employment Resources" earlier in this chapter) and decide to apply. Now you plan and compose your letter.

Use the introduction to get right to the point

In your brief introduction (five lines or fewer), do these things: Name the job and where you have seen it advertised; identify yourself and your background; and, if possible, establish a connection by naming a mutual acquaintance who encouraged you to apply—but only if that person has given you permission.

Use the body section to demonstrate your qualifications

In the body, spell out your case. Without merely repeating your résumé, relate your qualifications directly to this job. Also, be specific. Instead of referring to "much experience" or "increased sales," stipulate "three years of experience" or "a 35 percent increase in sales between June and October 2014." Support all claims with evidence. Instead of saying, "I have leadership skills," say, "I served as student senate president during my senior year and was captain of the lacrosse team."

Use the conclusion to restate interest

In the conclusion, restate your interest and emphasize your willingness to retrain or relocate if necessary. If the job is nearby, request an interview; otherwise, request a phone call or an email, suggesting a time you can be reached.

203 Elmwood Avenue
San Jose, CA 10462

April 22, 20XX

Sara Costanza
Personnel Director
Liberty International, Inc.
Lansdowne, PA 24153

Dear Ms. Costanza:

Please consider my application for a junior management position at your Lake
Geneva resort, as advertised on April 19 on Monster.com. I will graduate
from San Jose City College on May 30 with an Associate of Arts degree in hotel
and restaurant management. Dr. H. V. Garlid, my nutrition professor, described his
experience as a consultant for Liberty International and encouraged me to apply.

As you can see from my enclosed résumé, for two years I worked as a part-time
desk clerk, and I was promoted to manager, at a 200-unit resort. This experience,
combined with earlier customer relations work in a variety of situations, has given
me a clear and practical understanding of customers' needs and expectations.

As an amateur chef, I'm well aware of the effort, attention, and patience required
to prepare fine food. Moreover, my skiing and sailing background might be assets
to your resort's recreation program.

I have worked hard to hone my hospitality management skills. My experience,
education, and personality have prepared me to work well with others and to
respond creatively to challenges, crises, and added responsibilities.

If my background meets your needs, please phone me any weekday after 4:00 p.m.
at (555) 316-2419.

Sincerely,

James D. Purdy

James D. Purdy

Annotations (right margin):
- Writer identifies self and purpose
- Establishes a connection
- Relates qualifications to job opening
- Applies relevant interests to the job
- Expresses confidence and enthusiasm
- Makes follow-up easy for the reader

Figure 16.4 **A Solicited Application Letter**

Unsolicited Application Letters

Do not limit your job search to advertised openings. In fact, fewer than 20 percent of all job openings are advertised. Unsolicited application letters are a good way to uncover possibilities. They do have drawbacks, however: You may waste time writing to organizations that have no openings, and you cannot tailor your letter to advertised requirements. But there are also advantages: Even employers with no openings often welcome and file impressive unsolicited applications or pass them on to another employer who has an opening.

Use the introduction to spark reader interest

Because an unsolicited letter arrives unexpectedly, you need to get the reader's immediate attention. Don't begin, "I am writing to inquire about the possibility of obtaining a position with your company." Instead, open forcefully by establishing a connection with a mutual acquaintance, or by making a strong statement or asking a persuasive question as in the following example:

A forceful opening

> Does your hotel chain have a place for a junior manager with a degree in hospitality management, a proven commitment to quality service, and customer relations experience that extends far beyond textbooks? If so, please consider my application for a position.

Address your letter to the person most likely in charge of hiring. Consult company Web sites for names of company officers. Also, consider using a "Subject" line to attract a busy reader's attention and to announce the purpose of your letter, as in Figure 16.5.

> NOTE *Write to a specific person—not to a generic recipient such as "Director of Human Resources" or "Personnel Office." If you don't know who does the hiring, contact the company and ask for that person's name and title, and be sure you get the spelling right.*

Guidelines

for Application Letters

➤ **Develop an excellent prototype letter.** Presenting a clear and concise picture of who you are, what you have to offer, and what makes you special is arguably the hardest—but most essential—part of the application process. Revise this prototype, or model, until it represents you in the best possible light. Keep it to a single page, if possible.

➤ **Customize each letter for the specific job opening.** Although you can base letters to different employers on the same basic prototype—with appropriate changes—prepare each letter afresh. Don't look for shortcuts.

➤ **Use caution when adapting sample letters.** Plenty of free, online sample letters provide ideas for approaching your own situation. But never borrow them whole. Most employers are able to spot a "canned" letter immediately.

➤ **Create a dynamic tone with active voice and action verbs.** Instead of "Management responsibilities were steadily given to me," say "I steadily assumed management responsibilities." Be confident without seeming arrogant. (For more on tone, see Chapter 11, "Adjusting Your Tone.")

➤ **Never be vague.** Help readers visualize: Instead of saying, "I am familiar with the 1022 interactive database system and RUNOFF, the text-processing system," say "As a lab grader, I kept grading records on the 1022 database management system and composed lab procedures on the RUNOFF text-processing system."

➤ **Never exaggerate.** Liars get busted.

➤ **Convey some enthusiasm.** An enthusiastic attitude can sometimes be as important as your background (as in Figure 16.5).

➤ **Avoid flattery.** Don't say "I am greatly impressed by your remarkable company."

➤ **Be concise.** Review Chapter 11, "Editing for Conciseness." Limit your letters to one page, unless your discussion truly warrants the additional space.

➤ **Avoid being overly informal or overly stiff.** Avoid informal terms that sound unprofessional ("Your company sounds like a cool place to work") as well as stuffy language ("Hitherto, I request the honor of your acquaintance").

➤ **Never settle for a first draft—or even a second or third.** The application letter is your one chance to introduce yourself to a prospective employer. Make it perfect by trimming excess wording, double-checking the tone, and ensuring that you have connected your qualifications directly to the job. After you are satisfied with the content, proofread repeatedly to spot any factual errors or typos.

➤ **Never send a photocopied letter.**

Digital and Print Job Application Materials

16.5 Submit appropriately formatted digital application materials

Résumés and cover letters today are typically submitted as PDF documents, uploaded to the online job application site for a particular organization. Social networking sites like LinkedIn or Monster.com also allow you to upload your résumé and other materials and connect these to your online profile. Some employers may also ask you to submit your materials as email attachments, especially if you were solicited to apply. PDF files retain the look and feel (fonts, page breaks, line breaks) of the résumé you worked so hard to create, so be sure you know how to turn your Word or other word-processing document into PDF. (Hint: you can usually do so by selecting "PDF" in the "Save as" or similar menu.) PDF files also make it more difficult for a reader to accidentally insert a stray keystroke or change your formatting while reading your materials.

Using PDF files for digital résumés

A printed résumé may be necessary in certain situations. If you attend a campus career fair or similar in-person event (for instance, a company holds an information session or comes to campus to recruit), having print copies of your résumé ready to give to recruiters is invaluable. You can point out items of interest on the spot and make a lasting impression. Be sure to use high-quality white stationery paper, available at most copy shops. Ask for the recruiter's business card so you can send a follow-up PDF copy.

Print résumés are useful at career fairs

If you do hand out print copies of your materials, keep in mind that prospective employers will probably end up digitizing your résumé by scanning a copy for their job database systems. Digital résumés, especially at large organizations, become part of a much larger database of applicants (for current or later job openings). Digital résumés are much easier to search via computer, with employers looking for keywords to help them narrow down the application pool. Figure 16.6 shows a résumé suitable for scanning.

Scanned résumés

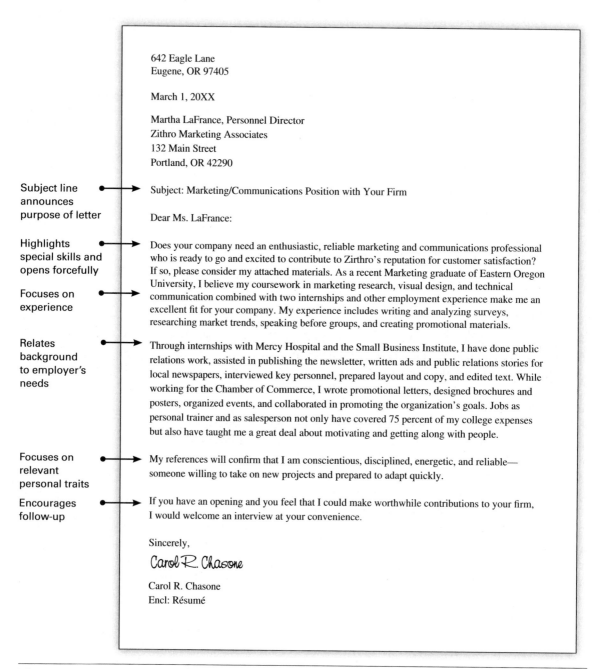

642 Eagle Lane
Eugene, OR 97405

March 1, 20XX

Martha LaFrance, Personnel Director
Zithro Marketing Associates
132 Main Street
Portland, OR 42290

Subject line announces purpose of letter → Subject: Marketing/Communications Position with Your Firm

Dear Ms. LaFrance:

Highlights special skills and opens forcefully → Does your company need an enthusiastic, reliable marketing and communications professional who is ready to go and excited to contribute to Zirthro's reputation for customer satisfaction? If so, please consider my attached materials. As a recent Marketing graduate of Eastern Oregon

Focuses on experience → University, I believe my coursework in marketing research, visual design, and technical communication combined with two internships and other employment experience make me an excellent fit for your company. My experience includes writing and analyzing surveys, researching market trends, speaking before groups, and creating promotional materials.

Relates background to employer's needs → Through internships with Mercy Hospital and the Small Business Institute, I have done public relations work, assisted in publishing the newsletter, written ads and public relations stories for local newspapers, interviewed key personnel, prepared layout and copy, and edited text. While working for the Chamber of Commerce, I wrote promotional letters, designed brochures and posters, organized events, and collaborated in promoting the organization's goals. Jobs as personal trainer and as salesperson not only have covered 75 percent of my college expenses but also have taught me a great deal about motivating and getting along with people.

Focuses on relevant personal traits → My references will confirm that I am conscientious, disciplined, energetic, and reliable—someone willing to take on new projects and prepared to adapt quickly.

Encourages follow-up → If you have an opening and you feel that I could make worthwhile contributions to your firm, I would welcome an interview at your convenience.

Sincerely,

Carol R. Chasone

Carol R. Chasone
Encl: Résumé

Figure 16.5   An Unsolicited Application Letter

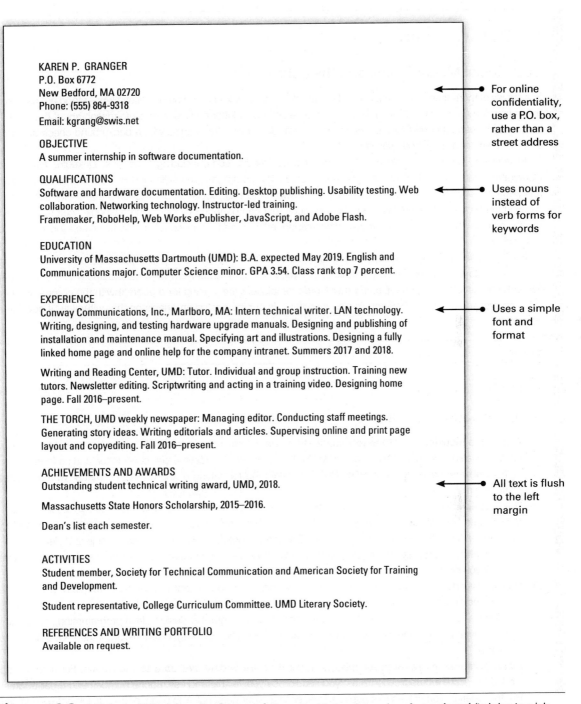

KAREN P. GRANGER
P.O. Box 6772
New Bedford, MA 02720
Phone: (555) 864-9318
Email: kgrang@swis.net

For online confidentiality, use a P.O. box, rather than a street address

OBJECTIVE
A summer internship in software documentation.

QUALIFICATIONS
Software and hardware documentation. Editing. Desktop publishing. Usability testing. Web collaboration. Networking technology. Instructor-led training.
Framemaker, RoboHelp, Web Works ePublisher, JavaScript, and Adobe Flash.

Uses nouns instead of verb forms for keywords

EDUCATION
University of Massachusetts Dartmouth (UMD): B.A. expected May 2019. English and Communications major. Computer Science minor. GPA 3.54. Class rank top 7 percent.

EXPERIENCE
Conway Communications, Inc., Marlboro, MA: Intern technical writer. LAN technology. Writing, designing, and testing hardware upgrade manuals. Designing and publishing of installation and maintenance manual. Specifying art and illustrations. Designing a fully linked home page and online help for the company intranet. Summers 2017 and 2018.

Uses a simple font and format

Writing and Reading Center, UMD: Tutor. Individual and group instruction. Training new tutors. Newsletter editing. Scriptwriting and acting in a training video. Designing home page. Fall 2016–present.

THE TORCH, UMD weekly newspaper: Managing editor. Conducting staff meetings. Generating story ideas. Writing editorials and articles. Supervising online and print page layout and copyediting. Fall 2016–present.

ACHIEVEMENTS AND AWARDS
Outstanding student technical writing award, UMD, 2018.

All text is flush to the left margin

Massachusetts State Honors Scholarship, 2015–2016.

Dean's list each semester.

ACTIVITIES
Student member, Society for Technical Communication and American Society for Training and Development.

Student representative, College Curriculum Committee. UMD Literary Society.

REFERENCES AND WRITING PORTFOLIO
Available on request.

Figure 16.6 A Résumé That Can Be Scanned If you decide to print copies of your résumé (to bring to a job fair or similar event), create one that is easy to scan, using sans serif font and avoiding fancy highlighting. To ensure that the scanner will recognize all characters, you may use all caps (rather than bold or italics) for headings and publications.

Consider This

Your Social Media Profile and the Job Search

In a survey of more than 2,000 employers, two qualities—personality and communication skills—stood out as essential when making a new hire (Barck). But your résumé and cover letter provide limited information about these areas, so employers use other methods to screen applicants. Screening often begins with a background check of education, employment history, and references.

Increasingly, however, employers also look at social media feeds, including Facebook, LinkedIn, Twitter, and Instagram to get a better feel for the applicant's personality, communication style, and fit. Note that in many states, it is illegal for employers to ask people for the passwords to their social media accounts (Kerr). But still, with so much information freely available online, there is much that can be found by just doing a Google search. According to one survey, 70 percent of employers who responded reported using social media to screen job candidates (CareerBuilder).

Social media postings, tweets, photos, and online comments create a digital persona, often reflecting the potential employee's values and communication style. In the end, employers want candidates who are well qualified *and* can get along with people. In other words, employers are looking for a good fit with the company: "what is taking precedence in the way companies recruit . . . is our increasing emphasis on the cultural fit of new hires" (Barck).

Guidelines

for Online Job Applications

➤ **Follow the instructions of the job application site.** Employer job sites will tell you what format to use, how large the file size can be, whether to upload just the résumé or the résumé and cover letter, and so forth.

➤ **Unless otherwise noted, use PDF.** PDF files retain the formatting of the original no matter what computer platform a reader is using.

➤ **Use a simple font.** Stick with those fonts that are easiest to scan, such as Times New Roman or Helvetica.

➤ **Use simple formatting.** Especially for print résumés that may be scanned, avoid fancy fonts, tables, and too much formatting. You may substitute ALL CAPS instead of boldface. Keep the design simple and clean.

➤ **Use templates carefully.** The Internet and most word-processing apps are full of résumé and cover letter templates. If you start off using a template, be sure to edit the formatting and layout to fit your needs.

➤ **Use keywords.** Use words that are likely to get hits if the document is searched. You may want to create a "qualifications" section at the top of your résumé. Include keywords for general skills (conflict management, report and proposal writing), specialized skills (graphic design, Java programming), credentials (B.S. in electrical engineering, Phi Beta Kappa), and job titles (manager, technician, intern). Use nouns for keywords.

➤ **Avoid personal information for job materials that are widely available to the public.** For résumés uploaded to secure job application sites, you should use your actual address, phone number, and email address. But for materials you upload to a public space, such as your personal Web page, or even to job networking sites where you can't predict who will see it, you may wish to avoid the potential for identity theft by leaving off your home address and phone number or using a P.O. box.

Dossiers, Portfolios, and E-Portfolios

16.6 Prepare an impressive dossier, portfolio, or e-portfolio

An employer impressed by your résumé and application letter will have further questions about your credentials and your past work. These questions will be answered, respectively, by your dossier and your portfolio (or e-portfolio). In some fields, such as technical writing (and other writing careers), architecture, design, photography, and graphics, a portfolio is an essential way to provide examples of your work.

Dossiers

Your dossier contains your credentials: college transcript, recommendation letters, and other items (such as a scholarship award or commendation letter) that offer evidence of your achievements. Prospective employers who decide to follow up on your application will request your dossier. By collecting recommendations in one folder, you spare your references from writing the same letter repeatedly.

What a dossier contains

Some college placement offices will keep the dossier (or placement folder) on file and, with permission from the job applicant, will provide copies to employers. Most of these dossier systems are digital; you have a login and password, and employers receive a secure link where they can access the materials. You might want to keep a print copy as well, including any nonconfidential recommendation letters. Then, if an employer requests your dossier, you can provide it, advising your recipient that the official placement copy is on the way, as dossiers are not always mailed immediately from a busy placement office.

Portfolios and E-portfolios

Whether print or digital, a portfolio (or e-portfolio, as in Figure 16.7) contains an introduction or mission statement explaining what you've included in your portfolio and why. Among the included items are your résumé, uploaded or scanned examples of your work, and anything else pertinent to your job search (such as copies of documents from your dossier). An organized, professional-looking portfolio or e-portfolio shows that you can apply your skills and helps you stand out as a candidate. It also gives you concrete material to discuss during job interviews.

What a portfolio contains

As you create your portfolio or e-portfolio, seek advice and feedback from professors in your major and from other people in the field. If you have a portfolio, indicate this on your résumé, followed by "Available on request." If you have a e-portfolio, provide the Web address on your résumé, but also bring printed copies of key items to your interview. Keep copies of these items on hand to leave with the interviewer if requested.

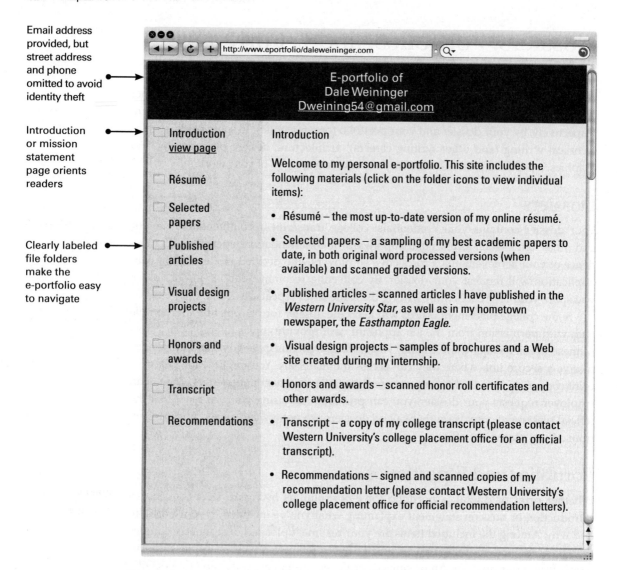

Email address provided, but street address and phone omitted to avoid identity theft

Introduction or mission statement page orients readers

Clearly labeled file folders make the e-portfolio easy to navigate

http://www.eportfolio/daleweininger.com

E-portfolio of
Dale Weininger
Dweining54@gmail.com

Introduction
view page

Résumé

Selected papers

Published articles

Visual design projects

Honors and awards

Transcript

Recommendations

Introduction

Welcome to my personal e-portfolio. This site includes the following materials (click on the folder icons to view individual items):

- Résumé – the most up-to-date version of my online résumé.

- Selected papers – a sampling of my best academic papers to date, in both original word processed versions (when available) and scanned graded versions.

- Published articles – scanned articles I have published in the *Western University Star*, as well as in my hometown newspaper, the *Easthampton Eagle*.

- Visual design projects – samples of brochures and a Web site created during my internship.

- Honors and awards – scanned honor roll certificates and other awards.

- Transcript – a copy of my college transcript (please contact Western University's college placement office for an official transcript).

- Recommendations – signed and scanned copies of my recommendation letter (please contact Western University's college placement office for official recommendation letters).

Figure 16.7 An E-portfolio Be sure that your e-portfolio is up to date. Check routinely to ensure that all links are functioning.

Guidelines

for Dossiers, Portfolios, and E-portfolios

➤ **Always provide an introduction or mission statement.** Place this page at the beginning, to introduce and explain the contents.

➤ **Collect relevant materials.** Gather documents or graphics you've prepared in school or on the job, presentations you've given, and projects or experiments you've worked on. Possible items: campus newspaper articles, reports on course projects, papers that earned an "A," examples of persuasive argument, documents from an internship, or visuals you've designed for an oral presentation.

➤ **Include copies of dossier materials.** Although they won't be official unless they go directly from your campus placement office to your prospective employer, post copies of your college transcript and recommendations on your e-portfolio.

➤ **Assemble your items.** Place your résumé first (after your introduction/mission statement) and use divider pages or electronic files to group related items. Follow the same structure for an e-portfolio. Aim for a professional look.

➤ **Omit irrelevant items.** Personal photographs and other items more appropriate for a Facebook page do not belong in a portfolio or e-portfolio.

Interviews and Follow-Up Communication

16.7 Succeed in a job interview and write professional follow-up correspondence

All your preparation leads to the last stages of the hiring process: the interview and follow-up to the interview.

Interviews

An employer who is impressed by your credentials will arrange an interview. The interview's purpose is to confirm the employer's impressions from your application letter, résumé, references, and dossier. You will likely be asked to present your portfolio (if you have one) at the interview, or you may be asked to give a short presentation. In addition to your original portfolio or the Web address for your e-portfolio, bring copies/printouts of the most relevant documents contained therein. Offer to leave these copies with the interviewer.

Purpose of interviews

Interviews come in various shapes and sizes. To narrow down the pool, some organizations start with *initial interviews*, which may take place on the phone or via Skype (or a similar video connection). Next comes an in-person interview. Here, you might meet with a single interviewer, a hiring committee, and/or several committees in succession. You might be interviewed alone or as part of a group of candidates. Interviews can last an hour or less, a full day, or several days. The interview can range from a pleasant chat to a grueling interrogation. Some interviewers may antagonize you deliberately to observe your reaction. Many times, the interview contains standard questions, which must be asked of all job candidates.

Types of interview situations

How to prepare

Careful preparation is the key to a productive interview. If you haven't already done so, learn all you can about the company from trade journals, business magazines (such as *Forbes, Fortune,* and *Business Week*), the company's Web site, and other online resources (start with a Google search of the organization, but be sure the sites you look at are credible and reliable). If possible, request company literature, including the most recent annual report. Speak with people who know about the company, or (well in advance) arrange an informational interview with someone at the company. Once you've done all this, ask yourself, "Does this job seem like a good fit?"

> NOTE *Taking the wrong job can be far worse than taking no job at all—especially for a recent graduate trying to build credentials.*

In addition to knowing about the company, be prepared in other ways. Dress and present yourself appropriately. Be psychologically prepared and confident (but not arrogant). Follow the rules of business etiquette. Unprepared interviewees make mistakes such as the following:

How people fail job interviews

- They know little about the company or what role they would play as an employee in this particular division or department.
- They have inflated ideas about their own worth.
- They have little idea of how their education prepares them for work.
- They dress inappropriately.
- They exhibit little or no self-confidence.
- They have only vague ideas of how they could benefit the employer.
- They inquire only about salary and benefits.
- They speak negatively of former employers or coworkers.

Practice answering interview questions

One important way to prepare for an interview is to practice answering typical questions. Think about how you would answer the following:

- Why does this job appeal to you?
- What do you know about our company? About this division or unit?

Questions to expect

- What do you know about our core values (for example, informal management structure, commitment to diversity or to the environment)?
- What do you know about the expectations and demands of this job?
- What are the major issues affecting this industry?
- How would you describe yourself as an employee?
- What do you see as your biggest weakness? Biggest strength?
- Can you describe an instance in which you came up with a new and better way of doing something?
- What are your short-term and long-term career goals?

Be sure to prepare your own list of well-researched questions about the job and the organization. You will be invited to ask questions, and what you ask can be as revealing as any answers you give.

Finally, tell the truth during the interview—doing so is both ethical and smart. Companies routinely verify an applicant's claims about education, prior employment, positions held, salary, and personal background. Perhaps you have some past infraction (such as a bad credit rating or a brush with the law), or some pressing personal commitment (such as caring for an elderly parent or a disabled child). Experts suggest that it's better to air these issues up front—before an employer learns from other sources. The employer will appreciate your honesty, and you will know exactly where you stand before accepting the job (Fisher, "Truth" 292).

Follow-Up Communication

There are two types of follow-up communication: thank you notes and, if you are offered the job, acceptance or refusal notes. A typed or handwritten thank you, acceptance, or refusal card or letter is considered the most professional and formal way to follow up after a job interview or job offer. However, if email has been the primary way you and the potential employer have communicated, you may decide to send your follow-up via email. If so, be sure that the email is professionally written, checked for spelling and grammar, and formal in tone. (See Chapters 14 and 15 for more on this topic.)

THANK YOU NOTES.   Within a day or so after the interview, send a thank you card, letter, or email to the person who interviewed you. If you were interviewed by multiple people, send your note to the primary interviewer and ask him or her to thank the others (typically, the recipient will forward your email to everyone else). Not only is this approach courteous, but it also reinforces a positive impression.

Open by thanking the interviewer and reemphasizing your interest in the position. Then refer to some details from the interview or some aspect of your visit that would help the recipient reconnect with the interview experience. If you forgot to mention something important during the interview, include it here—briefly. Finally, close with genuine enthusiasm, and provide your contact information again to make it easy for the interviewer to respond. Following is the text of a thank you email from James Purdy, the entry-level candidate in hotel-restaurant management, whose résumé and cover letter appeared earlier:

> Thank you for your hospitality during my Tuesday visit to Lake Geneva Resort. I am very interested in the restaurant-management position and was intrigued by our discussion about developing an eclectic regional cuisine.
>
> Everything about my tour was enjoyable, but I was especially impressed by the friendliness and professionalism of the resort staff. People seem to love working here, and it's not hard to see why.
>
> I'm convinced I would be a productive employee at Lake Geneva and would welcome the chance to prove my abilities. If you need additional information, please call me at (555) 316-2419.

Acceptance
letters are formal
and contractual

ACCEPTANCE OR REFUSAL LETTERS. You may receive a job offer by phone, letter, or email. If by phone, request a written offer and respond with a formal letter or email of acceptance. This correspondence may serve as part of your contract; spell out the terms you are accepting. Here is James Purdy's email of acceptance:

Accept an offer
with enthusiasm

> I am delighted to accept your offer of a position as assistant recreation supervisor at Liberty International's Lake Geneva Resort, with a starting salary of $44,500.
>
> As you requested, I will phone Elmer Druid in your Personnel Office for instructions on reporting date, physical exam, and employee orientation.
>
> I look forward to a long and satisfying career with Liberty International.

Refusal
letters are
courteous and
forward-thinking

You may also have to refuse a job offer. Even if you refuse by phone, write a prompt and cordial letter or email of refusal, explaining your reasons, and allowing for future possibilities. A courteous refusal and explanation can let the employer know why you have chosen a competing employer or why you have decided against taking the job for other reasons. Purdy handled one job refusal this way:

> Although I thoroughly enjoyed my visit to your company's headquarters, I have to decline your offer of a position as assistant desk manager of your London hotel.

Decline an offer
diplomatically

> I've decided to accept a position with Liberty International because the company has offered me the chance to participate in its manager-trainee program. Also, Liberty will provide tuition for courses in completing my B.A. degree in hospitality management.
>
> If any future openings should materialize at your Aspen resort, however, I would appreciate your considering me again as a candidate.
>
> Thank you for your confidence in me.

Guidelines

for Interviews and Follow-Up Communication

➤ **Confirm the interview's exact time and location.** Arrive early, but no more than 10 minutes.
   NOTE If you are offered a choice of interview times, choose mid-morning over late afternoon: According to an Accountemps survey of 1,400 managers, 69 percent prefer mid-morning for doing their hiring, whereas only 5 percent prefer late afternoon (Fisher, "My Company" 184).

➤ **Don't show up empty-handed.** Bring a briefcase, pen, and notepad. Have your own questions written out. Bring extra copies of your résumé (unfolded) and a portfolio (if appropriate).
   NOTE For a job interview, taking notes on paper is the best idea. A laptop's screen will compromise your ability to make eye contact; using your phone may give the appearance that you are checking email. Even a tablet might lose battery power.

➤ **Make a positive first impression.** Come dressed as if you already work for the company. Learn the name of your interviewer beforehand, so you can greet this person by name—but never by first name unless invited. Extend a firm handshake, smile, and look the interviewer in the eye. Wait to be asked to take a chair. Maintain eye contact much of the time, but don't stare. Do not fiddle with your face, hair, or other body parts.

➤ **Don't worry about having all the answers.** When you don't know the answer to a question, say so, and relax. Interviewers typically do most of the talking.

➤ **Avoid abrupt yes or no answers—as well as life stories.** Elaborate on your answers, but also keep them short and to the point.

➤ **Don't answer questions by merely repeating the material on your résumé.** Instead, explain how specific skills and types of experience could be assets to this particular employer. For concrete evidence refer to items in your portfolio whenever possible.

➤ **Remember to smile often and to be friendly and attentive throughout.** Qualifications are not the only reason a person gets hired. People often hire the candidate they *like* best.

➤ **Never criticize a previous employer.** Above all, interviewers like people who have positive attitudes.

➤ **Prepare to ask intelligent questions.** When questions are invited, focus on the nature of the job: travel involved, specific responsibilities, typical job assignments, opportunities for further training, types of clients, and so on. Avoid questions that could easily have been answered by your own prior research.

➤ **Take a hint.** At the interview's end, restate your interest, ask when a hiring decision will be made, say thank you, and leave.

➤ **Show some class.** If you are invited to lunch, don't order the most expensive dish on the menu; don't order an alcoholic beverage; don't salt your food before tasting it; don't eat too quickly; don't put your elbows on the table; don't speak with your mouth full; and don't order a huge dessert. And try to order last.

➤ **For a telephone or video interview, take notes.** Ask for the interviewer's name (spelled) and contact information. Organize all materials so you have them within easy reach. As the interview ends, encourage further contact by restating your interest in the position and your desire to visit and meet people in person.

➤ **Follow up as soon as possible.** Send a thank you note to the primary interviewer. Be sure to spell that person's name correctly.

Checklist

Résumés

Use the following Checklist when writing a reverse chronological or functional résumé.

Content

❑ Is all my contact information accurate? (See "Parts of a Résumé" in this chapter.)
❑ Does my statement of objective show a clear sense of purpose? (See "Parts of a Résumé" in this chapter.)
❑ If I am willing to relocate, have I so indicated? (See "Parts of a Résumé" in this chapter.)
❑ Did I include a summary of skills or qualifications, as needed? (See "Parts of a Résumé" in this chapter.)
❑ Is my educational background clear and complete? (See "Parts of a Résumé" in this chapter.)
❑ Did I accurately but briefly describe previous jobs? (See "Parts of a Résumé" in this chapter.)
❑ Did I list references or offer to provide them? (See "Parts of a Résumé" in this chapter.)
❑ Did I offer to provide a portfolio, as appropriate? (See "Parts of a Résumé" in this chapter.)
❑ Am I being scrupulously honest? (See "Résumés" in this chapter.)

Arrangement

❑ Did I place my strongest qualifications in positions of emphasis? (See "Parts of a Résumé" in this chapter.)
❑ Is the education versus experience content sequenced to highlight my strengths? (See "Parts of a Résumé" in this chapter.)
❑ Does my résumé's organization (reverse chronological or functional) put my best characteristics forward? (See "Organizing Résumé," and Figures 16.2 and 16.3 in this chapter.)

❑ In my scannable résumé, did I use keywords and basic formatting? (See "Parts of a Résumé" and Figure 16.6 in this chapter.)

Overall

❑ Did I limit the résumé to a single page, if possible? (See "Parts of a Résumé" and "Guidelines for Writing and Designing Your Résumé" in this chapter.)

❑ Is the résumé uncluttered and tasteful? (See "Résumés" and "Guidelines for Writing and Designing Your Résumé" in this chapter.)

❑ Did I use phrases instead of complete sentences? (See "Résumés" and "Guidelines for Writing and Designing Your Résumé" in this chapter.)

❑ Did I use action verbs and keywords? (See "Parts of a Résumé" and "Guidelines for Writing and Designing Your Résumé" in this chapter.)

❑ Are highlighting and punctuation consistent and simple? (See "Guidelines for Writing and Designing Your Résumé" in this chapter.)

❑ Have I proofread exhaustively? (See "Guidelines for Writing and Designing Your Résumé" in this chapter.)

Checklist

Application Letters

Use the following Checklist when writing a solicited or unsolicited application letter.

Content

❑ Is my letter addressed to a specifically named person? (See "Application Letters" and "Guidelines for Application Letters" in this chapter.)

❑ If my letter was solicited, did I indicate how I heard about the job? (See "Solicited Application Letters" in this chapter.)

❑ If my letter was unsolicited, does it have a forceful opening? (See "Unsolicited Application Letters" in this chapter.)

❑ Did I make my case without merely repeating my résumé? (See "Application Letters" and "Guidelines for Application Letters" in this chapter.)

❑ Did I support all claims with evidence? (See "Application Letters" in this chapter.)

❑ Am I being scrupulously honest? (See "Application Letters" in this chapter.)

Arrangement

❑ Does my introduction get directly to the point? (See "Application Letters" in this chapter.)

❑ Does the body section expand on qualifications sketched in my résumé? (See "Application Letters" in this chapter.)

❑ Does the conclusion restate my interest and request specific action? (See "Application Letters" in this chapter.)

Overall

❑ Did I limit the letter to a single page, whenever possible? (See "Application Letters" in this chapter.)

❑ Is my letter free of "canned" expressions? (See "Application Letters" and "Guidelines for Application Letters" in this chapter.)

❑ Is my tone appropriate? (See "Application Letters" and "Guidelines for Application Letters" in this chapter.)

❑ Did I convey enthusiasm and self-confidence without seeming arrogant? (See "Application Letters" and "Guidelines for Application Letters" in this chapter.)

❑ Have I proofread exhaustively? (See "Guidelines for Application Letters" in this chapter.)

Checklist

Supporting Materials

Use the following Checklist when engaging in other parts of the job search process.

❑ Is my dossier complete, (recommendation letters; evidence of achievements)? (See "Dossiers" and "Guidelines for Dossiers, Portfolios, and E-portfolios" in this chapter.)

❑ Is my portfolio or e-portfolio (if applicable) up-to-date, including functional links? (See "Portfolios and E-portfolios" and "Guidelines for Dossiers, Portfolios, and E-portfolios" in this chapter.)

❑ Have I prepared for interviews? (See "Interviews" and "Guidelines for Interviews and Follow-up Communication" in this chapter.)

❑ Have I sent the appropriate follow-up correspondence? (See "Follow-up Communication" and "Guidelines for Interviews and Follow-up Communication" in this chapter.)

Projects

For all projects, check with your instructor about whether to present your findings in class, bring drafts to class for discussion, upload your project to the class learning management system (LMS), and/or use the LMS forum or discussion boards to collaborate and review each activity below.

General

1. Create a résumé and application letter for a part-time or summer job in response to a specific ad. Choose an organization related to your career goals. Submit a copy of the ad along with your materials.

2. A friend has asked you for help with the following application letter. Rewrite it as needed.

Dear Ms. Brown,

Please consider my application for the position of assistant in the Engineering Department. I am a second-year student majoring in electrical engineering technology. I am presently an apprentice with your company and would like to continue my employment in the Engineering Department.

I have six years' experience in electronics, including two years of engineering studies. I am confident my background will enable me to assist the engineers, and I would appreciate the chance to improve my skills through their knowledge and experience.

I would appreciate the opportunity to discuss the possibilities and benefits of a position in the Engineering Department at Concord Electric. Please phone me any weekday after 3:00 p.m. at (555) 568–9867. I hope to hear from you soon.

Sincerely,

3. Write an unsolicited application letter to the human resources director of a company that interests you. Go to the company's Web site and social media feeds to research the various positions for which you may be qualified. Select one, and name that position in your letter. Also learn the name of the human resources director and address your letter to that person.

Team

In groups, search for Web sites that job seekers should visit for advice on résumés, cover letters, and e-portfolios. (If your school's career center has a Web site, include it in your search.) See if you can find information specific to certain careers. Are the guidelines and suggestions different based on a student's undergraduate major and career goals? Write a one-paragraph summary of the information you find on each site, then compare your findings with those of others in your class.

Digital and Social Media

Go to one or more of the following sites: Linkedin.com, Careerbuilders.com, Monster.com. Learn about how you would go about preparing a résumé to upload to one of these sites. If the site provides sample résumés, look at a few and compare these to the strategies provided in this chapter. Create a draft résumé for one of these sites and review it with your instructor. Find out how to create a profile and, if appropriate for you, create one and begin to review your connections and potentials for networking. Think carefully about how you word your profile description and qualifications.

Global

Assume that you and other students in your major would like to work in a particular country after graduation. Select a country and do some research on the economy, culture, and employment issues. Write a short memo that tells prospective students what they need to know about finding employment abroad.

Chapter 17
Technical Definitions

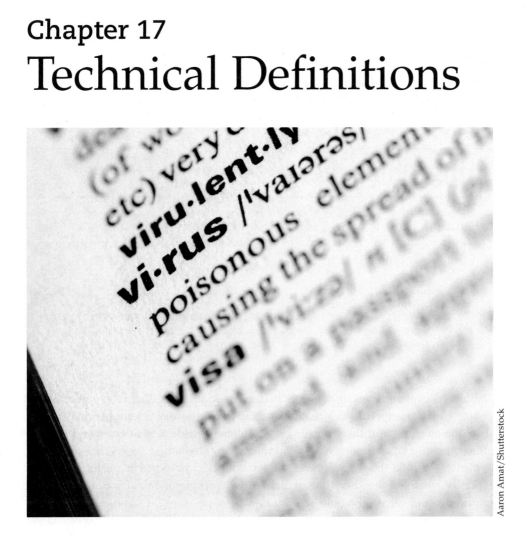

Aaron Amat/Shutterstock

❝As a nurse practitioner, much of my working day is spent defining specialized terms for patients and their families: medical conditions, treatments and surgical procedures, medications and side effects, and so on. Most of today's patients—especially those facing complex health decisions—have already done a lot of searching online and expect to be well-informed about medical issues that affect their lives. Clear, understandable definitions are an important first step in providing the information these people need.**❞**

—*Dana Ballinger, Nurse Practitioner in a primary-care clinic*

Learning Objectives

17.1 Determine when audience and purpose indicate the need for definition

17.2 Describe the legal, ethical, societal, and global implications of definitions

17.3 Differentiate between parenthetical, sentence, and expanded definitions

17.4 Identify the various ways to expand a definition

17.5 Place definitions effectively in your document

What definitions do

Definitions explain terms or concepts that are specialized and may be unfamiliar to people who lack expertise in a particular field. In many cases, a term may have more than one meaning or different meanings in different fields. Consider a word such as atmosphere: To an astronomer, it would refer to the envelope of gases that surrounds a planet ("the Earth's atmosphere"); to a politician or office manager, it would typically mean the mood of the country or the workplace ("an atmosphere of high hopes"); to a physicist, it would stand for a unit of pressure ("a standard atmosphere is 101,325 pascal"); and to a novelist, it would be associated with the mood of a novel ("a gothic atmosphere").

Why definitions must be precise

Precision is particularly important in specialized fields, in which field-specific terminology is common and undefined terms may prevent the overall document from making sense. Engineers talk about *elasticity* or *ductility*; bankers discuss *amortization* or *fiduciary relationships*. These terms must be defined if people both inside and outside of those fields are to understand the document as a whole. Imagine if a doctor continually used the words *myocardial infarction* in a patient brochure without ever defining the term. The fact that a myocardial infarction is the medical name for one form of *heart attack* would be lost on patients who need to have a clear understanding of the brochure.

Considering Audience and Purpose

17.1 Determine when audience and purpose indicate the need for definition

Definitions typically answer one of two questions: "What, exactly, is it?" or "What, exactly, does it entail?" The first question concerns what makes an item, concept, or process unique. For example, an engineering student needs to understand the distinction between *elasticity* and *ductility*. People in any audience have to grasp precisely what "makes a thing what it is and distinguishes that thing from all other things" (Corbett 38). The second question asks how readers are affected by the item being defined. For example, a person buying a new laptop needs to understand exactly what "manufacturer's guarantee" or "expandable memory" means in the context of that purchase. Unless you are certain that your audience already knows the exact meaning, always define a term the first time you use it.

Audience considerations

Consider the purpose of defining particular terms in your document by answering the question "Why does my audience need to understand this term?" The level of technicality you use must match the audience's background and experience. For a group of mechanical engineering students, your definition of a *solenoid*, for example, can use highly technical language, as in the following example:

Purpose considerations

> A solenoid is an inductance coil that serves as a tractive electromagnet.

A highly technical version

For general audiences, your definition will require language they can understand, as in the following example:

> A solenoid is a metal coil that converts electrical energy to magnetic energy capable of performing mechanical functions.

A nontechnical version

Finally, consider how your audience will read the definition: as a printed document, as a PDF, as a Web site, as a social media post, or in some combination. Digital documents can be updated and allow for the use of links to provide readers with more information. Online, definitions can be updated if there are any needed changes. But printed definitions may be needed in situations where readers can't count on Internet access (for instance, the mechanical rooms of many buildings are below ground and often don't get any signals; technicians would need materials in printed formats).

Legal, Ethical, Societal, and Global Implications

17.2 Describe the legal, ethical, societal, and global implications of definitions

Precise definition is essential because you (or the organization, if you write a document on its behalf) are legally responsible for that document. For example, contracts are detailed (and legally binding) definitions of the specific terms of an agreement. If you lease an apartment or a car, the contract will define both the *lessee's* and *lessor's* specific responsibilities. Likewise, an employment contract or employee handbook will spell

Definitions have legal implications

out responsibilities for both employer and employee. In preparing an employee handbook for your company, you would need to define such terms as *acceptable job performance, confidentiality, sexual harassment*, and *equal opportunity*.

Definitions have ethical implications, too. For example, the term *acceptable risk* had an ethical impact on January 28, 1986, when the space shuttle *Challenger* exploded 73 seconds after launch, killing all seven crew members. (Two rubber O-ring seals in a booster rocket had failed, allowing hot exhaust gases to escape and igniting the adjacent fuel tank.) Hours earlier—despite vehement objections from the engineers—management had decided that going ahead with the launch was a risk worth taking. In this case, management's definition of *acceptable risk* was based not on the engineering facts but rather on bureaucratic pressure to launch on schedule. Agreeing on meaning in such cases rarely is easy, but you are ethically bound to convey an accurate interpretation of the facts as you understand them.

Clear and accurate definitions help the public understand and evaluate complex technical and social issues. For example, as a first step in understanding the debate over the term *genetic engineering*, we need at least the following basic definition:

> Genetic engineering refers to [an experimental] technique through which genes can be isolated in a laboratory, manipulated, and then inserted stably into another organism. Gene insertion can be accomplished mechanically, chemically, or by using biological vectors such as viruses (U.S. Congress, Office of Technology Assessment 20).

Of course, to follow the debate, we would need increasingly detailed information (about specific procedures, risks, benefits, and so on). But the above definition gets us started on a healthy debate by enabling us to visualize the basic concept and to mutually agree on the basic meaning of the term.

Ongoing threats to our planet's environment, the prospect of nuclear proliferation, and the perils of terrorism top a list of complex issues that make definitions vital to global communication. For example, on the environmental front, world audiences need a realistic understanding of concepts such as *greenhouse effect* or *ozone depletion*. On the nuclear proliferation front, the survival of our species could well depend on the clear definition of terms such as *nuclear nonproliferation treaty* or *nonaggression pact* by the nations who are parties to such agreements. On the terrorism front, the public wants to understand such terms as *biological terrorism* or *weapons of mass destruction*.

In short, definition is more than an exercise in busy work. Figure 17.1 illustrates the informative power of an effective definition.

Types of Definition

17.3 Differentiate between parenthetical, sentence, and expanded definitions

Definitions fall into three distinct categories: *parenthetical, sentence*, and *expanded*. Decide how much detail your audience actually requires in order to grasp your exact meaning.

Parenthetical Definitions

Often, you can clarify the meaning of a word by using a more familiar synonym or a clarifying phrase in parentheses, as in these examples:

| The *leaching field* (sieve-like drainage area) requires crushed stone.

| The trees on the site are mostly *deciduous* (shedding foliage at season's end).

On a Web page, social media post, or online help system, these types of short definitions can be linked to the main word or phrase. Readers who click on *leaching field* would be taken to a window containing a brief definition.

In Figure 17.1, *allergenics* is defined as consisting of two different types of products (patch tests and extracts). Each individual definition is then carefully written using short sentences; technical phrases such as "allergic rhinitis" are defined in simpler language ("hay fever") for nonexpert readers. Because these definitions are available on a Web site, readers with different levels of expertise can search for more information, locate other definitions, and research any of the concepts within the definition.

When to use parenthetical definitions

Parenthetical definitions

Web and social media definitions

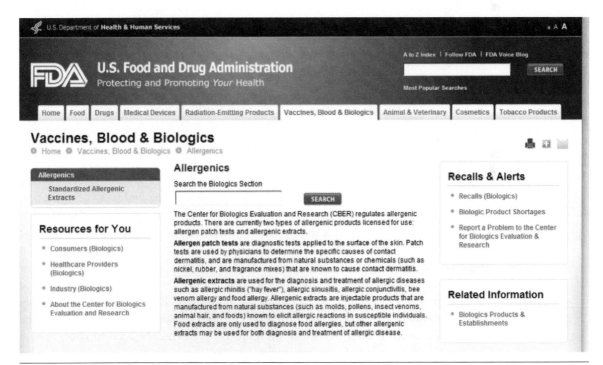

Figure 17.1 An Effective Definition
Source: U.S. Food and Drug Administration.

Sentence Definitions

When to use
sentence
definitions

When a term requires more elaboration than a parenthetical definition can offer, use a sentence definition. Begin by stating the term. Then indicate the broader class to which this item belongs, followed by the features that distinguish it from other items in that general class. Here are examples:

Sentence
definitions
(term-class-
features)

Term	Class	Distinguishing Features
Carburetor	a mixing device...	in gasoline engines that blends air and fuel into a vapor for combustion within the cylinders.
Diabetes	a metabolic disease...	caused by a disorder of the pituitary gland or pancreas and characterized by excessive urination, persistent thirst, and inability to metabolize sugar.
Stress	an applied force...	that strains or deforms a body.

The previous elements may be combined into one or more complete sentences, as in the following definition of *diabetes*:

A complete
sentence
definition

Diabetes is a metabolic disease caused by a disorder of the pituitary gland or pancreas. This disease is characterized by excessive urination, persistent thirst, and inability to metabolize sugar.

Sentence definitions are especially useful if you plan to use a term often and need to establish a working definition that you will not have to repeat throughout the document, as in the following definition *of disadvantaged student*:

A working
definition

Throughout this report, the term *disadvantaged student* will refer to all students who lack adequate funds to pay for on-campus housing, food services, and medical care, but who are able to pay for their coursework and books through scholarships and part-time work.

Expanded Definitions

When to use
expanded
definitions

Brief definitions are fine when your audience requires only a general understanding of a term. For example, the parenthetical definition of *leaching field* (see "Parenthetical Descriptions" in this chapter) might be adequate in a progress report to a client whose house you're building. But a document that requires more detail, such as a public health report on groundwater contamination from leaching fields, would call for an expanded definition.

Likewise, the nontechnical definition of "solenoid" (see "Considering Audience and Purpose" in this chapter) is adequate for a layperson who simply needs to know what a solenoid is. An instruction manual for mechanics, however, would define solenoid in much greater detail (as in Figure 17.3); mechanics need to know how a solenoid works and how to use and repair it.

Depending on audience and purpose, an expanded definition may be a short paragraph or may extend to several pages. For example, if a device, such as a digital dosimeter (used for measuring radiation exposure), is introduced to an audience who needs to understand how this instrument works, your definition would require at least several paragraphs, if not pages.

Methods for Expanding Definitions

17.4 Identify the various ways to expand a definition

An expanded definition can be created in any number of ways as described below. The method or methods you decide to use will depend on the questions you expect the audience will want answered, as illustrated in Figure 17.2.

As you read through the following sections, refer to the following sentence definition of the word *laser*, and consider how each expansion method provides detail in a different way:

| A laser is an electronic device that emits a highly concentrated beam of light.

Sentence definition

Etymology

Sometimes, a word's origin (its development and changing meanings), also known as the word's etymology, can help clarify its meaning. For example, *biometrics* (the statistical analysis of biological data) is a word derived from the Greek *bio*, meaning "life," and *metron*, meaning "measure." You can use a dictionary to learn the origins of most words. Not all words develop from Greek, Latin, or other roots, however. For

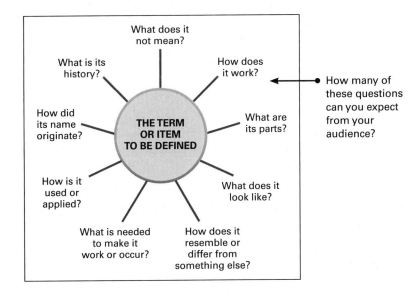

Figure 17.2 Questions for Expanding a Definition

example, some terms are acronyms, derived from the first letters or parts of several words. Such is the case with the word *laser* (derived from *light amplification by stimulated emission of radiation*); therefore, to expand the sentence definition of *laser*, you might phrase your definition as follows:

"How did its name originate?"

> The word *laser* is an acronym for *light amplification by stimulated emission of radiation*, and is the name for an electronic device that emits a highly concentrated beam of light.

History

In some cases, explaining the history of a term, concept, or procedure can be useful in expanding a definition. Specialized dictionaries and encyclopedias are good background sources. You might expand the definition of a laser by describing how the laser was invented, as in the following example:

"What is its history?"

> The early researchers in fiber optic communications were hampered by two principal difficulties—the lack of a sufficiently intense source of light and the absence of a medium which could transmit this light free from interference and with a minimum signal loss. Lasers emit a narrow beam of intense light, so their invention in 1960 solved the first problem. The development of a means to convey this signal was longer in coming, but scientists succeeded in developing the first communications-grade optical fiber of almost pure silica glass in 1970 (Stanton 28).

Negation

Some definitions can be clarified by an explanation of what the term *does not* mean. For example, the following definition of a laser eliminates any misconceptions an audience might already have about lasers:

"What does it not mean?"

> A laser is an electronic device that emits a highly concentrated beam of light. It is used for many beneficial purposes (including corrective eye and other surgeries), and not—as science fiction might tell you—as a transport medium to other dimensions.

Operating Principle

Anyone who wants to use a product correctly will need to know how it operates, as in the following example:

"How does it work?"

> Basically, a laser [uses electrical energy to produce] coherent light: light in which all the waves are in phase with each other, making the light hotter and more intense (Gartaganis 23).

Analysis of Parts

To create a complete picture, be sure to list all the parts. If necessary, define individual parts as well, as in the following passage:

"What are its parts?"

> A laser is an electronic device that emits a highly concentrated beam of light. To get a better idea of how a laser works, consider its three main parts:
>
> 1. [Lasers require] a source of energy, [such as] electric currents or even other lasers.
> 2. A resonant circuit...contains the lasing medium and has one fully reflecting end and one partially reflecting end. The medium—which can be a solid, liquid, or gas—absorbs the energy

and releases it as a stream of photons [electromagnetic particles that emit light]. The photons. . . . vibrate between the fully and partially reflecting ends of the resonant circuit, constantly accumulating energy—that is, they are amplified. After attaining a prescribed level of energy, the photons can pass through the partially reflecting surface as a beam of coherent light and encounter the optical elements.

3. Optical elements—lenses, prisms, and mirrors—modify size, shape, and other characteristics of the laser beam and direct it to its target (Gartaganis 23).

Visuals

Make sure any visual you use is well labeled. Always introduce and explain your visual and place it near your discussion. If the visual is borrowed, credit the source. The following visual accompanies the previous analysis of parts:

"What does it look like?"

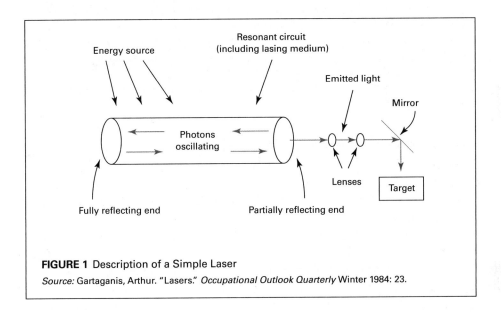

FIGURE 1 Description of a Simple Laser

Source: Gartaganis, Arthur. "Lasers." *Occupational Outlook Quarterly* Winter 1984: 23.

Comparison and Contrast

By comparing (showing similarities) or contrasting (showing differences) between new information and information your audience already understands, you help build a bridge between what people already know and what they don't. The following passage uses both comparison and contrast to expand upon a more basic definition of a laser:

Fiber optics technology results from the superior capacity of light waves to carry a communications signal. Sound waves, radio waves, and light waves can all carry signals; their capacity increases with their frequency. Voice frequencies carried by telephone operate at 1,000 cycles per second, or hertz. Television signals transmit at about 50 million hertz. Light waves, however, operate at frequencies in the hundreds of trillions of hertz (Stanton 28).

"How does it resemble or differ from something else?"

Required Conditions

Some items or processes need special materials and handling, or they may have other requirements or restrictions. An expanded definition incorporating required conditions should include the following important information:

"What is needed to make it work or occur?"

> In order to emit a highly concentrated beam of light, the laser must absorb energy through the reflecting end of a resonant circuit, amplify the photons produced between the reflecting and partially reflecting end of the resonant circuit, and release the photons as a beam of light via a set of lenses, prisms, and mirrors.

Examples

Examples are a powerful communication tool—as long as they are tailored to your audience's level of understanding. The following example shows how laser light is used in medical treatment:

"How is it used or applied?"

> Lasers are increasingly used to treat health problems. Thousands of eye operations involving cataracts and detached retinas are performed every year by ophthalmologists.... Dermatologists treat skin problems....Gynecologists treat problems of the reproductive system, and neurosurgeons even perform brain surgery—all using lasers transmitted through optical fibers (Gartaganis 24–25).

Using Multiple Expansion Methods

Use as many expansion methods as needed

Depending on the complexity of what you need to define, you may need to use multiple expansion methods. Whichever expansion strategies you use, be sure to document your information sources, as shown in Appendix A, "A Quick Guide to Documentation."

The two situations described below, and the related examples (Figures 17.3 and Figures 17.4) require expansion methods suitable for their respective audiences. Both definitions are unified and coherent but are written with different readers in mind. Each paragraph is developed around one main idea and logically connected to other paragraphs. Transitions emphasize the connection between ideas. Visuals are incorporated. Each definition displays a level of technicality appropriate for the intended audience. The authors of both definitions start by adapting the Audience and Use Profile (see Chapter 2, "Develop an Audience and Use Profile") in order to provide their audiences with exactly what they need.

AN EXPANDED DEFINITION FOR SEMITECHNICAL READERS.   Ron Vasile, a lab assistant in his college's Electronics Engineering Technology program, has been asked to contribute to a reference manual for the program's incoming students. His first assignment is to prepare a section defining basic solenoid technology. As a way of getting started on this assignment, Ron completes an Audience and Use Profile, then writes the following audience and purpose statement to guide his work:

The intended readers (future service technicians) are beginning student mechanics. Before they can repair a solenoid, they will need to know where the term *solenoid* comes from, what a solenoid looks like, how it works, how its parts operate, and how it is used. Diagrams will reinforce the explanations and enable readers to visualize this mechanism's parts and operating principle.

Ron's audience and purpose statement

Ron's definition (Figure 17.3) is designed as an *introduction,* and so it offers only a general view of the mechanism. Because the readers are not engineering students, they do *not* need electromagnetic or mechanical theory (e.g., equations or graphs illustrating voltage magnitudes, joules, lines of force).

AN EXPANDED DEFINITION FOR NONTECHNICAL READERS.   Amy Rogers has recently joined the public relations division of a government organization whose task is to explore the possible uses and applications of nanotechnology. She and other members of her division are helping to prepare a Web site that explains this complex topic to the general public. One of Amy's assignments is to prepare a definition of nanotechnology to be posted on the site. She first completes an Audience and Use Profile, targeting hi-tech investors and other readers interested in new and promising technologies, then writes the following audience and purpose statement to guide her work:

To understand *nanotechnology* and its implications, readers need an overview of what it is and how it developed, as well as its potential uses, present applications, health risks, and impact on the workforce. Question-type headings pose the questions in the same way readers are likely to ask them. Parenthetical definitions of *nanometer* and *micrometer* provide an essential sense of scale.

Amy's audience and purpose statement

This audience for Amy's definition (Figure 17.4) would have little interest in the physics or physical chemistry involved, such as *carbon nanotubes* (engineered nanoparticles), *nanolasers* (advanced applications), or *computational nanotechnology* (theoretical aspects). They simply need the broadest possible picture, including a diagram that compares the size of nanoparticles with the size of more familiar items.

Because Amy's definition will appear online, she chooses to place it in the FAQ (Frequently Asked Questions) section of the Web site, using a question and answer format. FAQs use a conversational style and help readers find what they are looking for in a quick and easy way. Consider creating FAQs if you have enough data and experience to know what questions people are most likely to ask.

Purpose of FAQs (Frequently Asked Questions)

Placing Definitions in a Document

17.5 Place definitions effectively in your document

Each time readers encounter an unfamiliar term or concept, that item should be defined. In most documents, you can place brief definitions in parentheses or in the document's margin, aligned with the terms being defined. Sentence definitions

Placing definitions within documents

1

SOLENOID

Formal
sentence
definition

A solenoid is an electrically energized coil that forms an electromagnet
capable of performing mechanical functions. The term "solenoid" is

Etymology

derived from the word "sole," which in reference to electrical equipment
means "a part of," or "contained inside, or with, other electrical
equipment." The Greek word *solenoides* means "channel," or "shaped
like a pipe."

Description
and analysis
of parts

A simple plunger-type solenoid consists of a coil of wire attached to
an electrical source and an iron rod, or plunger, that passes in and out of
the coil along the axis of the spiral. A return spring holds the rod outside
the coil when the current is deenergized, as shown in Figure 1.

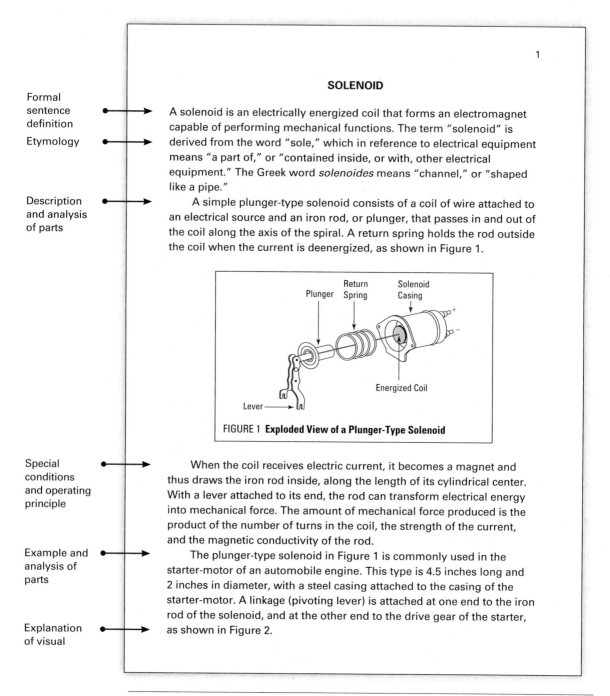

FIGURE 1 **Exploded View of a Plunger-Type Solenoid**

Special
conditions
and operating
principle

When the coil receives electric current, it becomes a magnet and
thus draws the iron rod inside, along the length of its cylindrical center.
With a lever attached to its end, the rod can transform electrical energy
into mechanical force. The amount of mechanical force produced is the
product of the number of turns in the coil, the strength of the current,
and the magnetic conductivity of the rod.

Example and
analysis of
parts

The plunger-type solenoid in Figure 1 is commonly used in the
starter-motor of an automobile engine. This type is 4.5 inches long and
2 inches in diameter, with a steel casing attached to the casing of the
starter-motor. A linkage (pivoting lever) is attached at one end to the iron
rod of the solenoid, and at the other end to the drive gear of the starter,

Explanation
of visual

as shown in Figure 2.

Figure 17.3 **An Expanded Definition for Semitechnical Readers**

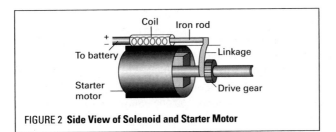

FIGURE 2 **Side View of Solenoid and Starter Motor**

When the ignition key is turned, current from the battery is supplied to the solenoid coil, and the iron rod is drawn inside the coil, thereby shifting the attached linkage. The linkage, in turn, engages the drive gear, activated by the starter-motor, with the flywheel (the main rotating gear of the engine).

Because of the solenoid's many uses, its size varies according to the work it must do. A small solenoid will have a small wire coil, hence a weak magnetic field. The larger the coil, the stronger the magnetic field; in this case, the rod in the solenoid can do harder work.

An electronic lock for a standard door would, for instance, require a much smaller solenoid than one for a bank vault.

Comparison
of sizes and
applications

Figure 17.3 **Continued**

Sentence definition

Parenthetical definitions

Comparison

A diagram comparing dimensions of nanoparticles with those of more familiar items

Contrast and operating principle

History

Examples

NANOTECHNOLOGY
FREQUENTLY ASKED QUESTIONS (FAQs)

What Is Nanotechnology?

Nanotechnology refers to the understanding and control of matter at dimensions of roughly 1 to 100 nanometers to produce new structures, materials, and devices. (A nanometer, μm, equals one-billionth of a meter; a sheet of paper is about 100,000 nanometers thick.) For further perspective, the diameter of DNA, our genetic material, is in the 2.5 nanometer range, while red blood cells are roughly 2.5 micrometers (a micrometer, mm, equals one-millionth of a meter), as shown in Figure 1.

Ant	Head of a pin	Red blood cells with white cell	DNA
5mm	1-2 mm	2-5 μm	2-12 μm diameter

FIGURE 1 **The Scale of Things—Nanometers and More**

Source: Adapted from U.S. Department of Energy, www.er.doe.gov/bes/scale_of_things.html

At the nanoscale level, the physical, chemical, and biological properties of materials differ from the properties of individual atoms and molecules or bulk matter. Nanotechnology research is directed toward understanding and creating improved materials, devices, and systems that exploit these new properties.

How did it develop?

Nanoscale science was enabled by advances in microscopy, most notably the electron, scanning-tunnel, and atomic-force microscopes, among others.

How is it used?

The use of nanoparticles is being researched and applied in many areas of technology and medicine, such as the following:

Figure 17.4 An Expanded Definition for Nontechnical Readers This type of question-and-answer format is often available on the Internet as a frequently asked questions (FAQ) document.

Source: Adapted from documents at the National Nanotechnology Initiative, www.nano.gov.

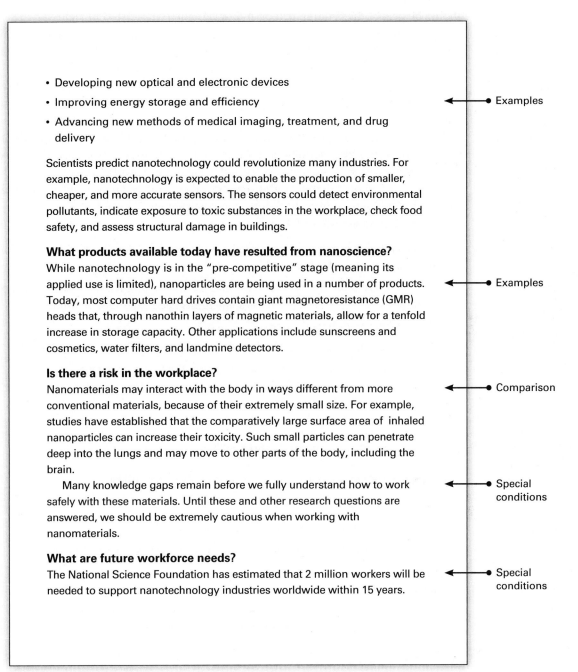

- Developing new optical and electronic devices
- Improving energy storage and efficiency ← ● **Examples**
- Advancing new methods of medical imaging, treatment, and drug delivery

Scientists predict nanotechnology could revolutionize many industries. For example, nanotechnology is expected to enable the production of smaller, cheaper, and more accurate sensors. The sensors could detect environmental pollutants, indicate exposure to toxic substances in the workplace, check food safety, and assess structural damage in buildings.

What products available today have resulted from nanoscience?
While nanotechnology is in the "pre-competitive" stage (meaning its applied use is limited), nanoparticles are being used in a number of products. ← ● **Examples**
Today, most computer hard drives contain giant magnetoresistance (GMR) heads that, through nanothin layers of magnetic materials, allow for a tenfold increase in storage capacity. Other applications include sunscreens and cosmetics, water filters, and landmine detectors.

Is there a risk in the workplace?
Nanomaterials may interact with the body in ways different from more ← ● **Comparison**
conventional materials, because of their extremely small size. For example, studies have established that the comparatively large surface area of inhaled nanoparticles can increase their toxicity. Such small particles can penetrate deep into the lungs and may move to other parts of the body, including the brain.

    Many knowledge gaps remain before we fully understand how to work ← ● **Special conditions**
safely with these materials. Until these and other research questions are answered, we should be extremely cautious when working with nanomaterials.

What are future workforce needs?
The National Science Foundation has estimated that 2 million workers will be ← ● **Special conditions**
needed to support nanotechnology industries worldwide within 15 years.

Figure 17.4 Continued

should be part of the running text or, if they are numerous, listed in a glossary. Place an expanded definition either near the beginning of a long document or in an appendix (see Chapter 21, "End Matter")—depending on whether the definition is essential to understanding the whole document or serves merely as a reference.

Using a glossary

A glossary alphabetically lists specialized terms and their definitions. It makes key definitions available to laypersons without interrupting technical readers. Use a glossary if your report contains numerous terms that may not be understood by all audience members. If fewer than five terms need defining, place them in the report introduction as working definitions, or use footnote definitions. If you use a glossary, announce its location: "(See the glossary at the end of this report)."

Figure 17.5 shows part of a glossary for a comparative analysis of two natural childbirth techniques, written by a nurse for expectant mothers.

Follow these suggestions for preparing a glossary:

How to prepare a glossary

- Define all terms unfamiliar to an intelligent layperson. When in doubt, overdefining is safer than underdefining.

- Define all terms by giving their class and distinguishing features (see "Sentence Definitions" in this chapter), unless some terms need expanded definitions.

- List all terms in alphabetical order.

- On first use, place an asterisk in the text by each item defined in the glossary.

- List your glossary and its first page number in the table of contents.

GLOSSARY

Analgesic: a medication given to relieve pain during the first stage of labor.

Cervix: the neck-shaped anatomical structure that forms the mouth of the uterus.

Dilation: cervical expansion occurring during the first stage of labor.

First stage of labor: the stage in which the cervix dilates and the baby remains in the uterus.

Induction: the stimulating of labor by puncturing the membranes around the baby or by giving an oxytoxic drug (uterine contractant), or by doing both.

Figure 17.5 **A Partial Glossary**

Guidelines
for Definitions

➤ **Decide on the level of detail you need**. Definitions vary greatly in length and detail, from a few words in parentheses to a multipage document. How much does this audience need in order to follow your explanation or grasp your point?

➤ **Classify the item precisely.** The narrower your class, the clearer your meaning. *Stress* is classified as an applied force; to say that stress "is what... or "takes place when... fails to denote a specific classification. Diabetes is precisely classified as a *metabolic disease,* not as a *medical term.*

➤ **Differentiate the item accurately.** If the distinguishing features are too broad, they will apply to more than the particular item you are defining. A definition of *brief* as a "legal document used in court" fails to differentiate *brief* from all other legal documents (*wills, affidavits,* and the like).

➤ **Avoid circular definitions.** Do not repeat, as part of the distinguishing feature, the word you are defining. "Stress is an applied force that places stress on a body" is a circular definition.

➤ **Expand your definition selectively.** Begin with a sentence definition and select the best combination of development strategies for your audience and purpose.

➤ **Use visuals to clarify your meaning.** No matter how clearly you explain, as the saying goes, a picture can be worth a thousand words—even more so when used with readable, accurate writing.

➤ **Know "how much is enough."** Don't insult people's intelligence by giving needless details or spelling out the obvious.

➤ **Consider the legal implications of your definition.** What does an "unsatisfactory job performance" mean in an evaluation of a company employee: that the employee should be fired, required to attend a training program, or given one or more chances to improve ("Performance Appraisal" 3–4)? Failure to spell out your meaning invites a lawsuit.

➤ **Consider the ethical implications of your definition.** Be sure your definition of a fuzzy or ambiguous term such as "safe levels of exposure," or "conservative investment," or "acceptable risk" is based on a fair and accurate interpretation of the facts. Consider, for example, a U.S. cigarette company's claim that cigarette smoking in the Czech Republic promoted "fiscal benefits," defined, in this case, by the fact that smokers die young, thus eliminating pension and health care costs for the elderly!

➤ **Place your definition in an appropriate location.** Allow readers to access the definition and then return to the main text with as little disruption as possible. On a Web site, don't kick people off the main page if they click on a definition link; instead, have the definition appear in a new tab or window.

➤ **Cite your sources as needed.** See Appendix A, "A Quick Guide to Documentation."

Checklist
Definitions

Use the following Checklist when writing definitions.

Content

❑ Is the type of definition (parenthetical, sentence, expanded) suited to its audience and purpose? (See "Types of Definition" in this chapter.)

❑ Does the definition adequately classify the item? (See "Guidelines for Definitions" in this chapter.)

❑ Does the definition adequately differentiate the item? (See "Guidelines for Definitions" in this chapter.)

❑ Will the level of technicality connect with the audience? (See "Considering Audience and Purpose" in this chapter.)

❑ Have circular definitions been avoided? (See "Guidelines for Definitions" in this chapter.)

❑ Is the expanded definition developed adequately for its audience? (See "Expanded Definitions" in this chapter.)

❑ Is the expanded definition free of needless details for its audience? (See "Expanded Definitions" in this chapter.)

❑ Are visuals used adequately and appropriately? (See "Visuals" in this chapter.)

❑ Are all information sources properly documented? (See "Guidelines for Definitions" in this chapter.)

❑ Is the definition ethically and legally acceptable? (See "Legal, Ethical, Societal, and Global Implications" in this chapter.)

Arrangement

❑ Is the expanded definition unified and coherent (like an essay)? (See "Using Multiple Expansion Methods" in this chapter.)

❑ Are transitions between ideas adequate? (See "Using Multiple Expansion Methods" in this chapter.)

❑ Is the definition appropriately located in the document? (See "Placing Definitions in a Document" in this chapter.)

Style and Page Design

❑ Is the definition in plain English? (See Chapter 11, "Global, Legal, and Ethical Implications of Style and Tone.")

❑ Are sentences clear, concise, and fluent? (See Chapter 11, "Editing for Clarity," "Editing for Conciseness," and "Editing for Fluency.")

❑ Is word choice precise? (See Chapter 11, "Finding the Exact Words.")

❑ Is the definition grammatical? (See Appendix B, "Grammar.")

❑ Is the page design inviting and accessible? (See Chapter 13, "Page Design in Print and Digital Workplace Documents.")

Projects

For all projects, check with your instructor about whether to present your findings in class, bring drafts to class for discussion, upload your project to the class learning management system (LMS), and/or use the LMS forum or discussion boards to collaborate and review each activity below.

General

Choose a situation and an audience, and prepare an expanded definition designed for this audience's level of technical understanding. Use at least four expansion strategies, including at least one visual. In preparing your expanded definition, consult no fewer than four outside references. Cite and document each source as shown in Appendix A, "A Quick Guide to Documentation."

Team

Divide into groups by majors or interests. Appoint one person as group manager. Decide on an item, concept, or process that would require an expanded definition for laypersons. Some examples follow:

From computer science: an algorithm, binary coding, or systems analysis
From nursing: a pacemaker, coronary bypass surgery, or natural childbirth

Complete an Audience and Use Profile (see Chapter 2, "Develop an Audience and Use Profile"). Once your group has decided on the appropriate expansion strategies, the group manager will assign each member to work on one or two specific strategies as part of the definition. As a group, edit and incorporate the collected material into an expanded definition, revising as often as needed. The group manager will assign one member to present the definition in class.

Digital and Social Media

Review your expanded definition from the General Project and consider what changes or additions could be made if the definition was converted to a Web page. For instance, where might you add hyperlinks so that readers who want more information can find it? Next, consider how you would revise the definition for a social media post (Hint: you might need to make the definition much shorter, then link to a longer version).

Global

Any definition you write may be read by someone for whom English is not a first language. Locate an expanded definition on Wikipedia that may be difficult for a nonnative English speaker to understand for some reason (use of idioms, use of abbreviations, use of American metaphors such as sports metaphors not used in other countries). Explain how the definition could be reworded so that most readers would understand it.

Chapter 18
Technical Descriptions, Specifications, and Marketing Materials

❝My company manufactures devices for the medical device industry. Most of my writing involves descriptions, specifications, and sometimes, technical marketing content. The people who use these documents range from highly technical (industrial designers, medical technicians, physicians, nurses) to legal (company lawyers, government regulators), to nontechnical (patients and their families). In all cases, I have to be sure that my materials serve the exact information needs of readers. Descriptions must be technically accurate and use language appropriate for the intended audience and how they need to use the document.❞

—*Zach Bowen, Biomedical Engineer*

∨ Learning Objectives

18.1 Understand the role of audience and
purpose in technical description

18.2 Appreciate the need for objectivity in
technical descriptions

18.3 Recognize the main components of
technical descriptions

18.4 Differentiate between product and
process descriptions

18.5 Write a set of specifications to ensure
safety and/or customer satisfaction

18.6 Write a technical marketing
document to sell a product or service

Description (creating a picture with words and images) is part of all writing. But a technical description conveys information about a product or mechanism to someone who will use it, operate it, assemble it, or manufacture it, or to someone who needs to know more about it. Any item can be visualized from countless different perspectives. Therefore, the way you describe something—your perspective—depends on your purpose and the audience's needs.

What
descriptions do

Considering Audience and Purpose

18.1 Understand the role of audience and purpose in technical description

Descriptions and definitions often go hand in hand and provide the foundation for many types of technical explanation. Definitions answer the questions such as "What is it?" or "What does it entail?" To help readers to visualize, descriptions answer additional questions that include "What does it look like?" "What are its parts?" "What does it do?" "How does it work?" or "How does it happen?"

Audience
considerations

Consider exactly what you want readers to know and why they need to know it. Before writing your description, you may want to jot down an audience and purpose

Purpose
considerations

statement such as: "The purpose of this description is to help first year medical students understand the parts of a stethoscope and how a stethoscope works."

As with definitions, be sure to consider how your audience will read a description: as a printed document, as a PDF, as a Web site, as a social media post, or in some combination. Descriptions in digital formats can be easily updated; and, online you can include links to provide readers with more information. But printed descriptions may be needed in situations where readers can't count on Internet access (for instance, mechanical rooms or areas in certain medical facilities may get poor or no signals; technicians and others would thus need materials in printed formats).

Objectivity in Technical Descriptions

18.2 Appreciate the need for objectivity in technical descriptions

Subjective versus objective descriptions

A description can be mainly *subjective* (based on feeling) or *objective* (based on fact). Subjective descriptions do more than simply convey factual information; subjective descriptions use sensory and judgmental expressions such as "The weather was miserable" or "The room was terribly messy." In contrast, objective descriptions present an impartial view, filtering out personal impressions and focusing on details any viewer could observe ("All day, we had freezing rain and gale-force winds").

Why descriptions should be objective

In general, descriptions should be objective. Pure objectivity is, of course, humanly impossible. Each writer filters the facts and their meaning through his or her own perspective, and therefore chooses what to include and what to omit. Also, technical marketing material is designed to help sell a product, so although the content must be accurate and clear, the document's purpose will tend to be a bit more on the persuasive side than other forms of descriptions.

Descriptions have ethical implications

When writing descriptions, you should communicate the facts as they are generally known and understood. Even positive claims made in marketing material (for example, "reliable," "rugged," "efficient") should be based on objective and verifiable evidence. Being objective does not mean forsaking personal evaluation, however, especially in cases where a product or process may have features that create safety concerns. An ethical communicator, in the words of one expert, "is obligated to express her or his opinions of products, as long as these opinions are based on objective and responsible research and observation" (MacKenzie 3).

How to remain objective

One way to maintain objectivity when writing descriptions is to provide details that are visual, not emotional. Ask yourself what any observer would recognize, or what a camera would record. For example, instead of saying, "His office has a depressing atmosphere" (not everyone would agree), say "His office has broken windows looking out on a brick wall, missing floorboards, broken chairs, and a ceiling with chunks of plaster missing."

A second way to maintain objectivity is to use precise and informative language. For instance, specify location and position, exact measurements, weights, and dimensions, instead of using inexact and subjective words like *large, long*, and *near*.

NOTE Never confuse precise language with overly complicated technical terms or needless jargon. For example, don't say "phlebotomy specimen" instead of "blood," "thermal attenuation" instead of "insulation," or "proactive neutralization" instead of "damage control." General readers prefer nontechnical language—as long as the simpler words do the job.

Elements of Descriptions

18.3 **Recognize the main components of technical descriptions**

The following elements are necessary in almost all technical descriptions. Please refer to Figure 18.1, "The Lightning Process," as you read through this section.

Clear and Limiting Title

An effective title promises exactly what the document will deliver—no more and no less. For example, the title "A Description of a Velo Ten-Speed Racing Bicycle" promises a comprehensive description. If you intend to describe the braking mechanism only, be sure your title indicates this focus: "A Description of the Velo's Center-Pull Caliper Braking Mechanism." The title of Figure 18.1 is extremely simple, indicating that the figure will do no more than show how the process works in the most basic way.

Give an immediate forecast

Appropriate Level of Detail and Technicality

Provide enough detail to convey a clear picture, but do not burden readers needlessly. Identify your audience and its reasons for using your description. Focus carefully on your purpose.

Give readers exactly and only what they need

The description of how lightning occurs, shown in Figure 18.1, is written for a nontechnical audience. The writer of this description assumes that the audience probably knows nothing about what causes lightning. Therefore, the document shows the process one step at a time, uses simple language for each step, and avoids any complex technical language. In contrast, a description of how lightning occurs for a semitechnical audience might use some technical language and add more complex detail about the process.

Visuals

Use drawings, diagrams, or photographs generously—with captions and labels that help readers interpret what they are seeing. Notice how the labeled illustrations in Figure 18.1 provide a simple but dynamic picture of a process in action. Economy in a visual often equals clarity. Avoid verbal and visual clutter.

Let the visual repeat, restate, or reinforce the prose

Sources for descriptive graphics include drawing or architectural drafting programs, clip art, electronic scans, and downloads from the Internet. (See Chapter 12 for a discussion of legal issues when using online graphics.)

Description begins with a clear and limiting title

Visuals show the process better than words alone

Level of detail and technicality is appropriate for a nonexpert audience

Description uses both functional and chronological sequence

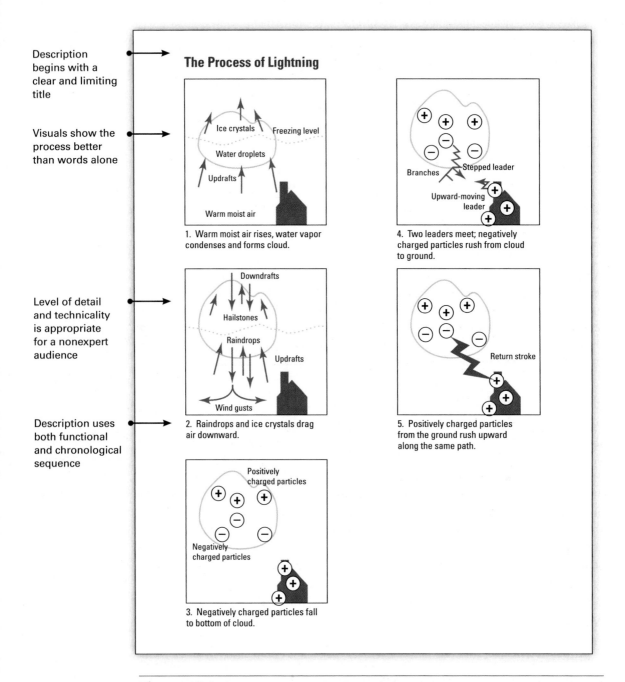

The Process of Lightning

1. Warm moist air rises, water vapor condenses and forms cloud.

2. Raindrops and ice crystals drag air downward.

3. Negatively charged particles fall to bottom of cloud.

4. Two leaders meet; negatively charged particles rush from cloud to ground.

5. Positively charged particles from the ground rush upward along the same path.

Figure 18.1 The Elements of Effective Descriptions

Source: From Richard E. Mayer et al., "When Less Is More: Meaningful Learning From Visual and Verbal Summaries of Science Textbook Lessons," *Journal of Educational Psychology,* Vol. 88, No. 1, 64-73. Copyright © 1996 by the American Psychological Association. Reprinted by permission of the American Psychological Association and Richard E. Mayer.

Clearest Descriptive Sequence

Any item or process usually has its own logic of organization, based on (1) the way it appears as a static object, (2) the way its parts operate in order, or (3) the way its parts are assembled. As a writer, you can describe these relationships, respectively, in spatial, functional, or chronological sequence.

Organize for the reader's understanding

SPATIAL SEQUENCE.   Part of all physical descriptions, a spatial sequence answers these questions: *What does it do? What does it look like? What parts and materials is it made of?* Use this sequence when you want readers to visualize a static item or a mechanism at rest (an office interior, the Statue of Liberty, a plot of land, a chainsaw, or a computer keyboard). Can readers best visualize this item from front to rear, left to right, top to bottom? (What logical path do the parts create?) A retractable pen, for example, would logically be viewed from outside to inside. The specifications in Figure 18.6 (see "Specifications" later in this chapter) proceed from the ground upward.

A spatial sequence parallels the reader's angle of vision in viewing the item

FUNCTIONAL SEQUENCE.   The functional sequence answers this question: *How does it work?* It is best used in describing a mechanism in action, such as a 35-millimeter camera, a smoke detector, or a car's cruise-control system. The logic of the item is reflected by the order in which its parts function.

A functional sequence parallels the order in which parts operate

CHRONOLOGICAL SEQUENCE.   A chronological sequence answers these questions: *How is it assembled? How does it work? How does it happen?* Use the chronological sequence for an item that is best visualized in terms of its order of assembly (such as a piece of furniture, a tent, or a prehung window or door unit). Architects might find a spatial sequence best for describing a proposed beach house to clients; however, they would use a chronological sequence (of blueprints) for specifying for the builder the prescribed dimensions, materials, and construction methods at each stage of the process.

A chronological sequence parallels the order in which parts are assembled or stages occur

Use more than one of these sequences as needed. For example, in describing an automobile jack (for a car owner's manual), you would employ a spatial sequence to help readers recognize this item, a functional sequence to show them how it works, and a chronological sequence to help them assemble and use the jack correctly. Notice how Figure 18.1 also uses more than one sequence in showing how lightning works. The process is both functional, in that various factors work together to create lightning, and chronological, in that the process occurs as these factors come together in a specific way over the course of time.

Types of Technical Descriptions

18.4 Differentiate between product and process descriptions

Technical descriptions divide into two basic types: *product* descriptions and *process* descriptions. Anyone learning to use a particular device (say, a stethoscope) relies on

Product versus process descriptions

A Description of the Standard Stethoscope

The stethoscope is a listening device that amplifies and transmits body sounds to aid in detecting physical abnormalities.

This instrument has evolved from the original wooden, funnel-shaped instrument invented by French physician R. T. Lennaec in 1819. Because of his female patients' modesty, he found it necessary to develop a device, other than his ear, for auscultation (listening to body sounds).

This description explains to the beginning paramedical or nursing student the structure, assembly, and operating principle of the stethoscope.

The standard stethoscope is roughly 24 inches long and weighs about 5 ounces. The instrument consists of a sensitive sound-detecting and amplifying device whose flat surface is pressed against a bodily area. This amplifying device is attached to rubber and metal tubing that transmits the body sound to a listening device inserted in the ear.

The stethoscope's Y-shaped structure contains seven interlocking pieces: (1) diaphragm contact piece, (2) lower tubing, (3) Y-shaped metal piece, (4) upper tubing, (5) U-shaped metal strip, (6) curved metal tubing, and (7) hollow ear plugs. These parts form a continuous unit (Figure 1).

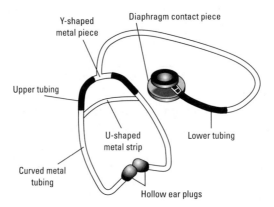

FIGURE 1 Stethoscope with Diaphragm Contact Piece (Front View)

The seven major parts of the stethoscope provide support for the instrument, flexibility of movement for the operator, and ease in use.

In an operating cycle, the diaphragm contact piece, placed against the skin, picks up sound impulses from the body's surface. These impulses cause the plastic diaphragm to vibrate. The amplified vibrations, in turn, are carried through a tube to a dividing point. From here, the amplified sound is carried through two separate but identical series of tubes to hollow ear plugs.

Figure 18.2 **A Product Description** Using both words and a visual, this description allows readers to see the basic stethoscope parts and understand the relationship among the parts.

product description. Anyone wanting to understand the steps or stages in a complex event (say, how contaminated soil is cleaned up) relies on process description.

For example, the product description in Figure 18.2 gives medical and nursing students a clear understanding of the overall device and its parts. The process description in Figure 18.3 begins with a descriptive introductory sentence but then explains the process for cleaning up contaminated soil.

While Figures 18.2 and 18.3 are relatively basic, workplace descriptions are often significantly more complex. The next two sections discuss more complex product and process descriptions and how to use outlining to help you write a thorough and accurate description.

A Complex Product Description

For all descriptions, it can be helpful to start by creating an outline. But outlines are especially important for planning and drafting descriptions of complex products (such as the standard flat-plate solar collector in Figure 18.4). The outline below is adaptable to many complex product descriptions. Note that in most descriptions, the subdivisions in the introduction or other sections can be combined and need not appear as individual headings in the document. You might modify, delete, or combine certain components of this outline to suit your subject, purpose, and audience.

I. Introduction: General Description   *Outlining a*
   A. Definition, Function, and Background of the Item   *complex product*
   B. Purpose (and Audience—for classroom only)   *description*
   C. Overall Description (with general visuals, if applicable)
   D. Principle of Operation (if applicable)
   E. Preview of Major Parts
II. Description and Function of Parts
   A. Part One in Your Descriptive Sequence
      1. Definition
      2. Shape, dimensions, material (with specific visuals)
      3. Subparts (if applicable)
      4. Function
      5. Relation to adjoining parts
      6. Mode of attachment (if applicable)
   B. Part Two in Your Descriptive Sequence (and so on)
III. Conclusion and Operating Description
   A. Summary (used only in a long, complex description)
   B. Interrelation of Parts
   C. One Complete Operating Cycle

The complex product description in Figure 18.4, aimed toward a general audience, adapts the previous outline model. The description was written by Roxanne Payton, a mechanical engineer who specializes in green energy technologies. Roxanne prepared this description as part of an informational booklet on solar energy systems distributed by her company, Eco-Solutions. In preparation for writing the description,

Writing a complex product description

The Lasagna Process

Dubbed the "lasagna" process because of its layers, this technology cleans up liquid-borne organic and inorganic contaminants in dense, claylike soils. Initial work is focused on removing chlorinated solvents.

Because clay is not very permeable, it holds groundwater and other liquids well. Traditional remediation for this type of site requires that the liquid in the soil (usually groundwater) be pumped out. The water brings many of the contaminants with it, then is chemically treated and replaced—a time-consuming and expensive solution.

The lasagna process, on the other hand, allows the soil to be remediated *in situ* (on site) by using low-voltage electric current to move contaminated groundwater through treatment zones in the soil. Depending on the characteristics of the individual site, the process can be done in either a vertical or horizontal configuration. (See figure below.)

The first step in the lasagna process is to "fracture" the soil, creating a series of zones. In a vertical configuration, a vertical borehole is drilled and a nozzle inserted; a highly pressurized mixture of water and sand (or another water/solid mix) is injected into the ground at various depths. The result: a stack of pancake-shaped, permeable zones in the denser, contaminated soil. The top and bottom zones are filled with carbon or graphite so they can conduct electricity. The zones between them are filled with treatment chemicals or microorganisms that will remediate the contaminants.

When electricity is applied to the carbon and graphite zones, they act as electrodes, creating an electric field. Within the field, the materials in the soil migrate toward either the positive or negative electrode. Along with the migrating materials, pollutants are carried into the treatment zones, where they are neutralized or destroyed.

The horizontal configuration works in much the same way, differing only in installation. Because the electrodes and treatment zones extend down from the surface, this configuration does not require the sophisticated hydraulic fracturing techniques that are used in the vertical configuration.

Schematic Diagram of the Lasagna Process

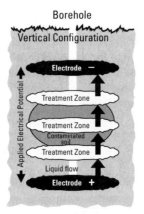
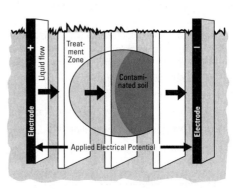

Figure 18.3 A Process Description This description uses words and visuals to help readers understand the process involved in cleaning up (remediating) contaminated soil.

Source: Adapted from Japikse, Catharina. "Lasagna in the Making." *EPA Journal*, 20.3.

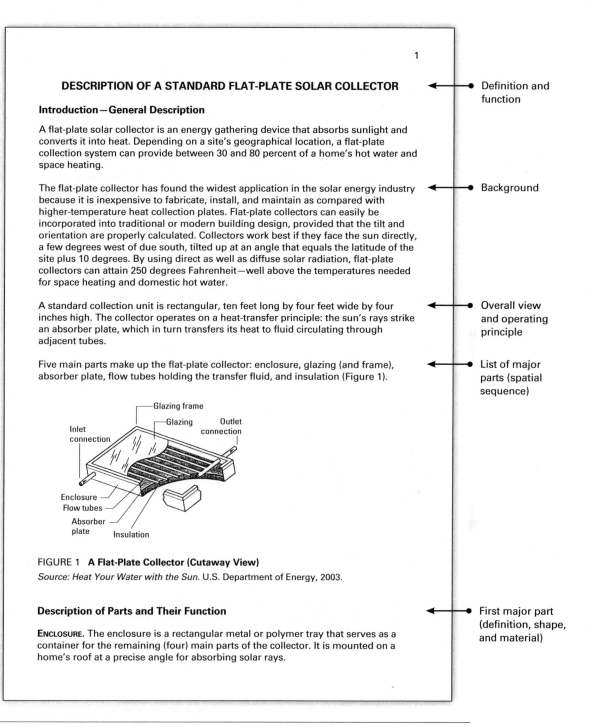

1

DESCRIPTION OF A STANDARD FLAT-PLATE SOLAR COLLECTOR

Definition and function

Introduction—General Description

A flat-plate solar collector is an energy gathering device that absorbs sunlight and converts it into heat. Depending on a site's geographical location, a flat-plate collection system can provide between 30 and 80 percent of a home's hot water and space heating.

The flat-plate collector has found the widest application in the solar energy industry because it is inexpensive to fabricate, install, and maintain as compared with higher-temperature heat collection plates. Flat-plate collectors can easily be incorporated into traditional or modern building design, provided that the tilt and orientation are properly calculated. Collectors work best if they face the sun directly, a few degrees west of due south, tilted up at an angle that equals the latitude of the site plus 10 degrees. By using direct as well as diffuse solar radiation, flat-plate collectors can attain 250 degrees Fahrenheit—well above the temperatures needed for space heating and domestic hot water.

Background

A standard collection unit is rectangular, ten feet long by four feet wide by four inches high. The collector operates on a heat-transfer principle: the sun's rays strike an absorber plate, which in turn transfers its heat to fluid circulating through adjacent tubes.

Overall view and operating principle

Five main parts make up the flat-plate collector: enclosure, glazing (and frame), absorber plate, flow tubes holding the transfer fluid, and insulation (Figure 1).

List of major parts (spatial sequence)

Glazing frame
Glazing
Outlet connection
Inlet connection
Enclosure
Flow tubes
Absorber plate
Insulation

FIGURE 1   A Flat-Plate Collector (Cutaway View)
Source: Heat Your Water with the Sun. U.S. Department of Energy, 2003.

Description of Parts and Their Function

First major part (definition, shape, and material)

ENCLOSURE. The enclosure is a rectangular metal or polymer tray that serves as a container for the remaining (four) main parts of the collector. It is mounted on a home's roof at a precise angle for absorbing solar rays.

Figure 18.4 A Complex Product Description for a Nontechnical Audience Readers are given only as much detail as they need to understand and visualize the item.

Other major parts, listed in order

GLAZING (AND FRAME). The glazing consists of one or more layers of transparent plastic or glass that allow the sun's rays to shine on the absorber plate. This part also provides a cover for the enclosure and serves as insulation by trapping the heat that has been absorbed. An insulated frame secures the glazing sheet to the enclosure.

ABSORBER PLATE. The metallic absorber plate, coated in black for maximum efficiency, absorbs solar radiation and converts it into heat energy. This plate is the heat source for the transfer fluid in the adjacent tubing.

FLOW TUBES AND TRANSFER FLUID. The captured solar heat is removed from the absorber by means of a transfer medium; generally, treated water. The transfer medium is heated as it passes through flow tubes attached to the absorbing plate and then transported to points of use in the home or to storage, depending on energy demand.

INSULATION. Polyurethane insulation surrounds the bottom, edges, and sides of the collector, to retain absorbed energy and limit heat loss.

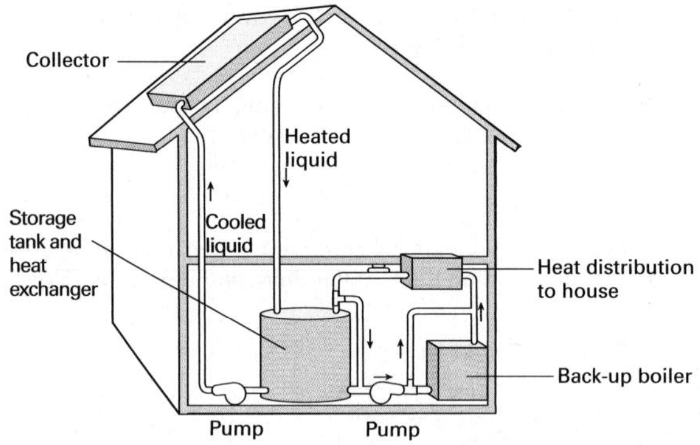

FIGURE 2 **How Solar Energy Is Captured and Distributed Throughout a Home**
Source: Heat Your Water with the Sun. U.S. Department of Energy, 2003.

Operating Description and Conclusion

One complete operating cycle (functional sequence)

In one operating cycle, solar rays penetrate the glazing to heat the absorber plate (Figure 2). Insulation helps retain the heat. The absorber plate, in turn, heats a liquid circulating through attached flow tubes, which is then pumped to a heat exchanger. The heat exchanger transfers the heat to the water in a storage tank, pumped to various uses in the home. The cooled liquid is then pumped back to the collector to be re-heated.

A conclusion emphasizing the collector's efficiency

The solar energy annually striking the roof of a typical house is ten times greater than its annual heat demand. Properly designed and installed, a flat-plate solar system can provide a large percentage of a house's space heating and domestic hot water.

Figure 18.4 Continued

Roxanne fills out an Audience and Use Profile, then writes the following audience and purpose statement to guide her work:

> The audience here will be homeowners or potential homeowners interested in incorporating solar flat-plate collectors as a heating source. Although many of these people probably lack technical expertise, they presumably have some general knowledge about active solar heating systems. Therefore, this description will focus on the collectors rather than on the entire system, while omitting specific technical data (for example, the heat conducting and corrosive properties of copper versus aluminum in the absorber plates). The team of engineers who designed the collector would include such data in research-and-development reports for the manufacturer. Informed laypersons, however, need only the information that will help them visualize and understand how a basic collector operates. Diagrams will be especially effective descriptive tools because they simplify the view by removing distracting features; labels will help readers interpret what they are seeing.

Roxanne's audience and purpose statement

A Complex Process Description

A complex process description divides the process into its parts or principles. Colleagues and clients need to know how stock and bond prices are governed, how your bank reviews a mortgage application, how an optical fiber conducts an impulse, and so on. A process description must be detailed enough to allow readers to follow the process step by step, as in the description of how acid rain develops, spreads, and destroys in Figure 18.5.

Much of your college writing explains how things work or happen. Your audience is your professor, who will evaluate what you have learned. Because this person knows *more* than you do about the subject, you often discuss only the main points, omitting the types of details that uninformed readers would require.

But your real challenge comes in describing a process for audiences who know *less* than you do, and who are neither willing nor able to fill in the blank spots; you then become the teacher, and the audience members become your students.

Introduce your description by telling what the process is, and why, when, and where it happens. In the body, tell how it happens, analyzing each stage in sequence. In the conclusion, summarize the stages and describe one full cycle of the process.

Sections from the following general outline can be adapted to any complex process description:

I. Introduction
   A. Definition, Background, and Purpose of the Process
   B. Intended Audience (usually omitted for workplace audiences)
   C. Prior Knowledge Needed to Understand the Process
   D. Brief Description of the Process
   E. Principle of Operation
   F. Special Conditions Needed for the Process to Occur
   G. Definitions of Special Terms
   H. Preview of Stages

II. Stages in the Process
   A. First Major Stage

Outlining a complex process description

1. Definition and purpose
2. Special conditions needed for the specific stage
3. Substages (if applicable)
   a.
   b.
   B. Second Stage (and so on)

III. Conclusion
   A. Summary of Major Stages
   B. One Complete Process Cycle

Writing a complex process description

The complex process description in Figure 18.5 adapts the above model. This description was written by Bill Kelly, who belongs to an environmental group studying the problem of acid rain in its Massachusetts community. (Massachusetts is among the states most affected by acid rain.) To gain community support, the environmentalists must educate citizens about the problem. Bill's group is publishing and mailing a series of brochures. The first brochure explains how acid rain is formed. Bill starts by completing an Audience and Use Profile, then writes the following audience and purpose statement to guide his work:

Bill's audience and purpose statement

Although some will be interested, others will have no awareness or interest; so, the topic is explained at a low level of technicality (no chemical formulas, equations). But the description does need to be vivid enough to appeal broadly. Visuals create interest and illustrate the situation, and dividing the content into three chronological steps—how acid rain develops, spreads, and destroys—helps readers understand the process.

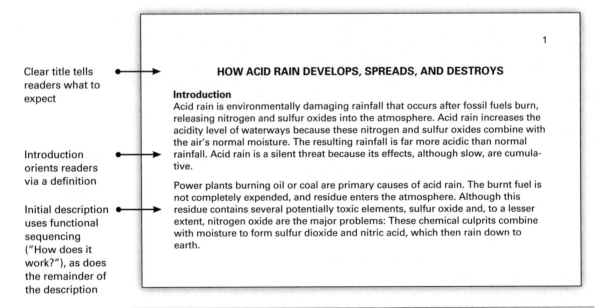

Clear title tells readers what to expect

Introduction orients readers via a definition

Initial description uses functional sequencing ("How does it work?"), as does the remainder of the description

1

HOW ACID RAIN DEVELOPS, SPREADS, AND DESTROYS

Introduction
Acid rain is environmentally damaging rainfall that occurs after fossil fuels burn, releasing nitrogen and sulfur oxides into the atmosphere. Acid rain increases the acidity level of waterways because these nitrogen and sulfur oxides combine with the air's normal moisture. The resulting rainfall is far more acidic than normal rainfall. Acid rain is a silent threat because its effects, although slow, are cumulative.

Power plants burning oil or coal are primary causes of acid rain. The burnt fuel is not completely expended, and residue enters the atmosphere. Although this residue contains several potentially toxic elements, sulfur oxide and, to a lesser extent, nitrogen oxide are the major problems: These chemical culprits combine with moisture to form sulfur dioxide and nitric acid, which then rain down to earth.

Figure 18.5 A Complex Process Description for a Nontechnical Audience Readers can follow the process as it unfolds.

2

The Process

HOW ACID RAIN DEVELOPS. Once fossil fuels have been burned, their usefulness ends. It is here that the acid rain problem begins (Figure 1).

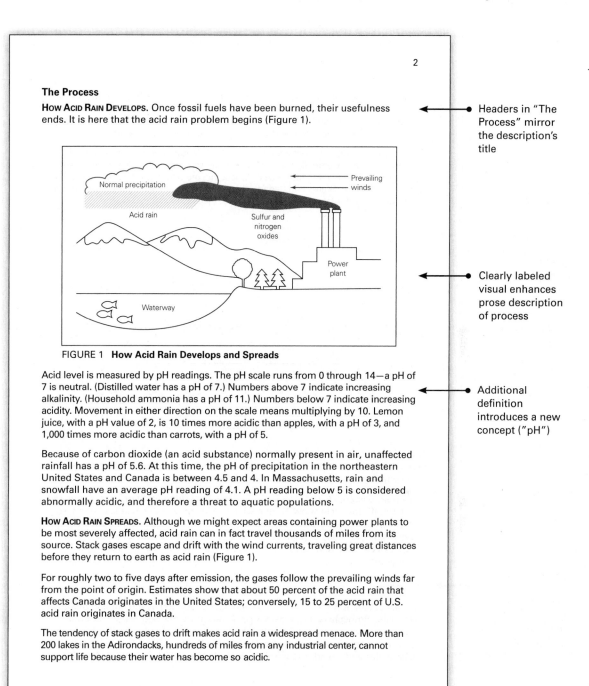

FIGURE 1 **How Acid Rain Develops and Spreads**

Acid level is measured by pH readings. The pH scale runs from 0 through 14—a pH of 7 is neutral. (Distilled water has a pH of 7.) Numbers above 7 indicate increasing alkalinity. (Household ammonia has a pH of 11.) Numbers below 7 indicate increasing acidity. Movement in either direction on the scale means multiplying by 10. Lemon juice, with a pH value of 2, is 10 times more acidic than apples, with a pH of 3, and 1,000 times more acidic than carrots, with a pH of 5.

Because of carbon dioxide (an acid substance) normally present in air, unaffected rainfall has a pH of 5.6. At this time, the pH of precipitation in the northeastern United States and Canada is between 4.5 and 4. In Massachusetts, rain and snowfall have an average pH reading of 4.1. A pH reading below 5 is considered abnormally acidic, and therefore a threat to aquatic populations.

HOW ACID RAIN SPREADS. Although we might expect areas containing power plants to be most severely affected, acid rain can in fact travel thousands of miles from its source. Stack gases escape and drift with the wind currents, traveling great distances before they return to earth as acid rain (Figure 1).

For roughly two to five days after emission, the gases follow the prevailing winds far from the point of origin. Estimates show that about 50 percent of the acid rain that affects Canada originates in the United States; conversely, 15 to 25 percent of U.S. acid rain originates in Canada.

The tendency of stack gases to drift makes acid rain a widespread menace. More than 200 lakes in the Adirondacks, hundreds of miles from any industrial center, cannot support life because their water has become so acidic.

Headers in "The Process" mirror the description's title

Clearly labeled visual enhances prose description of process

Additional definition introduces a new concept ("pH")

Figure 18.5 Continued

3

HOW ACID RAIN DESTROYS. Acid rain causes damage wherever it falls. It erodes various types of building rock such as limestone, marble, and mortar. Damage to buildings, houses, monuments, statues, and cars is widespread. Many priceless monuments have already been destroyed, and even trees of some varieties are dying.

More crucial is damage to waterways (Figure 2). Acid rain gradually lowers the pH in lakes and streams, eventually making a waterway so acidic that it dies. In areas with natural acid-buffering elements such as limestone, the dilute acid has less effect. The northeastern United States and Canada, however, lack this natural protection, and so are continually vulnerable.

Second labeled visual provides graphic illustration of the process

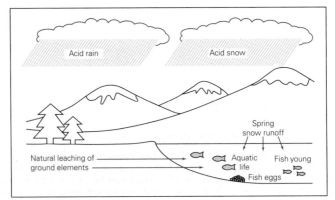

FIGURE 2 **How Acid Rain Infiltrates Waterways**

The pH level in an affected waterway drops so low that some species cease to reproduce. A pH of 5.1 to 5.4 means that entire fisheries are threatened: once a waterway reaches a pH of 4.5, fish reproduction ceases.

In the northeastern United States and Canada, the acidity problem is compounded by the runoff from acid snow. During winter, acid snow sits with little melting, so that by spring thaw, the acid released is greatly concentrated. Aluminum and other heavy metals normally present in soil are also released by acid rain and runoff. These concentrated toxins leach into waterways, affecting fish in all stages of development.

Conclusion summarizes the process and implicitly solicits audience support

Summary

Acid rain develops from nitrogen and sulfur oxides emitted by the burning of fossil fuels. In the atmosphere, these oxides combine with ozone and water to form precipitation with a low pH. This acid precipitation returns to earth miles from its source, damaging waterways that lack natural buffering agents. The northeastern United States and Canada are the most severely affected areas in North America.

Figure 18.5 **Continued**

Guidelines
for Descriptions

➤ **Take a look at the product or process.** Study your subject. For a product description, get your hands on the item if you can; weigh it, measure it, take it apart. For a process description, observe the process yourself, if possible.

➤ **Analyze your audience.** Determine your primary and secondary audiences. Then ask yourself exactly what your audience needs to know: "What does it look like?" "What are its parts?" "What does it do?" "How does it work?" or "How does it happen?" Decide upon the appropriate level of technicality for your audience.

➤ **Analyze your purpose.** Ask yourself why your audience needs this description.

➤ **Maintain objectivity.** Think in terms of visual (not emotional) details and specific language when describing location, measurements, weights, and dimensions.

➤ **Be concise.** Provide only what your audience needs, without distracting details.

➤ **Include all necessary parts.** Include a clear and limiting title, an orienting introduction, the appropriate sequence of topics (spatial, functional, chronological—or a combination), and a conclusion that brings readers full circle.

➤ **Incorporate visuals.** Enhance your verbal description using visuals, particularly if the product or process is too complicated to describe only in words. If your document will be viewed both online and in print, check to see how the visual looks in both formats.

Specifications

18.5 Write a set of specifications to ensure safety and/or customer satisfaction

Airplanes, bridges, smoke detectors, and countless other technologies are produced according to certain *specifications*. A particularly exacting type of description, specifications (or "specs") prescribe standards for performance, safety, and quality. For almost any product and process, specifications spell out the following:

> Specifications describe products and processes

- methods for manufacturing, building, or installing a product
- materials and equipment to be used
- size, shape, and weight of the product
- specific testing, maintenance, and inspection procedures

Specifications are often used to ensure compliance with a particular safety code, engineering standard, or government or legal ruling.

Because specifications define an "acceptable" level of quality, any product "below specifications" may provide grounds for a lawsuit. When injury or death results (as in a bridge collapse or an airline accident), the contractor, subcontractor, or supplier is criminally liable.

> Specifications have ethical and legal implications

Types of Specifications

Federal and state regulatory agencies routinely issue specifications to ensure safety. For example, federal or state consumer safety organizations may require power lawn

> Safety specifications

mowers to be equipped with a "kill switch" on the handle, a blade guard to prevent foot injuries, and a grass thrower that aims downward to prevent eye and facial injury. These same agencies issue specifications for baby products (such as the fire retardancy of pajama fabric) and many other products and services. Passenger airline specifications for aisle width, seat belt configurations, and emergency equipment are issued by the Federal Aviation Administration. State and local agencies also issue specifications in the form of building codes, fire codes, and other standards for safety and reliability.

Government departments (Defense, Interior, and so on) issue specifications for all types of military hardware and other equipment. A set of NASA specifications for spacecraft parts can be hundreds of pages long, prescribing the standards for even the smallest nuts and bolts, down to screw-thread depth and width in millimeters. Some government agencies even have specifications for how technical documents should be written, including the use of certain outline formats, numbering, and other features.

The private sector issues specifications for countless products or projects, to help ensure that customers get exactly what they want. Figure 18.6 shows partial specifications drawn up by an architect for a medical clinic building. This section of the specs covers only the structure's "shell." Other sections detail the requirements for plumbing, wiring, and interior finish work.

Considering Audience and Purpose

Specifications like those in Figure 18.6 must be clear enough for *identical* interpretation by variety of audiences with varied purposes (Glidden 258–59), such as the following:

- **The customer,** who has the big picture of what is needed and who wants the best product at the best price
- **The designer** (architect, engineer, computer scientist, other), who must translate the customer's wishes into the actual specification
- **The contractor or manufacturer,** who won the job by making the lowest bid, and so must preserve profit by doing only what is prescribed
- **The supplier,** who must provide the exact materials and equipment
- **The workforce,** who will do the actual assembly, construction, or installation (managers, supervisors, subcontractors, and workers—some working on only one part of the product, such as plumbing or electrical)
- **The inspectors** (such as building, plumbing, or electrical inspectors), who evaluate how well the product conforms to the specifications

Each of these parties needs to understand and agree on exactly *what* is to be done and *how* it is to be done. In the event of a lawsuit over failure to meet specifications, the readership broadens to include judges, lawyers, and jury.

In addition to guiding a product's design and construction, specifications can facilitate the product's use and maintenance. For instance, specifications in a computer manual include the product's performance limits, or *ratings*: its power requirements;

Ruger, Filstone, and Grant
Architects

MATERIAL SPECIFICATIONS FOR THE POWNAL CLINIC BUILDING

Foundation
    footings: 8" x 16" concrete (load-bearing capacity: 3,000 lbs. per sq. in.)
    frost walls: 8" x 4' @ 3,000 psi
    slab: 4" @ 3,000 psi, reinforced with wire mesh over vapor barrier

Exterior Walls
    frame: eastern pine #2 timber frame with exterior partitions set inside posts
    exterior partitions: 2" x 4" kiln-dried spruce set at 16" on center
    sheathing: 1/4" exterior-grade plywood
    siding: #1 red cedar with a 1/2" x 6" bevel
    trim: finished pine boards ranging from 1" x 4" to 1" x 10"
    painting: 2 coats of Clear Wood Finish on siding; trim primed and finished with one
          coat of bone white, oil base paint

Roof System
    framing: 2" x 12" kiln-dried spruce set at 24" on center
    sheathing: 5/8" exterior-grade plywood
    finish: 240 Celotex 20-year fiberglass shingles over #15 impregnated felt roofing paper
    flashing: copper

Windows
    Anderson casement and fixed-over-awning models, with white exterior cladding,
    insulating glass and screens, and wood interior frames

Landscape
    driveway: gravel base, with 3" traprock surface
    walks: timber defined, with traprock surface
    cleared areas: to be rough graded and covered with wood chips
    plantings: 10 assorted lawn plants along the road side of the building

Figure 18.6 Specifications for a Building Project (Partial) These specifications ensure that all parties agree on the specific materials to be used.

its processing and storage capacity; its operating environment requirements; the makeup of key parts; and so on. Product support literature for appliances, power tools, and other items routinely contains ratings to help customers select a good operating environment or replace worn or defective parts.

Guidelines
for Specifications

➤ **Analyze your audience.** Determine who will be reading the specs.
➤ **Know the minimum governmental and industry standards.** If your product is for specific customers, also consider the standards they expect you to meet.
➤ **Focus on consistency, quality, and safety.** Specifications fulfill all three purposes. Everyone who reads the document needs to be "on the same page" (consistency); the product you describe must satisfy quality requirements; and it must also meet safety requirements.
➤ **Use a standard format when applicable.** If your organization uses a standard format for specifications, follow that format.
➤ **Include a brief introduction or descriptive title.** Include some kind of overview, be it a one- or two-sentence introduction, a brief summary, or an abstract (see Chapter 9 for more on summaries and abstracts). For an audience completely familiar with the material, a clear and descriptive title will suffice.
➤ **List all parts and materials.** Group items into categories, if needed.
➤ **Refer to other documents or specs, as needed.** Often one set of specifications will refer to another set, or to government or industry standards. If your specifications are online, link to other specifications you refer to.
➤ **Use a consistent terminology.** Use the same terms for the same parts or materials throughout your specifications. If you refer to an "ergonomic adapter" in one section, do not substitute "iMac mouse adapter" in a later section. On a digital document (Web or PDF), consider adding links that define terms with one click.
➤ **Include retrieval aids.** Especially in longer specifications, some readers may be interested in only one portion. For example, someone working on a subset of a larger project may only want to look up technical details for that part of the project. Use clear headings and a table of contents.
➤ **Keep it simple.** People look at specifications because they want quick access to items, parts, technical requirements, and so on. If you can, limit your specs to short lists, using longer prose passages only as necessary.
➤ **Check your use of technical terms.** Use terms that are standard for the field.

Technical Marketing Materials

18.6 Write a technical marketing document to sell a product or service

Audience and purpose considerations

Technical marketing materials are designed to sell a product or service. Unlike proposals (Chapter 22), which also offer products or services, technical marketing materials tend to be less formal and more dynamic, colorful, and varied. A typical proposal is tailored to one client's specific needs and follows a fairly standard format, while marketing literature seeks to present the product in its best light for a broad array of audiences and needs. Also, technical marketing documents describe science and technology products and are often aimed at knowledgeable readers. A team of scientists looking to purchase a new electron microscope, for example, want specific technical information. Even when directed toward a general audience (say, home computer users), technical marketing materials must deal with specialized concepts.

Engineers and people in other technical disciplines who have a creative flair are often hired as technical marketing specialists. (For an intimate look at technical marketing as a career, see Richard Larkin's report in Chapter 21.)

Some situations that call for technical marketing materials include the following:

- *Cold calls*—sales representatives sending material to potential new customers
- *On-site visits*—sales representatives and technical experts visiting a customer to see if a new product or service might be of interest
- *Display booths*—booths at industry trade shows displaying engaging, interesting materials that people can take and read at their leisure
- *Online information*—Facebook and other social media pages, Twitter, and Web pages are often the primary place for information on a technical product or service

Common uses of technical marketing materials

Marketing documents range from simple "fact sheets" to brochures, booklets, or Web sites with colorful photographs and other visuals. Here are some common types:

Common formats for marketing documents

- **Social media pages.** Most companies use social media and the Web to display their marketing materials. The advantage of these sites over a printed document is that you can update price, specifications, or other features of the product. Social media also allows interactivity: Customers can provide feedback, request additional information, share tips and ideas, and rate your product or service.

- **Brochures.** Brochures are used to introduce a product or service, provide pricing information, and explain how customers can contact the company. A typical brochure is a standard-size page (8-½ ×11 inches) folded in thirds, but brochures can assume various shapes and sizes, depending on purpose, audience, and budget. Brochures can be made available to customers in PDF, downloadable from the company's Web site. You can create a professional looking brochure using any of today's word-processing or page design programs (most offer dozens of templates to choose from, which can be modified to meet your particular audience and purpose).

- **Fact sheets.** Fact sheets offer technical data and consumer information about the product or service being offered. Fact sheets are often one part of a more comprehensive brochure or Web page. Figure 18.7 shows one page from a longer brochure. This document, produced by the Vinyl Siding Institute, is included with the other marketing materials used by a building contractor to help customers understand the technology. The table format and visually appealing layout allow readers to compare two materials (vinyl and brick) based on key facts. Footnotes on this page are detailed on page four of the larger brochure.

Why America sides with vinyl

Beauty

Vinyl Siding	Brick
■ An impressive variety of profiles and shapes, with ideal choices to suit virtually any architectural style ■ An ever-increasing spectrum of colors, including darker options and period colors ■ Comprehensive architectural trim options and accessories in matching and complementary colors	■ While recognized for its attractive appearance, brick's limited variations and color options are not suited for popular architectural styles, including Queen Anne and Craftsman ■ Because accessories are not available in brick, many brick homes rely on high-maintenance trim (e.g., wood) to complete the look

Installation

Vinyl Siding	Brick
■ Lowest total installed cost of any exterior cladding[2] ■ An independent agency ensures that certified vinyl siding installers are trained and tested on ASTM-accepted application techniques	■ Labor-intensive process equates to higher installation cost; about 420 percent more expensive than vinyl siding[2] ■ No industry certification to verify installers are properly trained ■ Due to weight, more expensive to ship to jobsite

Environmental Impact

Vinyl Siding	Brick
■ Qualifies as an environmentally preferred product[3] ■ Per BEES software analysis, has lower environmental impact than brick in total embodied energy, global warming potential and criteria air pollutants ■ Produces less than 1.9 percent of the total construction waste by weight on a typical 2,000 sq. ft. home[4]	■ Per BEES software analysis, effects on global warming are three times greater than those of vinyl siding ■ Per BEES, uses more distribution energy, even when averaging about 140 miles less per shipment than vinyl siding ■ Per BEES, production of brick and mortar is responsible for generating almost 10 times the dioxin generated by the production of vinyl siding ■ Generates about 12.5 percent of the total construction waste by weight on a typical 2,000 sq. ft. home[4]

Value

Vinyl Siding	Brick
■ A cost-effective exterior cladding with the lowest total installed cost[2] ■ No long-term maintenance costs ■ No evidence of different appreciation rate of property value compared to other types of exterior cladding (see next page) ■ Lifetime, transferable warranties are available	■ Nearly 400 percent more expensive material cost than vinyl siding[2] ■ Maintenance to re-point joints adds to life cycle cost of the home

Figure 18.7 Technical Marketing Fact Sheet This fact sheet, one page of a technical marketing brochure, is included with materials given to potential customers by a building contractor.

Source: Courtesy of the Vinyl Siding Institute.

- **Letters.** Business letters are the most personal types of marketing documents. If a potential customer requests details about a product or service, you may send this information and include a brief cover letter. Thank the customer for his or her interest and point out specific features of your product or service that match this customer's needs. See, for example, how Rosemary Garrido's letter in Chapter 3, Figure 3.4 creates a persuasive connection with a potential customer. You can send the letter in the mail or, if you have the customer's contact information, as an email attachment.

- **Large color documents.** Some technical marketing materials are far more elaborate than a typical brochure. Consider the glossy booklets for a new car: The high-quality photography, slick color printing, and glossy feel are designed to evoke the feeling of owning such a car. These booklets include technical specifications, such as engine horsepower and wheel base size.

Guidelines

for Technical Marketing Materials

➤ **Research the background and experience of decision makers.** Your materials may be read by a range of people, but your main goal is to persuade those who make the final purchasing decisions. Gear the document toward their level of expertise and needs.

➤ **Situate your product in relation to others of its class.** Describe the product's main features as well as those that make it unique in relation to other such products.

➤ **Emphasize the special appeal of this product or service.** Briefly explain how this item fits the reader's exact needs, and support your claim with evidence.

➤ **Use upbeat, dynamic language.** Be careful not to overdo it, though. Technical people tend to dislike an obvious sales pitch.

➤ **Use visuals and color.** Visuals, especially diagrams and color photographs, are highly effective. Color images can convey the item's shape and feel while adding emotional and visual appeal (see Figure 18.7). If you create both print and online materials, make sure the visuals and color work in both formats. Coordinate your color choices to convey a consistent overall look and feel for your company and product.

➤ **Provide technical specifications, as needed.** Many types of technical marketing materials provide specifications such as product size, weight, and electrical requirements.

➤ **Consider including a FAQ list.** Some marketing materials anticipate customer questions with a "frequently asked questions" (FAQ) section.

Checklist

Technical Descriptions

Use the following Checklist when writing technical descriptions.

Content
❏ Is the description objective? (See "Objectivity in Technical Descriptions" in this chapter.)
❏ Does the title promise exactly what the description delivers? (See "Elements of Descriptions" in this chapter.)

❑ Are the item's overall features described, as well as each part? (See "Elements of Descriptions" in this chapter.)

❑ Is each part defined before it is discussed? (See "Elements of Descriptions" in this chapter.)

❑ Is the function of each part explained? (See "Elements of Descriptions" in this chapter.)

❑ Do visuals appear whenever they can provide clarification? (See "Elements of Descriptions" in this chapter.)

❑ Will readers be able to visualize the item? (See "Elements of Descriptions" in this chapter.)

❑ Are any details missing, needless, or confusing for this audience? (See "Elements of Descriptions" in this chapter.)

❑ Is the description ethically acceptable? (See "Objectivity in Technical Descriptions" in this chapter.)

Arrangement

❑ Does the description follow the clearest possible sequence? (See "Elements of Descriptions" in this chapter.)

❑ Are relationships among the parts clearly explained? (See "Elements of Descriptions" in this chapter.)

Style and Page Design

❑ Is the language informative and precise? (See "Objectivity in Technical Descriptions" in this chapter.)

❑ Is the level of technicality appropriate for the audience? (See "Elements of Descriptions" in this chapter.)

❑ Is the definition in plain English? (See Chapter 11, "Global, Legal, and Ethical Implications of Style and Tone.")

❑ Are sentences clear, concise, and fluent? (See Chapter 11, "Editing for Clarity," "Editing for Conciseness," and "Editing for Fluency.")

❑ Is word choice precise? (See Chapter 11, "Finding the Exact Words.")

❑ Is the definition grammatical? (See Appendix B, "Grammar.")

❑ Is the page design inviting and accessible? (See Chapter 13, "Page Design in Print and Digital Documents.")

Checklist

Specifications

Use the following Checklist when writing specifications.

Content

❑ Are the specifications appropriately detailed for the audience? (See "Specifications" in this chapter.)

❑ Do the specifications meet the requirements for consistency, quality, and safety? (See "Guidelines for Specifications" in this chapter.)

❑ Do the specifications adhere to prescribed standards? (See "Guidelines for Specifications" in this chapter.)

Arrangement

❑ Are other specifications or documents referred to, if needed? (See "Guidelines for Specifications" in this chapter.)

❑ Do the specifications follow a standard format, if applicable? (See "Guidelines for Specifications" in this chapter.)

❑ Is a brief introduction or descriptive title included? (See "Guidelines for Specifications" in this chapter.)

❑ Are the component parts or materials listed? (See "Guidelines for Specifications" in this chapter.)

Style and Page Design

❑ Is the terminology consistent? (See "Guidelines for Specifications" in this chapter.)

❑ Do short lists replace long prose passages wherever possible? (See "Guidelines for Specifications" in this chapter.)

❑ Are technical terms standard for the field? (See "Guidelines for Specifications" in this chapter.)

❑ Are the specifications easy to navigate, with clear headings and other retrieval aids? (See "Guidelines for Specifications" in this chapter.)

Checklist

Technical Marketing Materials

Use the following Checklist when writing technical marketing materials.

- ❏ Is the document based on a detailed audience and purpose analysis? (See "Technical Marketing Materials" in this chapter.)
- ❏ Is the material geared toward the final decision makers? (See "Guidelines for Technical Marketing Materials" in this chapter.)
- ❏ Is the format (brochure, Web page, letter, etc.) appropriate? (See "Technical Marketing Materials" in this chapter.)
- ❏ Is the product clearly situated in relation to others in its class? (See "Guidelines for Technical Marketing Materials" in this chapter.)
- ❏ Are the product's main characteristics as well as its unique features described? (See "Technical Marketing Materials" in this chapter.)
- ❏ Is the language dynamic and upbeat—without sounding like a sales pitch? (See "Guidelines for Technical Marketing Materials" in this chapter.)
- ❏ Does the document emphasize the special appeal of this product or service? (See "Guidelines for Technical Marketing Materials" in this chapter.)
- ❏ Do visuals and color convey an overall look and feel for the company and the product? (See "Guidelines for Technical Marketing Materials" in this chapter.)
- ❏ Are the product's technical specifications included, as needed? (See "Guidelines for Technical Marketing Materials" in this chapter.)
- ❏ Is a FAQ list included, as needed? (See "Guidelines for Technical Marketing Materials" in this chapter.)

Projects

For all projects, check with your instructor about whether to present your findings in class, bring drafts to class for discussion, upload your project to the class learning management system (LMS), and/or use the LMS forum or discussion boards to collaborate and review each activity below.

General

1. Choose a product requiring a description. Identify the audience and purpose, and prepare a description for this audience's level of technical understanding. As you prepare your description, refer to the "Guidelines for Descriptions" in this chapter.

2. Select a specialized process that you understand well and that has several distinct steps. Using the process description in Figure 18.3 as a model, explain this process to classmates who are unfamiliar with it.

3. The solar-collector description in this chapter is aimed toward a general audience. Evaluate its effectiveness for this audience. In one or two paragraphs, discuss your evaluation and suggest revisions.

4. How do specifications function in your workplace or home? Find one example of specifications used in your home or workplace and analyze it in terms of its usability. Are the specifications written for consistency, quality, and safety? How? What could the writers have done to improve the usability of this document?

Team

1. Divide into groups. Assume your group works in the product development division of a diversified manufacturing company. Your division has just thought of an idea for an inexpensive consumer item with a potentially vast market (choose a simple mechanism, such as nail clippers or a stapler). Your group's assignment is to prepare three descriptions of this invention:

   a. one for company executives who will decide whether to produce and market the item
   b. one for the engineers, machinists, and so on, who will design and manufacture the item
   c. one for the customers who might purchase and use the item

   Before writing for each audience, complete an Audience and Use Profile sheet (see Chapter 2, "Develop an Audience and Use Profile"). Appoint a group manager, who will assign tasks to members (visuals, typing, etc.). When the descriptions are prepared, the group manager will appoint one member to present the documents and explain their differences in class.

2. Select a specialized process you understand well or that you can learn about quickly by searching online (e.g., how gum disease develops, how an earthquake occurs, how steel is made, how a computer compiles and executes a program). Write a brief description of the process, incorporating at least one visual. Exchange your description with a classmate. Study your classmate's description for 15 minutes and then write the description in your own words. Now, evaluate your classmate's version of your original description. Does it show that your description was understood? If not, why not? Discuss your conclusions in a memo to your instructor, submitted with all samples. Be sure to document your information sources.

3. Assume your group is an architectural firm designing buildings at your college. Develop a set of specifications for duplicating the interior of the classroom in which this course is held. Focus only on materials, dimensions, and equipment (whiteboard, desk, etc.), and use visuals as appropriate. Your audience includes the firm that will construct the classroom, teachers, and school administrators. Use the same format as in Figure 18.6, or design a better one. Appoint one member to present the completed specifications in class. Compare versions from each group for accuracy and clarity.

Digital and Social Media

Go online and look for the Web page and social media sites of a local company that sells technical products or services. For instance, you might locate an installer of new energy-efficient windows or a company that offers radon abatement services. See how much technical marketing material is available on their Web site, and notice if they also use Facebook and Twitter. As a potential customer, can you get answers to basic questions? What other sort of marketing materials would be useful? What are other customers saying, and does the company respond to each and every customer post? Make recommendations for that company based on your observations. Share these recommendations with your instructor.

Global

Descriptions, especially those for nonspecialized or nontechnical audiences, are often written by comparing the item being described to an item that is already familiar to readers. For instance, the human heart is often described as a pump and compared to a pump people already know about (the pump for a swimming pool or a car's water pump). While descriptions based on these types of comparisons can be useful, they can also be problematic if your readers are from different countries and cultures. For instance, if you describe the human heart in relation to the pump for a swimming pool, people from parts of the United States or certain countries that don't have pools will not understand the comparison. Working with 2–3 other students, pick an item for which you'd like to write a description that will be distributed both within the United States and to at least one other country. Discuss whether or not to use a comparison and how you will decide which comparison(s) to use.

Chapter 19
Instructions and Procedures

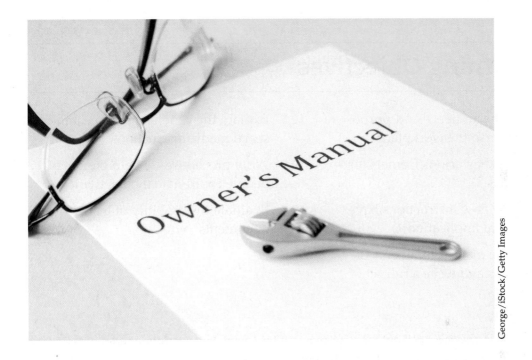

George/iStock/Getty Images

ffClear, accurate instructions and procedures are essential to the work we do in aerospace engineering. We need to ensure that the mechanics, ground control personnel, and pilots have the information they need to perform tasks and conduct safety and operations checks. These instructional documents can't have too much detail or be too wordy, and they need a clear list of steps. Hazard and warning material needs to show up easily, usually through the use of a visual. At our company, teams of engineers and technical writers work together to design, write, and evaluate all of our instructions, which we then print on quick-reference cards and make available on our Web site.**"**

—*Farid Akina, Aerospace Engineer at an international aerospace design firm*

Learning Objectives

19.1 Describe the audience and purpose of instructions in the workplace

19.2 Recognize the various formats for instructions

19.3 Appreciate how instructions have serious legal implications

19.4 Identify the main components of instructions and write a full set of instructions

19.5 Explain the benefits of online and social media instructions

19.6 Write procedures to help groups of people coordinate their activities

19.7 Evaluate the usability of instructional documents

What instructions do	*Instructions* spell out the steps required for completing a task or a series of tasks (say, installing printer software on your computer or operating an electron microscope). The audience for a set of instructions might be someone who doesn't know how to perform the task or someone who wants to perform it more effectively. In either case, effective instructions enable people to complete a job safely and efficiently.
What procedures do	*Procedures,* a special type of instructions, serve also as official guidelines. Procedures ensure that all members of a group (such as employees at the same company) follow the same steps to perform a particular task. For example, many companies have procedures in place that must be followed for evacuating a building or responding to emergencies.
The role of instructions on the job	Almost anyone with a responsible job writes and reads instructions. For example, you might instruct new employees on how to activate their voicemail system or advise a customer about shipping radioactive waste. An employee going on vacation typically writes instructions for the person filling in. When people buy a new computer, tablet, or any other electronic device, they turn to the instruction manual or quick reference card to get started.

Considering Audience and Purpose

19.1 Describe the audience and purpose of instructions in the workplace

Before preparing instructions, find out how much your audience already knows about the task(s) involved. For example, technicians who have done this procedure often (say, fixing a jammed photocopier) will need only basic guidelines rather than detailed explanations. But a more general audience (say, consumers trying to set up and use a new smart speaker or home security system) will need step-by-step guidance. A mixed audience (some experienced people and some novices) may require a layered approach; for instance, some initial basic information with a longer section later that has more details.

Audience considerations

The general purpose of instructions is to help people perform a task. The task may be simple (inserting a new toner cartridge in a printer) or complex (using an electron microscope). Whatever the task, people will have some basic questions:

Purpose considerations

- Why am I doing this?
- How do I do it?
- What materials and equipment will I need?
- Where do I begin?
- What do I do next?
- What could go wrong?

Because they focus squarely on the person who will "read" and then "do," instructions must meet the highest standards of excellence.

What people expect to learn from a set of instructions

Formats for Instructional Documents

19.2 Recognize the various formats for instructions

Instructional documents take various formats. Here are some of the most common ones:

- **Manuals** (Figure 19.1) are the most comprehensive form of instructions, often containing instructions for using the product along with descriptions, specifications, warnings, and troubleshooting advice. For a complex product (such as a 3D printer) or procedures (such as cleaning a hazardous waste site), manuals can be quite long (like a book).

Common formats for instructional documents

- **Quick reference materials** (Figures 19.2 and 19.6) typically fit on a single page or a small card. The instructions focus on basic steps for people who only need enough information to get started and perform the task.

- **Assembly guides** (Figure 19.4) are a common form of instructions found with consumer products (such as furniture, home repair items, and appliances) that describe how to assemble parts into a final product. These guides can be one page or several pages long and usually rely heavily on visuals (Figure 19.5).

Contents are grouped into logical categories

Frequently asked questions provide answers

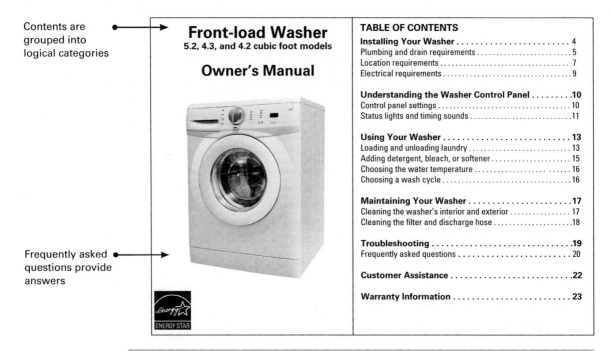

Front-load Washer
5.2, 4.3, and 4.2 cubic foot models

Owner's Manual

Figure 19.1 **Cover Page and Table of Contents from a User Manual**
Source: Energy Star Program; Photo: iStock/Getty Images Plus.

Opening paragraph is brief

Bullet items focus on basic steps

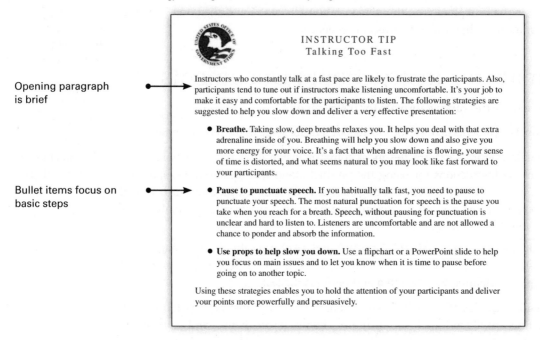

INSTRUCTOR TIP
Talking Too Fast

Instructors who constantly talk at a fast pace are likely to frustrate the participants. Also, participants tend to tune out if instructors make listening uncomfortable. It's your job to make it easy and comfortable for the participants to listen. The following strategies are suggested to help you slow down and deliver a very effective presentation:

- **Breathe.** Taking slow, deep breaths relaxes you. It helps you deal with that extra adrenaline inside of you. Breathing will help you slow down and also give you more energy for your voice. It's a fact that when adrenaline is flowing, your sense of time is distorted, and what seems natural to you may look like fast forward to your participants.

- **Pause to punctuate speech.** If you habitually talk fast, you need to pause to punctuate your speech. The most natural punctuation for speech is the pause you take when you reach for a breath. Speech, without pausing for punctuation is unclear and hard to listen to. Listeners are uncomfortable and are not allowed a chance to ponder and absorb the information.

- **Use props to help slow you down.** Use a flipchart or a PowerPoint slide to help you focus on main issues and to let you know when it is time to pause before going on to another topic.

Using these strategies enables you to hold the attention of your participants and deliver your points more powerfully and persuasively.

Figure 19.2 **A Brief Reference Card**
Source: Reprinted by permission of United States Office of Government Ethics, www.usoge.gov.

- **Web-based instructions** (Figure 19.3) allow readers to click on links to explore more information beyond the basic instructions on the main page.

- **Online help** (Figure 19.8) is part of most software packages. These instructions are "context sensitive"—that is, the help system recognizes what you are trying to accomplish (how to create a table in Microsoft Word, for example), provides a brief explanation if needed, and then guides you to that function.

Except for Web-based and online instructions, most other instructional documents are available in bothprint and PDF. Since consumers tend to lose the original manual that came with the product, PDF versions are usually available on the company

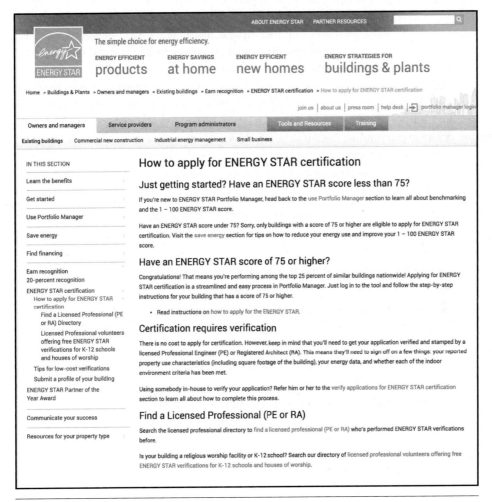

Figure 19.3 Web-based Instructions

Source: U.S. Environmental Protection Agency.

Web site. (See "Online and Social Media Instructions" later in this chapter for more information on PDF and instructional documents.) Regardless of its format, any set of instructions must meet the strict legal and usability requirements discussed on the following pages.

Faulty Instructions and Legal Liability

19.3 Appreciate how instructions have serious legal implications

Ethical implications of instructions

If you write, design, or are in any other way involved in researching and creating instructional documents, you need to remember that instructions have serious ethical and legal implications and that you, as part of the team, are responsible for making the material accurate and safe. Numerous workers are injured on the job each year, often due to faulty, unusable, or incomplete instructions. Countless injuries also result from misuse of consumer products such as power tools, car jacks, or household cleaners—types of misuse that are often caused by defective instructions.

Legal implications of instructions

Any person injured because of unclear, inaccurate, or incomplete instructions can sue the writer as well as the manufacturer. Courts have ruled that a writing defect in product support literature carries the same type of liability as a design or manufacturing defect in the product itself (Girill, "Technical Communication and Law" 37).

Those who prepare instructions are potentially liable for damage or injury resulting from information omissions such as the following (Caher 5–7; Manning 13; Nordenberg 7):

Examples of faulty instructions that create legal liability

- **Failure to instruct and caution readers in the proper use of a product:** for example, a medication's proper dosage or possible interaction with other drugs or possible side effects.

- **Failure to warn against hazards from proper use of a product:** for example, the risk of repetitive stress injury resulting from extended use of a keyboard.

- **Failure to warn against the possible misuses of a product:** for example, the danger of child suffocation posed by plastic bags or the danger of toxic fumes from spray-on oven cleaners.

- **Failure to explain a product's benefits and risks in language that average consumers can understand.**

- **Failure to convey the extent of risk with forceful language.**

- **Failure to display warnings prominently.**

Some legal experts argue that defects in the instructions carry even greater liability than defects in the product because such deficits are more easily demonstrated to a nontechnical jury (Bedford and Stearns 128).

> NOTE *Among all technical documents, instructions have the strictest requirements for giving readers precisely what they need precisely when they need it.*

Elements of Effective Instructions

19.4 Identify the main components of instructions and write a full set of instructions

Effective instructions typically contain several key elements including a title, accurate content, and appropriate visuals. These elements should be combined in ways that are most useful to your audience and the tasks they need to perform.

Clear and Limiting Title

Provide a clear and exact preview of the task. For example, the title "Instructions for Cleaning the DVD Drive of Your Laptop Computer" tells people what to expect: instructions for a specific procedure involving one selected part. But the title "Laptop Computer" gives no such forecast; a document so titled might contain a history of the laptop, a description of each part, or a wide range of related information.

Say no more or less than needed

Informed and Accurate Content

Make sure you know exactly what you are talking about. Ignorance, inexperience, or misinformation on your part makes you no less liable for faulty or inaccurate instructions:

Know the procedure

> If the author of [a car repair] manual had no experience with cars, yet provided faulty instructions on the repair of the car's brakes, the home mechanic who was injured when the brakes failed may recover [damages] from the author. (Walter and Marsteller 165)

Ignorance provides no legal excuse

Only write instructions when you completely understand the task and have performed the task often enough to understand all important details.

Visuals

Instructions often include a persuasive dimension: to promote interest, commitment, or action. In addition to showing what to do, visuals attract the reader's attention and help keep words to a minimum.

Enhance text with visuals

Types of visuals especially suited to instructions include icons, representational and schematic diagrams, flowcharts, photographs, and prose tables.

Visuals to accompany instructions can be created using a variety of software packages. Other sources for instructional graphics include clip art, scanning, and downloading from the Internet. (See Chapter 12, "Guidelines for Obtaining and Citing Visual Material," for information about using visuals you find online.)

To use visuals effectively, consider these suggestions:

- Illustrate any step that might be hard for readers to visualize. The less specialized your readers, the more visuals they are likely to need.

How to use instructional visuals

- Parallel the reader's angle of vision in performing the activity or operating the equipment. Name the angle (side view, top view) if you think people will have trouble figuring it out for themselves.

- Avoid illustrating any action simple enough for readers to visualize on their own, such as "PRESS ENTER" for anyone familiar with a keyboard.

Visuals can be used without words, too, especially for international audiences. Often called *wordless instructions*, these diagrams use clear, simple line drawings, arrows, and call-outs to let people see how to do something. Figure 19.4 shows one page from a set of wordless instructions for assembling a TV stand.

Figure 19.5 presents an array of visuals and their specific instructional functions. You may also require visuals for tasks such as how to repair something or how to understand key measures such as temperature or other measurements. Visuals like these are easily constructed, and some could be further enhanced, depending on your production budget and graphics capability.

Appropriate Level of Detail and Technicality

Unless you know your readers have the relevant background and skills, write for a general audience, and do three things:

Provide exactly and only what readers need

1. Give readers enough background to understand why they need to follow these instructions.
2. Give enough detail to show *what* to do.
3. Give enough examples so each step can be visualized clearly.

These three procedures are explained and illustrated on the following pages.

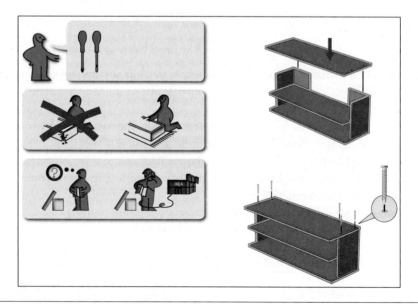

Figure 19.4 **Wordless Instructions**
Source: Used with the permission of Inter IKEA Systems B.V.

HOW TO LOCATE SOMETHING

Source: Adapted from Occupational and Safety Health Administration, www.osha.gov.

HOW TO POSITION SOMETHING

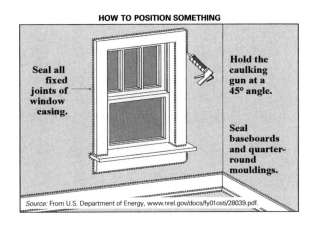

Source: From U.S. Department of Energy, www.nrel.gov/docs/fy01osti/28039.pdf.

HOW TO OPERATE SOMETHING

Source: Likoper/Shutterstock.

HOW TO BUILD SOMETHING

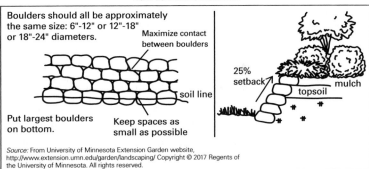

Source: From University of Minnesota Extension Garden website, http://www.extension.umn.edu/garden/landscaping/ Copyright © 2017 Regents of the University of Minnesota. All rights reserved.

Figure 19.5 Common Types of Instructional Visuals and Their Functions In addition to these examples, use visuals to show other tasks, such as how to perform a procedure (see Figure 19.6), how to assemble something, how to download something, how to repair something, and many more. Visuals can also provide important safety information (see "Notes and Hazard Notices" later in this chapter).

PROVIDE BACKGROUND. Begin by explaining the purpose of the task.

> You might easily lose information stored on a flash drive if
>
> - the drive is damaged by repeated use, moisture, or extreme temperature;
> - the drive is erased by a power surge, a computer malfunction, or a user error; or
> - the stored information is scrambled by a nearby magnet (telephone, computer terminal, or the like).
>
> Always use another back-up device, such as a Firewire hard drive, for important material.

Tell readers why they are doing this

Also, state your assumptions about your reader's level of technical understanding.

> To follow these instructions, you should be able to identify these parts of your iMac: computer, keyboard (wireless or USB), mouse, and external DVD drive.

Spell out what readers should already know

Define any specialized terms that appear in your instructions.

Tell readers what
each key term
means

> *Initialize:* Before you can store or retrieve information on a new CD, you must initialize the disk. Initializing creates a format that computers and CD players can understand—a directory of specific memory spaces on the disk where you can store information and retrieve it as needed.

When the reader understands *what* and *why*, you are ready to explain *how* he or she can complete the task.

Make
instructions
complete but not
excessive

PROVIDE ADEQUATE DETAIL. Include enough detail for people to understand and perform the task successfully. Omit general information that readers probably know, but if you are uncertain about their knowledge or experience level, do not overestimate the audience's background, as in the following example of inadequate detail:

Inadequate detail
for laypersons

First Aid for Electrical Shock

1. Check vital signs.
2. Establish an airway.
3. Administer CPR as needed.
4. Treat for shock.

These steps might be suitable for experts (such as paramedics or nurses), but terms such as "vital signs" and "CPR" are too technical for laypersons. Such instructions posted for workers in a high-voltage area would be useless. Illustrations and explanations are needed, as in the instructions in Figure 19.6 for item 3 above, administering CPR.

Don't assume that people know more than they really do, especially when you can perform the task almost automatically. (Think about when a relative or friend taught you to drive a car—or perhaps you tried to teach someone else.) When writing for a more general audience, always assume that your readers know less than you. A colleague will know at least a little less; a layperson will know a good deal less— maybe nothing—about this procedure.

Exactly how much information is enough? See "Guidelines for Providing Appropriate Detail in Instructions" following this section.

Give plenty of
examples

OFFER EXAMPLES. Instructions require specific examples (how to load a program, how to order a part) to help people follow the steps correctly:

To load your program, type this command:

Load "Style Editor"

Then press RETURN.

Like visuals, examples *show* readers what to do. Examples, in fact, often appear as visuals.

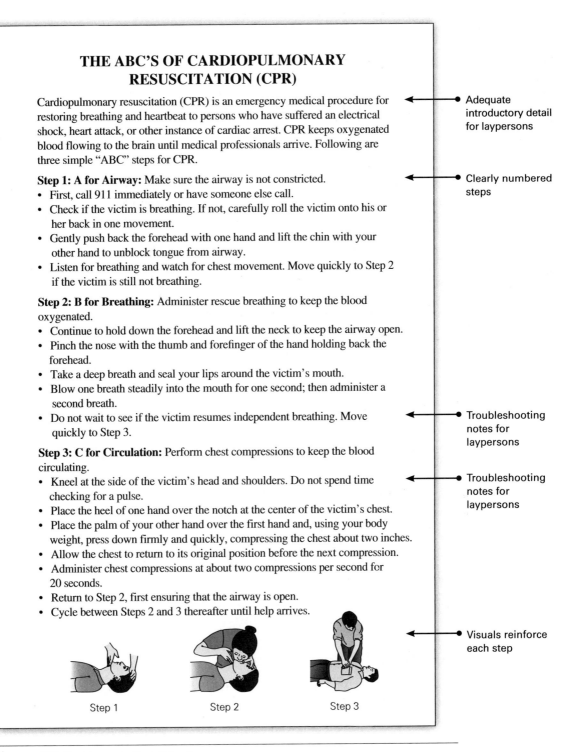

THE ABC'S OF CARDIOPULMONARY RESUSCITATION (CPR)

Cardiopulmonary resuscitation (CPR) is an emergency medical procedure for restoring breathing and heartbeat to persons who have suffered an electrical shock, heart attack, or other instance of cardiac arrest. CPR keeps oxygenated blood flowing to the brain until medical professionals arrive. Following are three simple "ABC" steps for CPR. ◀—— ● Adequate introductory detail for laypersons

Step 1: A for Airway: Make sure the airway is not constricted. ◀—— ● Clearly numbered steps
- First, call 911 immediately or have someone else call.
- Check if the victim is breathing. If not, carefully roll the victim onto his or her back in one movement.
- Gently push back the forehead with one hand and lift the chin with your other hand to unblock tongue from airway.
- Listen for breathing and watch for chest movement. Move quickly to Step 2 if the victim is still not breathing.

Step 2: B for Breathing: Administer rescue breathing to keep the blood oxygenated.
- Continue to hold down the forehead and lift the neck to keep the airway open.
- Pinch the nose with the thumb and forefinger of the hand holding back the forehead.
- Take a deep breath and seal your lips around the victim's mouth.
- Blow one breath steadily into the mouth for one second; then administer a second breath.
- Do not wait to see if the victim resumes independent breathing. Move quickly to Step 3. ◀—— ● Troubleshooting notes for laypersons

Step 3: C for Circulation: Perform chest compressions to keep the blood circulating.
- Kneel at the side of the victim's head and shoulders. Do not spend time checking for a pulse. ◀—— ● Troubleshooting notes for laypersons
- Place the heel of one hand over the notch at the center of the victim's chest.
- Place the palm of your other hand over the first hand and, using your body weight, press down firmly and quickly, compressing the chest about two inches.
- Allow the chest to return to its original position before the next compression.
- Administer chest compressions at about two compressions per second for 20 seconds.
- Return to Step 2, first ensuring that the airway is open.
- Cycle between Steps 2 and 3 thereafter until help arrives.

◀—— ● Visuals reinforce each step

Step 1   Step 2   Step 3

Figure 19.6 Adequate Detail for Laypersons

Guidelines
for Providing Appropriate Detail in Instructions

➤ **Provide *all* the necessary information.** The instructions must be able to stand alone.

➤ **Don't provide unnecessary information.** Give only what readers need. Don't tell them how to build a computer when they only need to know how to copy a file.

➤ **Instead of focusing on the *product*, focus on the *task*.** "How does it work?" "How do I use it?" or "How do I do it?" (Grice 132).

➤ **Omit steps that are obvious.** "Seat yourself at the computer," for example.

➤ **Divide the task into simple steps and substeps.** Allow people to focus on one step at a time.

➤ **Adjust the *information rate*.** This is "the amount of information presented in a given page" (Meyer 17), adjusted to the reader's background and the difficulty of the task. For complex or sensitive steps, slow the information rate. Don't make people do too much too fast.

➤ **Reinforce the prose with visuals.** Don't be afraid to repeat information if it saves readers from going back to look something up.

➤ **Keep it simple.** When writing instructions for consumer products, assume that your readers are not overly technical and that they have little to no experience with the product.

➤ **Recognize the persuasive dimension of the instructions.** Readers may need persuading that this procedure is necessary or beneficial, or that they can complete this procedure with relative ease and competence.

INCLUDE TROUBLESHOOTING ADVICE. Anticipate things that commonly go wrong when this task is performed—the paper jams in the printer, the tray of the DVD drive won't open, or some other malfunction. Explain the probable cause(s) and offer solutions, as in the following example:

Explain what to do when things go wrong

| NOTE: IF *X* doesn't work, first check *Y* and then do *Z*.

Logically Ordered Steps

Instructions are almost always arranged in chronological order, with warnings and precautions inserted for specific steps.

Show how the steps are connected

| You can't splice two wires to make an electrical connection until you have removed the insulation. To remove the insulation, you will need. . . .

Notes and Hazard Notices

Alert readers to special considerations

Following are the only items that normally should interrupt the steps in a set of instructions (Van Pelt 3):

• A *note* clarifies a point, emphasizes vital information, or describes options or alternatives.

This corrected version follows a logical sequence:

| Insert the DVD in the drive; then switch on the computer.   Logical

Simplify explanations by using a familiar-to-unfamiliar sequence. The following sentence is hard to follow, due to its unfamiliar-to-familiar sequence:

| You must initialize a blank CD before you can store information on it.   Hard

This corrected version follows an easier familiar-to-unfamiliar sequence:

| Before you can store information on a blank CD, you must initialize the CD.   Easier

USE PARALLEL PHRASING.   Parallelism is important in all writing but especially so in instructions, because repeating grammatical forms emphasizes the step-by-step organization. Parallelism also increases readability and lends continuity to the instructions. The following example is difficult to follow because the phrasing of the steps is not parallel:

| To connect to the server, follow these steps: Not parallel

  1. Switch the terminal to "on."
  2. The CONTROL key and C key are pressed simultaneously.
  3. Typing LOGON, and pressing the ESCAPE key.
  4. Type your user number, and then press the ESCAPE key.

All steps should be in identical grammatical form, as in the following easier-to-follow parallel example:

| To connect to the server, follow these steps: Parallel

  1. Switch the device to "on."
  2. Press the CONTROL key and C key simultaneously.
  3. Type LOGON, and then press the ESCAPE key.
  4. Type your user number, and then press the ESCAPE key.

PHRASE INSTRUCTIONS AFFIRMATIVELY.   Research shows that people respond more quickly and efficiently to instructions phrased affirmatively rather than negatively (Spyridakis and Wenger 205). The following sentence is phrased negatively, slowing readers down:

| Verify that your camera lens is not contaminated with dust.   Negative

This corrected version with affirmative phrasing is easier to grasp on a first reading:

| Examine your camera lens for dust.   Affirmative

USE TRANSITIONS TO MARK TIME AND SEQUENCE. Transitional expressions (see Appendix B, "Transitions") provide a bridge between related ideas. Some transitions ("first," "next," "meanwhile," "finally," "ten minutes later," "the next day," "immediately afterward") mark time and sequence. They help readers understand the step-by-step process, as in the next example. (Bold face is used to illustrate the transitions.)

Transitions enhance continuity

Preparing the Ground for a Tent

Begin by clearing and smoothing the area that will be under the tent. This step will prevent damage to the tent floor and eliminate the discomfort of sleeping on uneven ground. **First,** remove all large stones, branches, or other debris within a level area roughly 10 × 10 feet. Use your camping shovel to remove half-buried rocks that cannot easily be moved by hand. **Next,** fill in any large holes with soil or leaves. **Finally,** make several light surface passes with the shovel or a large, leafy branch to smooth the area.

Effective Design

Instructions rarely get undivided attention. The reader, in fact, is doing two things more or less at once: interpreting the instructions and performing the task. An effective instructional design conveys the sense that the task is within a qualified person's range of abilities. The more accessible and inviting the design, the more likely your readers will follow the instructions.

Guidelines
for Designing Instructions

➤ **Use informative headings.** Tell readers what to expect; emphasize what is most important; provide cues for navigation. A heading such as "How to Initialize Your Compact Disk" is more informative than "Compact Disk Initializing."

➤ **Arrange all steps in a numbered list.** Unless the procedure consists of simple steps (as in "Preparing the Ground for a Tent," above), list and number each step. Numbered steps not only announce the sequence of steps, but also help readers remember where they left off. (For more on using lists, see Appendix B, "Lists.")

➤ **Separate each step visually.** Single-space within steps and double-space between.

➤ **Double-space to signal a new paragraph, instead of indenting.**

➤ **Make warning, caution, and danger notices highly visible.** Use ruled boxes or highlighting, and plenty of white space.

➤ **Make visual and verbal information redundant.** Let the visual repeat, restate, or reinforce the prose.

➤ **Keep the visual and the step close together.** If room allows, place the visual right beside the step; if not, right after the step. Set off the visual with plenty of white space.

➤ **Consider a multicolumn design.** If steps are brief and straightforward and require back-and-forth reference from prose to visuals, consider multiple columns.

➤ **Keep it simple.** Readers can be overwhelmed by a page with excessive or inconsistent designs.

➤ **For lengthy instructions, consider a layered approach.** In a complex manual, for instance, you might add a "Quick Start Guide" for getting started, with cross-references to pages containing more detailed and technical information. PDF documents can provide a layered approach by using links.

➤ **For online instructions, use the appropriate software to format the information.** See "Online and Social Media Instructions" in this chapter.

For additional design considerations, see Chapter 13.

Introduction-Body-Conclusion Structure

As with all technical documents, instructions are typically organized using a standard introduction-body-conclusion structure, although a conclusion is not always required (for more information on conclusions in instructions, see below). For a longer set of instructions, you will also need to start with an outline. The following outline can be adapted to any instructions. Here are the possible components to include:

I. Introduction
   A. Definition, Benefits, and Purpose of the Procedure
   B. Intended Audience (often omitted for workplace audiences)
   C. Prior Knowledge and Skills Needed by the Audience
   D. Brief Overall Description of the Procedure
   E. Principle of Operation
   F. Materials, Equipment (in order of use), and Special Conditions
   G. Working Definitions (always in the introduction)
   H. Warnings, Cautions, Dangers (previewed here and spelled out at steps)
   I. List of Major Steps

II. Required Steps
   A. First Major Step
      1. Definition and purpose
      2. Materials, equipment, and special conditions for this step
      3. Substeps (if applicable)
         a. First substep
         b. Second substep (and so on)
   B. Second Major Step (and so on)

III. Conclusion
   A. Review of Major Steps (for a complex procedure only)
   B. Interrelation of Steps
   C. Troubleshooting or Follow-up Advice (as needed)

Outlining a longer set of instructions

This outline is only tentative; you might modify, delete, or combine some components, depending on your subject, purpose, audience, and workplace guidelines (for instance, in some companies, instructions must be written to conform with guidelines of relevant regulatory bodies, such as the Occupational Health and Safety Information (OSHA).

INTRODUCTION.   The introduction should help readers to begin "doing" as soon as they are able to proceed safely, effectively, and confidently (van der Meij and Carroll 245–46). Most people are interested primarily in how to use it or fix the product and will require only a general understanding of how it works. You don't want to bury your readers in a long introduction, nor do you want to set them loose on the procedure without adequate preparation. Know your audience—what they need and don't need.

BODY: REQUIRED STEPS.   In the body section (labeled Required Steps), give each step and substep in order. Insert warnings, cautions, and notes as needed. Begin each step with its definition or purpose or both. Readers who understand the reasons for a step will do a better job. A numbered list is an excellent way to segment the steps. Or, begin each sentence in a complex stage on a new line.

CONCLUSION.   The conclusion of a set of instructions has several possible functions:

- Summarize the major steps in a long and complex procedure, to help people review their performance.

- Describe the results of the procedure.

- Offer follow-up advice about what could be done next or refer the reader to further sources of documentation.

- Give advice about troubleshooting if anything goes wrong.

You might do all these things—or none of them. If your procedural section has provided all that is needed, omit the conclusion altogether.

Writing a
long set of
instructions

Figure 19.7 shows a complete set of instructions written for a nontechnical audience. These instructions follow the basic outline by offering an overview, a list of equipment needed, and simple numbered instructions. The design, which uses informative headings, numbered steps, and simple visual diagrams, is easy to use.

The writer of the instructions, Jill Sanning, has been told by the owner of her town's local hardware store that he is often asked the same questions by customers who want to make simple home repairs. One common question is about how to replace a worn faucet washer. He decides to hire Jill to write and design a simple yet effective set of instructions. He will offer print copies at the hardware store and also make the PDF file available on the store's Web site. Jill starts by filling out an Audience and Use Profile, then writing an audience and purpose statement to guide her work:

Jill's audience
and purpose
statement

These customers come from a wide range of backgrounds, but most of them are not engineers or plumbers. They are just regular homeowners who are confident in their ability to work with basic tools and comfortable trying a new task. They don't want a lot of detail about the history of faucets or the various types of faucets. They just want to know how to fix the problem. These people are busy—they have lots of chores to do on the weekend, and they want instructions that are easy to follow and use.

The results of Jill's work appear in Figure 19.7.

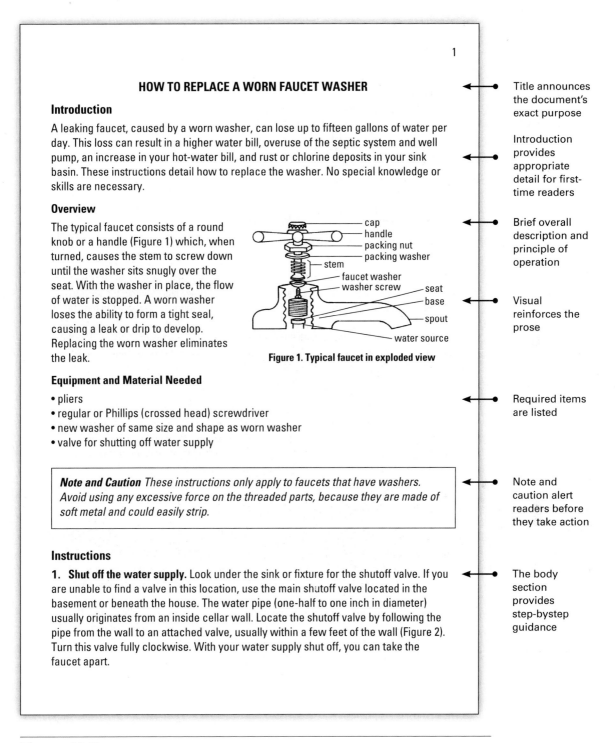

1

HOW TO REPLACE A WORN FAUCET WASHER

Introduction

A leaking faucet, caused by a worn washer, can lose up to fifteen gallons of water per day. This loss can result in a higher water bill, overuse of the septic system and well pump, an increase in your hot-water bill, and rust or chlorine deposits in your sink basin. These instructions detail how to replace the washer. No special knowledge or skills are necessary.

Overview

The typical faucet consists of a round knob or a handle (Figure 1) which, when turned, causes the stem to screw down until the washer sits snugly over the seat. With the washer in place, the flow of water is stopped. A worn washer loses the ability to form a tight seal, causing a leak or drip to develop. Replacing the worn washer eliminates the leak.

cap
handle
packing nut
packing washer
stem
faucet washer
washer screw
seat
base
spout
water source

Figure 1. Typical faucet in exploded view

Equipment and Material Needed

• pliers
• regular or Phillips (crossed head) screwdriver
• new washer of same size and shape as worn washer
• valve for shutting off water supply

Note and Caution These instructions only apply to faucets that have washers. Avoid using any excessive force on the threaded parts, because they are made of soft metal and could easily strip.

Instructions

1. Shut off the water supply. Look under the sink or fixture for the shutoff valve. If you are unable to find a valve in this location, use the main shutoff valve located in the basement or beneath the house. The water pipe (one-half to one inch in diameter) usually originates from an inside cellar wall. Locate the shutoff valve by following the pipe from the wall to an attached valve, usually within a few feet of the wall (Figure 2). Turn this valve fully clockwise. With your water supply shut off, you can take the faucet apart.

Title announces the document's exact purpose

Introduction provides appropriate detail for first-time readers

Brief overall description and principle of operation

Visual reinforces the prose

Required items are listed

Note and caution alert readers before they take action

The body section provides step-by-step guidance

Figure 19.7 A Complete Set of Instructions

2

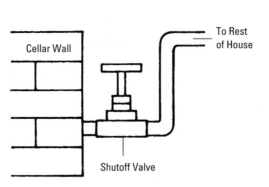

Figure 2. Location of main shutoff valve

2. Disassemble the faucet. Before removing the handle, open the faucet to allow remaining water in the pipe to escape. Using your screwdriver, remove the screw on top of the handle. If a cap covers the screw, pry it off (Figure 1). Remove the handle.

Next, remove the packing nut, using pliers to turn the nut counter-clockwise. The flat circular nut washer can now be lifted from the faucet base. With packing nut and washer removed, screw the stem out of the base by turning the faucet in the "open" direction. Lift the stem out of the base and proceed to step 3.

3. Replace the worn washer. Using your screwdriver, remove the screw holding the washer at the base of the stem (Figure 1). Remove the worn washer and replace it with a new one of the same size, using the washer screw to hold it in place. (Washers of various sizes can be purchased at any hardware store.) If the washer screw is worn, replace it. When the new washer is fixed in place, proceed to step 4.

Caution precedes step ●→ **4. Reassemble the faucet.** To reassemble your faucet, reverse the sequence described in step 2. *Caution: Do not overtighten the packing nut!*

Using pliers, screw the stem into the base until it ceases to turn. Next, place the packing washer and nut over the threads in the collar. Tighten the packing nut using the strength of one hand. Finally, secure the handle with your screwdriver. When the faucet is fully assembled, turn the handle to the "off" position and proceed to step 5.

Because the body section has given readers all they need, no conclusion is required ●→ **5. Turn on the water supply.** First, check to see that your faucet is fully closed. Next, turn the water on slowly (about one-half turn each time) until the shutoff valve is fully open. These slow turns prevent a sudden buildup of pressure, which could damage the pipes. Your faucet should now be as good as new.

Figure 19.7 Continued

Online and Social Media Instructions

19.5 **Explain the benefits of online and social media instructions**

Because of the costs involved in printing and updating instructions, especially lengthy user manuals, companies have shifted toward a combination of print and digital or have moved to digital entirely. Most products now come with a "Getting Started" quick reference guide and then direct readers to a Web site where they can access the longer, complete instruction manual. Companies also realize that people usually lose or misplace their original instructions or manual. By putting these documents on a Web site, these organizations save countless hours of time (answering phone calls or emails) and expense. Customers are also happy when they can find the manual they've been looking for and download it right when they need it.

Benefits of online instructions

The most common online format for instructional material is PDF, which retains the proper formatting no matter what computer or device is being used. For more information on PDF, see Chapter 13 "Adobe Acrobat and PDF files." Figure 19.7, the instructions for replacing a worn faucet washer, would be easy to convert to PDF using Microsoft Word or Apple Pages (just "Save as PDF").

Benefits of PDFs

Online Help

Unlike online user manuals and guides, online help is part of the product or software and provides quick answers, directs people to tasks they want to perform, and offers links to additional information. Figure 19.8 shows a typical online help screen for Microsoft Word. In this example, a person is trying to create a table; after entering "table" in the help search bar, the software guided her to the appropriate menu. More detailed help information on creating a table is also available by clicking on a link in the right-hand column of the screen.

As with Web pages (Chapter 24), online help should be written in short chunks. Experienced technical writers use special software, such as Adobe RoboHelp, to convert content into the appropriate format for online help. Online help can be coded to be *context sensitive*; that is, when people get stuck, the system recognizes what task they are trying to perform and offers help for that situation. You can also search using the help menu to locate information on specific tasks. Online help can be found in all kinds of apps and in many new consumer products such as cars, refrigerators, DVRs, and more. People appreciate being able to get help when they need it, rather than having to search around for the original user manual.

Benefits of online help

Social Media Instructions

Many people turn to social media for instructions, typically starting with a search on YouTube. Almost anyone with a cell phone can make an instructional video, and there can be dozens if not hundreds of videos posted on the same topic, each of differing qualities. For example, if you search on topics such as "how to do my own oil change,"

Cautions with using social media instructions

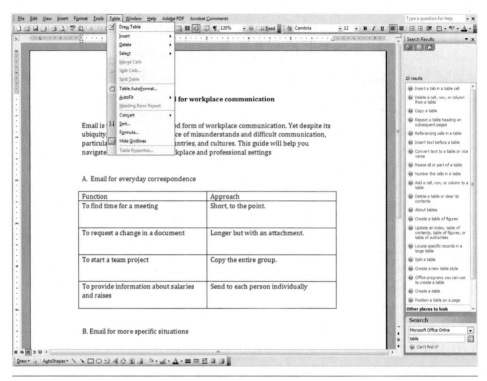

Figure 19.8 Online Help Screen
Source: Microsoft Word 2003.

"how to change the strings on a classical guitar," or "how to program the remote to my garage door opener," you will find more videos than you can possibly imagine, some of which are easy to follow, accurate, and thorough, but others that are almost unusable. Sorting through these videos and determining which ones are credible (and have been tested) and which ones are not can be incredibly time-consuming. Many of the people who make and upload instructional videos receive commercial compensation via ads, which is not necessarily bad but adds to the items that audiences must consider when choosing a credible, usable video for instructions.

Uses of organizational instructional videos

If you work for an organization that creates instructional videos, you may wish to put videos on your company's main Web site and Facebook page. The video can be run right from your server or from a YouTube link. Having these videos available on your official company page or social media feed provides customers with the confidence to know that these instructions are direct from the source and have been tested and authorized by the engineers, technicians, and others at your organization.

Advantages of video instructions

A big advantage of video instructions is the ability for customers to comment and rate the video (thumbs up or down). Customers are usually more than happy to point out places where the video is inaccurate or could use more detail. Remember that as with all forms of instructions, instructional videos need to take safety into account at all times. In a video, you may need to repeat the safety and caution information and include text frames that restate and summarize important safety considerations.

Chapter 25 discusses using social media for technical communication, including a complete section on creating video instructions and using sites such as YouTube, Vimeo, and others to post the video and interact with customers. As you work on the documents for this chapter, consider how these instructions might be converted to video, and save your ideas for projects in Chapter 25.

Procedures

19.6 Write procedures to help groups of people coordinate their activities

Instructions show an uninitiated person how to perform a task. *Procedures,* on the other hand, provide rules and guidance for people who usually know how to perform the task but who are required to follow accepted practice. To ensure that everyone does something in exactly the same way, procedures typically are aimed at groups of people who need to coordinate their activities so that everyone's performance meets a certain standard. Consider, for example, police procedures for properly gathering evidence from a crime scene: Strict rules stipulate how evidence should be collected and labeled and how it should be preserved, transported, and stored. Evidence shown to have been improperly handled is routinely discredited in a courtroom.

How instructions and procedures differ

Organizations need to follow strict safety procedures as defined by the U.S. Occupational Safety and Health Administration (OSHA) and other federal, state, and local government agencies. As laws and policies change, such procedures are often updated. The written procedures must be posted for employees to read. Figure 19.9 shows one page outlining OSHA regulations for evacuating high-rise buildings. This document is available in PDF on the OSHA Web site; it can be printed as well as shared with employees via email.

Procedures help keep everyone "on the same page"

Procedures are also useful in situations in which certain tasks need to be standardized. For example, if different people in your organization perform the same task at different times (say, monitoring groundwater pollution) with different equipment, or under different circumstances, this procedure may need to be standardized to ensure that all work is done with the same accuracy and precision. A document known as a *Standard Operating Procedure (SOP)* becomes the official guideline for that task. SOPs for complicated procedures may be many pages long, while those written for simpler procedures might only be a few pages in length. In many situations, such as manufacturing facilities, SOPs must be kept updated (usually by a team of engineers and technical writers) and be available for review during government safety inspections. See Figure 19.10 for an example of a Standard Operating Procedure.

Standard Operating Procedures help ensure safety

The steps in a procedure may or may not need to be numbered. This choice will depend on whether or not steps must be performed in strict sequence. Compare, for example, Figure 19.9 versus Figure 19.10.

Heading identifies this as a government (OSHA) procedure •────→

Questions that readers might have •────→

Bullets break out each step visually •────→

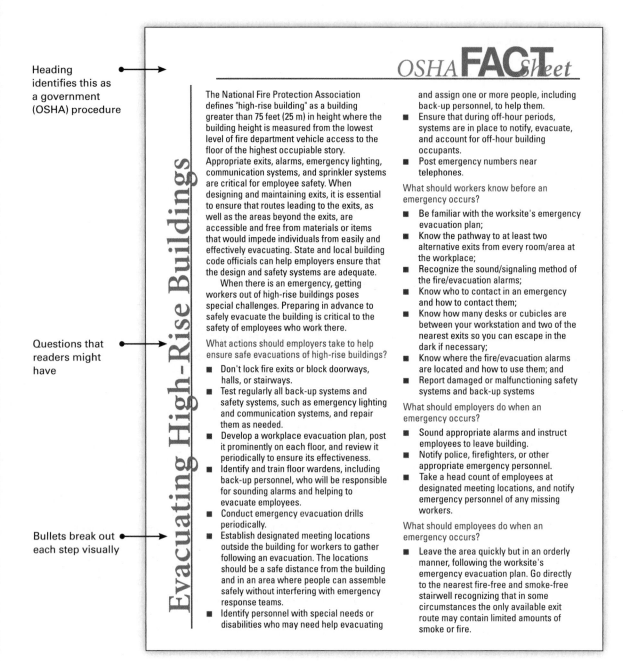

OSHA **FACT**Sheet

Evacuating High-Rise Buildings

The National Fire Protection Association defines "high-rise building" as a building greater than 75 feet (25 m) in height where the building height is measured from the lowest level of fire department vehicle access to the floor of the highest occupiable story. Appropriate exits, alarms, emergency lighting, communication systems, and sprinkler systems are critical for employee safety. When designing and maintaining exits, it is essential to ensure that routes leading to the exits, as well as the areas beyond the exits, are accessible and free from materials or items that would impede individuals from easily and effectively evacuating. State and local building code officials can help employers ensure that the design and safety systems are adequate.

When there is an emergency, getting workers out of high-rise buildings poses special challenges. Preparing in advance to safely evacuate the building is critical to the safety of employees who work there.

What actions should employers take to help ensure safe evacuations of high-rise buildings?

- Don't lock fire exits or block doorways, halls, or stairways.
- Test regularly all back-up systems and safety systems, such as emergency lighting and communication systems, and repair them as needed.
- Develop a workplace evacuation plan, post it prominently on each floor, and review it periodically to ensure its effectiveness.
- Identify and train floor wardens, including back-up personnel, who will be responsible for sounding alarms and helping to evacuate employees.
- Conduct emergency evacuation drills periodically.
- Establish designated meeting locations outside the building for workers to gather following an evacuation. The locations should be a safe distance from the building and in an area where people can assemble safely without interfering with emergency response teams.
- Identify personnel with special needs or disabilities who may need help evacuating

and assign one or more people, including back-up personnel, to help them.
- Ensure that during off-hour periods, systems are in place to notify, evacuate, and account for off-hour building occupants.
- Post emergency numbers near telephones.

What should workers know before an emergency occurs?

- Be familiar with the worksite's emergency evacuation plan;
- Know the pathway to at least two alternative exits from every room/area at the workplace;
- Recognize the sound/signaling method of the fire/evacuation alarms;
- Know who to contact in an emergency and how to contact them;
- Know how many desks or cubicles are between your workstation and two of the nearest exits so you can escape in the dark if necessary;
- Know where the fire/evacuation alarms are located and how to use them; and
- Report damaged or malfunctioning safety systems and back-up systems

What should employers do when an emergency occurs?

- Sound appropriate alarms and instruct employees to leave building.
- Notify police, firefighters, or other appropriate emergency personnel.
- Take a head count of employees at designated meeting locations, and notify emergency personnel of any missing workers.

What should employees do when an emergency occurs?

- Leave the area quickly but in an orderly manner, following the worksite's emergency evacuation plan. Go directly to the nearest fire-free and smoke-free stairwell recognizing that in some circumstances the only available exit route may contain limited amounts of smoke or fire.

Figure 19.9 Safety Procedures This page defines general safety and evacuation procedures to be followed by employers and employees. Each building in turn is required to have its own specific procedures, based on such variables as location, design, and state law.
Source: U.S. Occupational Safety and Health Administration, 2007, www.osha.gov.

January 2016

SOP No. 33

Standard Operating Procedure
for
Calibration of Weight Carts

1. Introduction

    1.1. This Standard Operation Procedure (SOP) describes the procedure to be followed for the calibration of weight carts used to test livestock and vehicle scales.

    1.2. Prerequisites

        1.2.1. Facility. Verify that the laboratory facilities meet the following minimum conditions to meet the expected uncertainty possible with this procedure.

Table 1. Environmental conditions.

Echelon	Temperature Requirements During a Calibration	Relative Humidity (%)
III	Lower and upper Limit: 18 °C to 27 °C Maximum changes: ± 5 °C/ 12 h and ± 3 °C/h	40 to 60 ± 20/4 h

    1.2.2. Balance/scale. Verify that the balance, scale, or load cell is in good operating condition with sufficiently small process standard deviation as verified by a valid control chart or preliminary experiments to ascertain its performance quality. The expanded uncertainty ($k = 2$) must be less than one-third of the applicable tolerance and balance operating characteristics must be evaluated against this requirement prior to calibration.

        1.2.2.1. If a scale or load cell is used for the calibration that is not a permanent piece of equipment in the calibration laboratory, appropriate verification and repeatability statistics must be obtained prior to a calibration to determine suitability and acceptability for calibration. Records must be maintained of this verification. Minimum verification includes an increasing and decreasing load test to at least the capacity of the weight cart, a shift test, a sensitivity test at the test load, and evaluation of the repeatability based on a minimum of 7 repeated weighings (following the same SOP to be used to test the weight cart).

Figure 19.10 Standard Operating Procedure Page one of a fifteen page SOP from the National Institutes of Standards and Technology describing the official procedure for calibrating weight carts (a device used to test large scales, such as those used for weighing livestock). This SOP uses an outline-style numbering system to list the required steps, starting with steps for balancing the scale. The format is simple and easy to read.

Source: https://www.nist.gov/pml/weights-and-measures/standard-operating-procedures.

Evaluating the Usability of Instructions and Procedures

19.7 Evaluate the usability of instructional documents

What a usable document does

A usable document enables readers to easily locate the information they need, understand this information immediately, and use it safely and effectively (Coe, *Human Factors* 193; Spencer 74). When you write and design any type of workplace or technical document, the end goal is for people to be able to *use* the document successfully. Although usability is an important feature of all documents, it is critical with instructions and procedures because of safety and liability concerns (see "Faulty Instructions and Legal Liability" earlier in this chapter).

Basic usability evaluations

Once you have created a first draft, you can conduct some basic usability evaluations to see how people use your document and to determine whether you need to revise the instructional material. Obtaining this feedback in the early stages enables you to correct errors or problems before the instructions are finalized.

Usability and the User Experience

User experience

Companies are increasingly interested in the overall "user experience" (UX). In many organizations, teams of technical writers, designers, engineers, and marketing specialists work together to ensure that customers have an outstanding experience with the product *and* its documentation. Instructions are key to the user experience: if instructional documents are too hard to understand, don't provide accurate information, contain poorly rendered drawings, or cause frustration, people will often return the entire product and purchase a different brand.

How usability testing works

One way to evaluate the user experience is to conduct some basic usability testing. You can do this by observing how people read, respond to, and work with your document. Begin by identifying the performance objectives—the precise tasks or goals readers must accomplish successfully, or the precise knowledge they must acquire (Carliner, "Physical" 564; Zibell 13). For example, the tasks involved in Figure 19.7 (instructions for replacing a worn faucet washer) are evident by the numbered steps. Readers must successfully complete each of these main steps as well as the tasks within each step. For instance, Step 2, "Disassemble the faucet," contains substeps (open the faucet; remove the screw; remove the handle; remove the packing nut). Your document is considered usable if readers are able to successfully complete these tasks, from start to finish.

Benefits of usability testing

Usability testing on documents that involve regulation (medical devices, for example, which are regulated by the Food and Drug Administration) or in other high-stakes settings is usually conducted by experts with training and experience in this area. Yet all instructional and procedural documents can benefit from basic usability testing. Conducting such testing and revising your document before publication can save your organization time, money, and potential legal challenges.

Basic Usability Survey

1. Briefly describe why this document is used. _____

   _____

2. Evaluate the *content:*
   • Identify any irrelevant information. _____

   • Indicate any gaps in the information. _____

   • Identify any information that seems inaccurate. _____

   • List other problems with the content. _____

3. Evaluate the *organization:*
   • Identify anything that is out of order or hard to locate or follow. ____

   • List other problems with the organization. _____

4. Evaluate the *style:*
   • Identify anything you misunderstood on first reading. _____

   • Identify anything you couldn't understand at all. _____

   • Identify expressions that seem wordy, inexact, or too complex. _____

   • List other problems with the style. _____

5. Evaluate the *design:*
   • Indicate any headings that are missing, confusing, or excessive. _____

   • Indicate any material that should be designed as a list. _____

   • Give examples of material that might be clarified by a visual. _____

   • Give examples of misleading or overly complex visuals. _____

   • List other problems with design. _____

6. Identify anything that seems misleading or that could create legal problems or cross-cultural misunderstanding. _____

7. Please suggest other ways of making this document easier to use. _____

   _____

Figure 19.11 A Basic Usability Survey Versions of these questions can serve as a basis for testing your document.
(*Source:* Based on Carliner, "Demonstrating Effectiveness" 258.)

NOTE In some companies, technical writing and usability jobs are often combined into positions called "UX engineers" or the like. If your college offers a class in usability, see if you can take it. Otherwise, look online for the Usability Professionals Association (UPA) and see if your campus has a student chapter.

Approaches for Evaluating a Document's Usability

First think about the intended use of the document

Using the Audience and Use Profile (Chapter 2, "Develop an Audience and Use Profile"), expand on the "Intended use of document" and "Information needs" items. What are the precise tasks readers need to accomplish using this document? How much time will readers typically have—a few minutes, or several hours? What other factors are key to your understanding of what will make your instructions or procedures usable?

Then analyze your audience and purpose

Once you have answered these questions, check your document using "Checklist: Analyzing Audience and Purpose" in Chapter 2, along with the specific Checklist for the given document (for example, "Checklist: Instructions and Procedures" in this chapter). Revise if necessary. Then, you can use two approaches to test the document's usability. Be sure to run your tests on people who represent the typical audience for the situation. For instance, if your audience for Figure 19.7 is homeowners who have experience using basic tools and doing simple home repairs, do not test your documents with people who have never used a pair of pliers.

Then revise your document

Revise your document based on what you learn from the usability evaluations. If time permits, see if you can set up a second usability test on the revised version, but don't run it on the same people from the first test.

How a think-aloud evaluation works

THINK-ALOUD EVALUATION.   In this approach, you will need three to five people but should test them one at a time. Each person is provided with your instructions and a way he or she can actually test the document. For instance, if your instructions explain how to connect a digital camera to a computer, provide a camera, cable, and laptop. Ask subjects to "think out loud" (talk about what they are doing) as they try to follow your instructions. Note those places where people are successful and where they get stuck. Don't coach them but do remind them to describe their thinking (sometimes when people are concentrating, they will stop talking, and you need to ask them "what are you thinking now?" or something along those lines). After the test is concluded, follow up with questions about places where your document seemed unclear or where people seemed to have particular problems. Take good notes and, at the end of your tests, compare findings to look for common themes.

How focus groups work

FOCUS GROUPS.   In this approach, eight to ten people are provided with the instructions and are asked to complete the task. Based on a targeted list of questions about the document's content, organization, style, and design (Basic Usability Survey, Figure 19.11), focus group members are asked to complete the survey and then describe (out loud) what information they think is missing or excessive, what they like or dislike, and what they find easy or hard to understand. They may also suggest revisions for graphics, format, word choice, platform (is the print copy easier to use than the online version? How about the phone app?), or level of technicality.

Checklist

Instructions and Procedures

Use the following Checklist when writing instructions or procedures.

Content

❏ Does the title promise exactly what the instructions deliver? (See "Clear and Limiting Title" in this chapter.)

❏ Is the background adequate for the intended audience? (See "Appropriate Level of Detail and Technicality" in this chapter.)

❏ Do explanations enable readers to understand what to do? (See "Appropriate Level of Detail and Technicality" in this chapter.)

❏ Do examples enable readers to see how to do it correctly? (See "Appropriate Level of Detail and Technicality" in this chapter.)

❏ Are the definition and purpose of each step given as needed? (See "Appropriate Level of Detail and Technicality" in this chapter.)

❏ Are all obvious steps and needless information omitted? (See "Appropriate Level of Detail and Technicality" in this chapter.)

❏ Do notes, cautions, or warnings appear before or with the step? (See "Notes and Hazard Notices" in this chapter.)

❏ Is the information rate appropriate for the reader's abilities and the difficulty of this procedure? (See "Appropriate Level of Detail and Technicality" in this chapter.)

❏ Are visuals adequate for clarifying the steps? (See "Visuals" in this chapter.)

❏ Do visuals repeat prose information whenever necessary? (See "Visuals" in this chapter.)

❏ Is everything accurate and based on your thorough knowledge? (See "Informed and Accurate Content" in this chapter.)

Organization

❏ Is the introduction adequate without being excessive? (See "Introduction-Body-Conclusion Structure" in this chapter.)

❏ Do the instructions follow the exact sequence of steps? (See "Logically Ordered Steps" in this chapter.)

❏ Is each step numbered, if appropriate? (See "Guidelines for Designing Instructions" in this chapter.)

❏ Is all the information for a particular step close together? (See "Guidelines for Designing Instructions" in this chapter.)

❏ For lengthy instructions, is a layered approach, with a brief reference card, more appropriate? (See "Guidelines for Designing Instructions" in this chapter.)

❏ Is the conclusion necessary and, if necessary, adequate? (See "Introduction-Body-Conclusion Structure" in this chapter.)

Style

❏ Does the familiar material appear *first* in each sentence? (See "Readability" in this chapter.)

❏ Do steps generally have short sentences? (See "Readability" in this chapter.)

❏ Does each step begin with an action verb? (See "Readability" in this chapter.)

❏ Are all steps in the active voice and imperative mood? (See "Readability" in this chapter.)

❑ Do all steps have parallel and affirmative phrasing? (See "Readability" in this chapter.)

❑ Are transitions adequate for marking time and sequence? (See "Readability" in this chapter.)

Page Design

❑ Does each heading clearly tell readers what to expect? (See "Guidelines for Designing Instructions" in this chapter.)

❑ Are steps single-spaced within, and double-spaced between? (See "Guidelines for Designing Instructions" in this chapter.)

❑ Is the overall design simple and accessible? (See "Guidelines for Designing Instructions" in this chapter.)

❑ Are notes, cautions, or warnings set off or highlighted? (See "Guidelines for Designing Instructions" in this chapter.)

❑ Are visuals beside or near the step, and set off by white space? (See "Guidelines for Designing Instructions" in this chapter.)

❑ Will the design work well in print and PDF? What changes need to be made if the document becomes a Web page? (See "Online and Social Media Instructions" in this chapter.)

Projects

For all projects, check with your instructor about whether to present your findings in class, bring drafts to class for discussion, upload your project to the class learning management system (LMS), and/or use the LMS forum or discussion boards to collaborate and review each activity below.

General

1. Improve readability by revising the style and design of these instructions.
    ### What to Do Before Jacking Up Your Car
    Whenever the misfortune of a flat tire occurs, some basic procedures should be followed before the car is jacked up. If possible, your car should be positioned on as firm and level a surface as is available. The engine has to be turned off; the parking brake should be set; and the automatic transmission shift lever must be placed in "park" or the manual transmission lever in "reverse." The wheel diagonally opposite the one to be removed should have a piece of wood placed beneath it to prevent the wheel from rolling. The spare wheel, jack, and lug wrench should be removed from the luggage compartment.

2. Select part of a technical manual in your field or instructions for a general audience and make a copy of the material. Using "Checklist: Instructions and Procedures" in this chapter, evaluate the sample's usability. In a memo to your instructor, discuss the strong and weak points of the instructions. Or explain your evaluation in class.

3. Assume that colleagues or classmates will be serving six months as volunteers in agriculture, education, or a similar capacity in a developing country. Do the research and create a set of procedures that will prepare individuals for avoiding diseases and dealing with medical issues in that specific country. Topics might include safe food and water, insect protection, vaccinations, medical emergencies, and the like. Be sure to provide background on the specific health risks that travelers will face. Design your instructions as a two-sided brief reference card, as a chapter to be included in a longer manual, a PDF page, or in some other format suggested by your instructor.

4. Select any one of the instructional visuals in Figure 19.4 and write a prose version of those instructions—without using visual illustrations or special page design. Bring your version to class and be prepared to discuss the conclusions you've derived from this exercise.

5. Find a set of instructions or some other technical document that is easy to use. Assume that you are Associate Director of Communications for the company that produced this document and you are doing a final review before the document is released. With "Checklist: Instructions and Procedures" as a guide, identify those features that make the document usable and prepare a memo to your boss that justifies your decision to release the document.

Following the identical scenario, find a document that is hard to use, and identify the features that need improving. Prepare a memo to your boss that spells out the needed improvements. Submit both memos and the examples to your instructor.

Team

1. Draw a map of the route from your classroom to your dorm, apartment, or home—whichever is closest. Be sure to include identifying landmarks. When your map is completed, write instructions for a classmate who will try to duplicate your map from the information given in your written instructions. Be sure your classmate does not see your map, and don't let either person use GPS. Exchange your instructions and try to duplicate your classmate's map. Compare your results with the original map. Discuss your conclusions about the usability of these instructions.

2. Divide into small groups and locate fairly brief instructions that could use revision for improved content, organization, style, or format. (Hint: look at the instructions that came with your most recent purchase.) Choose instructions for a procedure you are able to perform. Make a copy of the instructions, test them for usability, and revise as needed. Submit all materials to your instructor, along with a memo explaining the improvements. Or be prepared to discuss your revision in class.

3. Test the usability of a document prepared for this chapter or Chapter 18 by using the think-aloud evaluation (see "Approaches for Evaluating a Document's Usability" in this chapter). Begin by adapting the Audience and Use Profile Sheet (Chapter 2, "Develop an Audience and Use Profile") to include more questions about the intended use of the document. To become familiar with the technique, first practice the think-aloud evaluation on something simple (such as a one page set of instructions on how to get from your classroom to the nearest pizza shop). Then, run the test on your actual document. Take plenty of notes, and when the test is completed, use the Basic Usability Survey (Figure 19.11) to ask follow-up questions. Revise the document based on your observations.

Digital And Social Media

Think of a product you have at home, such as your coffee maker, dishwasher, Wi-Fi router, or DVD player. Imagine that the product is not working and you can't find the original instruction manual. Search online for the user manual for that specific product and model number. Did your search send you to the company's Web site, or did you find the manual on another site? How easy or difficult was it to locate? Is the manual in PDF and easy to download? Can you find the manual for all models or just newer ones? Write a one-page set of instructions to help others locate similar manuals online.

Global

In many cultures, the use of imperative mood is considered impolite or too direct. For instance, instructions that state "Place the disk into the disk drive" may sound bossy and inconsiderate. Find a set of instructions that use imperative mood (for example, Figure 19.7 in this chapter) and rewrite these for a cross-cultural audience where the imperative mood would be offensive. For instance, you might use an indirect imperative ("Be sure to insert the disk into the drive").

Chapter 20
Informal Reports

"In a large corporation like this one, short, informal reports keep vital day-to-day information moving along: Employees or groups report to managers or to one another; managers report to company department heads; department heads report to company officers, who, in turn report to the CEO. These reports address any conceivable topic, based on the need or crisis at the time. Depending on the urgency and importance of the issue, a report might be issued as an email, a PDF attachment, a paper document, or some combination. All such documents serve as a permanent record and are filed for later reference and retrieval, as needed."

—*Elaine Hering, Division Chief for a major auto-parts supplier*

∨ Learning Objectives

20.1 Differentiate between informal and formal reports

20.2 Differentiate between informational and analytical reports

20.3 Plan and write progress, periodic activity, and trip reports, as well as meeting minutes

20.4 Plan and write feasibility, recommendation, justification, and peer review reports

Informal Versus Formal Reports

20.1 Differentiate between informal and formal reports

When you think of reports, you probably picture long, formal documents that contain many pages of research, information, charts and graphs, footnotes, and other details. Yet in the workplace, for every long, formal report that is written, there are countless specific shorter reports created. These *informal reports* help people make decisions on matters as diverse as the most comfortable office chairs to buy or the best recruit to hire for management training. Unlike long formal reports (discussed in Chapter 21), most informal reports require no extended

How informal and formal reports differ

planning, are prepared quickly, contain little or no background information, and have no *front* or *end matter* supplements (title page, table of contents, glossary, works cited, index).

Informal report formats vary depending on the particular organization where you work. Routine, short reports are frequently written as the body of an email message; for example, in some companies, employees are required to submit to their manager a weekly status report (a short form of a periodic activity report) via email. But less frequent, longer reports may be written as memos, which can be printed or attached to email as a PDF document.

Especially when an informal report is from a superior (such as a vice president) or from an official unit within the organization (such as human resources), the report may be in memo format and placed on company letterhead. Usually these reports would then be sent out via email as an attachment. When an informal report is for an external client, you would typically use a letter instead (see Chapter 15). In all cases, make sure your document is easy to navigate: Provide clear headings and use numbered or bulleted lists whenever possible.

Informal report formats vary by organization

When to use letterhead

Informational Versus Analytical Reports

20.2 Differentiate between informational and analytical reports

Informational versus analytical reports

In the professional world, decision makers rely on two types of informal reports. Some reports focus primarily on information ("what we're doing now," "what we did last month," "what our customer survey found," "what went on at the department meeting"). Other reports also include analysis ("what this information means for us," "what courses of action should be considered," "what we recommend, and why").

Types of informational reports

Informational reports answer basic questions such as how much progress has been made on a project (progress reports), what activity transpired during a given period (activity reports), what activity took place during a business trip (trip reports), or what discussions occurred in a meeting (meeting minutes). These reports help keep an organization running from day to day by providing short, timely updates.

Types of analytical reports

Analytical reports offer information as well as interpretations and conclusions based on the information. Analysis is the heart of technical communication. Analysis involves evaluating information, interpreting it accurately, drawing valid conclusions, and making persuasive recommendations. Although gathering and reporting information are essential workplace skills, analysis is ultimately what professionals do to earn their pay. Analytical reports evaluate whether a project or situation is feasible (feasibility reports), make recommendations on how to proceed (recommendation reports), constructively critique the work of others (peer review reports), or justify the writer's position on an issue (justification reports).

Types of Informational Reports

20.3 Plan and write progress, periodic activity, and trip reports, as well as meeting minutes

Following are the major types of informational reports, with strategies for how to plan and write each type.

Progress Reports

Organizations depend on *progress reports* (also called status reports) to monitor progress and problems on various projects. Progress reports may be written either for internal personnel or outside clients. In the case of internal audiences, managers use progress reports to evaluate projects, monitor employees, decide how to allocate funds, and keep track of delays or expense overages that could dramatically affect outcomes and project costs. In the case of external audiences, progress reports explain to clients how time and money are being spent and how difficulties have been overcome. The reports can therefore be used to assure the client that the project will be completed on schedule and within the budget.

> Audience and purpose considerations

Many contracts stipulate the dates and stages when progress will be reported. Failing to report on time may invoke contractual penalties. Some organizations require regular progress reports (daily, weekly, monthly), whereas others only use progress reports as needed, such as when a project milestone has been reached.

Figure 20.1 shows a progress report from a training manager to a company vice president. Because the audience is internal, the writer has chosen a memo format. (A report like this to an external client might also be written in letter format and on company letterhead as a legal record.) Notice also how the report provides only those details essential to the reader.

As you work on a longer report or term project, your instructor may require a progress report. In Figure 20.2, Tina Fitzgerald documents progress made on her term project: a causal analysis examining the impact of a recently formed town committee. Her instructor indicated that the progress report should be written as a memo and sent as an email attachment; for such a short document, an instructor might also tell students to send the report as the body of an email message or as a post to a learning management system discussion forum.

Subject line states the report's purpose

Summarizes first achievement

Summarizes second achievement

Describes work remaining

Bulleted list breaks up dense information

Concludes with request for approval of next phase

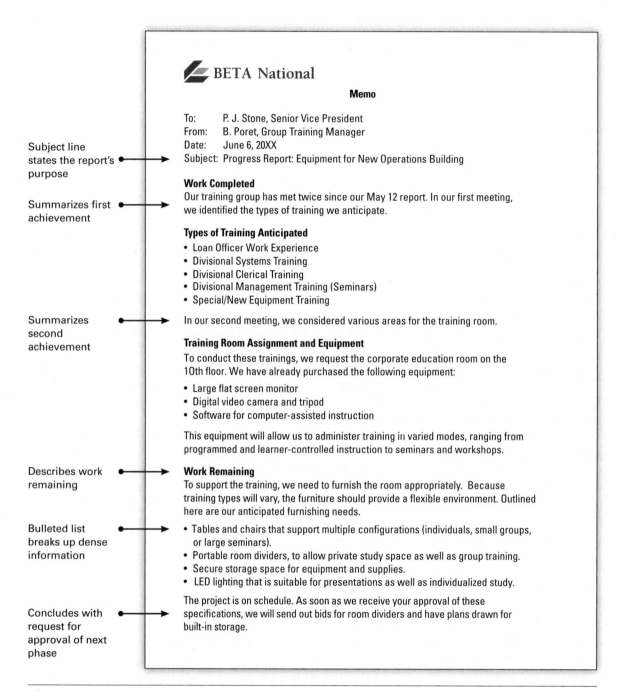

BETA National

Memo

To: P. J. Stone, Senior Vice President
From: B. Poret, Group Training Manager
Date: June 6, 20XX
Subject: Progress Report: Equipment for New Operations Building

Work Completed
Our training group has met twice since our May 12 report. In our first meeting, we identified the types of training we anticipate.

Types of Training Anticipated
- Loan Officer Work Experience
- Divisional Systems Training
- Divisional Clerical Training
- Divisional Management Training (Seminars)
- Special/New Equipment Training

In our second meeting, we considered various areas for the training room.

Training Room Assignment and Equipment
To conduct these trainings, we request the corporate education room on the 10th floor. We have already purchased the following equipment:

- Large flat screen monitor
- Digital video camera and tripod
- Software for computer-assisted instruction

This equipment will allow us to administer training in varied modes, ranging from programmed and learner-controlled instruction to seminars and workshops.

Work Remaining
To support the training, we need to furnish the room appropriately. Because training types will vary, the furniture should provide a flexible environment. Outlined here are our anticipated furnishing needs.

- Tables and chairs that support multiple configurations (individuals, small groups, or large seminars).
- Portable room dividers, to allow private study space as well as group training.
- Secure storage space for equipment and supplies.
- LED lighting that is suitable for presentations as well as individualized study.

The project is on schedule. As soon as we receive your approval of these specifications, we will send out bids for room dividers and have plans drawn for built-in storage.

Figure 20.1 A Progress Report

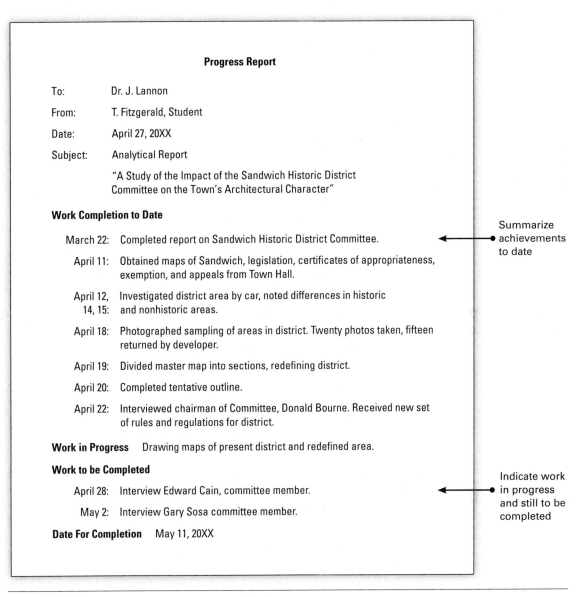

Progress Report

To: Dr. J. Lannon

From: T. Fitzgerald, Student

Date: April 27, 20XX

Subject: Analytical Report

           "A Study of the Impact of the Sandwich Historic District
           Committee on the Town's Architectural Character"

Work Completion to Date

March 22:   Completed report on Sandwich Historic District Committee.   *Summarize achievements to date*

April 11:   Obtained maps of Sandwich, legislation, certificates of appropriateness, exemption, and appeals from Town Hall.

April 12, 14, 15:   Investigated district area by car, noted differences in historic and nonhistoric areas.

April 18:   Photographed sampling of areas in district. Twenty photos taken, fifteen returned by developer.

April 19:   Divided master map into sections, redefining district.

April 20:   Completed tentative outline.

April 22:   Interviewed chairman of Committee, Donald Bourne. Received new set of rules and regulations for district.

Work in Progress   Drawing maps of present district and redefined area.

Work to be Completed

April 28:   Interview Edward Cain, committee member.   *Indicate work in progress and still to be completed*

May 2:   Interview Gary Sosa committee member.

Date For Completion   May 11, 20XX

Figure 20.2 **Progress Report on a Term Project**

Guidelines

for Progress Reports

➤ **Provide a clear subject line**. All progress reports should clearly identify their purpose in the subject line of a memo, email, or letter.

➤ **Present information efficiently**. Because readers of progress reports are managers or clients who want the bottom line information as quickly as possible, chunk the information into logically headed sections and use bulleted or numbered lists. Leave out information readers will already know, but do not omit anything that they might need to know.

➤ **Use a timeline structure to answer the anticipated questions**. First, identify what has been accomplished since the last report. Then, discuss any important details such as outcomes of meetings, problems encountered and solutions implemented, deadlines met or missed, and resources/materials needed. Conclude with steps to be completed in time for the next report, with specific dates, if available.

Periodic Activity Reports

Audience and purpose considerations

Periodic activity reports resemble progress reports in that they summarize activities over a specified period. But unlike progress reports, which summarize specific accomplishments on a particular project, periodic activity reports summarize general activities during a particular period. Periodic activity reports are almost always internal, written by employees to update their supervisors on their activities as a whole, in order to help managers monitor workload and project status. These reports, usually written weekly or monthly, are often called "status reports."

Fran DeWitt's report (Figure 20.3) answers her boss's primary question: *What did you accomplish last month?* Her response has to be detailed and informative.

> NOTE *Both progress reports and periodic activity reports inform management and clients about what employees are doing and how well they are doing it. Therefore, accuracy, clarity, and appropriate detail are essential, as is the ethical dimension.*

Guidelines

for Periodic Activity Reports

➤ **Provide a clear subject line.** Identify the exact purpose and timeframe of the report in a subject line, such as "Monthly activity report for August."

➤ **Present information efficiently.** Report readers just want the bottom line. Omit minor details but include all the essentials. Use headings to chunk information and bulleted and numbered lists to break up dense prose for easier reading.

➤ **Make sure your report answers the expected questions.** Describe your key accomplishments during the period (such as progress on an ongoing project). Then, describe other relevant activity since the last report (such as completion of one-time projects, attendance at meetings, problems encountered, or changes).

■ Mammon Trust

MEMORANDUM
Date: 6/18/XX
To: N. Morgan, Assistant Vice President
From: F. C. DeWitt
Subject: Online Learning Activity Report for May 20XX

◄── ● Subject line announces the topic

Overview
For the past month, I've been working on a cooperative project with the Banking Administration Institute, Online Training Associates (OTA), and several banks. My purpose has been to develop online training programs for bank employees. We focus on three areas: Proof/Encoding Training for entry-level personnel, Productivity Skills for Management, and Banking Principles for Supervisors.

◄── ● Gives overview of recent activities, and their purpose

Meetings Hosted
I hosted two meetings for this task force. On June 6, we discussed Proof/ Encoding Training, and on June 7, Productivity Training. The objective for the Proof/Encoding meeting was to compare ideas, information, and available training packages. We are now designing a course.

◄── ● Gives details

The objective for the Productivity meeting was to explore methods for increasing Banking Operations skills. Discussion included instances in which computer-assisted instruction is appropriate. OTA also described software applications used to teach productivity. Other banks outlined their experiences with similar applications.

Meeting Attended
On June 10, I attended a meeting in Washington, D.C., to design a course in basic banking principles for high-level clerical/supervisory-level employees. We also discussed the feasibility of adapting this course to online learning modules. This type of training, not currently available through Corporate Education, would meet a definite supervisor/management need in the division.

Benefits
My involvement in these meetings has two benefits. First, structured discussions with trainers in the banking industry provide an exchange of ideas, methods, and experiences. This involvement expedites development of our training programs because it saves me time on research. Second, with a working knowledge of these systems and applications, I am able to assist my group in designing programs specific to our needs.

◄── ● Explains the benefits of these activities

Figure 20.3 **Periodic Activity Report**

Trip Reports

Trip reports focus on business-related travel during a given period. Employers who pay travel expenses need to know that the company is getting its money's worth. Employers also need to know what employees learn through their travels.

Figure 20.4 shows a trip report from an employee who has traveled to a branch office to inspect the site and interview the office staff about absenteeism and hiring problems at that branch.

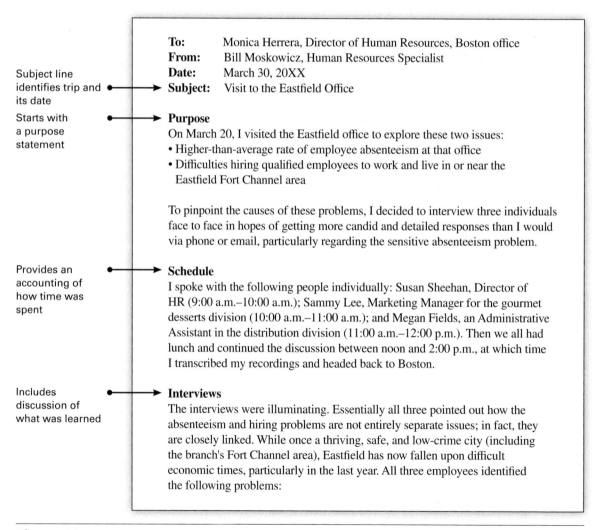

Subject line
identifies trip and
its date

Starts with
a purpose
statement

Provides an
accounting of
how time was
spent

Includes
discussion of
what was learned

To: Monica Herrera, Director of Human Resources, Boston office
From: Bill Moskowicz, Human Resources Specialist
Date: March 30, 20XX
Subject: Visit to the Eastfield Office

Purpose
On March 20, I visited the Eastfield office to explore these two issues:
• Higher-than-average rate of employee absenteeism at that office
• Difficulties hiring qualified employees to work and live in or near the
  Eastfield Fort Channel area

To pinpoint the causes of these problems, I decided to interview three individuals face to face in hopes of getting more candid and detailed responses than I would via phone or email, particularly regarding the sensitive absenteeism problem.

Schedule
I spoke with the following people individually: Susan Sheehan, Director of HR (9:00 a.m.–10:00 a.m.); Sammy Lee, Marketing Manager for the gourmet desserts division (10:00 a.m.–11:00 a.m.); and Megan Fields, an Administrative Assistant in the distribution division (11:00 a.m.–12:00 p.m.). Then we all had lunch and continued the discussion between noon and 2:00 p.m., at which time I transcribed my recordings and headed back to Boston.

Interviews
The interviews were illuminating. Essentially all three pointed out how the absenteeism and hiring problems are not entirely separate issues; in fact, they are closely linked. While once a thriving, safe, and low-crime city (including the branch's Fort Channel area), Eastfield has now fallen upon difficult economic times, particularly in the last year. All three employees identified the following problems:

Figure 20.4 A Trip Report

- Increased gang activity and vagrancy in the Fort Channel area
- Lack of safe public transportation to and from the office or a secure, on-site parking facility
- An increasing decline in amenities, especially lunch venues or safe public areas nearby
- Decreased security resulting from a busy police force and high security guard turnover

Uses bulleted list (and headings) to break up text and identify key points

While Susan and Sammy can both afford to occasionally take cabs and leave the Fort Channel area for lunch, Megan pointed out her limitations due to salary and an insufficient reimbursement program.

The Fort Channel area's problems have all had a direct impact on employee hiring in recent months. Both Susan and Sammy reported losing promising potential employees following interviews. According to Susan, while United Foods manufactures and distributes gourmet products, attracting those who wish to work in an upscale environment, potential employees are often frightened away, preferring employment at either the Boston office or with competitors located elsewhere in the state. At the same time, very few have opted to relocate from the Boston to Eastfield offices to help account for imbalances.

Conclusion

While the goal of this trip was not for me to make recommendations, I have some specific ideas. I would be glad to speak with you further to answer any questions and to share those ideas with you.

Provides an offer to follow up

Figure 20.4 Continued

Guidelines
for Trip Reports

➤ **Choose the right delivery format**. Many organizations have templates or online forms that combine trip reimbursements with trip reports. Even so, managers usually appreciate a more detailed trip report that is written just for them and sent via email.

➤ **Take accurate notes.** If you must interview people, either record the conversation—with their permission—or take careful notes. Transcribe interviews immediately, while your memory is fresh. If you investigate a location or site, take careful notes.

➤ **Begin with a clear subject line and purpose statement.** The subject line identifies the exact trip, including date(s), while the purpose statement prevents any possible confusion about your reasons for the trip.

➤ **Record the names of people and places.** Specify whom you spoke with (spelling names correctly and getting job titles right) and places visited.

➤ **Account for times and locations.** If you visit more than one location, be clear about what occurred at each site.

➤ **Describe findings completely and objectively.** Include all vital information and omit irrelevant material. Do not insert personal impressions—stick to the facts.

➤ **Offer to follow up.** Because trip reports may evoke further questions from a supervisor, indicate that you are available to help provide clarification. Unless you have been specifically asked, do not make recommendations.

Meeting Minutes

Audience and purpose considerations

Many team or project meetings require someone to record the proceedings. *Meeting minutes* are the records of such meetings. Copies of minutes usually are distributed (often via email) to all members and interested parties to track the proceedings and to remind members about their responsibilities. Usually one person is appointed to record the minutes. Often, the person assigned will type minutes on a laptop and email them out as soon as the meeting is completed.

Figure 20.5 shows minutes from a personnel managers' meeting.

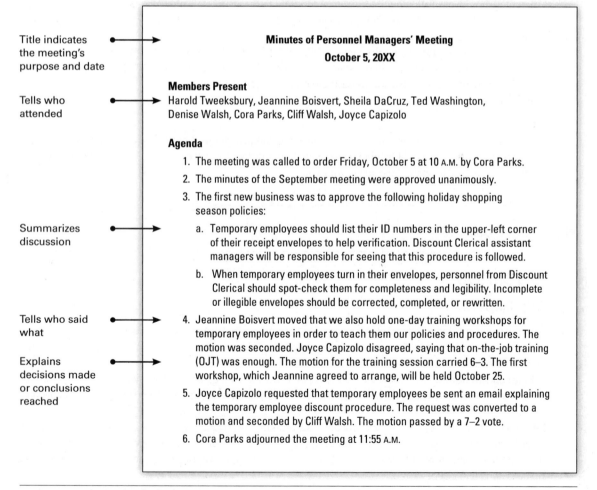

Title indicates the meeting's purpose and date

Minutes of Personnel Managers' Meeting
October 5, 20XX

Members Present

Tells who attended

Harold Tweeksbury, Jeannine Boisvert, Sheila DaCruz, Ted Washington, Denise Walsh, Cora Parks, Cliff Walsh, Joyce Capizolo

Agenda

1. The meeting was called to order Friday, October 5 at 10 A.M. by Cora Parks.
2. The minutes of the September meeting were approved unanimously.
3. The first new business was to approve the following holiday shopping season policies:

Summarizes discussion

   a. Temporary employees should list their ID numbers in the upper-left corner of their receipt envelopes to help verification. Discount Clerical assistant managers will be responsible for seeing that this procedure is followed.

   b. When temporary employees turn in their envelopes, personnel from Discount Clerical should spot-check them for completeness and legibility. Incomplete or illegible envelopes should be corrected, completed, or rewritten.

Tells who said what

Explains decisions made or conclusions reached

4. Jeannine Boisvert moved that we also hold one-day training workshops for temporary employees in order to teach them our policies and procedures. The motion was seconded. Joyce Capizolo disagreed, saying that on-the-job training (OJT) was enough. The motion for the training session carried 6–3. The first workshop, which Jeannine agreed to arrange, will be held October 25.

5. Joyce Capizolo requested that temporary employees be sent an email explaining the temporary employee discount procedure. The request was converted to a motion and seconded by Cliff Walsh. The motion passed by a 7–2 vote.

6. Cora Parks adjourned the meeting at 11:55 A.M.

Figure 20.5 Meeting Minutes

Guidelines

for Meeting Minutes

➤ **Take good notes during the meeting**. Don't rely on your memory, even if you write up the minutes immediately after the meeting. Write down who said what, especially if an important point was raised.

➤ **Complete the minutes immediately after the meeting**. Even if you take good notes, you may forget the context of a particular point if you wait too long.

➤ **Include a clear title and the meeting date**. Indicate the meeting's exact purpose ("sales conference planning meeting"), and include the date to prevent confusion with similar meetings.

➤ **List all attendees**. If the meeting was chaired or moderated, indicate by whom.

➤ **Describe all agenda items**. Make sure all topics discussed are recorded. Your own memory (and the memories of other meeting attendees) of meeting topics is usually fleeting.

➤ **Record all decisions or conclusions**. If everyone agreed on a point, or if a vote was taken, include those outcomes.

➤ **Make the minutes easy to navigate**. Use headings, lists, and other helpful design features.

➤ **Make the minutes precise and clear**. Describe each topic fully yet concisely. Don't omit important nuances, but stick to the facts.

➤ **Keep personal commentary, humor, and "sidebar" comments out of meeting minutes**. Comments ("As usual, Ms. Jones disagreed with the committee") or judgmental expressions ("good," "poor," "irrelevant") are not appropriate.

➤ **Proofread**. Check the spelling of everything but especially, all attendees' names.

➤ **Try to anticipate any unintended consequences**. Consider how you have described any politically sensitive or confidential issues before sending out minutes. Remember, email can travel widely.

➤ **Consider using a blog or other in-house app to post meeting minutes**. You can make meeting minutes easy to access and locate if you use a blog or similar app, saving readers from having to search back through old emails.

Types of Analytical Reports

20.4 **Plan and write feasibility, recommendation, justification, and peer review reports**

Following are the major types of analytical reports, with guidelines for how to plan and write each type.

Feasibility Reports

Feasibility reports help decision makers assess whether an idea, plan, or course of action is realistic and practical. An idea that seems to make perfect sense from one perspective may not be feasible for one reason or another, perhaps, say, because of timing. For example, a manufacturing company may want to switch to a more automated process, which would reduce costs over the long term. But in the short term, the resulting layoffs would have a negative impact on morale.

Audience and purpose considerations

A feasibility report provides answers to questions such as these:

Typical questions about feasibility

- What is the problem or situation, and how should we deal with it?
- Is this course of action likely to succeed?
- Do the benefits outweigh the drawbacks or risks?
- What are the pros and cons, and the alternatives?
- Should anything be done at all? Should we wait? Is the timing right?

The answers are based on the writer's careful research and analysis.

Managers and other decision makers are the primary audience for feasibility reports. These busy people expect the recommendation or answer at or near the beginning of the document, followed by supporting evidence and reasoning. A feasibility report generally uses the "bottom line first" organizing pattern: here is the situation; here is our recommendation; and here is why this approach is feasible at this time.

Figure 20.6 shows a report in which a securities analyst for a state pension fund reports to the fund's manager on the feasibility of investing in a rapidly growing computer manufacturer. Feasibility reports can be short, as in this example, or much longer, with each heading becoming an entire section. When longer, more detailed reports are required for your audience and situation, you would follow the advice and guidelines given in Chapter 21.

Guidelines
for Feasibility Reports

- ➤ **Make the subject line clear**. Always use the word "feasibility" (or a synonym) in the title ("Subject: Feasibility of . . .").
- ➤ **Provide background if needed**. Your readers may not always need background information, but if they do, make it brief.
- ➤ **Offer the recommendation early**. State the "bottom line" explicitly at the beginning of the report, or just after the background information.
- ➤ **Follow up with details, data, and criteria**. Include supporting data, such as costs, equipment needed, and results expected. Include only the directly relevant details that will persuade readers to support your recommendation.
- ➤ **Explain why your recommendation is the most feasible among all the choices**. Your readers may prefer other options. Persuade them that your suggested course of action is the best.
- ➤ **End with a call to action.**

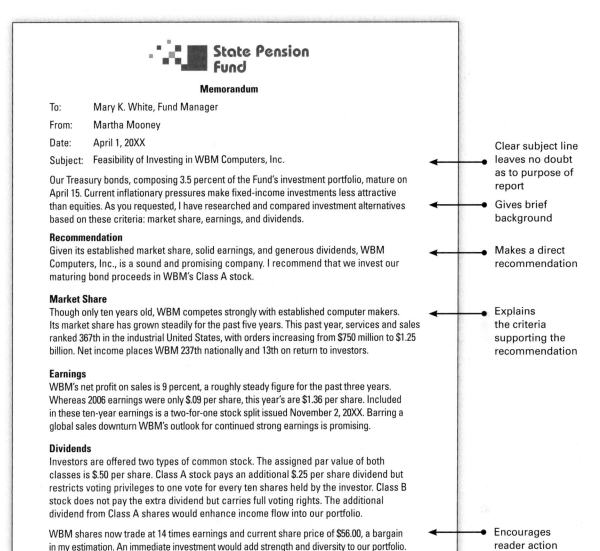

State Pension Fund

Memorandum

To: Mary K. White, Fund Manager

From: Martha Mooney

Date: April 1, 20XX

Subject: Feasibility of Investing in WBM Computers, Inc.   ← Clear subject line leaves no doubt as to purpose of report

Our Treasury bonds, composing 3.5 percent of the Fund's investment portfolio, mature on April 15. Current inflationary pressures make fixed-income investments less attractive than equities. As you requested, I have researched and compared investment alternatives based on these criteria: market share, earnings, and dividends.   ← Gives brief background

Recommendation

Given its established market share, solid earnings, and generous dividends, WBM Computers, Inc., is a sound and promising company. I recommend that we invest our maturing bond proceeds in WBM's Class A stock.   ← Makes a direct recommendation

Market Share

Though only ten years old, WBM competes strongly with established computer makers. Its market share has grown steadily for the past five years. This past year, services and sales ranked 367th in the industrial United States, with orders increasing from $750 million to $1.25 billion. Net income places WBM 237th nationally and 13th on return to investors.   ← Explains the criteria supporting the recommendation

Earnings

WBM's net profit on sales is 9 percent, a roughly steady figure for the past three years. Whereas 2006 earnings were only $.09 per share, this year's are $1.36 per share. Included in these ten-year earnings is a two-for-one stock split issued November 2, 20XX. Barring a global sales downturn WBM's outlook for continued strong earnings is promising.

Dividends

Investors are offered two types of common stock. The assigned par value of both classes is $.50 per share. Class A stock pays an additional $.25 per share dividend but restricts voting privileges to one vote for every ten shares held by the investor. Class B stock does not pay the extra dividend but carries full voting rights. The additional dividend from Class A shares would enhance income flow into our portfolio.

WBM shares now trade at 14 times earnings and current share price of $56.00, a bargain in my estimation. An immediate investment would add strength and diversity to our portfolio.   ← Encourages reader action

Figure 20.6 A Feasibility Report

Recommendation Reports

Audience
and purpose
considerations Whereas a feasibility report sets out to prove that a particular course of action is the right one, a *recommendation report* shortens or even skips the feasibility analysis (since it has already occurred or been discussed) and gets right to the recommendation. Recommendation reports, like feasibility reports, may include supporting data, but they also state an affirmative position ("Here's what we should do and why") rather than examining whether the approach will work ("Should we do it?"). Like feasibility reports, recommendation reports are for the eyes of decision makers, and these documents take a direct stance.

Figure 20.7 shows a recommendation report from a health and safety officer at an airline company to a vice president, regarding workstation comfort of reservation and booking agents. After receiving numerous employee complaints about chronic discomfort, this writer's boss has asked him to study the problems and recommend improvements in the work environment. Like feasibility reports, recommendation reports can be short, as in this example, or much longer. Longer documents would require more research and detail for each section (each heading in Figure 20.7). When longer, more detailed reports are required for your audience and situation, you would follow the advice and guidelines given in Chapter 21.

> NOTE *Before making any recommendation, be sure you've gathered the right information. For the report in Figure 20.7, the writer did enough research to rule out one cause of the problem (computer screens as the cause of employee headaches) before settling on the actual cause (excessive glare on display screens from background lighting).*

Guidelines

for Recommendation Reports

➤ **Provide a clear subject line.** Make sure the subject line announces the purpose of the report.
➤ **Keep the background brief.** But do discuss how feasibility has already been determined.
➤ **Summarize the problem or situation prior to making recommendations.** Outline the issue that the recommended actions will resolve. Then discuss the recommendations in as much detail as necessary.
➤ **Use an authoritative tone.** Take a strong stance and write with confidence, knowing that the recommendation already is considered feasible.
➤ **Use informative headings.** Instead of using the vague heading "Problem," be specific ("Causes of Agents' Discomfort"). Remember that you are writing to a "bottom line" audience.
➤ **End with a list of benefits for taking action.** Rather than appealing for action, assume it will be taken. Reemphasize the benefits.

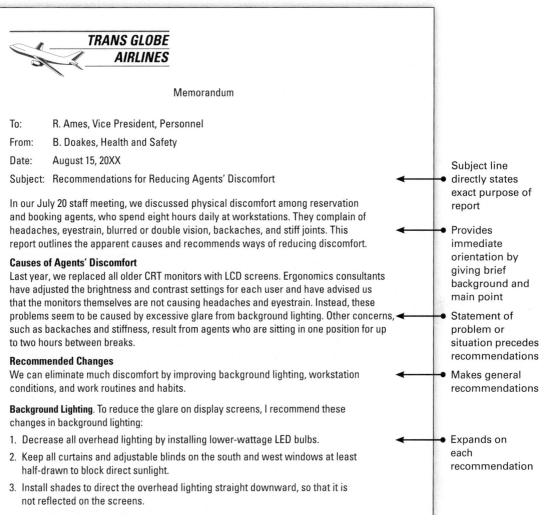

TRANS GLOBE AIRLINES

Memorandum

To: R. Ames, Vice President, Personnel
From: B. Doakes, Health and Safety
Date: August 15, 20XX
Subject: Recommendations for Reducing Agents' Discomfort   ◀————● Subject line directly states exact purpose of report

In our July 20 staff meeting, we discussed physical discomfort among reservation and booking agents, who spend eight hours daily at workstations. They complain of headaches, eyestrain, blurred or double vision, backaches, and stiff joints. This report outlines the apparent causes and recommends ways of reducing discomfort.   ◀————● Provides immediate orientation by giving brief background and main point

Causes of Agents' Discomfort
Last year, we replaced all older CRT monitors with LCD screens. Ergonomics consultants have adjusted the brightness and contrast settings for each user and have advised us that the monitors themselves are not causing headaches and eyestrain. Instead, these problems seem to be caused by excessive glare from background lighting. Other concerns,   ◀————● Statement of problem or situation precedes recommendations
such as backaches and stiffness, result from agents who are sitting in one position for up to two hours between breaks.

Recommended Changes
We can eliminate much discomfort by improving background lighting, workstation   ◀————● Makes general recommendations
conditions, and work routines and habits.

Background Lighting. To reduce the glare on display screens, I recommend these changes in background lighting:

1. Decrease all overhead lighting by installing lower-wattage LED bulbs.   ◀————● Expands on each recommendation

2. Keep all curtains and adjustable blinds on the south and west windows at least half-drawn to block direct sunlight.

3. Install shades to direct the overhead lighting straight downward, so that it is not reflected on the screens.

Workstation Conditions. I recommend the following changes in the workstations:

1. Reposition all screens so light sources are neither at front nor back.

2. Clean the surface of each screen weekly.

3. Adjust each screen so the top is slightly below the operator's eye level.

4. Adjust all keyboards so they are 27 inches from the floor.

5. Replace all fixed chairs with pneumatic, multi-task chairs.

<continued>

Figure 20.7 **A Recommendation Report**

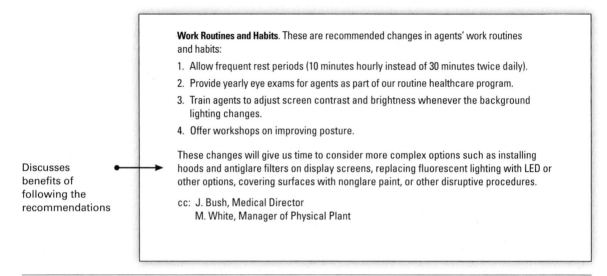

Discusses benefits of following the recommendations

> **Work Routines and Habits.** These are recommended changes in agents' work routines and habits:
>
> 1. Allow frequent rest periods (10 minutes hourly instead of 30 minutes twice daily).
> 2. Provide yearly eye exams for agents as part of our routine healthcare program.
> 3. Train agents to adjust screen contrast and brightness whenever the background lighting changes.
> 4. Offer workshops on improving posture.
>
> These changes will give us time to consider more complex options such as installing hoods and antiglare filters on display screens, replacing fluorescent lighting with LED or other options, covering surfaces with nonglare paint, or other disruptive procedures.
>
> cc: J. Bush, Medical Director
> M. White, Manager of Physical Plant

Figure 20.7 Continued

Justification Reports

Audience and purpose considerations

Many recommendation reports respond to reader requests for a solution to a problem (as in Figure 20.7); others originate with the writer, who has recognized a problem and has devised a solution. This latter type is often called a *justification report*; such reports justify the writer's position by answering this key question for recipients: *Why should we follow your recommendation?*

Unsolicited recommendations—no matter how sound the reasoning behind them—present a complex persuasive challenge and they *always* carry the possibility of inviting a hostile or defensive response from a surprised or offended recipient. ("Who asked for your two cents worth?") Never come across as presumptuous. Give readers notice beforehand; feel them out on the issue. Notice how the writer of the report in Figure 20.8 approaches the subject with confidence but stays professional in tone.

Guidelines

for Justification Reports

➤ **State the problem and your recommended solution.** Unless you expect total resistance to your idea, get to the point quickly and make your case by using some version of the direct organizational plan described in Chapter 15, "Memo Tone."

➤ **Highlight the benefits of your plan before presenting the costs.** An expensive "bottom line" is often an audience deterrent.

➤ **If needed, explain how your plan can be implemented.**

➤ **Conclude by encouraging the reader to act.**

Global Biosolutions, Inc.

MEMORANDUM

To: D. Spring, President

From: M. Marks, Chief, Biology Division

Date: April 18, 20XX

Subject: The Need to Hire Additional Personnel

Introduction and Recommendation

With 56 active employees, GBI has been unable to keep up with scheduled contracts. As a result, we have a contract backlog of roughly $1,225,000. This backlog is caused by understaffing in the biology and chemistry divisions.   ← Opens with the problem

To increase production and ease the workload, I recommend that we hire three general laboratory assistants.   ← Recommends a solution

The lab assistants would be responsible for cleaning glassware and general equipment; feeding and monitoring fish stocks; preparing yeast, algae, and shrimp cultures; preparing stock solutions; and assisting scientists in various procedures.   ← Expands on the recommendation

Benefits and Costs

Three full-time lab assistants would have a positive effect on overall productivity: ← Shows how benefits would offset costs

- Dirty glassware would no longer pile up.

- Because other employees would no longer need to work more than forty hours weekly, morale would improve.

- Research scientists would be freed from general maintenance work.

- With our backlog eliminated, clients would no longer be impatient.

Costs: Initial yearly salaries (at $21.56/hour) for three lab assistants would come to $134,534. Even with health insurance and other benefits factored in, our chief financial officer has assured me that this expenditure would more than offset anticipated revenue growth.

Conclusion

Increased productivity at GBI is essential to maintaining good client relations. These additional personnel would allow us to continue a reputation of prompt and efficient service, thus ensuring our steady growth and development.   ← Encourages acceptance of the recommendation

Could we meet to discuss this request in detail? I will contact you Monday.

Figure 20.8 A Justification Report The tone here is confident yet diplomatic, appropriate for an unsolicited recommendation to an executive. For more on connecting with a reluctant audience, see Chapter 3.

Peer Review Reports

Audience
and purpose
considerations

Peer review reports provide a way for people (peers) to give each other constructive criticism and feedback. Because such reports are shared by colleagues wishing to preserve good workplace relationships, they must be written very tactfully. Figure 20.9 can serve as a model for reviewing the work of other students in the classroom. (For more on peer review, see Chapter 5, "Reviewing and Editing Others' Work" and "Guidelines for Peer Review and Editing.")

Guidelines
for Peer Review Reports

➤ **Start with the positives.** Briefly enumerate the good points of the reviewed document as a way of leading into and balancing against the criticisms.

➤ **Organize by topic area.** Provide separate sections for suggestions about the reviewed document's writing style, design, use of visuals, and other elements.

➤ **Always provide constructive criticism.** Remember that you are reviewing the work of a colleague or colleagues. Review their work tactfully, as you would want yours reviewed. However, don't ignore problem areas.

➤ **Support your critique with examples and advice.** Point to particular examples in the reviewed document as you discuss various elements. Suggest alternatives, if possible. Point to helpful resources, if applicable.

➤ **Close positively.** End with a friendly, encouraging tone. If the reviewed document is particularly problematic, state as much—but diplomatically.

Checklist
Informal Reports

Use the following Checklist as well as the appropriate Guidelines box when writing an informal report.

❏ Have I determined the right report type (status report, recommendation report, other) for this situation? (See "Informational versus Analytical Reports" in this chapter.)

❏ For a progress or status report, have I determined how much detail to include? (See "Progress Reports" in this chapter.)

❏ When taking minutes for a meeting, have I included all the key decisions and discussion items and omitted unnecessary detail? (See "Meeting Minutes" in this chapter.)

❏ If I have decided to write a recommendation or justification report, has the feasibility of the project or idea already been determined, or is a feasibility study and report more appropriate at this time? (See "Feasibility Reports" and "Justification Reports" in this chapter.)

❏ Depending on the audience, have I used a memo, email, or letter format? (See "Informal versus Formal Reports" in this chapter.)

❏ Is the subject line clear? (See all Guidelines boxes in this chapter.)

❏ Is the tone professional and polite but direct? (See all Figures in this chapter.)

❏ Have I done enough research so that my position is credible? (See "Feasibility Reports" in this chapter.)

❏ Have I determined whether to deliver the report as the body of an email message or as a PDF attachment? (See "Informal versus Formal Reports" in this chapter.)

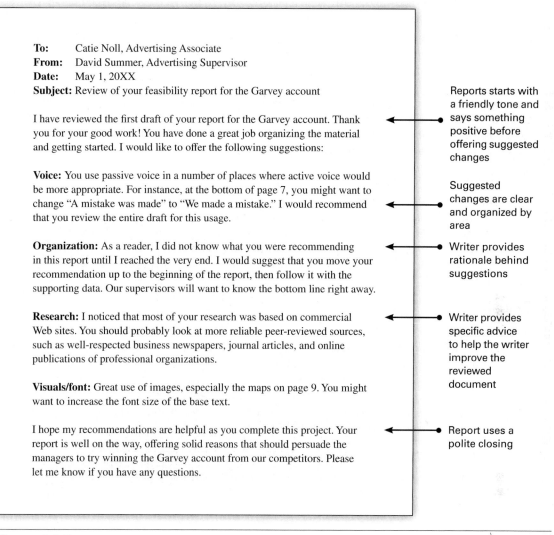

To: Catie Noll, Advertising Associate
From: David Summer, Advertising Supervisor
Date: May 1, 20XX
Subject: Review of your feasibility report for the Garvey account

I have reviewed the first draft of your report for the Garvey account. Thank you for your good work! You have done a great job organizing the material and getting started. I would like to offer the following suggestions:

Voice: You use passive voice in a number of places where active voice would be more appropriate. For instance, at the bottom of page 7, you might want to change "A mistake was made" to "We made a mistake." I would recommend that you review the entire draft for this usage.

Organization: As a reader, I did not know what you were recommending in this report until I reached the very end. I would suggest that you move your recommendation up to the beginning of the report, then follow it with the supporting data. Our supervisors will want to know the bottom line right away.

Research: I noticed that most of your research was based on commercial Web sites. You should probably look at more reliable peer-reviewed sources, such as well-respected business newspapers, journal articles, and online publications of professional organizations.

Visuals/font: Great use of images, especially the maps on page 9. You might want to increase the font size of the base text.

I hope my recommendations are helpful as you complete this project. Your report is well on the way, offering solid reasons that should persuade the managers to try winning the Garvey account from our competitors. Please let me know if you have any questions.

Reports starts with a friendly tone and says something positive before offering suggested changes

Suggested changes are clear and organized by area

Writer provides rationale behind suggestions

Writer provides specific advice to help the writer improve the reviewed document

Report uses a polite closing

Figure 20.9 A Peer Review Report

Projects

For all projects, check with your instructor about whether to present your findings in class, bring drafts to class for discussion, upload your project to the class learning management system (LMS), and/or use the LMS forum or discussion boards to collaborate and review each activity below.

General

1. For this or another class, write a progress report to your instructor, describing the progress you've made on a project (your final project, for instance). Use the guidelines suggested in this chapter. Or, write a peer review report of a classmate's draft of an upcoming assignment.

2. For a current or past job or internship, write a periodic activity report in which you report on the past week's activities (actual or typical). Follow the guidelines in this chapter.

3. Take notes at the next group event or meeting that you attend. Write up minutes for the meeting, thinking about how much detail to include and what kinds of information will be useful to the other members of the team.

Team

With other classmates, identify a dangerous or inconvenient area or situation on campus or in your community (endless cafeteria lines, an unsafe walkway, slippery stairs, a bad campus intersection). Observe the problem for several hours during a peak use period. Write a recommendation report to a specifically identified decision maker (the head of campus security, for example) describing the problem, listing your observations, and making a recommendation.

Digital and Social Media

As the examples in this chapter illustrate, most informal reports are written for internal audiences. Yet with the rise of social media, organizations are making some informal reports available to the public. Look at the Twitter feed and Facebook pages for agencies like NASA, the Department of Agriculture (USDA), or a public health or environmental agency at your state or local level. Find links to one or two short, informal reports. Compare these reports to the examples in this chapter. What are the differences in formatting, tone, and style when these reports are written for external audiences?

Global

Assume you are a team leader of a company based in the United States but with offices worldwide. You have been assigned a new manager who is a citizen of another country. This week you need to write a report recommending that all the staff on your team receive bonuses. You know the tone you would take if your manager were from the United States, but you are uncertain about what tone or level of politeness is appropriate for this new manager. Pick a country and do some research online about tone, style, and politeness in different cultures, remembering that not everyone in a country or culture is exactly alike. Write a short memo to your instructor explaining what you have learned and how you would shape your recommendation report in this situation.

Chapter 21
Formal Analytical Reports

Pressmaster/Shutterstock

❝Our clients make investment decisions based on the feasibility of designing and marketing innovative new technical products or services. Our job is to research consumer interest in these potential products (for example, a new banking app or an updated version of a digital camera). We use online surveys and in-person focus groups to learn as much as possible, then we analyze what customers tell us in relation to what our clients have in mind. I always do a bit more research, usually by studying recent trade publications and credible online review sites, then I write a report that makes accurate, research-based interpretations and recommendations for our clients.❞

—*James North, Senior Project Manager, technical marketing research firm*

Considering Audience and Purpose

Typical Analytical Problems

Elements of an Effective Analysis

Structuring a Formal Report

Front Matter and End Matter Supplements

A Situation Requiring an Analytical Report

Guidelines for Reasoning through an Analytical Problem

Checklist: Analytical Reports

Projects

 ## Learning Objectives

21.1 Understand the role of audience and purpose in formal reports

21.2 Identify three major types of analyses: causal, comparative, and feasibility

21.3 Describe the elements of an effective analysis

21.4 Follow an introduction/body/conclusion structure in a formal report

21.5 Identify the front and end matter elements that accompany a formal report

21.6 Write a formal analytical report

What formal analytical reports do

The formal analytical report, like some of the shorter versions discussed in Chapter 20, usually leads to recommendations. The formal report replaces the memo when the topic requires lengthy discussion. Formal reports generally include a title page, table of contents, a system of headings, a list of references or works cited, and other front-matter and end-matter supplements.

An essential component of workplace problem solving, analytical reports are designed to answer these questions:

What readers of an analytical report want to know

- Based on the information gathered about this issue, what do we know?
- What conclusions can we draw?
- What should we do or not do?

Assume, for example, that you receive this assignment from your supervisor:

A typical analytical problem

Recommend the best method for removing the heavy metal contamination from our company dump site.

First, you need to learn all you can about the problem. Then you have to compare the advantages and disadvantages of various options based on the criteria you are using to assess feasibility: say, cost-effectiveness, time required versus time available for completion, potential risk to the public and the environment. For example, the cheapest option might also pose the greatest environmental risk and could result in heavy fines or criminal charges. But the safest option might simply be too expensive for this struggling company to afford. Or perhaps the Environmental Protection Agency (or similar state and local agencies) has imposed a legal deadline for the cleanup. In making your recommendation, you will need to weigh all the criteria (cost, safety, time) very carefully, or your report could cause more harm than good.

Recommendations have legal and ethical implications

The above situation calls for critical thinking (Chapter 9) and research (Chapters 7–9). Besides interviewing legal and environmental experts, you would need to do a very thorough search of the scientific literature. You would also want to conduct a systematic online search, starting with Google (or another search engine) but then narrowing down your findings to the most factual and updated ones. From these sources, you can discover whether anyone has been able to solve a problem like yours, and you can also learn about the newest technologies for toxic waste cleanup. Then you will have to decide how much, if any, of what others have done applies to your situation. (For more on analytical reasoning, see "Elements of an Effective Analysis" in this chapter.)

Considering Audience and Purpose

21.1 Understand the role of audience and purpose in formal reports

On the job, you may be assigned to evaluate a new assembly technique on the production line or to locate and purchase the best equipment at the best price. You might have to identify the cause behind a monthly drop in sales, the reasons for low employee morale, the causes of an accident, or the reasons for equipment failure. You might need to assess the feasibility of a proposal for a company's expansion or merger or investment. You will present your findings in a formal report.

Using analysis on the job

Because of their major impact on the decision-making process, formal reports are almost always written for an audience of decision makers such as government officials, corporate managers, purchasing officers, and others in such positions. You need to know whom you are writing for and whether the report will be read primarily by an individual, a team, or a series of individuals with differing roles in the company.

Audience considerations

To determine the purpose of the report, consider what question or questions it will ultimately answer. Also, consider why this particular topic is timely and useful to the intended audience. Use the Audience and Use Profile Sheet on (Figure 2.7) to begin mapping out your audience and purpose for the report.

Purpose considerations

Typical Analytical Problems

21.2 Identify three major types of analyses: causal, comparative, and feasibility

Far more than an encyclopedic presentation of information, the analytical report traces your inquiry, your evidence, and your reasoning to show exactly how you arrived at your conclusions and recommendations. Workplace problem solving calls for skills in

Three types of analysis

three broad categories: *causal analysis, comparative analysis,* and *feasibility analysis.* Each approach relies on its own type of reasoning.

Causal Analysis: "Why Does X Happen?"

Designed to reveal a problem at its source, the causal analysis answers questions such as this: *Why do so many apparently healthy people have sudden heart attacks?*

Case

The Reasoning Process in Causal Analysis

Identify the problem

Medical researchers at the world-renowned Hanford Health Institute found that 20 to 30 percent of deaths from sudden heart attacks occur in people who have none of the established risk factors (weight gain, smoking, diabetes, lack of exercise, high blood pressure, or family history).

Examine possible causes

To better identify people at risk, researchers continue to look for new and powerful risk factors such as bacteria, viruses, genes, stress, anger, and depression.

Recommend solutions

Once researchers identify these factors and their mechanisms, they can recommend preventive steps such as careful monitoring, lifestyle and diet changes, drug treatment, or psychotherapy (Lewis 39–43).

A different version of causal analysis employs reasoning from effect to cause, to answer questions such as: *What are the health effects of exposure to electromagnetic radiation?* For more on causal reasoning, see Chapter 8, "Faulty Causal Reasoning."

> NOTE *Keep in mind that faulty causal reasoning is extremely common, especially when we ignore other possible causes or we confuse correlation with causation (see Chapter 8, "Faulty Statistical Analysis").*

Comparative Analysis: "Is X OR Y Better for Our Needs?"

Designed to rate competing items on the basis of specific criteria, the comparative analysis answers questions such as: *Which type of security (firewall/encryption) program should we install on our company's computer system?*

Case

The Reasoning Process in Comparative Analysis

Identify the criteria

XYZ Corporation needs to identify exactly what information (personnel files, financial records) or functions (in-house communication, file transfer) it wants to protect from whom. Does it need both virus and tamper protection? Does it wish to restrict network access or encrypt email and computer files so they become unreadable to unauthorized persons? Does it wish to restrict access to or from the Web? In addition to the level of protection, how important are ease of maintenance and user friendliness?

After identifying their specific criteria, XYZ decision makers need to rank them in order of importance (for example, 1. tamper protection, 2. user-friendliness, 3. secure financial records, and so on).

On the basis of these ranked criteria, XYZ will assess relative strengths and weaknesses of competing security programs and recommend the best one (Schafer 93–94).

Rank the criteria

Compare items and recommend the best one

For more on comparative analysis, see "Guidelines for Reasoning through an Analytical Problem" in this chapter.

Feasibility Analysis: "Is This a Good Idea?"

Designed to assess the practicality of an idea or a plan, the feasibility analysis answers questions such as this: *Should healthy young adults be encouraged to receive genetic testing to measure their susceptibility to various diseases?*

Case

The Reasoning Process in Feasibility Analysis

As a step toward "[translating] genetic research into health care," a National Institutes of Health (NIH) study investigated the feasibility of offering genetic testing to healthy young adults at little or no cost. The testing focused on diseases such as melanoma, diabetes, heart disease, and lung cancer. This study measured the target audience's interest and how those people tested "will interpret and use the results in making their own health care decisions in the future."

Arguments in favor of testing included the following:

➤ Young people with a higher risk for a particular disorder might consider preventive treatments.

➤ Early diagnosis and treatment could be "personalized," tailored to an individual's genetic profile.

➤ Low-risk test results might "inspire healthy people to stay healthy" by taking precautions such as limiting sun exposure or changing their dietary, exercise, and smoking habits.

Consider the strength of supporting reasons

Arguments against testing included the following:

➤ The benefits of early diagnosis and treatment for a small population might not justify the expense of testing the population at large.

➤ False positive results are always traumatic, and false negative results could be disastrous.

➤ Any positive result could potentially subject a currently healthy person to disqualification for disability insurance or life insurance (National Institutes of Health; Notkins 74, 79).

Consider the strength of opposing reasons

After assessing the benefits and drawbacks of testing in this situation, NIH decision makers were able to make the appropriate recommendations.

Weigh the pros and cons, and recommend a course of action

For more on feasibility analysis, see "Guidelines for Reasoning through an Analytical Problem" in this chapter.

Combining Types of Analysis

Analytical categories overlap considerably. Any one study may in fact require answers to two or more of the previous questions. The sample report in Figure 21.3 later in this chapter is both a feasibility analysis and a comparative analysis. It is designed to answer these questions: *Is technical marketing the right career for me? If so, which is my best option for entering the field?*

Elements of an Effective Analysis

21.3 Describe the elements of an effective analysis

The formal analytical report incorporates many elements from documents in earlier chapters, along with the suggestions that follow.

Clearly Identified Problem or Purpose

Define your purpose

To solve any problem or achieve any goal, you must first identify the issues precisely. Always begin by defining the main questions and thinking through any subordinate questions they may imply. Only then can you determine what to look for, where to look, and how much information you will need.

Your employer, for example, might pose this question: Will a low-impact exercise program significantly reduce stress among my employees? The question obviously requires answers to three other questions: What are the therapeutic claims for low-impact exercise? Are these claims valid? Will this kind of exercise work in this situation? With the main questions identified, you can formulate an audience and purpose statement:

Audience and purpose statement

> My goal is to examine and evaluate claims about the therapeutic benefits of low-impact exercise and to recommend a course of action to my employer.

Words such as *examine* and *evaluate* (or *compare, identify, determine, measure, describe,* and so on) help readers understand the specific analytical activity that forms the subject of the report. (For more on asking the right questions, see "Typical Analytical Problems" in this chapter.)

Adequate but Not Excessive Data

Decide how much is enough

A superficial analysis is basically worthless. Worthwhile analysis, in contrast, examines an issue in depth. In reporting on your analysis, however, you filter that material for the audience's understanding, deciding what to include and what to leave out. "Do decision makers in this situation need a closer look or am I presenting excessive detail when only general information is needed?" Is it possible to have too much information? In some cases, yes—as behavioral expert Dietrich Dörner explains:

> The more we know, the more clearly we realize what we don't know. This probably explains why . . . organizations tend to [separate] their information-gathering and decision-making branches. A business executive has an office manager; presidents have . . . advisers; military commanders have chiefs of staff. The point of this separation may well be to provide decision makers with only the bare outlines of all the available information so they will not be hobbled by excessive detail when they are obliged to render decisions. Anyone who is fully informed will see much more than the bare outlines and will therefore find it extremely difficult to reach a clear decision. (99)

Excessive information hampers decision making

Confusing the issue with excessive information is no better than recommending hasty action on the basis of inadequate information (Dörner 104).

As you research the issue, you may also want to filter material for your own understanding. Your decision about whether to rely on the abstract or summary or to read the complete text of a specialized article or report depends on the question you're trying to answer and the level of technical detail your readers expect. If you are an expert in the field, writing for other experts, you probably want to read the entire document in order to assess the methods and reasoning behind a given study. But if you are less than expert, a summary or abstract of this study's findings might suffice. The fact sheet in Figure 21.1, for example, summarizes a detailed feasibility study for a general reading audience. Readers seeking more details, including nonclassified elements of the complete report, could visit the Transportation Security Administration's Web site.

When you might consult an abstract or summary instead of the complete work

> NOTE Online, we tend to read quickly, often reading only the abstract or summary in our haste to get to the next item in our search. Be sure to consider whether you need to read more deeply, into the body of the report. If you have relied only on the abstract or summary instead of the full article, be sure to indicate this point ("Abstract," "Press Release," or the like) when you cite the source in your report (as shown in Appendix A, "MLA Works Cited Entries" and "APA Reference List Entries").

Accurate and Balanced Data

Avoid stacking the evidence to support a preconceived point of view. Assume, for example, that you are asked to recommend the best chainsaw brand for a logging company. Reviewing test reports, you come across this information:

Give readers all they need to make an informed judgment

> Of all six brands tested, the Bomarc chainsaw proved easiest to operate. However, this brand also offers the fewest safety features.

In citing these equivocal findings, you need to present both of them accurately, and not simply the first—even though the Bomarc brand may be your favorite. Then argue for the feature (ease of use or safety) you think should receive priority. (Refer to Chapter 7, "Exploring a Balance of Views" and Chapter 8, "Evaluate the Evidence" for more on exploring and presenting balanced and reasonable evidence.)

FACT SHEET: Train and Rail Inspection Pilot, Phase I

U.S. DEPARTMENT OF HOMELAND SECURITY
Transportation Security Administration
FOR IMMEDIATE RELEASE – June 7, 2014
TSA Press Office: (571) 227-2829

Objective:
Implement a pilot program to determine the feasibility of screening passengers, luggage, and carry-on bags for explosives in the rail environment.

TRIP I Background:
- Homeland Security Secretary announced TRIP on March 22, 2014, to test new technologies and screening concepts.
- The program is conducted in partnership with the Department of Transportation, Amtrak, Maryland Rail Commuter, and Washington, D.C.'s Metro.
- The New Carrollton, MD, station was selected because it serves multiple types of rail operations and is located close to Washington, D.C.

TRIP I Facts:
- Screening for Phase I of TRIP began on May 4 and was completed on May 26, 2014.
- A total of 8,835 passengers and 9,875 pieces of baggage were screened during the test.
- The average time to wait in line and move through the screening process was less than 2 minutes.
- Customer Feedback cards reflect a 93 percent satisfaction rate with both the screening process and the professional demeanor of TSA personnel.

Lessons Learned:
- Results indicate efficient checkpoints throughout with minimal customer inconvenience.
- Passengers were overwhelmingly receptive to the screening process.
- Providing a customer service representative on-site during all screening operations helped Amtrak ensure passengers received outstanding customer service.
- Skilled TSA screeners from the agency's National Screening Force were able to quickly transition to screening in the rail environment.
- Most importantly, Phase I showed that currently available technology could be utilized to screen for explosives in the rail environment.

Figure 21.1 A Summary Description of a Feasibility Study
Notice that the criteria for assessing feasibility include passenger wait times, passenger receptiveness to screening, and—most important—effectiveness of screening equipment in this environment.
Source: Transportation Security Administration.

Fully Interpreted Data

Interpretation shows the audience "what is important and what is unimportant, what belongs together and what does not" (Dörner 44). For example, you might interpret the above chainsaw data in this way:

> Our logging crews often work suspended by harness, high above the ground. Also, much work is in remote areas. Safety features therefore should be our first requirement in a chainsaw. Despite its ease of operation, the Bomarc saw does not meet our safety needs.

By saying "therefore," you engage in analysis—not just information sharing. Don't merely list your findings—explain what they mean.

Explain the significance of your data

Explain the meaning of your evidence

Subordination of Personal Bias

To arrive at the *truth* of the matter (see Chapter 8, "Identify Your Level of Certainty"), you need to see clearly. Don't let your biases fog up the real picture. Each stage of analysis requires decisions about what to record, what to exclude, and where to go next. You must evaluate your data (Is this reliable and important?), interpret your evidence (What does it mean?), and make recommendations (What action is needed?). An ethically sound analysis presents a balanced and reasonable assessment of the evidence. Do not force viewpoints that are not supported by dependable evidence. Make sure your evidence is from credible, reliable sources, especially if such evidence is from online sources that might appear credible but are actually quite biased against scientific information, professional interpretations, and the sound judgement that goes along with peer review, editorial oversight, and sound reasoning. (Refer to Chapter 7, "Evaluating and Interpreting Your Findings" and Chapter 8, "Interpret Your Findings".)

Evaluate and interpret evidence impartially

Appropriate Visuals

Provide as much visual information as appropriate to make complex statistics and numeric data easy for readers to understand. Graphs are especially useful for analyzing trends (rising or falling sales, employment trends). Tables and charts are helpful for comparing data. Illustrations, diagrams, and photographs are excellent ways to show a component or special feature. Place the visual close to its accompanying text; be sure your text explains the visual in the context of the larger analysis. (See Chapter 12 for more information on using visuals.)

Use visuals appropriately

NOTE *The simplicity of a visual such as the one used by writer Richard Larkin, on page 2 of his report (within Figure 21.3 in this chapter), illustrates that a visual does not need to be overly fancy or complex, nor its story long and involved. Sometimes, less is more.*

Valid Conclusions and Recommendations

Along with the informative abstract (see Chapter 9, "Special Types of Summaries"), conclusions and recommendations are the sections of a long report that receive the most

Be clear about what the audience should think and do

audience attention. The goal of analysis is to reach a valid conclusion—an overall judgment about what all the material means (that *X* is better than *Y*, that *B* failed because of *C*, that *A* is a good plan of action). The following example shows the conclusion of a report on the feasibility of installing an active solar heating system in a large building.

Offer a final
judgment

1. Active solar space heating for our new research building is technically feasible because the site orientation will allow for a sloping roof facing due south, with plenty of unshaded space.
2. It is legally feasible because we are able to obtain an access easement on the adjoining property to ensure that no buildings or trees will be permitted to shade the solar collectors once they are installed.
3. It is economically feasible because our sunny, cold climate means high fuel savings and faster payback (15 years maximum) with solar heating. The long-term fuel savings justify our short-term installation costs (already minimal because the solar system can be incorporated during the building's construction—without renovations).

Conclusions are valid when they are logically derived from accurate interpretations.

Having explained *what it all means*, you then recommend *what should be done*. Taking all possible alternatives into account, your recommendations urge the most feasible option (to invest in *A* instead of *B*, to replace *C* immediately, to follow plan *A*, or the like). Here are the recommendations based on the previous conclusions:

Tell what should
be done

1. I recommend that we install an active solar heating system in our new research building.
2. We should arrange an immediate meeting with our architect, building contractor, and solar heating contractor. In this way, we can make all necessary design changes before construction begins in two weeks.
3. We should instruct our legal department to obtain the appropriate permits and easements immediately.

Recommendations are valid when they propose an appropriate response to the problem or question and when such response is based on unbiased and careful research and analysis.

Because they culminate your research and analysis, recommendations challenge your imagination, your creativity, and—above all—your critical thinking skills. What strikes one person as a brilliant suggestion might be seen by others as irresponsible, offensive, or dangerous. (Figure 21.2 depicts the kinds of decisions writers encounter in formulating, evaluating, and refining their recommendations.)

> NOTE Keep in mind that solving one problem might create new and worse problems—or unintended consequences. For example, to prevent crop damage by rodents, a pest control specialist might recommend trapping and poisoning. While rodent eradication may increase crop yield temporarily, it also increases the insects these rodents feed on—leading eventually to even greater crop damage. In short, before settling on any recommendation, try to anticipate its "side effects and long-term repercussions" (Dörner 15).

Use a confident
tone to state your
conclusions

When you do achieve definite conclusions and recommendations, express them with assurance and authority. Unless you have reason to be unsure, avoid noncommittal statements ("It would seem that" or "It looks as if"). Be direct and assertive

Consider All the Details

- What exactly should be done—if anything at all?
- How exactly should it be done?
- When should it begin and be completed?
- Who will do it, and how willing are they?
- What equipment, materials, or resources are needed?
- Are any special conditions required?
- What will this cost, and where will the money come from?
- What consequences are possible?
- Whom do I have to persuade?
- How should I order my list (priority, urgency, etc.)?

Locate the Weak Spots

- Is anything unclear or hard to follow?
- Is this course of action unrealistic?
- Is it risky or dangerous?
- Is it too complicated or confusing?
- Is anything about it illegal or unethical?
- Will it cost too much?
- Will it take too long?
- Could anything go wrong?
- Who might object or be offended?
- What objections might be raised?

Make Improvements

- Can I rephrase anything?
- Can I change anything?
- Should I consider alternatives?
- Should I reorder my list?
- Can I overcome objections?
- Should I get advice or feedback before I submit this?

Figure 21.2 How to Think Critically about Your Recommendations

Source: Ruggiero, Vincent R. *The Art of Thinking: A Guide to Critical and Creative Thought,* 8th Ed., (c) 2007, 188–89.
Adapted and electronically reproduced by permission of Pearson Education, Inc., Upper Saddle River, New Jersey.

("The earthquake danger at the reactor site is acute," or "I recommend an immediate investment"). Announce where you stand.

If, however, your analysis yields nothing definite, do not force a simplistic conclusion on your material. Instead, explain the limitations ("The contradictory responses to our consumer survey prevent me from reaching a definite conclusion. Before we make any decision about this product, I recommend a full-scale market analysis"). The wrong recommendation is far worse than no recommendation at all. (Chapter 8 offers helpful guidelines for evaluating and interpreting information.)

Self-Assessment

Assess your analysis continuously

Things that might go wrong with your analysis

The more we are involved in a project, the larger our stake is in its outcome—making self-criticism less likely just when it is needed most! For example, it is hard to admit that we might need to backtrack, or even start over, in instances like these (Dörner 46):

- During research you find that your goal isn't clear enough to indicate exactly what information you need.
- As you review your findings, you discover that the information you have is not the information you need.
- After making a recommendation, you discover that what seemed like the right course of action turns out to be the wrong one.

If you meet such obstacles, acknowledge them immediately, and revise your approach as needed.

Structuring a Formal Report

21.4 Follow an introduction/body/conclusion structure in a formal report

Whether you outline earlier or later, the finished report depends on a good outline. This model outline can be adapted to most analytical reports.

I. Introduction
    A. Definition, Description, and Background
    B. Purpose of the Report, and Intended Audience
    C. Method of Inquiry
    D. Limitations of the Study
    E. Working Definitions (here or in a glossary)
    F. Scope of the Inquiry (topics listed in logical order)
    G. Conclusion(s) of the Inquiry (briefly stated)
II. Collected Data
    A. First Topic for Investigation
        1. Definition
        2. Findings
        3. Interpretation of findings

   B. Second Topic for Investigation
       1. First subtopic
           a. Definition
           b. Findings
           c. Interpretation of findings
       2. Second subtopic (and so on)
III. Conclusion
   A. Summary of Findings
   B. Overall Interpretation of Findings (as needed)
   C. Recommendations (as needed and feasible)

(This outline is only tentative. Modify the components as necessary.)

Two sample reports in this chapter follow the model outline. The first one, "Children Exposed to Electromagnetic Radiation: A Risk Assessment" (minus the front matter and end matter that ordinarily accompany a long report), begins below. The second report, "Feasibility Analysis of a Career in Technical Marketing," appears in Figure 21.3.

Each report responds to slightly different questions. The first tackles these questions: *What are the effects of* X *and what should we do about them?* The second tackles two questions: *Is* X *feasible, and which version of* X *is better for my purposes?* At least one of these reports should serve as a model for your own analysis.

Introduction

The introduction engages and orients the audience and provides background as briefly as possible for the situation. Often, writers are tempted to write long introductions because they have a lot of background knowledge. But readers generally don't need long history lessons on the subject.

Identify your topic's origin and significance, define or describe the problem or issue, and explain the report's purpose. (Generally, stipulate your audience only in the version your instructor will read and only if you don't attach an audience and use profile.) Briefly identify your research methods (interviews, literature searches, and so on) and explain any limitations or omissions (person unavailable for interview, research still in progress, and so on). List working definitions, but if you have more than two or three, use a glossary. List the topics you have researched. Finally, briefly preview your conclusion; don't make readers wade through the entire report to find out what you recommend or advise.

> NOTE *Not all reports require every component. Give readers only what they need and expect.*

As you read the following introduction, think about the elements designed to engage and orient the audience (i.e., local citizens), and evaluate their effectiveness. (Review the Case Study in Chapter 7, "Asking the Right Questions" for the situation that gave rise to this report.)

CHILDREN EXPOSED TO ELECTROMAGNETIC RADIATION: A RISK ASSESSMENT

LAURIE A. SIMONEAU

INTRODUCTION

Definition and background of the problem

Wherever electricity flows—through the largest transmission line or the smallest appliance—it emits varying intensities of charged waves: an *electromagnetic field* (EMF). Some medical studies have linked human exposure to EMFs with definite physiologic changes and possible illness including cancer, miscarriage, and depression.

Experts disagree over the health risk, if any, from EMFs. Some question whether EMF risk is greater from high-voltage transmission lines, the smaller distribution lines strung on utility poles, or household appliances. Conclusive research may take years; meanwhile, concerned citizens worry about avoiding potential risks.

Description of the problem

In Bocaville, four sets of transmission lines—two at 115 Kilovolts (kV) and two at 500 kV—cross residential neighborhoods and public property. The Adams elementary school is less than 100 feet from this power line corridor. EMF risks—whatever they may be—are thought to increase with proximity.

Purpose and methods of this inquiry

Based on a review of key research as well as interviews with local authorities, this report assesses whether potential health risks from EMFs seem significant enough for Bocaville to (a) increase public awareness, (b) divert the transmission lines that run adjacent to the elementary school, and (c) implement widespread precautions in the transmission and distribution of electrical power throughout Bocaville.

Scope of this inquiry

This report covers five major topics: what we know about various EMF sources, what research indicates about physiologic and health effects, how experts differ in evaluating the research, what the power industry and the public have to say, and what actions are being taken locally and nationwide to avoid risk.

Conclusions of the inquiry (briefly stated)

The report concludes by acknowledging the ongoing conflict among EMF research findings and by recommending immediate and inexpensive precautionary steps for our community.

Body

The body section (or data section) describes and explains your findings. Present a clear and detailed picture of the evidence, interpretations, and reasoning on which you will base your conclusion. Divide topics into subtopics and use informative headings as aids to navigation.

> *NOTE Remember your ethical responsibility for presenting a fair and balanced treatment of the material, instead of "loading" the report with only those findings that support your viewpoint. Also, keep in mind the body section can have many variations, depending on the audience, topic, purpose, and situation.*

As you read the following section, evaluate how effectively it informs readers, keeps them on track, reveals a clear line of reasoning, and presents an impartial analysis.

DATA SECTION

Sources of EMF Exposure

Electromagnetic intensity is measured in *milligauss* (mG), a unit of electrical measurement. The higher the mG reading, the stronger the field. Studies suggest that consistent exposure above 1–2 mG may increase cancer risk significantly, but no scientific evidence concludes that exposure even below 2.5 mG is safe.

Definition

Table 1 gives the EMF intensities from electric power lines at varying distances during average and peak usage.

Findings

Table 1 EMF Emissions from Power Lines (in milligauss)

Types of Transmission Lines	Maximum on Right-of-Way	Distance from lines			
		50'	100'	200'	300'
115 Kilovolts (kV)					
Average usage	30	7	2	0.4	0.2
Peak usage	63	14	4	0.9	0.4
230 Kilovolts (kV)					
Average usage	58	20	7	1.8	0.8
Peak usage	118	40	15	3.6	1.6
500 Kilovolts (kV)					
Average usage	87	29	13	3.2	1.4
Peak usage	183	62	27	6.7	3.0

Source: United States, Environmental Protection Agency, Office of Radiation and Indoor Air. *EMF in Your Environment.* Government Printing Office, 1992, p. 24.

As Table 1 indicates, EMF intensity drops substantially as distance from the power lines increases.

Interpretation

Although the EMF controversy has focused on 2 million miles of power lines criss-crossing the country, potentially harmful waves are also emitted by household wiring, appliances, computer terminals—and even from the earth's natural magnetic field. The background magnetic field (at a safe distance from any electrical appliance) in the average American home varies from 0.5 to 4.0 mG (United States, Environmental Protection Agency, Office of Radiation and Indoor Air 10). Table 2 compares intensities of various sources.

EMF intensity from certain appliances tends to be higher than from transmission lines because of the amount of current involved.

Interpretation

Voltage measures the speed and pressure of electricity in wires, but *current* measures the volume of electricity passing through wires. Current (measured in *amperage*) is what produces electromagnetic fields. The current flowing through a transmission line typically ranges from 200 to 400 amps. Most homes have a 200-amp service, which means that if every electrical item in the house were turned on at the same time, the house could run about 200 amps—almost as high as the transmission line. Consumers then have the ability to put 200 amps of current-flow into their homes, while transmission lines carrying 200 to 400 amps are at least 50 feet away (Miltane).

Definitions

Finding

Table 2 EMF Emissions from Selected Sources (in milligauss)

Source	Range[a,b]
Earth's magnetic field	0.1–2.5
Blowdryer	60–1,400
Four in. from TV screen	40–100
Four ft. from TV screen	0.7–9
Fluorescent lights	10–12
Electric razor	1,200–1,600
Electric blanket	2–25
Computer terminal (12 inches away)	3–15
Toaster	10–60

[a]Data from Miltane, John. Personal Interview. 5 Apr. 2013.; United States, National Institutes of Health, National Institute of Environmental Health Sciences. *EMF Electric and Magnetic Fields Associated with the Use of Electric Power.* Government Printing Office, 2002, pp. 32–35.

[b]Readings are made with a gaussmeter, and vary with technique, proximity of gaussmeter to source, its direction of aim, and other random factors.

Proximity and duration of exposure, however, are other risk factors. People are exposed to EMFs from home appliances at close proximity, but appliances run only periodically: exposure is therefore sporadic, and intensity diminishes sharply within a few feet (Figure 1).

As Figure 1 indicates, EMF intensity drops dramatically over very short distances from the typical appliance.

Finding

Power line exposure, on the other hand, is at a greater distance (usually 50 feet or more), but it is constant. Moreover, its intensity can remain strong well beyond 100 feet (Miltane).

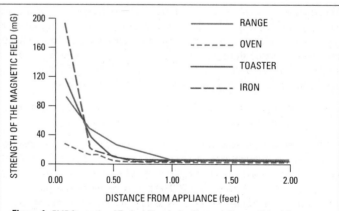

Figure 1 EMF Strengths of Typical Electric Appliances *Source:* United States, Environmental Protection Agency, Office of Radiation and Indoor Air. *EMF In Your Environment.* Government Printing Office, 1992, p. 11.

Research has yet to determine which type of exposure might be more harmful: briefly, to higher intensities or constantly, to lower intensities. In any case, proximity seems most significant because EMF intensity drops rapidly with distance.

Interpretation

Physiologic Effects and Health Risks from EMF Exposure

Research on EMF exposure falls into two categories: epidemiologic studies and laboratory studies. The findings are sometimes controversial and inconclusive, but also disturbing.

Second major
topic

Epidemiologic Studies. Epidemiologic studies look for statistical correlations between EMF exposure and human illness or disorders. Of 77 such studies in recent decades, over 70 percent suggest that EMF exposure increases the incidence of the following conditions (Pinsky 155–215):

First subtopic

Definition

- cancer, especially leukemia and brain tumors
- miscarriage
- stress and depression
- learning disabilities
- heart attacks

General findings

Following, for example, are summaries of several noted epidemiologic studies implicating EMFs in occurrences of cancer.

A Landmark Study of the EMF/Cancer Connection. A 1979 Denver study by Wertheimer and Leeper was the first to implicate EMFs as a cause of cancer. Researchers compared hundreds of homes in which children had developed cancer with similar homes in which children were cancer free. Victims were two to three times as likely to live in "high-current homes" (within 130 feet of a transmission line or 50 feet of a distribution line).

Detailed findings

This study has been criticized because (1) it was not "blind" (researchers knew which homes cancer victims were living in), and (2) researchers never took gaussmeter readings to verify their designation of "high-current" homes (Pinsky 160–62; Taubes 96).

Critiques of
findings

Follow-up Studies. Since then, several major studies of the EMF/cancer connection have confirmed Wertheimer's findings:

Detailed findings

- In 1988, Savitz studied hundreds of Denver houses and found that children with cancer were 1.7 times as likely to live in high-current homes. Unlike his predecessors, Savitz did not know whether a cancer victim lived in the home being measured, and he took gaussmeter readings to verify that houses could be designated "high-current" (Pinsky 162–63).
- In 1990, London and Peters found that Los Angeles children had 2.5 times more risk of leukemia if they lived near power lines (Brodeur 115).
- In 1992, a massive Swedish study found that children in houses with average intensities greater than 1 mG had twice the normal leukemia risk; at greater than 2 mG, the risk nearly tripled; at greater than 3 mG, it nearly quadrupled (Brodeur 115).
- In 2002, British researchers evaluated findings from 34 studies of power line EMF effects (a *meta-analysis*). This study found "a degree of consistency in the evidence suggesting adverse health effects of living near high voltage power lines" (Henshaw et al. 1).

Workplace Studies. More than 80 percent of 51 studies over a 13 year period—most notably a landmark 1992 Swedish study—concluded that electricians, electrical engineers, and power line workers constantly exposed to an average of 1.5 to 4.0 mG had a significantly elevated cancer risk (Brodeur 115; Pinsky 177–209).

Two additional workplace studies seem to support or even amplify the above findings.

- A University of North Carolina study of 138,905 electric utility workers concluded that occupational EMF exposure roughly doubles brain cancer risk. This study, however, found no increased leukemia risk (Cavanaugh 8; Moore 16).
- A Canadian study of electrical-power employees indicates that those who had worked in strong electric fields for more than 20 years had "an eight- to tenfold increase in the risk of leukemia," along with a significantly elevated risk of lymphoma ("Strong Electric Fields" 1–2).

Interpretation	Although none of the above studies can be said to "prove" a direct cause-effect relationship, their strikingly similar results suggest a conceivable link between prolonged EMF exposure and illness.
Second subtopic	**Laboratory Studies.** Laboratory studies assess cellular, metabolic, and behavioral effects of EMFs on humans and animals. EMFs directly cause the following physiologic changes (Brodeur 88; Pinsky 24–29; Raloff, "EMFs'" 30):
General findings	• reduced heart rate • altered brain waves • impaired immune system • interference with the synthesis of genetic material • disrupted regulation of cell growth • interaction with the biochemistry of cancer cells • altered hormonal activity • disrupted sleep patterns
Detailed findings	These changes are documented in the following summaries of several significant laboratory studies. **EMF Effects on Cell Chemistry.** Other studies have demonstrated previously unrecognized effects on cell growth and division. Most notably, a study by Michigan State University found that EMFs equal to the intensity that occurs "within a few feet" of outdoor power lines caused cells with cancer-related genetic mutations to multiply rapidly (Sivitz 196). **EMF Effects on Hormones.** Studies have found that EMF exposure (for example, from an electric blanket) inhibits production of melatonin, a hormone that fights cancer and depression, stimulates the immune system, and regulates bodily rhythms. A study at the Lawrence National Laboratory found that EMF exposure can suppress both melatonin and the hormone-like, anticancer drug Tamoxifen (Raloff, "EMFs'" 30). Similarly, physiologist Charles Graham found that EMFs elevate female estrogen levels and depress male testosterone levels—hormone alterations associated with risk of breast or testicular cancer, respectively (Raloff, "EMFs'" 30).
Interpretation	Although laboratory studies seem more conclusive than the epidemiologic studies, what these findings *mean* is debatable.
	Debate over Quality, Cost, and Status of EMF Research
Third major topic	Experts differ over the meaning of EMF research findings largely because of the following limitations attributed to various studies.

Limitations of Various EMF Studies. Epidemiologic studies are criticized for overstating evidence. For example, some critics claim that so-called EMF cancer links are produced by "data dredging" (making countless comparisons between cancers and EMF sources until random correlations appear) (Taubes 99). Other critics argue that news media distort the issue by publicizing positive findings while often ignoring negative or ambiguous findings (Goodman). Some studies are also accused of mistaking *coincidence* for *correlation,* without exploring "confounding factors" (e.g., exposure to toxins or to other adverse conditions—including the earth's natural magnetic field) (Moore 16).

First subtopic

Critiques of population studies

Supporters of EMF research respond that the sheer volume of epidemiologic evidence seems overwhelming (Kirkpatrick 81, 83). Moreover, the Swedish studies cited earlier seem to invalidate the above criticisms (Brodeur 115).

Response to critiques

Laboratory studies are criticized—even by scientists who conduct them—because effects on an isolated culture of cells or on experimental animals do not always equal effects on the total human organism (Jauchem 190–94).

Critiques of lab studies

Until recently, critics argued that no scientist had offered a reasonable hypothesis to explain the possible health effects of EMFs (Palfreman 26). However, a University of Washington study showed that a weak electromagnetic field can break DNA strands and lead to brain cell death in rats, presumably because of cell-damaging agents known as free radicals (Lai and Singh).

Response to critiques

Costs of EMF Research. Critics claim that research and publicity about EMFs are becoming a profit venture, spawning "a new growth industry among researchers as well as marketers of EMF monitors" ("Electrophobia" I). Environmental expert Keith Florig identifies adverse economic effects of the EMF debate that include decreased property values, frivolous lawsuits, expensive but needless "low field" consumer appliances, and costly modifications to schools and public buildings (Monmonier 190).

Cost objections

Present Status of EMF Research. In July 1998, an editor at *The New England Journal of Medicine* called for the ending of EMF/cancer research. He cited studies from the National Cancer Institute and other respected sources that showed "little evidence" of any causal connection. In a parallel development, federal and industry funding for EMF research has been reduced drastically (Stix 33). But, in August 1998, experts from the Energy Department and the National Institute of Environmental Health Sciences (NIEHS) proposed that EMFs should be officially designated a "possible human carcinogen" (Gross 30).

Conflicting scientific opinions

However, one year later, in a report based on its seven-year review of EMF research, NIEHS concluded that "the scientific evidence suggesting that. . .EMF exposures pose any health risk is weak." But the report also conceded that such exposure "cannot be recognized at this time as entirely safe" (United States, National Institutes of Health, National Institute of Environmental Health Sciences, *Health* 1–2). In 2005, the National Cancer Institute fueled the controversy, concluding that the EMF–cancer connection is supported by only "limited evidence" and "inconsistent associations" ("Magnetic"). In a 2009 update, NIEHS announced "that the overall pattern of results suggests a weak association between exposure to EMFs and increased risk of childhood leukemia" ("Electric"). Recently, noted epidemiologist Daniel Wartenberg has testified in support of the "Precautionary Principle," arguing that policy decisions should be based on "the possibility of risk." Specifically, Wartenberg advocates "prudently lowering exposures of greatest concern [i.e., of children] in case the possible risk is shown eventually to be true" (6).

Interpretation

In short, after more than twenty-five years of study, the EMF/illness debate continues, even among respected experts. While most scientists agree that EMFs exert measurable effects on the human body, they disagree about whether a real hazard exists. Given the drastic cuts in research funding, definite answers are unlikely to appear any time soon.

Views from the Power Industry and the Public

Fourth major
topic
First subtopic

While the experts continue their debate, other viewpoints are also worth considering.

The Power Industry's Views. The Electrical Power Research Institute (EPRI), the research arm of the nation's electric utilities, claims that recent EMF studies have provided valuable but inconclusive data that warrant further study (Moore 17). What does our local power company think about the alleged EMF risk? Marianne Halloran-Barney, Energy Service Advisor for County Electric, expressed this view in an email correspondence:

Findings

> There are definitely some links, but we don't know, really, what the effects are or what to do about them. . . .There are so many variables in EMF research that it's a question of whether the studies were even done correctly. . . .Maybe in a few years there will be really definite answers.

Echoing Halloran-Barney's views, John Miltane, Chief Engineer for County Electric, added this political insight:

> The public needs and demands electricity, but in regard to the negative effects of generation and transmission, the pervasive attitude seems to be "not in my back yard!" Utilities in general are scared to death of the EMF issue, but at County Electric we're trying to do the best job we can while providing reliable electricity to 24,000 customers.

Miltane stresses that County Electric takes the EMF issue very seriously: Whenever possible, new distribution lines are run underground and configured to diminish EMF intensity.

Second subtopic

Public Perception. Industry views seem to parallel the national perspective among the broader population: Informed people are genuinely concerned, but remain unsure about what level of anxiety is warranted or what exactly should be done. A survey by the Edison Electric Institute did reveal that EMFs are considered a serious health threat by 33 percent of the American public (Stix 33).

Risk-Avoidance Measures Being Taken

Fifth major topic

Although conclusive answers may require decades of research, concerned citizens are already taking action against potential EMF hazards.

First subtopic

Risk Avoidance Nationwide. Following are examples of steps taken by various communities to protect schoolchildren from EMF exposure:

Findings

- Hundreds of individuals and community groups have taken legal action to block proposed construction of new power lines. A single Washington law firm has defended roughly 140 utilities in cases related to EMFs (Dana and Turner 32).
- Houston schools "forced a utility company to remove a transmission line that ran within 300 feet of three schools. Cost: $8 million" (Kirkpatrick 85).

- California parents and teachers are pressuring reluctant school and public health officials to investigate cancer rates in the roughly 1,000 schools located within 300 feet of transmission lines, and to close at least one school (within 100 feet) in which cancer rates far exceed normal (Brodeur 118).

Although critics argue that the questionable risks fail to justify the costs of such measures, widespread concern about EMF exposure continues to grow.

Risk Avoidance Locally. Local awareness of the EMF issue seems low. The main public concern seems to be with property values. According to Halloran-Barney, County Electric receives one or two calls monthly from concerned customers, including people buying homes near power lines. The lack of public awareness adds another dimension to the EMF problem: People can't avoid a health threat that they don't know exists.

Before risk avoidance can be considered on a broader community level, the public must first be informed about EMFs and the associated risks of exposure.

Second subtopic

Interpretation

Conclusion

The conclusion is likely to interest readers most because it answers the questions that originally sparked the analysis.

> NOTE Many workplace reports are submitted with the conclusion preceding the introduction and body sections.

In the conclusion, you summarize, interpret, and recommend. Although you have interpreted evidence at each stage of your analysis, your conclusion presents a broad interpretation and suggests a course of action, where appropriate. The summary and interpretations should lead logically to your recommendations.

- The summary accurately reflects the body of the report.
- The overall interpretation is consistent with the findings in the summary.
- The recommendations are consistent with the purpose of the report, the evidence presented, and the interpretations given.

Elements
of a logical
conclusion

> NOTE Don't introduce any new facts, ideas, or statistics in the conclusion.

As you read the following conclusion, evaluate how effectively it provides a clear and consistent perspective on the whole document.

CONCLUSION

Summary and Overall Interpretation of Findings

Electromagnetic fields exist wherever electricity flows; the stronger the current, the higher the EMF intensity. While no "safe" EMF level has been identified, long-term exposure to intensities

Review of major
findings

greater than 2.5 milligauss is considered dangerous. Although home appliances can generate high EMFs during use, power lines can generate constant EMFs, typically at 2 to 3 milligauss in buildings within 150 feet. Our elementary school is less than 100 feet from a high-voltage power line corridor.

Notable epidemiologic studies implicate EMFs in increased rates of medical disorders such as cancer, miscarriage, stress, depression, and learning disabilities—all directly related to intensity and duration of exposure. Laboratory studies show that EMFs cause the kinds of cellular and metabolic changes that could produce these disorders.

An overall judgment about what the findings mean

Though still controversial and inconclusive, most of the various findings are strikingly similar and they underscore the need for more research and for risk avoidance, especially as far as children are concerned.

Concerned citizens nationwide have begun to prevail over resistant school and health officials and utility companies in reducing EMF risk to schoolchildren. And even though our local power company is taking reasonable risk-avoidance steps, our community can do more to learn about the issues and diminish potential risk.

Recommendations

In light of the conflicting evidence and interpretations, any type of government regulation any time soon seems unlikely. Also, considering the limitations of what we know, drastic and enormously expensive actions (such as burying all the town's power lines or increasing the height of utility towers) seem inadvisable. In fact, these might turn out to be the wrong actions.

Despite this climate of uncertainty, however, our community still can take some immediate and inexpensive steps to address possible EMF risk. Please consider the following recommendations:

Feasible and realistic course of action

- Relocate the school playground to the side of the school most distant from the power lines.
- Discourage children from playing near any power lines.
- Distribute a version of this report to all Bocaville residents.
- Ask our school board to hire a licensed contractor to take milligauss readings throughout the elementary school, to determine the extent of the problem, and to suggest reasonable corrective measures.
- Ask our Town Council to meet with County Electric Company representatives to explore options and costs for rerouting or burying the segment of the power lines near the school.
- Hold a town meeting to answer citizens' questions and to solicit opinions.
- Appoint a committee (consisting of at least one physician, one engineer, and other experts) to review emerging research as it relates to our school and town.

A call to action

As we await conclusive answers, we need to learn all we can about the EMF issue, and to do all we can to diminish this potentially significant health issue.

WORKS CITED

[The Works Cited section for the preceding report appears in Appendix A, "MLA Sample Works Cited Pages." This author uses MLA documentation style.]

Front Matter and End Matter Supplements

21.5 Identify the front and end matter elements that accompany a formal report

Most formal reports or proposals consist of the front matter, the text of the report or proposal, and the end matter. (Some parts of the front and end matter may be optional.) Submit your completed document with these supplements, in this order:

- letter of transmittal
- title page
- table of contents
- list of tables and figures
- abstract
- text of the report (introduction, body, conclusion)
- glossary (as needed)
- appendices (as needed)
- Works Cited page (or alphabetical or numbered list of references)

Front matter precedes the report

End matter follows the report

For discussion of the above supplements, see below. For examples in a formal report or proposal, see Figures 21.3 and 22.4.

Front Matter

Preceding the text of the report is the front matter: letter of transmittal, title page, table of contents, list of tables and figures (if appropriate), and abstract or executive summary. See the pages preceding the Introduction to Figure 21.3 for examples of these items.

LETTER OF TRANSMITTAL. Many formal reports or proposals include a letter of transmittal, addressed to a specific reader or readers, which precedes the document. This letter might acknowledge people that helped with the document, refer readers to sections of special interest, discuss any limitations of the study or any problems in gathering data, offer personal (or off-the-record) observations, or urge readers to take immediate action. Even if the report is sent as a PDF attachment to an email, don't use email for the transmittal information. Just write a short email stating that the report is attached, then include the full letter of transmittal with the report itself.

TITLE PAGE. The title page provides the document title, the names of all authors and their affiliations (and/or the name of the organization that commissioned the report), and the date the report was submitted. The title announces the report's purpose and subject by using descriptive words such as *analysis, comparison, feasibility,* or *recommendation.* Be sure the title fully describes your report, but avoid an overly long and involved title. Make the title the most prominent item, highest on the page, followed by the name of the recipient(s), the author(s), and the date of submission.

TABLE OF CONTENTS. For any long document, the table of contents helps readers by listing the page number for each major section, including any front matter that falls after the table of contents. (Do not include the letter of transmittal, title page, or the table of contents itself, but do include the list of tables and figures, along with the abstract or executive summary.)

Indicate page numbers for front matter in lowercase Roman numerals (i, ii, iii). Note that the title page, though not numbered itself or listed on the table of contents, is counted as page i. Number the report text pages using Arabic numerals (1, 2, 3), starting with the first page of the report. Number end matter using Arabic numerals continuing from the end of the report's text.

Make sure headings and subheadings in the table of contents match exactly the headings and subheadings in the document. Indicate headings of different levels (a-level, b-level, c-level) using different type styles or indentations. Use *leader lines* (........) to connect headings with their page numbers.

LIST OF TABLES AND FIGURES. On a separate page following the table of contents (or at the end of the table of contents, if it fits), list the tables and figures in the report. If the report contains only one or two tables and figures, you may skip this list.

ABSTRACT OR EXECUTIVE SUMMARY. Instead of reading an entire formal report, readers interested only in the big picture may consult the abstract or executive summary that commonly precedes the report proper (see Chapter 9). The purpose of this summary is to explain the issue, describe how you researched it, and state your conclusions (and, in the case of an executive summary, indicate what action the conclusions suggest). Busy readers can then flip through the document to locate sections important to them.

Make the abstract or executive summary as brief as possible. Summarize the report without adding new information or leaving out crucial information. Write for a general audience and follow a sequence that moves from the reason the report was written to the report's major findings, to conclusions and recommendations.

Text of the Report

The text of the report consists of the introduction, the body, and the conclusion, as discussed and illustrated in Figure 21.3.

End Matter

Following the report text (as needed) is the end matter, which may include a glossary, appendices, and/or a list of references cited in your report. Readers can refer to any of these supplements or skip them altogether, according to their needs.

GLOSSARY. Use a glossary if your report contains more than five technical terms that may not be understood by all intended readers. If five or fewer terms need defining, place them in the report's introduction as working definitions, or use footnote definitions. If you do include a separate glossary, announce its location when you introduce technical terms defined there ("see the glossary at the end of this report"). Chapter 17, "Placing Definitions in a Document," shows a glossary.

APPENDICES. If you have large blocks of material or other documents that are relevant but will bog readers down, place these in an appendix. For example, if your report on the cost of electricity at your company refers to another report issued by the local utility company, you may wish to include this second report as an appendix. Other items that belong in an appendix include complex formulas, interview questions and responses, maps, photographs, questionnaires and tabulated responses, and texts of laws and regulations.

Do not stuff appendices with needless information or use them to bury bad or embarrassing news that belongs in the report itself. Title each appendix clearly: "Appendix A: Projected Costs." Mention the appendixes early in the introduction and refer to these appendixes at appropriate points in the report: ("see Appendix A"). The sample proposal in Chapter 22, Figure 22.4 includes an appendix.

REFERENCES OR WORKS CITED LIST. If you have used outside sources in your report (and typically you should), you must provide a list of References (per APA style) or of Works Cited (per MLA style). For detailed advice on documenting sources using APA or MLA style, see A Quick Guide to Documentation. The report in Figure 21.3 uses APA style and includes a References list, which appears in Appendix A, "APA Sample References List." The report in the "Structuring a Formal Report" section of this chapter uses MLA style and includes a Works Cited list, which is shown in Appendix A, "MLA Sample Works Cited Pages."

A Situation Requiring an Analytical Report

21.6 Write a formal analytical report

The formal report that follows, patterned after the model outline (at the beginning of "Structuring a Formal Report" in this chapter), combines a feasibility analysis with a comparative analysis.

The Situation

Richard Larkin, author of the following report, has a work-study job fifteen hours weekly in his school's placement office. His boss, John Fitton (placement director), likes to keep up with trends in various fields. Larkin, an engineering major, has developed an interest in technical marketing and sales. Needing a report topic for his writing course, Larkin offers to analyze the feasibility of a technical marketing and sales career, both for his own decision making and for technical and science graduates in general. Fitton accepts Larkin's offer, looking forward to having the final report in his reference file for use by students choosing careers. Larkin wants his report to be useful in three ways: (1) to satisfy a course requirement, (2) to help him in choosing his own career, and (3) to help other students with their career choices.

With his topic approved, Larkin begins gathering his primary data, using interviews, letters of inquiry, telephone inquiries, and lecture notes. He supplements these primary sources with articles in recent publications. He will document his findings in APA (author-date) style.

As a guide for designing his final report (Figure 21.3), Larkin completes the following audience and use profile (based on the worksheet in Chapter 2, "Develop an Audience and Use Profile").

Audience and Use Profile

The primary audience consists of John Fitton, Placement Director, and the students who will refer to the report. The secondary audience is the writing instructor.

Because he is familiar with the marketing field, Fitton will need very little background to understand the report. Many student readers, however, will have questions like these:

- What, exactly, is technical marketing and sales?
- What are the requirements for this career?
- What are the pros and cons of this career?
- Could this be the right career for me?
- How do I enter the field?

Readers affected by this document are primarily students making career choices. Readers' attitudes likely will vary:

- Some readers should have a good deal of interest, especially those seeking a people-oriented career.
- Others might be only casually interested as they investigate a range of possible careers.
- Some readers might be skeptical about something written by a fellow student instead of by some expert. To connect with all these people, this report needs to persuade them that its conclusions are based on reliable information and careful reasoning.

All readers expect things spelled out, but concisely. Visuals will help compress and emphasize material throughout.

Essential information will include an expanded definition of technical marketing and sales, the skills and attitudes needed for success, the career's advantages and drawbacks, and a description of various paths for entering the career.

This report combines feasibility and comparative analysis, so the structure of the report must reveal a clear line of reasoning: in the feasibility section, reasons for and reasons against; in the comparison section, a block structure and a table that compares the four entry paths point by point. The report will close with recommendations based on solid conclusions.

For various readers who might not wish to read the entire report, an informative abstract will be included.

NOTE This report's end matter (list of references) is shown in Appendix A, Figure A.5 and discussed in Appendix A, "Discussion of Figure A.5."

165 Hammond Way
Hyannis, MA 02457

April 29, 20XX

John Fitton
Placement Director
University of Massachusetts
North Dartmouth, MA 02747

Dear Mr. Fitton:

Here is my report, Feasibility Analysis of a Career in Technical Marketing. In preparing
this report, I've learned a great deal about the requirements and modes of access to
this career, and I believe my information will help other students as well. Thank you for
your guidance and encouragement throughout this process.

Although committed to their specialties, some technical and science graduates seem
interested in careers in which they can apply their technical knowledge to customer
and business problems. Technical marketing may be an attractive choice of career for
those who know their field, who can relate to different personalities, and who
communicate well.

Technical marketing is competitive and demanding, but highly rewarding. In fact, it is
an excellent route to upper-management and executive positions. Specifically,
marketing work enables one to develop a sound technical knowledge of a company's
products, to understand how these products fit into the marketplace, and to perfect
sales techniques and interpersonal skills. This is precisely the kind of background that
paves the way to top-level jobs.

I've enjoyed my work on this project, and would be happy to answer any questions.
Please phone at 690-555-1122 or email at larkin@com.net anytime.

Sincerely,

Richard B. Larkin

Richard B. Larkin, Jr.

Letter of
transmittal
targets and
thanks
specific reader
and provides
additional
context

Figure 21.3 A Formal Report

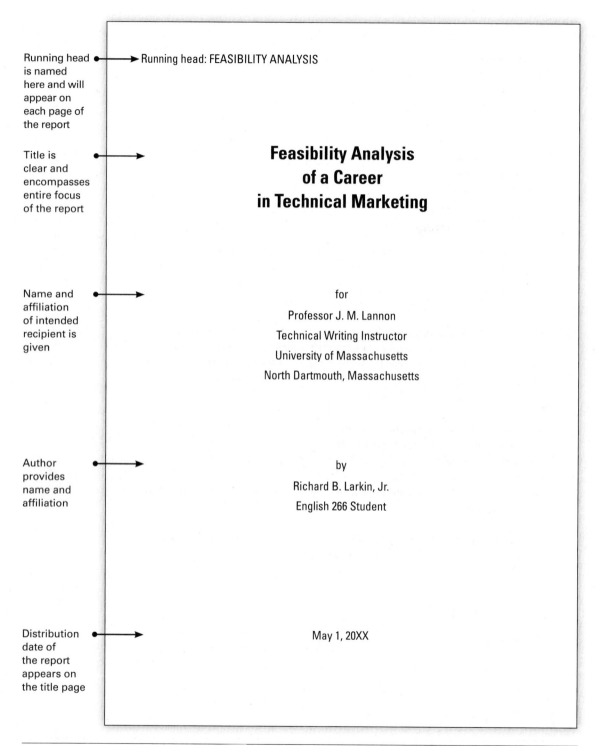

Running head is named here and will appear on each page of the report

Running head: FEASIBILITY ANALYSIS

Title is clear and encompasses entire focus of the report

Feasibility Analysis
of a Career
in Technical Marketing

Name and affiliation of intended recipient is given

for

Professor J. M. Lannon

Technical Writing Instructor

University of Massachusetts

North Dartmouth, Massachusetts

Author provides name and affiliation

by

Richard B. Larkin, Jr.

English 266 Student

Distribution date of the report appears on the title page

May 1, 20XX

Figure 21.3 **Continued**

FEASIBILITY ANALYSIS

Table of Contents

Figures and Tables

Table of contents helps readers find information and visualize the structure of the report

This section makes visuals easy to locate

Figure 21.3 Continued

Abstract
fully
summarizes
the content
of the
report

Abstract

The feasibility of technical marketing as a career is based on a college graduate's interests, abilities, and expectations, as well as on possible entry options.

Technical marketing is a feasible career for anyone who is motivated, who can communicate well, and who knows how to get along. Although this career offers job diversity and potential for excellent income, it entails almost constant travel, competition, and stress.

College graduates enter technical marketing through one of four options: entry-level positions that offer hands-on experience, formal training programs in large companies, prior experience in one's specialty, or graduate programs. The relative advantages and disadvantages of each option can be measured in resulting immediacy of income, rapidity of advancement, and long-term potential.

Anyone considering a technical marketing career should follow these recommendations:

- Speak with people who work in the field.
- Weigh the implications of each entry option carefully.
- Consider combining two or more options.
- Choose options for personal as well as professional benefits.

Figure 21.3 Continued

FEASIBILITY ANALYSIS   1

Introduction

In today's global business climate, graduates in science and engineering face narrowing career opportunities because of "offshoring" of hi-tech jobs to low-wage countries. Government research indicates that more than two-thirds of the 40 occupations "most prone to offshoring" are in science and engineering (Bureau of Labor Statistics [BLS], 2006, p. 14). Experts Hira and Hira have suggested that the offshoring situation threatens the livelihood of some of the best-paid workers in America (2005, p. 12). University career counselor Troy Behrens offers the disturbing fact that the U.S. graduated 30,000 engineers in 2006, whereas India and China graduated 3 million (cited in Jacobs, 2007).

— • Introduction identifies the problem

Given such bleak prospects, graduates might consider alternative careers. Technical marketing is one field that combines science and engineering expertise with "people skills"— those least likely to be offshored (BLS, 2006, p. 12). Engineers, for example, might seek jobs as *sales engineers*, specially trained professionals who market and sell highly technical products and services (BLS, 2009a, p. 1).

— • Proposes a possible solution

What specific type of work do technical marketers perform? *The Occupational Outlook Handbook* offers this job description:

> They [technical marketing specialists] possess extensive knowledge of [technologically and scientifically advanced] products, including…the components, functions, and scientific processes that make them work. They use their technical skills to explain the benefit of their products to potential customers and to demonstrate how their products are better than the products of their competitors. Often they modify and adjust products to meet customers' specific needs (BLS, 2009a, p.1).

— • Definition

(For a more detailed job description, refer to "The Technical Marketing Process," on page 2.)

Undergraduates interested in technical marketing need answers to the following basic questions:

— • Clear purpose statement leads into the body of the report

- *Is this the right career for me?*
- *If so, how do I enter the field?*

To help answer these questions, this report analyzes information gathered from professionals as well as from the literature. After defining *technical marketing,* the analysis examines employment outlook, required skills and personal qualities, career benefits and drawbacks, and entry options.

Figure 21.3 Continued

Data Section

Key Factors in a Technical Marketing Career

Anyone considering technical marketing needs to assess whether this career fits his or her interests, abilities, and aspirations.

The technical marketing process. The classic process (identifying, reaching, and selling to customers) entails six key activities (Cornelius & Lewis, 1983, p. 44):

1. *Market research:* assessing size and character of the target market.
2. *Product development and management:* producing the goods to fill a need.
3. *Cost determination and pricing:* measuring every expense in the product's production, distribution, advertising, and sales to determine its price.
4. *Advertising and promotion:* developing strategies for reaching customers.
5. *Product distribution:* coordinating all elements of a technical product or service, from conception through final delivery to the customer.
6. *Sales and technical support:* creating and maintaining customer accounts, and servicing and upgrading products.

Engaged in all these activities, the technical marketing professional gains detailed understanding of the industry, the product, and the customer's needs (Figure 1).

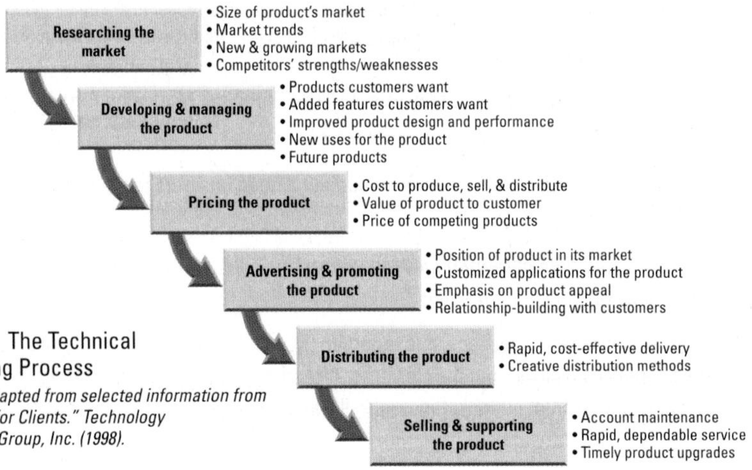

Researching the market
- Size of product's market
- Market trends
- New & growing markets
- Competitors' strengths/weaknesses

Developing & managing the product
- Products customers want
- Added features customers want
- Improved product design and performance
- New uses for the product
- Future products

Pricing the product
- Cost to produce, sell, & distribute
- Value of product to customer
- Price of competing products

Advertising & promoting the product
- Position of product in its market
- Customized applications for the product
- Emphasis on product appeal
- Relationship-building with customers

Distributing the product
- Rapid, cost-effective delivery
- Creative distribution methods

Selling & supporting the product
- Account maintenance
- Rapid, dependable service
- Timely product upgrades

Figure 1 The Technical Marketing Process

Source: Adapted from selected information from "Services for Clients." Technology Marketing Group, Inc. (1998).

Figure 21.3 **Continued**

FEASIBILITY ANALYSIS   3

Employment outlook. For graduates with the right combination of technical and personal qualifications, the outlook for technical marketing (and management) is excellent. Most engineering jobs will increase at less than average for jobs requiring a Bachelor's degree, while marketing and marketing management jobs will exceed the average rate (Figure 2).

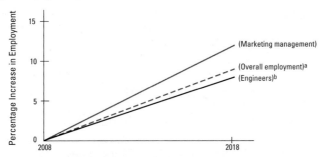

Visual provides instant comparison

Figure 2 The Employment Outlook for Technical Marketing
[a]Jobs requiring a Bachelor's degree.
[b]Excluding outlying rates for specialties at the positive end of the spectrum (environmental engineers: +31%; biomedical: +72%; civil: +24%; petroleum: +18%).
Source: Data from U.S. Department of Labor. Bureau of Labor Statistics. (2009).
http://www.bls.gov/oco/ocos027.htm

Although highly competitive, these marketing positions call for the very kinds of technical, analytical, and problem-solving skills that engineers can offer—especially in an automated environment.

Technical skills required. Interactive Web sites and social media marketing will increasingly influence the way products are advertised and sold. Also, marketing representatives increasingly work from a "virtual office." Using laptops, smartphones, and other such devices, representatives out in the field have real-time access to digital catalogs of product lines, multimedia presentations, pricing for customized products, inventory data, product distribution channels, and sales contacts (Tolland, 2013).

With their rich background in computer, technical, and problem-solving skills, engineering graduates are ideally suited for (a) working in automated environments, and (b) implementing and troubleshooting these complex and often sensitive electronic systems.

Figure 21.3 Continued

FEASIBILITY ANALYSIS   4

Other skills and qualities required. *Business Week*'s Peter Coy offers this
distinction between routine versus non routine work:

> The jobs that will pay well in the future will be ones that are hard to reduce to a recipe. These
> attractive jobs—from factory floor management to sales to teaching to the professions—require
> flexibility, creativity, and lifelong learning. They generally also require subtle and frequent
> interactions with other people, often face to face. (2004, p. 50)

For emphasis, select sources are quoted directly

Technical marketing is just such a job: it involves few "cookbook-type" tasks and
requires "people skills." Besides a strong technical background, success in this field
calls for a generous blend of those traits summarized in Figure 3.

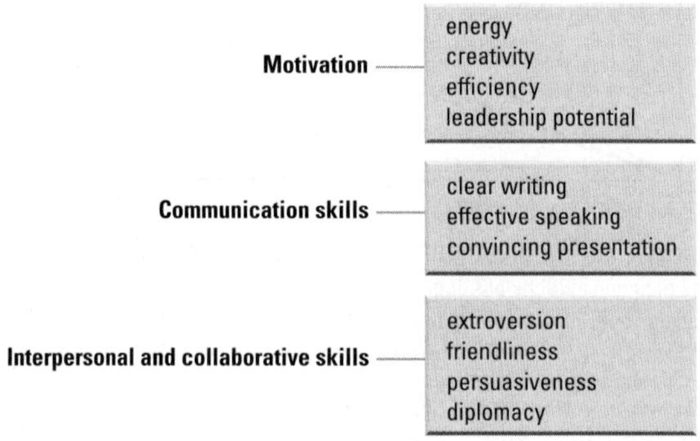

Motivation — energy / creativity / efficiency / leadership potential

Communication skills — clear writing / effective speaking / convincing presentation

Interpersonal and collaborative skills — extroversion / friendliness / persuasiveness / diplomacy

Figure 3 Required People Skills for a Technical Marketing Career

Motivation is essential in marketing. Professionals must be energetic and able to
function with minimal supervision. Career counselor Phil Hawkins describes the ideal
candidates as people who can plan and program their own tasks, can manage
their time, and have no fear of hard work (personal interview, February 11, 2013).
Leadership potential, as demonstrated by extracurricular activities, is an asset.

Report is based on primary as well as secondary research

Motivation alone, however, provides no guarantee of success. Marketing professionals
are paid to communicate the value of their products and services, orally, online, on
paper, and face to face. They routinely prepare such documents as sales proposals,
product descriptions, and user manuals. Successful job candidates typically have
taken courses in advertising, public speaking, technical communication, and—
increasingly— foreign language (BLS, 2009a, pp. 2-3).

Figure 21.3 Continued

Skilled oral presentation is vital to any sales effort, as Phil Hawkins points out. Technical marketing professionals need to speak confidently and persuasively—to represent their products and services in the best light (personal interview, February 11, 2013). Sales presentations often involve public speaking at conventions and trade shows.

The ultimate requirement for success in marketing is interpersonal and collaborative skills: "tact, good judgement, and exceptional ability to establish and maintain relationships with supervisory and professional staff and client firms" (BLS, 2009b, p. 4).

Advantages of the career. As shown in Figure 1, technical marketing offers experience in every phase of a company's operation, from a product's design to its sales and service. Such broad exposure provides excellent preparation for upper-management positions. In fact, experienced sales engineers often open their own businesses as freelance "manufacturers' agents" representing a variety of companies who have no marketing staff. In effect their own bosses, manufacturers' agents are free to choose, from among many offers, the products they wish to represent (Tolland, 2013).

Offers balanced coverage: advantages versus drawbacks (below)

Another career benefit is the attractive salary. In addition to typically receiving a base pay plus commissions, marketing professionals are reimbursed for business expenses. Other employee benefits often include health insurance, a pension plan, and a company car. In 2008, the median annual earnings for sales engineers was $83,100. The highest 10 percent earned more than $136,000 annually (BLS, 2008, p.1).

The interpersonal and communication skills that marketing professionals develop are highly portable. This is vital in our rapidly shifting economy, in which job security is disappearing in the face of more and more temporary positions (Tolland, 2013).

Drawbacks of the career. Technical marketing is by no means a career for every engineer or technology professional. Sales engineer Roger Cayer cautions that personnel might spend most of their time traveling to meet potential customers. Success requires hard work over long hours, evenings, and occasional weekends. Above all, the job entails constant pressure to meet sales quotas (phone interview, February 8, 2013). Anyone considering this career should be able to work and thrive in a competitive environment. The Bureau of Labor Statistics (2009a, p. 2) adds that the expanding global economy means that "international travel, to secure contracts with foreign customers, is becoming more common"—placing more pressure on an already hectic schedule.

Provides a realistic view

Figure 21.3 Continued

A Comparison of Entry Options

Each option in the comparison is previewed →

Engineers and other technical graduates enter this field through one of four options. Some join small companies and learn their trade directly on the job. Others join companies that offer formal training programs. Some begin by getting experience in their technical specialty. Others earn a graduate degree beforehand. These options are compared below.

Option 1: Entry-level marketing with on-the-job training. Smaller manufacturers offer marketing positions in which people learn on the job. Elaine Carto, president of ABCO Electronics, believes small companies offer a unique opportunity; entry-level salespersons learn about all facets of an organization, and often enjoy rapid advancement (personal interview, February 10, 2013). Career counselor Phil Hawkins says, "It's all a matter of whether you prefer to be a big fish in a small pond or a small fish in a big pond" (personal interview, February 11, 2013).

Here and below, interpretations clarify the pros and cons of each entry option →

Entry-level marketing offers immediate income and a chance for early promotion. But one disadvantage might be the loss of any technical edge acquired in college.

Option 2: A marketing and sales training program. Formal training programs offer the most popular entry. Larger companies offer two formats: (a) a product-specific program, focused on a particular product or product line, or (b) a rotational program, in which trainees learn about an array of products and work in the various positions outlined in Figure 1. Programs last from weeks to months. Intel Corporation, for example, offers 30-month training programs titled "Sales and Marketing Rotation," to prepare new graduates for positions as technical sales engineer, marketing technical engineer, and technical applications engineer.

Like direct entry, this option offers the advantage of immediate income and early promotion. With no chance to practice in their specialty, however, trainees might eventually find their technical expertise compromised.

Option 3: Prior experience in one's technical specialty. Instead of directly entering marketing, some candidates first gain experience in their specialty. This option combines direct exposure to the workplace with the chance to sharpen technical skills in practical applications. In addition, some companies, such as Roger Cayer's, will offer marketing and sales positions to outstanding staff engineers as a step toward upper management (phone interview, February 8, 2013).

Figure 21.3 Continued

FEASIBILITY ANALYSIS 7

Although the prior-experience option delays entry into technical marketing, industry experts consider direct workplace and technical experience key assets for career growth in any field. Also, work experience becomes an asset for applicants to top MBA programs (Shelley, 1997, pp. 30–31).

Option 4: Graduate program. Instead of direct entry, some people choose to pursue an MS in their specialty or an MBA. According to engineering professor Mary McClane, MS degrees are usually unnecessary for technical marketing unless the particular products are highly complex (personal interview, April 2, 2013).

In general, jobseekers with an MBA have a competitive advantage. Also, new MBAs with a technical bachelor's degree and one to two years of experience command salaries from 10 to 30 percent higher than MBAs who lack work experience and a technical bachelor's degree. In fact, no more than 3 percent of candidates offer a "techno-MBA" specialty, making this unique group highly desirable to employers (Shelley, 1997, p. 30). A motivated student might combine graduate degrees. Dora Anson, president of Susimo Systems, sees the MS/MBA combination as ideal preparation for technical marketing (2013).

One disadvantage of a full-time graduate program is lost salary, compounded by school expenses. These costs must be weighed against the prospect of promotion and monetary rewards later in one's career.

Interprets evidence impartially

An overall comparison by relative advantage. Table 1 compares the four entry options on the basis of three criteria: immediate income, rate of advancement, and long-term potential.

Table 1 Relative Advantages Among Four Technical-Marketing Entry Options

Table summarizes the prior information, for instant comparisons

Option	Early, immediate income	Greatest advancement in marketing	Long-term potential
Entry level, no experience	yes	yes	no
Training program	yes	yes	no
Practical experience	yes	no	yes
Graduate program	no	no	yes

Figure 21.3 Continued

Summary accurately and concisely reflects the report's body section

Overall interpretation explains what the findings mean

Recommendations are clear about what the audience should think and do

Conclusion

Summary of Findings

Technical marketing and sales requires solid technical background, motivation, communication skills, and interpersonal skills. This career offers job diversity and excellent income potential, balanced against relentless pressure to perform.

Graduates interested in this field confront four entry options: (1) direct entry with on-the-job training, (2) a formal training program, (3) prior technical experience, and (4) graduate programs. Each option has benefits and drawbacks based on immediacy of income, rate of advancement, and long-term potential.

Interpretation of Findings

For graduates with strong technical backgrounds and the right skills and motivation, technical marketing offers attractive prospects. Anyone contemplating this career, however, needs to enjoy customer contact and thrive in a competitive environment.

Those who decide that technical marketing is for them have various entry options:

- For hands-on experience, direct entry is the logical option.
- For intensive sales training, a formal program with a large company is best.
- For sharpening technical skills, prior work in one's specialty is invaluable.
- If immediate income is not vital, graduate school is an attractive option.

Recommendations

If your interests and abilities match the requirements, consider these suggestions:

1. For a firsthand view, seek the advice and opinions of people in the field. You might begin by contacting professional organizations such as the Manufacturers' Agents National Association at www.manaonline.org.
2. Before settling on an entry option, consider its benefits and drawbacks and decide whether this option best coincides with your career goals.
3. When making any career decision, consider career counselor Phil Hawkins' s advice: "Listen to your brain and your heart" (personal interview, February 11, 2013). Seek not only professional advancement but also personal satisfaction.

References

The References section for this report appears in Appendix A, "A Quick Guide to Documentation," Figure A.5. This author uses APA documentation style.

Figure 21.3 **Continued**

Guidelines

for Reasoning through an Analytical Problem

Audiences approach an analytical report with this basic question:

| *Is this analysis based on sound reasoning?*

Whether your report documents a causal, comparative, or feasibility analysis (or some combination), you need to trace your line of reasoning so that readers can follow it clearly. As you prepare your report, observe the following guidelines:

For Causal Analysis

➤ **Be sure the cause fits the effect.** Keep in mind that faulty causal reasoning is extremely common, especially when we ignore other possible causes or when we confuse mere coincidence with causation. For more on this topic, refer to Chapter 8, "Faulty Causal Reasoning."

➤ **Make the links between effect and cause clear.** Identify the immediate cause (the one most closely related to the effect) as well as the distant cause(s) (the ones that often precede the immediate cause). For example, the immediate cause of a particular airplane crash might be a fuel tank explosion, caused by a short circuit in frayed wiring, caused by faulty design or poor quality control by the airplane manufacturer. Discussing only the immediate cause often just scratches the surface of the problem.

➤ **Clearly distinguish between possible, probable, and definite causes.** Unless the cause is obvious, limit your assertions by using *perhaps, probably, maybe, most likely, could, seems to, appears to,* or similar qualifiers that prevent you from making an insupportable claim. Keep in mind that "certainty" is elusive, especially in causal relationships.

For Comparative Analysis

➤ **Rest the comparison on clear and definite criteria: costs, uses, benefits/drawbacks, appearance, results.** In evaluating the merits of competing items, identify your specific criteria (cost, ease of use, durability, and so on) and rank these criteria in order of importance.

➤ **Give each item balanced treatment.** Discuss points of comparison for each item in identical order.

➤ **Support and clarify the comparison or contrast through credible examples.** Use research, if necessary, for examples that readers can visualize.

➤ **Follow either a block pattern or a point-by-point pattern.** In the block pattern, first one item is discussed fully, then the next. Choose a block pattern when the overall picture is more important than the individual points. In the point-by-point pattern, one point about both items is discussed, then the next point, and so on. Choose a point-by-point pattern when specific points might be hard to remember unless placed side by side.

Block pattern	**Point-by-point pattern**
Item A	first point of A/first point of B, etc.
first point	
second point, etc.	
Item B	second point of A/second point of B, etc.
first point	
second point, etc.	

➤ **Order your points for greatest emphasis.** Try ordering your points from least to most important or dramatic or useful or reasonable. Placing the most striking point last emphasizes it best.

> ➤ **In an evaluative comparison ("X is better than Y"), offer your final judgment.** Base your judgment squarely on the criteria presented.

For Feasibility Analysis

> ➤ **Consider the strength of supporting reasons.** Choose the best reasons for supporting the action or decision being considered—based on your collected evidence.

> ➤ **Consider the strength of opposing reasons.** Remember that people—including ourselves—usually see only what they want to see. Avoid the temptation to overlook or downplay opposing reasons, especially for an action or decision that you have been promoting. Consider alternate points of view; examine and evaluate all the evidence. But be sure that all information is from respected, credible sources.

> ➤ **Recommend a realistic course of action.** After weighing all the pros and cons, make your recommendation— but be prepared to reconsider if you discover that what seemed like the right course of action turns out to be the wrong one.

Checklist

Analytical Reports

Use the following Checklist when writing a formal analytic report.

Content

❑ Does the report address a clearly identified problem or purpose? (See "Clearly Identified Problem or Purpose" in this chapter.)

❑ Are the report's length and detail appropriate for the subject? (See "Adequate but Not Excessive Data" in this chapter.)

❑ Is there enough information for readers to make an informed decision? (See "Accurate and Balanced Data" in this chapter.)

❑ Are all data fully interpreted? (See "Fully Interpreted Data" in this chapter.)

❑ Is the information accurate, unbiased, and complete? (See "Subordination of Personal Bias" in this chapter.)

❑ Are visuals used whenever possible to aid communication? (See "Appropriate Visuals" in this chapter.)

❑ Are the conclusions logically derived from accurate interpretation? (See "Valid Conclusions and Recommdations" in this chapter.)

❑ Do the recommendations constitute an appropriate and reasonable response to the question or problem? (See "Valid Conclusions and Recommdations" in this chapter.)

❑ Are all limitations of the analysis clearly acknowledged? (See "Valid Conclusions and Recommdations" in this chapter.)

❑ Are all needed front and end matter supplements included? (See "Front Matter and End Matter Supplements" in this chapter.)

❑ Is each source and contribution properly cited? (See "End Matter" in this chapter.)

Arrangement

❑ Is there a distinct introduction, body, and conclusion? (See "Structuring a Formal Report" in this chapter.)

❑ Does the introduction provide sufficient orientation to the issue or problem? (See "Introduction" in this chapter.)

❏ Does the body section present a clear picture of the evidence and reasoning? (See "Body" in this chapter.)

❏ Does the conclusion answer the question that originally sparked the analysis? (See "Conclusion" in this chapter.)

❏ Are there clear transitions between related ideas? (See Appendix B, "Transitions.")

Style and Page Design

❏ Is the level of technicality appropriate for the primary audience? (See Chapter 2, "Assess the Audience's Technical Background.")

❏ Are headings informative and adequate? (See Chapter 13, "Using Headings for Access and Orientation.")

❏ Is the writing clear, concise, and fluent? (See Chapter 11, "Editing for Clarity," "Editing for Conciseness," and "Editing for Fluency.")

❏ Is the language precise, and informative? (See Chapter 11, "Finding the Exact Words.")

❏ Is the report grammatical? (See Appendix B, "Grammar.")

❏ Is the page design inviting and accessible? (See Chapter 13, "Page Design in Print and Digital Workplace Documents.")

❏ Will the report be delivered in digital format, in print, or both? (See Chapter 13, "Page Design in Print and Digital Workplace Documents.")

Projects

For all projects, check with your instructor about whether to present your findings in class, bring drafts to class for discussion, upload your project to the class learning management system (LMS), and/or use the LMS forum or discussion boards to collaborate and review each activity below.

General

Prepare an analytical report, using this procedure:

a. Choose a problem or question for analysis from your major or a subject of interest.

b. Restate the main question as a declarative sentence in your audience and purpose statement.

c. Identify an audience—other than your instructor—that will use your information for a specific purpose.

d. Hold a private brainstorming session to generate major topics and subtopics.

e. Use the topics to make a working outline based on the model outline in this chapter.

f. Make a tentative list of all sources (primary and secondary) that you will investigate. Verify that adequate sources are Ḳable available.

g. In a proposal memo to your instructor (see Chapter 23, "Research Proposals"), describe the topic and your plan for analysis. Attach a tentative bibliography.

h. Use your working outline as a guide to research.

i. Submit a progress report (see Chapter 20, "Progress Reports") to your instructor describing work completed, problems encountered, and work remaining.

j. Compose an audience and use profile. (See Chapter 2, "Develop an Audience and Use Profile.")

k. Write the report for your stated audience. Work from a clear statement of audience and purpose, and be sure your reasoning is shown clearly. Verify that your evidence, conclusions, and recommendations are consistent. Be especially careful that your recommendations observe the critical-thinking guidelines in Figure 21.2.

l. After writing your first draft, make any needed changes in the outline and revise your report according to "Checklist: Analytical Reports." Include front matter and end matter.

m. Exchange reports with a classmate for further suggestions for revision.

n. Prepare an oral report of your findings for the class as a whole.

Team

1. Divide into small groups. Choose a topic for group analysis—preferably a campus issue—and brainstorm. Draw up a working outline that could be used as an analytical report on this subject.

2. In the workplace, it is common for reports to be written in teams. Think of an idea for a report for this class that might be a team project. How would you divide the tasks and ensure that the work was being done fairly? What are the advantages and disadvantages of a team approach to a report? In groups of two to three, discuss these issues. If your instructor indicates that your report is to be team-based, write a memo (as a team) to your instructor indicating the role of each team member and the timeline.

Digital and Social Media

Many reports, particularly government reports, are created as PDF documents and made available online. When you have a draft of your report, turn it into a PDF file. (Use the "Save As" function in Microsoft Word or similar word processing apps.) Now, open the file using Adobe Acrobat (you can download a free version) and read the report on a computer. Are there places in the report where you might consider adding links to external Web sites, so readers can obtain more information? How would you decide where and when to add links? Also, consider how you might make your report available on social media: as a link within a post to a Facebook page? As a tweet? How would you summarize your report in your social media post in a way that makes the topic inviting for readers to explore further? Make some notes about this and discuss with at least one other student. Write a brief memo to your classmates explaining your observations.

Global

Many formal analytical reports are written in English but are also for readers who speak many different languages and come from many different countries and cultures. Find the Web sites and social media feeds for organizations that have an international audience. Organizations of interest might include the following (you can add to this list with other organizations related to your major or career):

- Association of Computing Machinery (ACM)
- Doctors Without Borders
- Institute of Electrical and Electronics Engineers (IEEE)
- International Association for Bridge and Structural Engineering
- Society for Technical Communication (STC)
- United Nations
- World Health Organization (WHO)

Look for the "publications" tab or other links that give you access to the organization's published reports. Select several reports and analyze these based on the material from this chapter. Can you determine the type of analysis (causal, comparative, feasibility, or some combination)? Why do you think the writer(s) chose this approach for an international audience? In addition, analyze the report for features such as structure and organization, word choice, formatting, use of visuals, levels of politeness, and format (PDF, hyperlinked Web page, or other). Write a memo to your instructor with your findings. What conclusions can you draw about effective writing and design practices for formal reports that have an international audience?

Chapter 22
Proposals

Alix Minde/PhotoAlto/Getty Images

"My work is all about proposal writing—it's my full-time job. While some of our museum's financial support comes from private donors, most of it is generated from grants. I mainly write for panels or committees reviewing grant proposals to decide whether to fund them: for example, corporate officers making a decision on a corporate gift, trustees of private foundations deciding on a foundation gift, and so on. Any such proposal (prepared for, say, The National Endowment for the Arts) usually has to be submitted on a yearly basis in order for funding to continue. And each proposal must compete yearly with proposals from similar institutions. It's a continuous challenge!**"**

—*Patrick Welch, Grant Writer for a major museum*

Learning Objectives

22.1 Understand that audiences expect to be persuaded by a proposal

22.2 Describe the proposal process

22.3 Differentiate between the various types of proposals

22.4 Identify the elements that are typically part of an effective proposal

22.5 Follow an introduction/body/ conclusion structure in a proposal

22.6 Write a formal proposal

What proposals do

Proposals attempt to *persuade* an audience to take some form of action: to authorize a project, accept a service or product, or support a specific plan for solving a problem or improving a situation.

You might write as a proposal a letter to your school board to suggest changes in the English curriculum; you might write a memo to your firm's vice president to request funding for a training program for new employees; or you might work on a team preparing an extensive document to bid on a Defense Department contract (competing with proposals from other firms). As a student or as an intern at a non-profit agency, you might submit a *grant proposal*, requesting financial support for a research or community project.

You might work alone or collaboratively. Developing and writing the proposal might take hours or months. If your job depends on funding from outside sources, proposals might be the most important documents you produce.

Considering Audience and Purpose

22.1 Understand that audiences expect to be persuaded by a proposal

Audience considerations

In science, business, government, or education, proposals are written for decision makers: managers, executives, directors, clients, board members, or community leaders. Inside or outside your organization, these people review various proposals and then decide whether a specific plan is worthwhile, whether the project will materialize, or whether the service or product is useful.

Before accepting a particular proposal, reviewers look for persuasive answers to these basic questions:

- What exactly is the problem or need, and why is this such a big deal?
- Why should we spend time, money, and effort on this?
- What exactly is your plan, and how do we know it is feasible?
- Why should we accept the items that seem costly about your plan?
- What action are we supposed to take?

What proposal reviewers want to know

Connect with your audience by addressing the previous questions early and systematically. Here are the persuasive tasks involved:

A proposal involves these basic persuasive tasks

1. **Spell out the problem (and its causes) clearly and convincingly.** Supply enough detail for your audience to appreciate the problem's importance.
2. **Point out the benefits of solving the problem.** Explain specifically what your readers stand to gain.
3. **Offer a realistic, cost-effective solution.** Stick to claims or assertions you can support. (For more on feasibility, see Chapter 20, "Feasibility Reports" and Chapter 21, "Feasibility Analysis: 'Is This a Good Idea?'")
4. **Address anticipated objections to your solution.** Consider carefully your audience's level of skepticism about this issue.
5. **Convince your audience to act.** Decide exactly what you want your readers to do and give reasons why they should be the ones to take action.

The "Elements of a Persuasive Proposal" section and "Guidelines for Proposals" in this chapter can help you meet your audience's needs. The singular purpose of your proposal is to convince your audience to accept your plan and get them to say "Yes. Let's move ahead on this."

Purpose considerations

While they may contain many of the same basic elements as a report, proposals have a primarily *persuasive* purpose. Of course, reports can also contain persuasive elements, as in recommending a specific course of action or justifying an equipment purchase. But reports typically focus more on *informative* purposes—such as keeping track of progress, explaining why something happened, or predicting an outcome.

How proposals and reports differ in purpose

A report often precedes a proposal. For example, a report on high levels of chemical pollution in a major waterway typically leads to various proposals for cleaning up that waterway. In short, once the report has *explored* a particular need, a proposal will be developed to *sell* the idea for meeting that need.

The Proposal Process

22.2 Describe the proposal process

The basic proposal process can be summarized like this: Someone offers a plan for something that needs to be done. This process has three stages:

Proposals in the commercial sector

1. Client *X* needs a service or product.

Stages in the proposal process

2. Firms *A*, *B*, and *C* propose a plan for meeting the need.
3. Client *X* awards the job to the firm offering the best proposal.

Following is a typical scenario.

Case

Submitting a Competitive Proposal

You manage a mining engineering firm in Tulsa, Oklahoma. You regularly read *Commerce Business Daily*, an essential online reference tool for anyone whose firm seeks government contracts. This publication lists the government's latest needs for services (salvage, engineering, maintenance) and for supplies, equipment, and materials (guided missiles, engine parts, and so on). On Wednesday, February 19, you spot this announcement:

> **Development of Alternative Solutions to Acid Mine Water Contamination from Abandoned Lead and Zinc Mines** near Tar Creek, Neosho River, Ground Lake, and the Boone and Roubidoux aquifers in northeastern Oklahoma. This project will include assessment of environmental effects of mine drainage followed by development and evaluation of alternate solutions to alleviate acid mine drainage in receiving streams. An optional portion of the contract to be bid on as an add-on and awarded at the discretion of the OWRB will be to prepare an Environmental Impact Assessment for each of three alternative solutions as selected by the OWRB. The project is expected to take six months to accomplish, with an anticipated completion date of September 30, 20XX. The projected effort for the required task is thirty person-months. The request for proposal is available at www.owrb.gov. Proposals are due March 1.
>
> Oklahoma Water Resources Board
> P.O. Box 53585
> 1000 Northeast 10th Street
> Oklahoma City, OK 73151
> (405) 555–2541

Your firm has the personnel, experience, and time to do the job, so you decide to compete for the contract. Because the March 1 deadline is fast approaching, you immediately download the request for proposal (RFP). The RFP will give you the guidelines for developing and submitting the proposal—guidelines for spelling out your plan to solve the problem (methods, timetables, costs).

You then get right to work with the two staff engineers you have appointed to your proposal team. Because the credentials of your staff could affect the client's acceptance of the proposal, you ask team members to update their résumés for inclusion in an appendix to the proposal.

In situations like the one above, the client will award the contract to the firm submitting the best proposal, based on the following criteria (and perhaps others):

Criteria by which reviewers evaluate proposals

- understanding of the client's needs, as described in the RFP
- clarity and feasibility of the plan being offered
- quality of the project's organization and management

- ability to complete the job by deadline
- ability to control costs
- firm's experience on similar projects
- qualifications of staff to be assigned to the project
- firm's performance record on similar projects

A client's specific evaluation criteria are often listed (in order of importance or on a point scale) in the RFP. Although these criteria may vary, every client expects a proposal that is clear, informative, and realistic.

In contrast to proposals prepared for commercial purposes, museums, community service groups, and other nonprofit organizations prepare *grant proposals* that request financial support for worthwhile causes. Government and charitable granting agencies such as the Department of Health and Human Services, the Department of Agriculture, the Pew Charitable Trust, the MacArthur Foundation, and others solicit proposals for funding in areas such as medical research, educational TV programming, and rural development. Submission and review of grant proposals follow the same basic process used for commercial proposals. *Proposals in the nonprofit sector*

In both the commercial and nonprofit sectors, the proposal process typically occurs online. The National Science Foundation's *Fastlane* Web site, for example, allows grant applicants to submit proposals in electronic format. This process enables applicants to include sophisticated graphics, to revise budget estimates, to update other aspects of the plan as needed, and to maintain real-time contact with the granting agency while the proposal is being reviewed. *Submitting online proposals*

Types of Proposals

22.3 Differentiate between the various types of proposals

Proposals may be either *solicited* or *unsolicited*. Solicited proposals are those that have been requested by a manager, client, or customer (see the Case Study in "The Proposal Process" section in this chapter). Unsolicited proposals are those that have not been requested. If you are a new advertising agency in town, for example, you might decide to send unsolicited proposals to local radio stations to suggest they use your agency for their advertising. *Solicited and unsolicited proposals*

Because the audience for a solicited proposal has made a specific request, you will not need to spend time introducing yourself or providing background on the product or service. For an unsolicited proposal (sometimes termed a "cold call" in sales), you will need to catch readers' attention quickly and provide incentives for them to continue reading— perhaps by printing a price comparison of your fees on the first page, for example.

Proposals may also be *informal* or *formal*. Informal proposals can take the form of an email or memo (if distributed within an organization), or a letter (when sent outside of an organization). Formal proposals, meanwhile, take on the same format as formal reports (see Chapter 21), including front matter, proposal text, and end matter, if applicable. *Informal and formal proposals*

Both solicited and unsolicited proposals, whether informal or formal, fall into three categories: planning proposals, research proposals, and sales proposals.

Planning Proposals

The role of planning proposals

Planning proposals offer solutions to a problem or suggestions for improvement. A planning proposal might be a request for funding to expand the campus newspaper (as in the formal proposal in this chapter), an architectural plan for new facilities at a ski area, or a plan to develop energy alternatives to fossil fuels.

Figure 22.1 is the first page of a solicited, informal planning proposal. Architects from Brewster Architectural, Inc. have been meeting with personnel from Southeastern Massachusetts University to discuss installing an elevator in an older campus building. Because the project is complicated (need to maintain the building's historic elements but install an elevator fully compliant with the Americans with Disabilities Act), the architects need to persuade the client that their analysis and research are sound and their methods will succeed. Because this proposal is addressed to an external reader, it is cast as a letter. To save space, only the salutation and first page are shown here. (Some planning proposals for far less complex situations are purposely brief, such as the proposal from Leverett Land & Timber Company, shown in Chapter 15, Figure 15.6.)

Notice that word choice ("Chris," "we've," "thanks") creates a friendly, yet still professional tone—appropriate for this document since the architects and their clients have spent many hours together already. Notice also the "Limitations" section, where the architects are careful to remind their clients that any construction project may have unanticipated surprises.

This proposal does not need a lengthy introduction or problem statement: Southeastern Massachusetts University, having solicited the proposal, is aware of the problem.

Research Proposals

The role of research proposals

Research (or grant) proposals request approval (and often funding) for some type of study. For example, a university chemist might address a research proposal to the state environmental agency requesting funds to help identify toxic contaminants in local groundwater. Research proposals are solicited by many agencies, including the National Science Foundation and the National Institutes of Health. Each agency has its own requirements and guidelines for proposal format and content. Successful research proposals follow those guidelines and carefully articulate the goals of the project. In these cases, proposal readers will generally be other scientists; therefore, writers can use language that is appropriate for other experts.

Other research proposals might be submitted by students requesting funds or approval for projects such as independent studies or thesis projects. A technical writing student usually submits an informal research proposal that will lead to the term project. For example, in the following research proposal (Figure 22.2), Tom Dewoody

Dear Chris and team:

Thanks for meeting over the past few weeks to help us gather the necessary ◄——● States the
information so we can provide a plan for the elevator project in Bass Hall.   document's
                                                                                  purpose

Background

To make the building fully accessible, Bass Hall needs an elevator. Last month, ◄——● Summarizes
you contacted us requesting a proposal for this project. Although Bass Hall is not   the background
an officially designated historic building, it has distinct historic features   and problem
that you wish to retain. In addition to historic preservation issues, the timing of this
project is very important for the building occupants.

Needs assessment

We've met with project managers, the University building committee,   ◄——● Outlines the
and the building's tenants (Department of Chemistry). Based on those meetings   client's needs
and a review of blueprints and other documents, we have identified
the following needs as important:

- Maintain the historic east-facing facade of Bass Hall
- Locate the elevator shaft so it begins in the underground parking garage
- Avoid any physical changes to the Department of Chemistry's student lab
- Ensure the fewest disruptions during the academic semester
- Use code compliant elevator doors that preserve the historic look of other entrances
- Avoid disturbing the marble portico at the building entrance

Proposed Plan

Based on the needs identified above, and on the overall University master building ◄——● Proposes a
plan, city and state codes and guidelines, and the timeline you have outlined, we   solution
propose to install a handicap-accessible elevator, starting at the underground parking
garage level and terminating on the 4th floor, with the following conditions:

- Situate the elevator shaft in the middle of the building, thus allowing it to terminate
  in the underground parking garage
- Locate the elevator on the north side of the hallway. Although this location will
  require relocating one faculty office, using the north side avoids the student
  chemistry lab, marble portico, and other important building features
- Work with our historic preservation team to create custom entrance doors and
  hardware
- Begin demolition on June 1, so that the most disruptive work will be completed
  before the fall semester begins and the final project can be finished by December

Limitations

Older buildings bring unanticipated issues. For instance, although Bass Hall   ◄——● Sets realistic
recently underwent asbestos abatement, should any additional asbestos be located   expectations
during initial testing, we would need to reconsider the timing of this proposal to
allow for removal.

Please see the next several pages for schematic drawings and other details.

Figure 22.1 A Planning Proposal The complete proposal contains several schematic drawings (two pages additional)
for the client to review. (To save space, only the salutation and first page are shown here.)

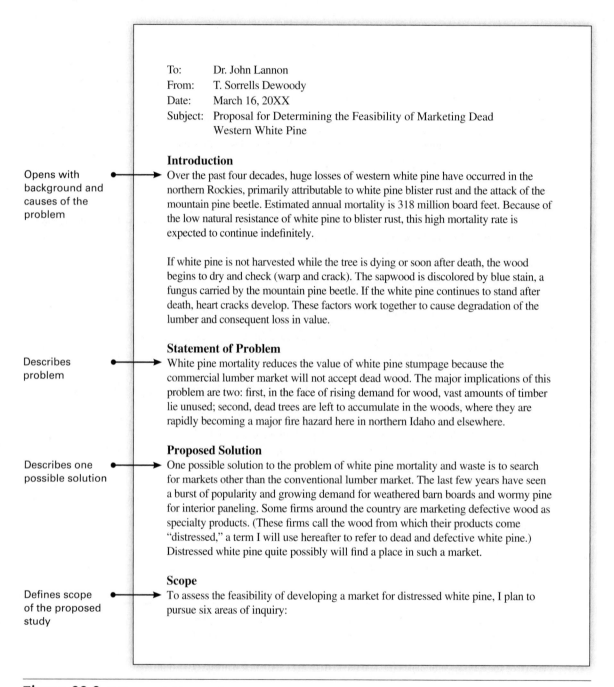

To: Dr. John Lannon
From: T. Sorrells Dewoody
Date: March 16, 20XX
Subject: Proposal for Determining the Feasibility of Marketing Dead
          Western White Pine

Introduction

Opens with background and causes of the problem ▸ Over the past four decades, huge losses of western white pine have occurred in the northern Rockies, primarily attributable to white pine blister rust and the attack of the mountain pine beetle. Estimated annual mortality is 318 million board feet. Because of the low natural resistance of white pine to blister rust, this high mortality rate is expected to continue indefinitely.

If white pine is not harvested while the tree is dying or soon after death, the wood begins to dry and check (warp and crack). The sapwood is discolored by blue stain, a fungus carried by the mountain pine beetle. If the white pine continues to stand after death, heart cracks develop. These factors work together to cause degradation of the lumber and consequent loss in value.

Statement of Problem

Describes problem ▸ White pine mortality reduces the value of white pine stumpage because the commercial lumber market will not accept dead wood. The major implications of this problem are two: first, in the face of rising demand for wood, vast amounts of timber lie unused; second, dead trees are left to accumulate in the woods, where they are rapidly becoming a major fire hazard here in northern Idaho and elsewhere.

Proposed Solution

Describes one possible solution ▸ One possible solution to the problem of white pine mortality and waste is to search for markets other than the conventional lumber market. The last few years have seen a burst of popularity and growing demand for weathered barn boards and wormy pine for interior paneling. Some firms around the country are marketing defective wood as specialty products. (These firms call the wood from which their products come "distressed," a term I will use hereafter to refer to dead and defective white pine.) Distressed white pine quite possibly will find a place in such a market.

Scope

Defines scope of the proposed study ▸ To assess the feasibility of developing a market for distressed white pine, I plan to pursue six areas of inquiry:

Figure 22.2 **A Research Proposal**

1. What products presently are being produced from dead wood, and what are the approximate costs of production?
2. How large is the demand for distressed-wood products?
3. Can distressed white pine meet this demand as well as other species meet it?
4. Does the market contain room for distressed white pine?
5. What are the costs of retrieving and milling distressed white pine?
6. What prices for the products can the market bear?

Methods

My primary data sources will include consultations with Dr. James Hill, Professor of Wood Utilization, and Dr. Sven Bergman, Forest Economist—both members of the College of Forestry, Wildlife, and Range. I will also inspect decks of dead white pine at several locations and visit a processing mill to evaluate it as a possible base of operations. I will round out my primary research with a letter and telephone survey of processors and wholesalers of distressed material.

Describes how study will be done

Secondary sources will include publications on the uses of dead timber, and a review of a study by Dr. Hill on the uses of dead white pine.

Mentions literature review

My Qualifications

I have been following Dr. Hill's study on dead white pine for two years. In June of this year I will receive my B.S. in forest management. I am familiar with wood milling processes and have firsthand experience at logging. My association with Drs. Hill and Bergman gives me the opportunity for an in-depth feasibility study.

Cites a major reference and gives the writer's qualifications for this project

Conclusion

Clearly, action is needed to reduce the vast accumulations of dead white pine in our forests—among the most productive forests in northern Idaho. By addressing the six areas of inquiry mentioned earlier, I can determine the feasibility of directing capital and labor to the production of distressed white pine products. With your approval I will begin research at once.

Encourages reader acceptance

Figure 22.2 **Continued**

requests his instructor's authorization to do a feasibility study (Chapter 21) that will produce an analytical report for potential investors. Dewoody's proposal clearly answers the questions about *what, why, how, when,* and *where.* Because this proposal is addressed to an internal reader, it is cast as a memo. The instructor might request the memo as an email attachment or uploaded to the class learning management system.

Sales Proposals

The role of sales proposals

Sales proposals offer services or products and may be either solicited or unsolicited. If the proposal is solicited, several firms may be competing for the contract; in such cases, submitted proposals may be ranked by a committee.

A successful sales proposal persuades customers that your product or service surpasses those of competitors. In the following solicited proposal (Figure 22.3), the writer explains why her machinery is best for the job, how the job can be done efficiently, what qualifications her company can offer, and what costs are involved. To protect herself, she points out possible causes of increased costs. Because this document is addressed to an external reader, it is cast as a letter.

> NOTE *Never underestimate costs by failing to account for and acknowledge all variables—a sure way to lose money or clients.*

The proposal categories (planning, research, and sales) discussed in this section are not mutually exclusive. A research proposal, for example, may request funds for a study that will lead to a planning proposal. The Vista proposal partially shown here combines planning and sales features; if clients accept the preliminary plan, they will hire the firm to install the automated system.

> NOTE *Proposals can be cast as letters if the situation calls for the proposal to be brief. If the situation requires a longer proposal, you may need to include most or all of the components listed in the proposal outline (see "Structuring a Proposal" in this chapter) as well as supplements (see "Supplements Tailored for a Diverse Audience" in this chapter and Chapter 21, "Front and End Matter Supplements").*

Elements of a Persuasive Proposal

22.4 Identify the elements that are typically part of an effective proposal

Proposal reviewers expect a clear, informative, and realistic presentation. They will evaluate your proposal on the basis of the following quality indicators. (See also the criteria listed in "The Proposal Process" earlier in this chapter).

A Forecasting Title or Subject Line

Provide a clear forecast

Announce the proposal's purpose and content with an informative title such as "Recommended Wastewater Treatment System for the Mudpie Resorts and Spa" (for a formal proposal) or with a subject line (in an informal proposal). Instead of a vague title

Modern Landscaping
23–44 18th Street
Sunnyside, NY 11104

October 4, 20XX

Martin Haver
35–66 114th Avenue
Jamaica, NY 11107

Subject: Proposal to dig a trench and move boulders at Bliss site

Dear Mr. Haver:

I've inspected your property and would be happy to undertake the landscaping ◄——————● Describes the
project necessary for the development of your farm.   subject and
                                                                                                                          purpose

The backhoe I use cuts a span 3 feet wide and can dig as deep as 18 feet—more
than an adequate depth for the mainline pipe you wish to lay. Because this backhoe is on
tracks rather than tires and is hydraulically operated, it is particularly efficient in moving ◄——● Gives the
rocks. I have more than twelve years of experience with backhoe work and have   writer's
completed many jobs similar to this one.   qualifications

After examining the huge boulders that block access to your property, I am ◄—————————● Explains how
convinced they can be moved only if I dig out underneath and exert upward   the job will be
pressure with the hydraulic ram while you push forward on the boulders with your   done
D-9 Caterpillar. With this method, we can move enough rock to enable you to farm ◄————● Maintains a
that now inaccessible tract. Because of its power, my larger backhoe will save you   confident tone
both time and money in the long run.   throughout

This job should take 12 to 15 hours, unless we encounter subsurface ledge ◄——————● Gives a
formations. My fee is $200 per hour. The fact that I provide my own dynamiting crew   qualified cost
at no extra charge should be an advantage to you because you have so much rock   estimate
to be moved.

Please contact me any time for more information (sharon@mlandscape.com or ◄————● Encourages
814-551-2222). I'm sure we can do the job economically and efficiently.   reader acceptance
                                                                                                                          by emphasizing
                                                                                                                          economy and
Sincerely yours,   efficiency
Sharon Ingram
Sharon Ingram

Figure 22.3 **A Sales Proposal**

such as "Proposed Office Procedures for Vista Freight Company," be specific: "A Proposal for Automating Vista's Freight Billing System." An overworked reviewer facing a stack of proposals might very well decide that the proposal lacking a clear, focused title or subject line probably also will be unclear and unfocused in its content—and set it aside.

Background Information

A background section can be brief or long. If the reader is familiar with the project, a quick reminder of the context is sufficient:

Brief background, for readers familiar with the context

> **Background**
>
> Vista provides two services: (1) They locate freight carriers for their clients. The carriers, in turn, pay Vista a 6 percent commission for each referral. (2) Vista also handles all shipping paperwork for its clients. For this auditing service, clients pay Vista a monthly retainer.

In an unsolicited proposal, you may need to provide a longer introduction. If the topic warrants, the background section may take up several pages.

Statement of the Problem

The problem and its resolution form the backbone of any proposal. Show that you clearly understand your clients' problems and their expectations, and then offer an appropriate solution.

Describes problem and its effects

> **Statement of the Problem**
>
> Although Vista's business has increased steadily for the past three years, record keeping, accounting, and other paperwork are still done almost entirely by hand and on paper. These inefficient procedures have caused a number of problems, including late billings, lost commissions, and poor account maintenance. Updated office procedures and technologies seem crucial to competitiveness and continued growth.

Description of Solution

The proposal audience wants specific suggestions for meeting their specific needs. Their biggest question is: "What will this plan do for me?" In the following proposal for automating office procedures at Vista, Inc., Gerald Beaulieu begins with a clear assessment of needs and then moves quickly into a proposed plan of action.

Describes plan to solve the problem

> **Objective**
>
> This proposal offers a realistic and effective plan for streamlining Vista's office procedures. We first identify the burden imposed on your staff by the current system, and then we show how to reduce inefficiency, eliminate client complaints, and improve your cash flow by automating most office procedures.

A Clear Focus on Benefits

Conduct a detailed audience and use analysis to identify readers' major concerns and to anticipate likely questions and objections. Show that you understand what readers (or their organization) will gain by adopting your plan. The following list spells the exact tasks Vista employees will be able to accomplish once the proposed plan is implemented.

> Once your digital, automated system is operational, you will be able to
>
> - identify cost-effective carriers
> - coordinate shipments (which will ensure substantial client discounts)
> - deliver commission bills via print or secure shared files
> - track shipments online, by weight, miles, fuel costs, and destination
> - provide clients with secure online access to weekly audit reports on their shipments
> - bill clients on a 25-day cycle
> - produce weekly or monthly reports
>
> Additional benefits include eliminating repetitive tasks, improving cash flow, and increasing productivity.

Relates benefits directly to client's needs

(Each of these benefits will be described at length later in the "Plan" section.)

Honest and Supportable Claims

Because they typically involve expenditures of large sums of money as well as contractual obligations, proposals require a solid ethical and legal foundation. Clients in these situations often have doubts or objections about time and financial costs and a host of other risks involved whenever any important project is undertaken. Your proposal needs to address these issues openly and honestly. For example, if you are proposing to install customized virus-protection software, be clear about what this software cannot accomplish under certain circumstances. False or exaggerated promises not only damage reputations, but also invite lawsuits. (For more on supporting your claims, see Chapter 3, "Support Your Claims Convincingly.")

Promise only what you can deliver

Here is how the Vista proposal qualifies its promises:

> As countless firms have learned, imposing automated procedures on employees can create severe morale problems—particularly among senior staff who feel coerced and often marginalized. To diminish employee resistance, we suggest that your entire staff be invited to comment on this proposal. To help avoid hardware and software problems once the system is operational, we have included recommendations and a budget for staff training. (Adequate training is essential to the automation process.)

Anticipates a major objection and offers a realistic approach

If the best available solutions have limitations, say so. Notice how the above solutions are qualified ("diminish" and "help avoid" instead of "eliminate") so as not to promise more than the plan can achieve.

A proposal can be judged fraudulent if it misleads potential clients by

- making unsupported claims,
- ignoring anticipated technical problems, or
- knowingly underestimating costs or time requirements.

Major ethical and legal violations in a proposal

For a project involving complex tasks or phases, provide a realistic timetable (perhaps using a Gantt chart, as discussed in Chapter 12, "Gantt and PERT Charts" and

Figure 12.19) to show when each major phase will begin and end. Also provide a realistic, accurate budget, with a detailed cost breakdown (for supplies and equipment, travel, research costs, outside contractors, or the like) to show clients exactly how the money is being spent. For a sample breakdown of costs, see the construction repair proposal (Chapter 15, Figure 15.6).

> NOTE *Be certain that you spend every dollar according to the allocations that have been stipulated. For example, if a grant award allocates a certain amount for "a research assistant," be sure to spend that exact amount for that exact purpose—unless you receive written permission from the granting agency to divert funds for other purposes. Keep strict accounting of all the money you spend. Proposal experts Friedland and Folt remind us that "[f]inancial misconduct is never tolerated, regardless of intent" (161). Even an innocent mistake or accounting lapse on your part can lead to charges of fraud.*

Appropriate Detail

Vagueness in a proposal is fatal. Spell everything out. Instead of writing, "We will install state-of-the-art equipment," enumerate the products or services to be provided.

Spells out what will be provided

> To meet your requirements, we will install 12 iMac desktop computers, each with 2 TB hard drives. The system will be configured for secure file transfer between office locations. The plan also includes network printers with four HP LaserJet CP2020 color printers and one HP DesignJet T120 printer/scanner.

To avoid misunderstandings that could produce legal complications, a proposal must elicit *one* interpretation only.

Place support material (maps, blueprints, specifications, calculations) in an appendix so as not to interrupt the discussion.

> NOTE *While concrete and specific detail is vital, never overburden reviewers with needless material. A precise audience and use analysis (Chapter 2) can pinpoint specific information needs.*

Readability

A readable proposal is straightforward, easy to follow, and understandable. Avoid language that is overblown or too technical for your audience. Review Chapter 11 for style strategies.

A Tone That Connects with Readers

Your proposal should move people to action. Review Chapter 4 for persuasion guidelines. Keep your tone confident and encouraging, not bossy and critical. For more on tone, see Chapter 11, "Adjusting Your Tone."

Visuals

Emphasize key points in your proposal with relevant tables, flowcharts, and other visuals (Chapter 12), properly introduced and discussed.

As the flowchart (Figure 1) illustrates, Vista's manual routing and billing system creates redundant work for your staff. The routing sheet alone is handled at least six times. Such extensive handling leads to errors, misplaced paperwork, and late billing.

Visual repeats, restates, or reinforces the prose

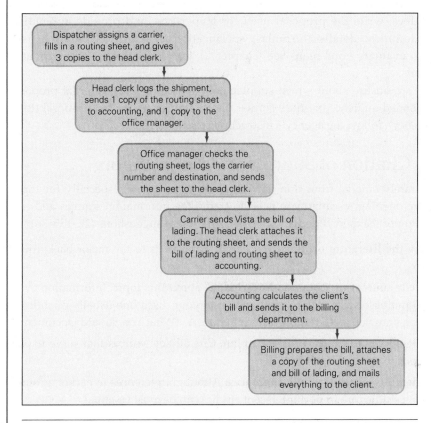

Figure 1 **Flowchart of Vista's Manual Routing and Billing System**

Accessible Page Design

Yours might be one of several proposals being reviewed. Help the audience to find what they need quickly by creating a document that's easy to navigate and makes information accessible to all readers. Review Chapter 13 for page and document design strategies.

Supplements Tailored for a Diverse Audience

A single proposal often addresses a diverse audience: executives, managers, technical experts, attorneys, politicians, and so on. Various reviewers are interested in different parts of your proposal. Experts look for the technical details. Others might be interested in the recommendations, costs, timetable, or expected results, but they will also need an explanation of technical details.

Analyze the specific needs and interests of each major reviewer

Give each major reviewer what he or she expects

If the primary audience is expert or informed, you can write the proposal text using technical language. For uninformed secondary reviewers (if any), you should provide an informative abstract, a glossary, and appendices explaining specialized information. If the primary audience has no expertise and the secondary audience does, write the proposal itself for laypersons, and provide appendices with the technical details (formulas, specifications, calculations) that experts will use to evaluate your plan. See Chapter 21 for front-matter and end-matter supplements.

If you are unsure about which supplements to include in an internal proposal, ask the intended audience or study similar proposals. For a solicited proposal (to an outside agency), follow the agency's instructions exactly.

Proper Citation of Sources and Contributors

Proposals rarely emerge from thin air. Whenever appropriate, especially for topics that involve ongoing research, you need to credit key information sources and contributors. Proposal experts Friedland and Folt offer these suggestions (22, 135–36):

How to cite sources and contributors

- **Review the literature on this topic.** Limit your focus to the major background studies.

- **Don't cite sources of "common knowledge" about this topic.** Information available in multiple sources or readily known in your discipline usually qualifies as common knowledge. (For more, see Appendix A, "What You Should Document.")

- **Provide adequate support for your plan.** Cite all key sources that serve to confirm your plan's feasibility.

- **Provide up-to-date principal references.** Although references to earlier, groundbreaking studies are important, recent studies can be most essential.

- **Present a balanced, unbiased view.** Acknowledge sources that differ from or oppose your point of view; explain the key differences among the various viewpoints before making your case.

- **Give credit to all contributors**. Recognize everyone who has worked on or helped with this proposal: for example, coauthors, editors, data gatherers, and people who contributed ideas.

Proper citation is not only an ethical requirement, but also an indicator of your proposal's feasibility. See Appendix A, "A Quick Guide to Documentation" for more on citation techniques.

Structuring a Proposal

22.5 Follow an introduction/body/conclusion structure in a proposal

Depending on a proposal's complexity, each section contains some or all of the components listed in the following general outline:

I. Introduction
  A. Statement of Problem and Objective/Project Overview
  B. Background and Review of the Literature (as needed)
  C. Need
  D. Benefits
  E. Qualifications of Personnel
  F. Data Sources
  G. Limitations and Contingencies
  H. Scope
II. Plan
  A. Objectives and Methods
  B. Timetable
  C. Materials and Equipment
  D. Personnel
  E. Available Facilities
  F. Needed Facilities
  G. Cost and Budget
  H. Expected Results
  I. Feasibility
III. Conclusion
  A. Summary of Key Points
  B. Request for Action
IV. Works Cited

These components can be rearranged, combined, divided, or deleted as needed. Not every proposal will contain all components; however, each major section must persuasively address specific information needs as illustrated in the sample proposal in the "Structuring a Proposal" section in this chapter.

Introduction

From the beginning, your goal is to *sell your idea*—to demonstrate the need for the project, your qualifications for tackling the project, and your clear understanding of what needs to be done and how to proceed. Readers quickly lose interest in a wordy, evasive, or vague introduction.

Following is the introduction for a planning proposal titled "Proposal for Solving the Noise Problem in the University Library." Jill Sanders, a library work-study student, addresses her proposal to the chief librarian and the administrative staff. Because this proposal is unsolicited, it must first make the problem vivid through details that arouse concern and interest. This introduction is longer than it would be in a solicited proposal, whose audience would already agree on the severity of the problem.

> NOTE Title page, informative abstract, table of contents, and other front-matter and end-matter supplements that ordinarily accompany long proposals of this type are omitted here to save space. See Chapter 21 for discussion and examples of each type of supplement. Also see Figure 21.3 in this chapter.

INTRODUCTION

Statement of Problem

During the October 20XX Convocation at Margate University, students and faculty members complained about noise in the library. Soon afterward, areas were designated for "quiet study," but complaints about noise continue. To create a scholarly atmosphere, the library should take immediate action to decrease noise.

Objective

This proposal examines the noise problem from the viewpoint of students, faculty, and library staff. It then offers a plan to make areas of the library quiet enough for serious study and research.

Sources

My data come from a university-wide questionnaire; interviews with students, faculty, and library staff; inquiry letters to other college libraries; and my own observations for three years on the library staff.

Details of the Problem

This subsection examines the severity and causes of the noise.

Severity. Since the 20XX Convocation, the library's fourth and fifth floors have been reserved for quiet study, but students hold group study sessions at the large tables and disturb others working alone. The constant use of digital devices (mostly laptops and phones) on both floors adds to the noise, especially when students talk and discuss. Moreover, people often are talking as they enter or leave study areas.

On the second and third floors, designed for reference, staff help patrons locate materials, causing constant shuffling of people and books, as well as loud conversation. At the information technology service desk on the third floor, conferences between students and instructors create more noise.

The most frequently voiced complaint from the faculty members interviewed was about the second floor, where people using the Reference and Government Documents services converse loudly. Students complain about the lack of a quiet spot to study, especially in the evening, when even the "quiet" floors are as noisy as the dorms.

More than 80 percent of respondents (530 undergraduates, 30 faculty, 22 graduate students) to an online university-wide questionnaire (Appendix A) insisted that excessive noise discourages them from using the library as often as they would prefer. Of the student respondents, 430 cited quiet study as their primary reason for wishing to use the library.

The library staff recognizes the problem but has insufficient personnel. Because all staff members have assigned tasks, they have no time to monitor noise in their sections.

Causes. Respondents complained specifically about these causes of noise (in descending order of frequency):

1. Loud study groups that often lapse into social discussions
2. General disrespect for the library, with some students' attitudes characterized as "rude," "inconsiderate," or "immature"
3. On all five floors, the constant sound of people typing on laptops and talking (including using voice commands for searching on their phones)
4. Vacuuming by the evening custodians

Concise descriptions of problem and objective immediately alert the readers

This section comes early because it is referred to in the next section

Details help readers understand the problem

Shows how campus feels about problem

Shows concern is widespread and pervasive

Identifies specific causes

All complaints converged on lack of enforcement by library staff. Because the day staff works on the first three floors, quiet-study rules are not enforced on the fourth and fifth floors. Work-study students on these floors have no authority to enforce rules not enforced by the regular staff. Small, black-and-white "Quiet Please" signs posted on all floors go unnoticed, and the evening security guard provides no deterrent.

Needs

Excessive noise in the library is keeping patrons away. By addressing this problem immediately, we can help restore the library's credibility and utility as a campus resource. We must reduce noise on the lower floors and eliminate it from the quiet-study floors.

Needs statement evolves logically and persuasively from earlier evidence

Scope

The proposed plan includes a detailed assessment of methods, costs and materials, personnel requirements, feasibility, and expected results.

Previews the plan

Body

The body (or plan section) of your proposal will receive the most audience attention. The main goal of this section is to prove your plan will work. Here you spell out your plan in enough detail for the audience to evaluate its soundness. If this section is vague, your proposal stands no chance of being accepted. Be sure that your plan is realistic and promises no more than you can deliver.

PROPOSED PLAN

This plan takes into account the needs and wishes of our campus community, as well as the available facilities in our library.

Phases of the Plan

Noise in the library can be reduced in three complementary phases: (1) improving publicity, (2) shutting down and modifying our facilities, and (3) enforcing the quiet rules.

Tells how plan will be implemented

Improving Publicity.   First, the library must publicize the noise problem. This assertive move will demonstrate the staff's interest. Publicity could include articles by staff members in the campus newspaper, leaflets distributed on campus, and a freshman library orientation acknowledging the noise problem and asking for cooperation from new students. All forms of publicity should detail the steps being taken by the library to solve the problem.

Describes first phase

Shutting Down and Modifying Facilities.   After notifying campus and local newspapers, you should close the library for one week. To minimize disruption, the shutdown should occur between the end of summer school and the beginning of the fall term.

During this period, you can convert the fixed tables on the fourth and fifth floors to cubicles with temporary partitions (six cubicles per table). You could later convert the cubicles to shelves as the need increases.

Then you can take all unfixed tables from the upper floors to the first floor, and set up a space for group study. Plans are already under way for removing the computer terminals from the fourth and fifth floors.

Describes second phase

Describes third
phase

Enforcing the Quiet Rules.
Enforcement is the essential long-term element in this plan. No one of any age is likely to follow all the rules all the time—unless the rules are enforced.

First, you can make new "Quiet" posters to replace the present, innocuous notices. A visual-design student can be hired to draw up large, colorful posters that attract attention. Either the design student or the university print shop can take charge of poster production.

Next, through publicity, library patrons can be encouraged to demand quiet from noisy people. To support such patron demands, the library staff can begin monitoring the fourth and fifth floors, asking study groups to move to the first floor, and revoking library privileges of those who refuse. Patrons on the second and third floors can be asked to speak in whispers. Staff members should set an example by regulating their own voices. Students who need to work in groups should be directed to the group study areas in the student learning commons.

Costs and Materials

Estimates costs
and materials
needed

- The major cost would be for salaries of new staff members who would help monitor. Next year's library budget, however, will include an allocation for four new staff members.
- A design student has offered to make up four different posters for $200. The university printing office can reproduce as many posters as needed at no additional cost.
- Prefabricated cubicles for 26 tables sell for $150 apiece, for a total cost of $3,900.
- Rearrangement on various floors can be handled by the library's custodians.
- The student learning commons is already available and so there is no cost to redirecting students to that space.

The Student Fee Allocations Committee and the Student Senate routinely reserve funds for improving student facilities. A request to these organizations would presumably yield at least partial funding for the plan.

Personnel

Describes
personnel
needed

The success of this plan ultimately depends on the willingness of the library administration to implement it. You can run the program itself by committees made up of students, staff, and faculty. This is yet another area where publicity is essential to persuade people that the problem is severe and that you need their help. To recruit committee members from among students, you can offer Contract Learning credits.

The proposed committees include an Anti-noise Committee overseeing the program, a Public Relations Committee, a Poster Committee, and an Enforcement Committee.

Feasibility

Assesses
probability of
success

On March 15, 20XX, I surveyed twenty-five New England colleges, inquiring about their methods for coping with noise in the library. Among the respondents, sixteen stated that publicity and the administration's attitude toward enforcement were main elements in their success.

Improved publicity and enforcement could also work for us. And slight modifications in our facilities, to concentrate group study on the busiest floors, would automatically lighten the burden of enforcement.

Benefits

Offers a realistic
and persuasive
forecast of
benefits

Publicity will improve communication between the library and the campus. An assertive approach will show that the library is aware of its patrons' needs and is willing to meet those needs. Offering the program for public inspection will draw the entire community into improvement efforts. Publicity, begun now, will pave the way for the formation of committees.

The library shutdown will have a dual effect: It will dramatize the problem to the community, and it will provide time for the physical changes. (An anti-noise program begun with carpentry noise in the quiet areas would hardly be effective.) The shutdown will be both a symbolic and a concrete measure, leading to the reopening of the library with a new philosophy and a new image.

Continued strict enforcement will be the backbone of the program. It will prove that staff members care enough about the atmosphere to jeopardize their friendly image in the eyes of some users, and that the library is not afraid to enforce its rules.

Conclusion

The conclusion reaffirms the need for the project and induces the audience to act. End on a strong note, with a conclusion that is assertive, confident, and encouraging—and keep it short.

CONCLUSION AND RECOMMENDATION

The noise in Margate University Library has become embarrassing and annoying to the whole campus. Forceful steps are needed to restore the academic atmosphere.

Aside from the intangible question of image, close inspection of the proposed plan will show that it will work if the recommended steps are taken and—most important—if daily enforcement of quiet rules becomes a part of library policy.

Reemphasizes need and feasibility and encourages action

In long, formal proposals, especially those beginning with a comprehensive abstract, the conclusion can be omitted.

A Situation Requiring a Formal Proposal

22.6 Write a formal proposal

The proposal that follows is essentially a grant proposal since it requests funding for a nonprofit enterprise. Notice how it adapts elements from the sample outline (see the beginning of "Structuring a Proposal" in this chapter). As in any funding proposal, a precise, realistic plan and an itemized budget provide the justification for the requested financial support.

The Situation

Southeastern Massachusetts University's newspaper, the SMU *Torch*, is struggling to meet rising production costs. The paper's yearly budget is funded by the Student Fee Allocation Committee, which disburses money to various campus organizations. Drastic budget cuts have resulted in reduced funding for all state schools. As a result, the newspaper has received no funding increase for the last three years. Meanwhile, production costs keep rising.

Bill Trippe, the *Torch*'s business manager, has to justify a requested increase of 17.3 percent for the coming year's budget. Before drafting his proposal (Figure 22.4), Bill constructs an audience and use profile (based on the worksheet in Chapter 2, "Develop an Audience and Use Profile").

Audience and Use Profile

The primary audience includes all members of the Student Fee Allocation Committee. The secondary audience is the newspaper staff, who will implement the proposed plan—if it is approved by the committee.

The primary audience will use this document as perhaps the sole basis for deciding whether to grant the additional funds. Most of these readers have overseen the newspaper budget for years, and so they already know quite a bit about the newspaper's overall operation. But they still need an item-by-item explanation of the conditions created by problems with funding and ever-increasing costs. Probable questions Bill can anticipate:

- Why should the paper receive priority over other campus organizations?
- Just how crucial is the problem?
- Are present funds being used efficiently?
- Can any expenses be reduced?
- How would additional funds be spent?
- How much will this increase cost?
- Will the benefits justify the cost?

The primary audience often has expressed interest in this topic. But they are likely to object to any request for more money by arguing that everyone has to economize in these difficult times. Thus their attitude could be characterized as both receptive and hesitant. (Almost every campus organization is trying to make a case for additional funds.)

Bill knows most of the committee members pretty well, and he senses that they respect his management skills. But he still needs to spell out the problem and propose a realistic plan, showing that the newspaper staff is sincere in its intention to eliminate nonessential operating costs. At a time when everyone is expected to make do with less, this proposal needs to make an especially strong case for salary increases (to attract talented personnel).

The primary audience has solicited this proposal, and so it is likely to be carefully read—but also scrutinized and evaluated for its soundness. Especially in a budget request, this audience expects no shortcuts: Every expense will have to be itemized, and the Costs section should be the longest part of the proposal.

To further justify the budget request, the proposal needs to demonstrate just how well the newspaper manages its present funds. In the Feasibility section, Bill provides

SMU *Torch*

Old Westport Road
North Dartmouth, Massachusetts 02747

May 1, 20XX

Charles Marcus, Chair
Student Fee Allocation Committee
Southeastern Massachusetts University
North Dartmouth, MA 02747

Dear Dean Marcus:

No one needs to be reminded about the effects of increased costs on our campus community. We are all faced with having to make do with less.

Accordingly, we at the *Torch* have spent long hours devising a plan to cope with increased production costs—without compromising the newspaper's tradition of quality service. I think you and your colleagues will agree that our plan is realistic and feasible. Even the "bare-bones" operation that will result from our proposed spending cuts, however, will call for a $6,478.57 increase in next year's budget.

We have received no funding increase in three years. Our present need is absolute. Without additional funds, the *Torch* simply cannot continue to function as a professional newspaper. I therefore submit the following budget proposal for your consideration.

Respectfully,

William Trippe

William Trippe
Business Manager, SMU *Torch*

— Letter of transmittal provides additional context and persuasion

Figure 22.4 **A Formal Proposal**

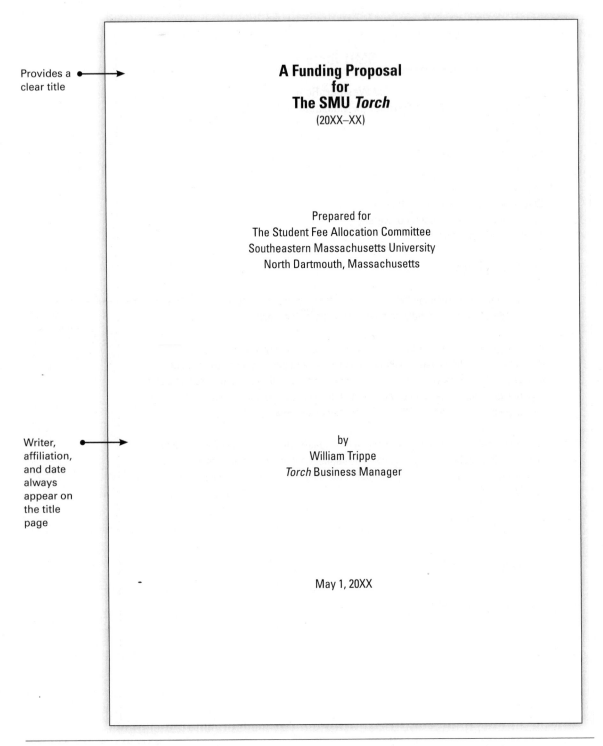

Provides a clear title ●——————▶

**A Funding Proposal
for
The SMU *Torch***
(20XX–XX)

Prepared for
The Student Fee Allocation Committee
Southeastern Massachusetts University
North Dartmouth, Massachusetts

Writer, affiliation, and date always appear on the title page ●——————▶

by
William Trippe
Torch Business Manager

May 1, 20XX

Figure 22.4 **Continued**

ii

Table of Contents   ← Table of
contents
orients
PAGE   readers and
demonstrates
proposal's

Figure 22.4 Continued

Informative Abstract

Abstract
accurately
encapsulates
entire
document

The SMU *Torch*, the student newspaper at Southeastern Massachusetts University, is crippled by inadequate funding, having received no budget increase in three years. Increased costs and inadequate funding are the major problems facing the *Torch*. Increases in costs of technology upgrades and in printing have called for cutbacks in production. Moreover, our low staff salaries are inadequate to attract and retain qualified personnel. A nominal pay increase would make salaries more competitive.

Our staff plans to cut costs by reducing page count and by hiring a new press for the *Torch*'s printing work. The only proposed cost increase (for staff salaries) is essential.

A detailed breakdown of projected costs establishes the need for a $6,478.57 budget increase to keep the paper a weekly publication with adequate page count to serve our campus.

Compared with similar newspapers at other colleges, the *Torch* makes much better use of its money. The comparison figures in the Appendix illustrate the cost-effectiveness of our proposal.

Figure 22.4 Continued

Introduction

Overview

Our campus newspaper faces the contradictory challenge of surviving ever-growing production costs while maintaining its reputation for quality. The following proposal addresses that crisis. This plan's ultimate success, however, depends on the Allocation Committee's willingness to approve a long-overdue increase in the *Torch*'s upcoming yearly budget. ◀— ● Opens with an overview of the situation

Background

In ten years, the *Torch* has grown in size, scope, and quality. Roughly 6,000 copies (24 pages/issue) are printed weekly for each fourteen-week semester. Each week, the *Torch* prints national and local press releases, features, editorials, sports articles, announcements, notices, classified ads, a calendar column, and letters to the editor. A vital part of university life, our newspaper provides a forum for information, ideas, and opinions—all with the highest professionalism. For several years, we have published an online version as well, but the print paper, available across campus and at many local coffee shops, is still a popular and important news vehicle for our campus community. ◀— ● Provides relevant background

Statement of Problem

With much of its staff about to graduate, the *Torch* faces next year with rising costs in every phase of production, and the need to replace outdated and worn equipment. ◀— ● Describes the problem concisely

Our newspaper also suffers from a lack of student involvement. Students gain valuable experience and potential career credentials and understand that our jobs are not paid at the same level as a comparable professional position. But our salaries are not even on par with campus-based student worker positions; therefore, students who want to work for us are forced to look elsewhere. Since more and more SMU students must work part-time, the *Torch* will have to make its salaries more competitive.

The newspaper's operating expenses can be divided into four categories: hardware and software upgrades, salaries, printing costs, and miscellaneous (office supplies, mail, and so on). The first three categories account for nearly 90 percent of the budget. Over the past year, costs in all categories have increased: from as little as 2 percent for miscellaneous expenses to as much as 19 percent for technology upgrades. Printing costs (roughly one-third of our total budget) rose 9 percent in the past year, and another price hike of 10 percent has just been announced.

Figure 22.4 Continued

Need

States the
need in
a clear,
succinct
manner

Despite growing production costs, the *Torch* has received no increase in its yearly budget allocation ($37,400) in three years. Inadequate funding is virtually crippling our newspaper.

Scope

Previews
the plan
before
getting
into details

The following plan includes
1. Methods for reducing production costs while maintaining the quality of our staff
2. Projected costs for technology upgrades, salaries, and services during the upcoming year
3. A demonstration of feasibility, showing our cost-effectiveness
4. A summary of attitudes shared by our personnel

Proposed Plan

This plan is designed to trim operating costs without compromising quality.

Methods

Itemizes
realistic
ways to
save
money
and retain
staff

We can overcome our budget and staffing crisis by taking these steps:

Reducing Page Count. By condensing free notices for campus organizations, abolishing "personal" notices, and limiting press releases to one page, we can reduce page count per issue from 24 to 20, saving nearly 17 percent in production costs. (Items deleted from hard copy could be linked as add-ons in the *Torch's* online version.)

Reducing Hard-Copy Circulation. Reducing circulation from 6,000 to 5,000 copies barely will cover the number of full-time students, but will save 17 percent in printing costs. We continue to track access to our online version, and although the numbers are rising, our print paper is still far more popular.

Hiring a New Press. We can save money by hiring Arrow Press for printing. Other presses (including our present printer) bid at least 25 percent higher than Arrow. With its state-of-the art production equipment, Arrow will import our "camera-ready" digital files to produce the hard-copy version. Moreover, no other company offers the rapid turnover time (from submission to finished product) that Arrow promises.

Figure 22.4 Continued

Upgrading Our Desktop Publishing Technology. To meet Arrow's specifications for submitting digital files, we must upgrade our equipment. Upgrade costs will be largely offset the first year by reduced printing costs. Also, this technology will increase efficiency and reduce labor costs, resulting in substantial payback on investment.

Increasing Staff Salaries. Although we seek talented students who expect little money and much experience, salaries for all positions must increase by an average of 25 percent. Otherwise, any of our staff could earn as much money elsewhere by working only a little more than half the time. In fact, many students could exceed the minimum wage by working for local newspapers. To illustrate: The *Standard Beacon* pays $60 to $90 per news article and $30 per photo; the *Torch* pays nothing for articles and $6 per photo.

A striking example of low salaries is the rate we pay our desktop publishing staff. Our present desktop publishing cost of $3,038 could be as much as $7,000 or even higher if we had this service done by an outside firm, as many colleges do. Without a nominal salary increase, we cannot possibly attract qualified personnel.

Costs

Our proposed budget is itemized in Table 1, but the main point is clear: If the *Torch* is to remain viable, increased funding is essential for meeting our projected costs.

Table 1 Projected Costs and Requested Funding for Next Year's *Torch* Budget

PROJECTED COSTS

Hardware/Software Upgrades

Apple iMac w/8 GB memory, 1TB HD	$1,162.00
HP Pavilion 2709m 27" (second monitor)	355.00
Seagate 8 TB external hard drive (for backups)	128.99
Olympus Stylus SH-1 digital camera	299.98
HP Scanjet 5000 sheet-feed scanner	799.00
Microsoft Office	499.00
Adobe Creative Suite	2,450.00
Subtotal	**$5,693.97**

Provides detailed breakdown of costs— the central issue in the situation

Figure 22.4 **Continued**

Wages and Salaries

Production staff	$5,880.00
Editor-in-Chief	3,150.00
News Editor	1,890.00
Features Editor	1,890.00
Advertising Manager	2,350.00
Advertising Designer	1,575.00
Webmaster	2,520.00
Layout Editor	1,890.00
Art Director	1,260.00
Photo Editor	1,890.00
Business Manager	1,890.00
Distributor	560.00
Subtotal	**$26,745.00**

Miscellaneous Costs

Graphics by SMU art students	$ 840.00
Mailing	1,100.00
Telephone	1,000.00
Campus print shop services	400.00
Copier fees	100.00
Subtotal	**$3,440.00**

Fixed Printing Costs (5,000 copies/wk x 28/wk)	**$24,799.60**

TOTAL YEARLY COSTS	**$60,678.57**
Expected Advertising Revenue ($600/wk x 28 wks)	**($16,800.00)**
Total Costs Minus Advertising Revenue	**$45,109.39**

TOTAL FUNDING REQUEST	**$43,878.57**

Figure 22.4 **Continued**

Feasibility

Beyond exhibiting our need, we feel that the feasibility of this proposal can be measured through an objective evaluation of our cost-effectiveness: Compared with newspapers at similar schools, how well does the *Torch* use its funding?

In a survey of the four area college newspapers, we found that the *Torch*—by a sometimes huge margin—makes the best use of its money per page. Table 1A in the Appendix shows that, of the five newspapers, the *Torch* costs students the least, runs the most pages weekly, and spends the least money per page, *despite a circulation two to three times the size of the other papers*.

The *Torch* has the lowest yearly cost of all five newspapers, despite having the largest circulation. With the requested budget increase, the cost would rise by only $0.88, for a yearly cost of $9.00 to each student. Although Alden College's newspaper costs each student $8.58, it is published only every third week, averages 12 pages per issue, and costs more than $71.00 yearly per page to print—in contrast to our yearly printing cost of $55.65 per page. As the figures in the Appendix demonstrate, our cost management is responsible and effective.

→ Assesses probability of success

Personnel

The *Torch* staff is determined to maintain the highest professionalism. Many are planning careers in journalism, writing, editing, advertising, photography, Web design, or public relations. In any *Torch* issue, the balanced, enlightened coverage is evidence of our judicious selection and treatment of articles and our shared concern for quality.

→ Addresses important issue of personnel

Conclusion

As a broad forum for ideas and opinions, the *Torch* continues to reflect a seriousness of purpose and a commitment to free and responsible expression. Its role in campus life is more vital than ever during these troubled times.

→ Reemphasizes need and encourages action

Every year, allocations to student organizations increase or decrease based on need. Last year, for example, eight allocations increased by an average of $4,332. The *Torch* has received no increase in three years.

Presumably, increases are prompted by special circumstances. For the *Torch*, these circumstances derive from increasing production costs and the need to update vital equipment. We respectfully urge the Committee to respond to the *Torch*'s legitimate needs by increasing next year's allocation to $43,878.57.

Figure 22.4 Continued

Appendix (Comparative Performance)

Table 1A Allocations and Performance of Five Local College Newspapers

Appendix provides detailed breakdown of cost comparisons

	Stonehorse College	Alden College	Simms University	Fallow State	SMU
Enrollment	1,600	1,400	3,000	3,000	5,000
Fee paid (per year)	$65.00	$85.00	$35.00	$50.00	$65.00
Total fee budget	$104,000	$119,000	$105,000	$150,000	$325,000
Newspaper budget	$18,300	$8,580	$36,179	$52,910	$37,392
					$45,109[a]
Yearly cost per student	$12.50	$8.58	$16.86	$24.66	$8.12
					$9.00[a]
Publication rate	Weekly	Every third week	Weekly	Weekly	Weekly
Average no. of pages	8	12	18	12	24
Average total pages	224	120	504	336	672
					560[a]
Yearly cost per page	$81.60	$71.50	$71.78	$157.47	$55.65
					$67.12[a]

[a]These figures are next year's costs for the SMU *Torch*.

Source: Figures were quoted by newspaper business managers in April 20XX.

Figure 22.4 Continued

a detailed comparison of funding, expenditures, and the size of the *Torch* in relation to the size of the newspapers of the four other local colleges. These are the facts most likely to persuade readers that the plan is cost-effective.

To organize his document, Bill (1) identifies the problem, (2) establishes the need, (3) proposes a solution, (4) shows that the plan is cost-effective, and (5) concludes with a request for action.

This audience expects a confident and businesslike—but not stuffy—tone.

Guidelines

for Proposals

➤ **Understand the audience's needs.** Demonstrate a clear understanding of the audience's problem, and then offer an appropriate solution.

➤ **Perform research as needed.** For example, you might research the very latest technology for solving a problem; compare the costs, benefits, and drawbacks of various approaches; contact others in your field for their suggestions; or find out what competitors are up to.

➤ **Credit all information sources and contributors.** If anything in your proposal represents the work or input of others, document the sources.

➤ **Use an appropriate format.** For an informal proposal distributed internally, use email or memo format. For an informal proposal distributed externally, use letter format. For a formal proposal, include all the required front matter and end matter.

➤ **Provide a clear title or subject line and background information.** Tell readers what to expect, and orient them with the appropriate background information.

➤ **Spell out the problem (and its causes).** Answer the implied question, "Why is this such a big deal?"

➤ **Point out the benefits of solving the problem.** Answer the implied question, "Why should we spend time, money, and effort to do this?"

➤ **Offer a realistic solution.** Stick to claims or assertions you can support. Answer the implied question, "How do we know this will work?" If the solution involves accounting for costs, budgeting time, or proving your qualifications, include this information.

➤ **Address anticipated objections to your plan.** Decision makers typically approach a proposal with skepticism, especially if the project will cost them money and time. Answer the implied question, "Why should we accept the items that seem costly with your plan?"

➤ **Include all necessary details, but don't overload.** Include as much supporting detail as you need to induce readers to say yes. Leave nothing to guesswork. At the same time, don't overload readers with irrelevant information.

➤ **Write clearly and concisely.** Use action verbs and plain English. Avoid terms that are too technical for your audience. If necessary for a mixed audience with differing technical levels, include a glossary.

➤ **Express confidence.** You are trying to sell yourself, your ideas, or your services to a skeptical audience. Offer the supporting facts ("For the third year in a row, our firm has been ranked as the number one architecture firm in the Midwest") and state your case directly ("We know you will be satisfied with the results").

➤ **Make honest and supportable claims.** If the solutions you offer have limitations, make sure you say so.

➤ **Induce readers to act.** Decide exactly what you want readers to do, and give reasons why they should be the ones to act. In your conclusion, answer the implied question, "What action am I supposed to take?"

Checklist

Proposals

Use the following Checklist when writing a proposal.

Content

❏ Are all required proposal elements included? (See "Elements of a Persuasive Proposal" in this chapter.)

❏ Does the title or subject line provide a clear forecast? (See "A Forecasting Title or Subject Line" in this chapter.)

❏ Is the background section appropriate for this audience's needs? (See "Background Information" in this chapter.)

❏ Is the problem clearly identified? (See "Statement of the Problem" in this chapter.)

❏ Is the objective clearly identified? (See "Description of Solution" in this chapter.)

❏ Does the proposal maintain a clear focus on benefits? (See "A Clear Focus on Benefits" in this chapter.)

❏ Are the claims honest and supportable? (See "Honest and Supportable Claims" in this chapter.)

❏ Does it address anticipated objections? (See "Honest and Supportable Claims" in this chapter.)

❏ Are the proposed solutions feasible and realistic? (See "Honest and Supportable Claims" in this chapter.)

❏ Are all foreseeable limitations and contingencies identified? (See "Honest and Supportable Claims" in this chapter.)

❏ Is the cost and budget section accurate and easy to understand? (See "Honest and Supportable Claims" in this chapter.)

❏ Is every *relevant* detail spelled out? (See "Appropriate Detail" in this chapter.)

❏ Are visuals used effectively? (See "Visuals" in this chapter.)

❏ Is each source and contribution properly cited? (See "Proper Citation of Sources and Contributors" in this chapter.)

Arrangement

❏ Does the introduction spell out the problem and preview the plan? (See "Introduction" in this chapter.)

❏ Does the body section explain how, where, when, and how much? (See "Body" in this chapter.)

❏ Does the conclusion encourage acceptance of the proposal? (See "Conclusion" in this chapter.)

❏ Is the informal proposal cast as a memo or letter, as appropriate? (See "Types of Proposals" in this chapter.)

❏ Does the formal proposal have adequate front matter and end matter supplements to serve the needs of different readers? (See "Types of Proposals" in this chapter.)

Style and Page Design

❏ Is the level of technicality appropriate for primary readers? (See Chapter 2, "Assess the Audience's Technical Background.")

❏ Does the tone encourage acceptance of the proposal? (See "A Tone That Connects with Readers" in this chapter and Chapter 11, "Adjusting Your Tone.")

❏ Is the writing clear, concise, and fluent? (See "Readability" in this chapter and Chapter 11, "Editing for Clarity," "Editing for Conciseness," and "Editing for Fluency.")

❏ Is the language precise? (See Chapter 11, "Finding the Exact Words.")

❏ Is the proposal grammatical? (See Appendix B, "Grammar.")

❏ Is the page design inviting and accessible? (See "Accessible Page Design" in this chapter and Chapter 13, "Page Design in Print and Digital Workplace Documents.")

Projects

For all projects, check with your instructor about whether to present your findings in class, bring drafts to class for discussion, upload your project to the class learning management system (LMS), and/or use the LMS forum or discussion boards to collaborate and review each activity below.

General

1. After identifying your primary and secondary audience, write a short planning proposal for improving an unsatisfactry situation in the classroom, on the job, or in your dorm or apartment (e.g., poor lighting, drab atmosphere, health hazards, poor seating arrangements). Choose a problem or situation whose solution or resolution is more a matter of common sense and lucid observation than of intensive research. Be sure to (a) identify the problem clearly, give a brief background, and stimulate interest; (b) clearly state the methods proposed to solve the problem; and (c) conclude with a statement designed to gain audience support.

2. Write a research proposal to your instructor (or an interested third party) requesting approval for your final term project (a formal analytical report or formal proposal). Verify that adequate primary and secondary sources are available. Convince your audience of the soundness and usefulness of the project.

3. As an alternate term project to the formal analytical report (Chapter 21), develop a long proposal for solving a problem, improving a situation, or satisfying a need in your school, community, or job. Choose a subject sufficiently complex to justify a formal proposal, a topic requiring research (mostly primary). Identify an audience (other than your instructor) who will use your proposal for a specific purpose. Complete an audience and use profile.

Team

Working in groups of four, develop an unsolicited planning proposal for solving a problem, improving a situation, or satisfying a need in your school, community, or workplace. Begin by brainstorming as a group to come up with a list of possible issues or problems to address in your proposal. Narrow your list, and work as a group to focus on a specific issue or idea. Your proposal should address a clearly identified audience of decision makers and stakeholders in the given issue. Complete an audience and use profile.

Digital and Social Media

As noted earlier, solicited proposals are usually written in response to a formal "request for proposal" (RFP). Many RFPs are available online; for instance, the National Science Foundation (NSF) provides RFPs for numerous funding opportunities. Go to the NSF Facebook page and look for topics that interest you. Then, go to their Web site and search for specific grants. For instance, you might want to focus on their Research Experiences for Undergraduates (REU) program and its different program areas (for example, astronomy; chemistry; computer science; ethics and values studies; physics). The Web site also offers examples of previously funded REU projects. Write a memo to your instructor explaining what you learned and what focus your proposal might take. Based on the information in this chapter, how would you make your best case?

Global

Compare the sorts of proposals regularly done in the United States with those created in other countries. For example, is the format the same? Are there more proposals for certain purposes in the United States than in another country? You can learn about this topic by interviewing an expert in international business (someone you meet on the job, during an internship, or through your adviser). You can also search online for information on international technical communication. Describe your findings in an informal report (memo format) that you will share with your classmates.

Chapter 23
Oral Presentations and Video Conferencing

Monkey Business Images/Shutterstock

❝In any particular week, I might need to give two or three presentations. Sometimes these are brief and informal, designed to provide my team an update on our project's status. Other times, the presentation needs to be quite detailed and formal—for instance, when I present to my manager and other division heads. Since it's never clear who will be attending in person and who will be connecting online, I always make sure my visuals are easy to understand and not too cluttered. Regardless of the technology, my primary goal is to make sure that I'm providing just the right amount of information for my audience and focusing on content that they need to know about.**❞**

—*Nicole Hillman, Industrial Engineer for a medical device company*

Learning Objectives

23.1 Identify the advantages and drawbacks of oral presentations

23.2 Follow the steps to plan your oral presentation

23.3 Prepare your oral presentation systematically

23.4 Use visuals to enhance your presentation and choose the right media format

23.5 Create audience-friendly slides using presentation apps

23.6 Deliver an effective oral presentation

23.7 Determine how and when to use video conferencing

In the workplace, professionals communicate their ideas via written documents but may also need to present their ideas by giving an oral presentation. These presentations vary in style, complexity, and formality; they may include convention speeches, reports at national meetings, technical briefings for colleagues, sales presentations, project updates, and speeches to community groups. Workplace presentations may be designed to *inform* (e.g., to describe new government safety requirements); to *instruct* (e.g., to show volunteers how to safely clean up an oil spill); to *persuade* (e.g., to induce company officers to vote for a pay raise); or to achieve some combination of all three of the above. The higher your status, on the job or in the community, the more you can expect to give oral presentations.

Most workplace presentations are designed and given using PowerPoint or other presentation software. In addition to PowerPoint slides, you may also need to use a whiteboard or flip chart to sketch out some ideas or to write down audience questions and suggestions. This chapter assumes in general the use of PowerPoint slides, but Figure 23.3 and other sections of this chapter offer ideas for using other media formats for workplace presentations. For all presentations, it's important to consider when to give handouts (before, during, or after the presentation). This topic is addresses later in this chapter.

Advantages and Drawbacks of Oral Presentations

23.1 Identify the advantages and drawbacks of oral presentations

Advantages

Unlike written documents, oral presentations are highly interactive. When you speak to people in person (in the same room or over a video connection or both), you use body language (vocal tone, eye contact, gestures) to establish credibility and rapport with your audience. A likable personality can have a powerful effect on how an audience understands and trusts the presentation and its contents. Also, unlike most written documents, oral presentations provide for give-and-take between you and your audience. As you listen and watch for audience reactions, you can adjust your presentation on the spot, and audience members can get their questions answered right away. Depending on the type of meeting, some audience members will be in the room, while others will join the meeting via an online video connection.

Drawbacks

In a written report you generally have to think about what you're saying and how you're saying it, and to revise the report until the message is just right. For an oral report, one attempt is basically all you get, and that pressure makes it easier to stumble. (People consistently rank fear of public speaking higher than fear of dying!) Also, an oral report is limited in the amount and complexity of information it can present. When people read a written report, they can follow at their own pace and direction, going back and forth, skimming some sections and studying others, creating bookmarks and coming back to the document later. In an oral presentation, on the other hand, you establish the pace and the information flow, thereby creating the risk of "losing" or boring the listeners.

Avoiding Presentation Pitfalls

Importance of avoiding presentation pitfalls

An oral presentation is only the tip of a pyramid built from many earlier labors. But such presentations often serve as the concrete measure of your overall job performance. In short, your audience's only basis for judgment may be the brief moments during which you stand before them.

Common presentation pitfalls

Oral presentations are an important part of workplace communication, and yet they often turn out to be boring, confusing, unconvincing, or too long. Many are delivered ineptly, with the presenter losing her or his place, fumbling through notes, apologizing for forgetting something, or reading word-for-word the bullet points on the screen. Table 23.1 lists some of the features that can go wrong. Avoid these difficulties by carefully analyzing, planning, preparing, and practicing.

Table 23.1 Common Pitfalls in Oral Presentations

Speaker ●●●	Visuals/Slides* * *	Setting ■■■
• makes no eye contact	* are nonexistent	■ is too noisy
• seems like a robot	* are hard to see	■ is too hot or cold
• hides behind the lectern	* are hard to interpret	■ is too large or small
• speaks too softly/loudly	* are out of sequence	■ is too bright for visuals
• sways, fidgets, paces	* are shown too rapidly	■ is too dark for notes
• rambles or loses her/his place	* are shown too slowly	■ has equipment missing
• never gets to the point	* have typos/errors	■ has broken equipment
• fumbles with notes	* are word-filled	
• has too much material		

Planning Your Presentation

23.2 Follow the steps to plan your oral presentation

Planning your presentation involves carefully analyzing your audience and purpose, analyzing your speaking situation, selecting a type of presentation, and selecting a delivery method. The following paragraphs explore these planning considerations in further depth.

Analyze Your Audience and Purpose

When analyzing audience and purpose for oral presentations, you'll need to consider a few items in addition to those you focused on when doing the same sort of analysis for written documents. Use the Audience and Purpose Profile Sheet in Figure 23.1 to help you get started; you can modify this document to fit your specific situation.

Start with an audience and use profile

Do all you can to find out exactly who will be attending your presentation and how many will be in the room versus connecting online. Determine the roles within the organization for each attendee. Learn about their attitudes and experiences regarding your subject. For example, managers may focus on the bottom line, while engineers may care more about the technical difficulties involved in the project you are proposing. Oral presentations are often delivered to a mixed group, so consider the attitudes of the group as a whole and speak to the needs of various factions. The group is also likely to consist of people from different cultural and linguistic backgrounds, so be sure to account for these differences (see Chapter 5 and the Consider This box later in this chapter).

Audience considerations

Take a few minutes and write out an audience and purpose statement. Who are your listeners? What do you want listeners to think, know, or do? (This statement can also serve as the introduction to your presentation.)

Assume, for example, that you represent an environmental engineering firm that has completed a study of groundwater quality in your area. The organization that

AUDIENCE

Primary audience members:_____
(names, titles)

Secondary audience members:_____

_____

Are most attendees of this presentation members of the same team or group?
Y   N

Relationship with audience members:_____
(client, employer, other)

Technical background of audience:_____
(layperson, expert, other)

Any international or non-native speakers of English in the audience?_____

How many people will attend in person?_____

How many connecting via Webinar or video?_____

PURPOSE

Primary purpose:_____(to inform or instruct;
to persuade; to propose an action; to sell something)

Secondary purpose(s):_____

ROOM or LOCATION for presentation

Conference table, theater style, other. Describe: _____

_____

**TECHNOLOGY (use space below to make notes about these and
related items)**

If using PowerPoint, do I need my own laptop?

(If not, how will I access my presentation: flash drive? network drive?)

Where is the projector located? Does the room have speakers?

Do I need anything else, such as a flip chart, whiteboard, or poster?

_____

_____

Figure 23.1 Audience and Purpose Profile Sheet for an Oral Presentation

sponsored your study has asked you to present an oral version of your written report, titled "Pollution Threats to Local Groundwater," at a town meeting. After careful thought, you settle on this statement:

> **Audience and Purpose:** By informing Cape Cod residents about the dangers to the Cape's freshwater supply posed by rapid population growth, this report is intended to increase local interest in the problem.

Audience and purpose statement

Now you are prepared to focus on the listeners and the speaking situation by asking these questions:

- Who are my listeners (strangers, peers, superiors, clients)?
- What is their attitude toward me or the topic (hostile, indifferent, needy, friendly)?
- Why are they here (they want to be here, are forced to be, are curious)?
- What kind of presentation do they expect (brief, informal; long, detailed; lecture)?
- What do these listeners already know (nothing, a little, a lot)?
- What do they need or want to know (overview, bottom line, nitty-gritty)?
- How large is their stake in this topic (about layoffs, new policies, pay raises)?
- Do I want to motivate, mollify, inform, instruct, or warn my listeners?
- What are their biggest concerns or objections about this topic?
- What do I want them to think, know, or do?

Questions for analyzing your listeners and purpose

Analyze Your Speaking Situation

The more you can discover about the circumstances, the setting, and the constraints for your presentation, the more deliberately you will be able to prepare.

Ask yourself these questions:

- How much time will I have to speak?
- Will other people be speaking before or after me?
- How formal or informal is the setting?
- How large is the audience?
- How large is the room?
- How bright and adjustable is the lighting?
- What equipment (laptop? projector? flip chart or white board?) is available?
- How much time do I have to prepare?

Questions for analyzing your speaking situation

Later parts of this chapter explain how to incorporate your answers to the above questions into the planning and preparation of your presentation.

Select a Type of Presentation

Your primary purpose determines the type of presentation required: informative presentation, training/instructional presentation, persuasive presentation, action plan

Types of oral presentations

presentation, or sales presentation. Keep in mind that all presentations have a primary purpose but that presentations may have secondary purposes that blend some of the types listed below. For example, a presentation about an unsafe work situation may be primarily persuasive (to convince listeners that the situation is unsafe) but may end with a few slides in the action plan mode (to present a plan for addressing these safety issues).

Informative presentations provide facts and explanations

INFORMATIVE PRESENTATIONS.   Informative presentations are often given at conferences, product update meetings, briefings, or class lectures. Your goal is to be as impartial as possible and to provide the best information you can. If your primary purpose is informative, observe these criteria:

- Keep the presentation title clear and factual.
- Stipulate at the outset that your purpose is simply to provide information.
- Be clear about the sources of information that you present or draw upon.

Training/ instructional presentations show how to perform a task

TRAINING/INSTRUCTIONAL PRESENTATIONS.   Training (or instructional) presentations can cover such topics as how to ensure on-the-job safety, how to use a specific software application, or how to exit a capsized kayak. Some technical communicators specialize in giving training presentations. If your primary purpose is instructional, observe these criteria:

- Use a title that indicates the training purpose of the presentation.
- Provide an overview of the learning outcomes—what participants can expect to learn from the presentation.
- Create slides or a handout that participants can reference later, when they are trying to perform the task(s) on their own.

Persuasive presentations attempt to gain support or change an opinion

PERSUASIVE PRESENTATIONS.   To influence people's thinking, give a persuasive presentation. For example, an engineer at a manufacturing plant may wish to persuade her peers that a standard procedure is unsafe and should be changed. In a persuasive situation, you need to perform adequate research so that you are well informed on all sides of the issue. If your primary purpose is persuasive, observe these criteria:

- Be clear from the start that you are promoting a particular point of view on the subject.
- Use research and visual data (charts, graphs, photographs) to support your stance.
- Consider and address counterarguments in advance ("Some might say that this approach won't work, but here is why it will") and be prepared to take questions that challenge your view.

Action plan presentations motivate people to take action

ACTION PLAN PRESENTATIONS.   To get something done, give an action plan presentation. For example, if you wanted your company to address a design flaw in one of its products, you would give a presentation that outlined the problem, presented a specific solution, and then encouraged audience members to implement the solution. If your primary purpose is to move people to take action, observe these criteria:

- Be clear up front about your purpose ("My primary purpose today is to ask you to act on this matter").
- Present the research to back up the need for your plan.
- Show that you have considered other plans but that yours is the most effective.
- In closing, restate what you want your audience to do.

SALES PRESENTATIONS.   Technical sales presentations need to be well researched. (At many high-tech companies, technical sales representatives are often scientists or engineers who understand the product's complexities and are effective communicators.) If your primary purpose is to sell something, observe these criteria:

Sales presentations inform and persuade

- Let the facts tell the story. Use examples to help explain why your product or service is the right one.
- Know your product or service—and those of your competitors—inside and out. Thorough knowledge of your own product or service and a well-researched competitive analysis will make your presentation that much more persuasive.
- Display sincere interest in the needs and concerns of your customers.
- Provide plenty of time for questions. You may think you have made an airtight case for your product or service, but you will need extra time to field questions and convince skeptical audience members.

Select a Delivery Method

Your presentation's effectiveness will depend largely on *how* the presentation connects with listeners. Different types of delivery create different connections.

Four types of delivery methods

THE MEMORIZED DELIVERY.   This type of delivery takes a long time to prepare, offers no chance for revision during the presentation, and spells disaster if you lose your place. Avoid a memorized delivery in most workplace settings.

THE IMPROMPTU DELIVERY.   An impromptu (off-the-cuff) delivery can be a natural way of connecting with listeners—but only when you really know your material, feel comfortable with your audience, and are in an informal speaking situation (group brainstorming, or responding to a question: "Tell us about your team's progress on the automation project"). Avoid impromptu deliveries for complex information, no matter how well you know the material. If you have minimal warning beforehand, at least jot down a few notes about what you want to say.

THE SCRIPTED DELIVERY.   For a complex technical presentation, a conference paper, or a formal speech, you may want to read some material verbatim from a prepared script. Scripted presentations work well if you have many details to present or if the content is extremely technical. Yet although a scripted delivery helps you control the content, it offers little chance for audience interaction and it can be boring. If you *do* have material that you need to read aloud, intersperse this reading within an *extemporaneous* presentation (see next item): for instance, if you have 15 slides in your presentation, you might have a few spots where you show a key quote from the written material but read the entire page out loud. Keep in mind that when reading from a prepared text, you should plan on roughly two minutes per double-spaced page. Rehearse until you are able to glance up from the page periodically without losing your place.

THE EXTEMPORANEOUS DELIVERY.   The most common kind of workplace presentation, an extemporaneous delivery is carefully planned, practiced, and based on notes that keep you on track. In this natural way of addressing an audience, you glance at your material and speak in a conversational style. Extemporaneous delivery is based on key ideas in sentence or topic outline form, often projected as PowerPoint slides (or similar apps).

The dangers in extemporaneous delivery are that you might lose track of your material, forget something important, say something unclearly, or exceed your time limit. Careful preparation is the key.

Table 23.2 summarizes the various uses and drawbacks of the most common types of delivery. In many instances, some combination of methods can be effective.

Table 23.2 A Comparison of Oral Presentation Methods

Delivery Method	* Main Uses *	• Main Drawbacks •
IMPROMPTU (inventing as you speak)	* in-house meetings	• offers no chance to prepare
	* small, intimate groups	• speaker might ramble or misstate key technical data
	* simple topics	• speaker might lose track
SCRIPTED (reading verbatim from a written work)	* formal speeches	• takes a long time to prepare
	* sections with highly technical content	• speaker can't move around
	* strict time limit	• limits human contact
	* cross-cultural audiences where translation might be an issue	• can appear stiff and unnatural
	* highly nervous speaker who is more comfortable reading straight from a script	• might bore listeners
		• makes working with visuals difficult
EXTEMPORANEOUS (speaking from an outline of key points)	* face-to-face presentations	• speaker might just read the slide bullet points
	* medium-sized, familiar groups	• speaker might leave something out
	* moderately complex topics	• slide after slide can be boring
	* somewhat flexible time limit	• speaker might exceed time limit
	* visually based presentations	• speaker might fumble with notes, visuals, or equipment

For example, in an orientation for new employees, you might prefer the flexibility of an extemporaneous format but also read a brief passage aloud from time to time (e.g., excerpts from the company's formal code of ethics). Most workplace presentations are extemporaneous.

Preparing Your Presentation

23.3 Prepare your oral presentation systematically

To stay in control and build confidence, plan your presentation systematically by researching your topic thoroughly, aiming for simplicity and conciseness, anticipating audience questions, and outlining your presentation.

Steps in preparing an oral presentation

Research Your Topic

Anyone can open a PowerPoint slide deck and create a presentation. But as with all technical communication, the most effective presentations are based on solid, credible research. Be prepared to support each assertion, opinion, conclusion, and recommendation with evidence and reason. Even if the details don't make it into your slides, have that material available to you (or memorized) in case listeners question your assumptions. Check your facts for accuracy. Begin gathering material well ahead of time. Use summarizing techniques from Chapter 10. If your presentation is based on a previously written report, you can expand your outline for the written report into an outline for the presentation.

Research your topic thoroughly

Aim for Simplicity and Conciseness

Boil the material down to a few main points. A typical attention span is about twenty minutes. Time yourself in practice sessions and trim as needed. (If your situation requires a lengthy presentation, plan a short break, about halfway.) Remember that at work, people are subjected to hours and hours of PowerPoint presentations, so the more concise and clear you can make yours, the better chance people will remember your key points.

Make your presentation easy to understand and remember

Anticipate Audience Questions

Consider those parts of your presentation that listeners might question or challenge. You might need to clarify or justify information that is new, controversial, disappointing, or surprising.

Be prepared for audience challenges

Outline Your Presentation

As noted above, if your presentation is based on a written report or other written document, you can use the outline from that document (or the document itself) to help shape the presentation (see also Chapter 10 for more on outlining). For most presentations, follow an introduction-body-conclusion outline format.

Use an introduction-body-conclusion format

Set the stage

INTRODUCTION.   The introduction should accomplish three things:

1. Capture your audience's attention by telling a quick story, asking a question, or relating your topic to a current event or to something else the audience cares about.
2. Establish credibility by stating your credentials or explaining where you obtained your information.
3. Preview your presentation by listing the main points and the overall conclusion.

You might try doing the first two items without a slide (just put up your first slide) so you can establish a personal connection with your listeners. Then, for the third item, create a simple slide listing your key points.

Use small chunks and transitions in the body

BODY.   Readers who get confused or want to know the scope of a print document can look back at the headings, table of contents, or previous pages. But oral presentations offer no such options. To make your presentation easy to follow, structure the material into small chunks. To signal that you are moving from one main point to another, use a transition statement such as, "Now that I've explained how to separate good information from bad, let me suggest how you can contribute to medical discussions on the Internet."

Tie everything together

CONCLUSION.   Your conclusion should return full circle to your introduction. Remind your audience of the big picture, restate the main points you've just covered, and leave listeners with some final advice or tips for locating more information.

The example below illustrates a complete presentation outline using the introduction-body-conclusion format. Notice how each sentence is a topic sentence for a paragraph that a well-prepared speaker can develop in detail.

Pollution Threats to Local Groundwater

Arnold Borthwick

Introduction

I. Introduction to the Problem
   A. Do you know what you are drinking when you turn on the tap and fill a glass?
   B. The quality of our water is good but not guaranteed to last forever.
   C. Cape Cod's rapid population growth poses a serious threat to our freshwater supply.
   D. Measurable pollution in some town water supplies has already occurred.
   E. What are the major causes and consequences of this problem and what can we do about it?

Body, divided into sections

II. Description of the Aquifer
   A. The groundwater is collected and held in an aquifer.
      1. This porous rock formation creates a broad, continuous arch beneath the entire Cape.
      2. The lighter freshwater flows on top of the heavier saltwater.
   B. This type of natural storage facility, combined with rapid population growth, creates potential for disaster.

III. Hazards from Sewage and Landfills
   A. With increasing population, sewage and solid waste from landfill dumps increasingly invade the aquifer.

B. The Cape's sandy soil promotes rapid seepage of wastes into the groundwater.

C. As wastes flow naturally toward the sea, they can invade the drawing radii of town wells.

IV. Hazards from Saltwater Intrusion

A. Increased population also causes overdraw on some town wells, resulting in saltwater intrusion.

B. Salt and calcium used in snow removal add to the problem by seeping into the aquifer from surface runoff.

V. Long-Term Environmental and Economic Consequences

A. The environmental effects of continuing pollution of our water table will be far-reaching.

    1. Drinking water will have to be piped in more than 100 miles from Quabbin Reservoir.

    2. The Cape's beautiful freshwater ponds will be unfit for swimming.

    3. Aquatic and aviary marsh life will be threatened.

    4. The Cape's sensitive ecology might well be damaged beyond repair.

B. Such environmental damage would, in turn, spell economic disaster for Cape Cod's major industry—tourism.

VI. Conclusion and Recommendations   Conclusion

A. This problem is becoming more real than theoretical.

B. The conclusion is obvious: If the Cape is to survive ecologically and financially, we must take immediate action to preserve our *only* water supply.

C. These recommendations offer a starting point for action.

    1. Restrict population density in all Cape towns by creating larger building lot requirements.

    2. Keep strict watch on proposed high-density apartment and condominium projects.

    3. Create a committee in each town to educate residents about water conservation.

    4. Prohibit salt, calcium, and other additives in sand spread on snow-covered roads.

    5. Explore alternatives to landfills for solid waste disposal.

D. Given its potential effects on our quality of life, such a crucial issue deserves the active involvement of every Cape resident.

If you will be using PowerPoint, along with a storyboard for visuals (see next section), your outline can become the framework for creating the slides. If you plan to give the presentation without slides (using, for example, a flip chart), transfer your outline to index cards and use these to help guide you through your presentation. But note that for most public presentations, people expect slides, especially because those slides can be uploaded to a Web site for viewing by those who connect via video or who can't attend the actual presentation.

Planning Your Visuals and Choosing a Media Format

23.4 Use visuals to enhance your presentation and choose the right media format

Using visuals as part of your presentation is critical. Slide after slide filled only with text is boring and unengaging. To help your audience stay focused and interested and to help them understand and retain the material, use visuals appropriately. The trick   Importance of using visuals is not to overuse visuals. Select those that will clarify and enhance your talk without making you or your main points fade into the background. Focus on the following decisions to help you use visuals effectively.

Decide Which Visuals to Use and Where to Use Them

Choose the most appropriate visuals

Should you use tables (numbers or text), graphics, charts, illustrations, or diagrams? How complex or simple should the visuals be for this audience? Chapter 12 can help you answer these questions. Then, determine how a visual will look as a PowerPoint slide. A nice, detailed table from your report might work great in a written document or PDF, but on a slide, the fonts might be too small for anyone to read.

Balance visuals with text

Next, decide *where* to use these visuals. Visuals are best used to emphasize a point, or any place where *showing* is more effective than just *telling*. You might have a text-only slide showing key data but then follow that slide with, say, a pie chart that shows the percentage distributions of that data in relation to the whole. Decide how many visuals are appropriate. Of the sample slides in Figure 23.4, two contain visuals. The photograph of a glass of water (Slide 2) creates visual interest in a slide that would otherwise be all text. The saltwater intrusion diagram (Slide 5) provides a visual understanding of the key points made in the previous slide (Slide 4). Keep in mind that for international audiences, visuals may be better than text. But you need to strike a balance: A good ratio would be one visual slide for every two to three text slides.

Create a Storyboard

Format for a storyboard

A presentation storyboard is a double-column format in which your discussion is outlined in the left column and aligned with the specific supporting visuals in the right column (Figure 23.2). Even though you can just open up PowerPoint and start creating slides without any prior planning, it's a good idea to create a quick storyboard to help you think through the structure and use of visuals beforehand. Using a storyboard lets you see the big picture before you get into the slide-by-slide approach when working with PowerPoint or similar apps.

Decide Which Visuals You Can Realistically Create

Fit each visual to the situation

Fit each visual to the situation. The visuals you select will depend on the room, the equipment, and the production resources available. Ask yourself how large the room is and how the room is set up. Some visuals work well in small rooms but not large ones and vice versa. How well can the room be darkened? Which lights can be left on? Can the lighting be adjusted selectively? What size should visuals be, to be seen clearly by the whole room? (A smaller, intimate room is usually better than a room that is too big.) Visuals that look great on your laptop computer may be unreadable when projected onto a large screen. Create a few samples and find a place to test out your slides before you create the entire presentation. As you prepare visuals, focus on economy, clarity, and simplicity.

Pollution Threats to Local Groundwater

I. Introduce the Problem   (slide: *forecast of presentation*)

    A.   Do you know what you are drinking when   (slide: *showing opening question and*
          you turn on the tap and fill the glass?   *photo of glass of water*)

    B.   The quality of our water is good but not
          guaranteed to last forever.

    C.   Cape Cod's rapid population growth poses a   (poster: *a line graph showing twenty-*
          serious threat to our freshwater supply.   *year population growth*)

    D.   Measurable pollution in some town water   (poster: *two side-by-side tables showing*
          supplies has already occurred.   *twenty-year increases in nitrate and chloride*
                                         concentrations in three town wells)

    E.   What are the major causes and consequences   (slide: *a list that previews my five subtopics*)
          of this problem and what can we do about it?

Figure 23.2 A Partial Storyboard This storyboard is based on a Presentation Outline.

BE SELECTIVE. Use a visual only when it truly serves a purpose. Use restraint in choosing what to highlight with visuals. Try not to begin or end the presentation with a visual. At those times, listeners' attention should be focused on the presenter, not a visual.

Prepare visuals that are clear and simple

MAKE VISUALS EASY TO READ AND UNDERSTAND. Think of each visual as an image that flashes before your listeners. They will not have the luxury of studying the visual at leisure. Listeners need to know at a glance what they are looking at and what it means. In addition to being able to *read* the visual, listeners need to *understand* it. Follow these suggestions:

- Make visuals large enough to be read anywhere in the room.
- Don't cram too many words, ideas, designs, or type styles into a single visual.
- Distill the message into the fewest words and simplest images possible.
- Chunk material into small sections.
- Summarize with key words, phrases, or short sentences.
- Check to see how your visual looks on a big screen (versus on your laptop).
- Give each visual a title that announces the topic.
- Use color sparingly to highlight key words, facts, trends, or the bottom line.
- Use the brightest color for what is most important.

- Label each part of a diagram or illustration.
- Proofread each visual carefully.

USE THE RIGHT APP TO PREPARE YOUR VISUALS Spreadsheets (such as Microsoft Excel or Apple Numbers) can generate charts and graphs based on numeric data sets. Word-processing programs (such as Microsoft Word or Apple Pages) let you create prose tables and simple graphics easily. Many people use the "draw" tool in PowerPoint or similar apps to create graphs, charts, and simple diagrams right within the slide deck. More sophisticated programs such as Adobe Illustrator can be used for line drawings and diagrams. You might also be able to take a simple photograph using your phone or a digital camera, or find a copyright-free photograph on the Internet. For more on creating visuals, see Chapter 12.

Choose the Right Media Format

Consider all the options

As discussed previously, in almost all workplace settings, PowerPoint is the standard presentation software. However, you may want to consider other media options, either instead of PowerPoint or in addition to your slides. Which format or combination is best for your topic, setting, and audience? What do listeners expect in this particular setting? Will some listeners be connecting online (such as Skype or an in-house system); if so, what format will allow them to participate, too? (See Figure 23.3 for a list of different media formats and the "Video Conferencing" section later in this chapter.)

Fit the format to the situation

- For weekly team meetings where everyone typically attends in person, and where the focus is project updates (and some brainstorming about next steps), writing ideas out on a whiteboard might suffice. The team leader can take a picture of the whiteboard and attach it to a follow-up email or post it to the appropriate wiki, blog, or discussion forum.
- Handouts can be useful, but whether to give these out before, during, or after the presentation is always a question. See "When and How to Use Handouts" later in this chapter for more on handouts.
- For immediate orientation, you might begin with a poster or flip chart that lists key ideas or themes to which you will refer repeatedly.
- For ideas best conveyed in film or video, try embedding a small bit of video directly into your slide set. (Or, you can use YouTube, Vimeo, or another video-sharing site.)

Figure 23.3 presents the various media that you can use for a presentation.

NOTE Keep in mind that the more technology you use in your presentation, the more prepared you must be for things to fail. Always have Plan B. Bring a flash drive of your presentation and have a hard copy outline in case you need to present without technology.

Whiteboard/Chalkboard

Uses
- Interactive settings with lots of ideas being generated
- recording audience responses
- informal settings
- small, well-lighted rooms

Tips
- Write out long material in advance
- make it legible and visible to all
- use washable markers
- speak to the listeners—not to the board

Poster

Uses
- overviews, previews, emphasis
- recurring themes
- small, well-lighted rooms

Tips
- use 20" × 30" posterboard (or larger)
- use intense, washable colors
- keep each poster simple and uncrowded
- arrange/display posters in advance
- point to what you are discussing

PowerPoint and other presentation apps

Uses
- incorporate text, images, sound
- provide slide deck for those who can't attend in person
- formal or informal settings
- dark or semi-dark room required

Tips
- create an outline and storyboard
- find out if you need to bring your own laptop
- don't just read the slides; engage your audience
- have a backup plan in case the technology fails

Handouts

Uses
- present complex material
- help listeners follow along, take notes, and remember

Tips
- staple or bind the packet
- number the pages
- try saving for the end
- may need to make available in advance for people connecting via video

Flip Chart

Uses
- a sequence of visuals
- back-and-forth movement
- small, well-lighted rooms

Tips
- use an easel pad and easel
- use intense, washable colors
- check your sequence beforehand
- point to what you are discussing

Movies

Uses
- necessary display of moving images
- coordinated sound and visual images

Tips
- introduce segment to be shown; tell viewers what to expect
- show only relevant segments
- if the segment is complex, replay it where needed, using slow-motion replay
- practice beforehand

Figure 23.3 **Media Choices for Visual Presentations**

Using Powerpoint and Other Presentation Apps

23.5 Create audience-friendly slides using presentation apps

Overview of presentation software

Workplace presentations rely heavily on presentation software, mainly PowerPoint. Depending on what's being used at your job, other choices are available, such as Keynote (Apple) and Slides (Google). Prezi is another popular choice; it offers many creative options for animation and transitions. Most presentation apps have a cloud-based option that works well for collaboration since everyone on the team can access the same file in real time.

Whatever you choose, remember that good presentations are based on the qualities discussed previously in this chapter. They must be well researched, carefully planned, well organized, and correctly timed. Excessive material crammed on one slide might help you as you speak but will frustrate and bore your audience. Also, while presentation software helps structure a speaker's story or argument, an overreliance on bulleted items can oversimplify complex issues because each bullet may look equal in importance, especially after an audience is forced to sit through slide after slide of text-heavy material. See the "Case: PowerPoint and the Space Shuttle Columbia Disaster," later in this chapter for an example of a serious ethical situation that has been blamed, in part, on the use of slides.

PowerPoint and similar apps usually offer a range of colorful and professionally designed templates. You can also create a custom template for your company or brand. Make sure the template fits with your overall message by choosing a color and theme suitable for the message you wish to convey. Avoid overly fancy fonts and colors that might not appear clearly on a projection screen. Modify templates to fit your specific audience and purpose.

As presentation software becomes the standard for meetings, employers are now finding that meeting upon meeting filled with endless slide decks is not an effective way to communicate. The CEO of Amazon, Jeff Bezos, recently banned the use of PowerPoint in all executive meetings, opting instead for the use of a short memo (like an executive summary) followed by discussion (C. Gallo). They key take-away is not to avoid presentation software but to use these apps as tools, not as the brains of the operation. Keep audience and purpose in mind; do your research; create a storyboard and use appropriate visuals; keep the slides interesting and concise.

Presentation apps offer many technical features, only a small percentage of which are used by the average writer. Take a tutorial or attend a workshop on using some of the more high-end features of PowerPoint, including the following:

Selected design and display features of presentation software

- Create slide designs in various colors, shading, and textures.
- Use templates effectively, including the ability to create a template for your company or brand.
- Create drawings or graphs and import clip art, photos, and video.

- Create animated text and images such as bullets that flash one-at-a-time on the screen or bars and lines on a graph that are highlighted individually, to emphasize specific characteristics of the data.

- Create dynamic transitions between each slide, such as having one slide dissolve toward the right side of the screen as the following slide uncovers from the left.

- Amplify each slide with speaker notes that are invisible to the audience.

- Time your entire presentation with precision.

Figure 23.4 shows parts of a sample presentation based in part on the outline earlier in this chapter.

When and How to Use Handouts

Handouts can be useful in helping your audience retain key information from your presentation. But there is always the question of when to distribute handouts: before, during, or after the presentation? Some basic suggestions include the following:

- In most situations, avoid emailing or giving out your entire slide deck as a handout before the presentation. You don't want people reading or skipping ahead when they should be paying attention to you.

- For sections of the presentation where information is too detailed or complicated for a slide, consider providing a handout that you distribute at that point in the presentation. Or, if there are more than, say, 10 people in the room, offer these handouts in advance but ask people to look at these only when directed.

- People who connect to your presentation via a video link may be able to see your slides as you present them (in real time) if they are connected via a webinar. But for people who are connecting on a video stream that only shows the presenter (you), it may be necessary to provide your slides in advance; do so only to those online audience members.

- After the presentation, you can upload your slide deck and all handouts to a common file space (a shared folder or the company file server). You might also be asked to send your materials to all attendees via email. Follow the appropriate practices for your workplace. If you are a contractor or consultant, you may want to consider how much of your intellectual property you wish to share beyond the presentation itself. Your work might be considered "works for hire," where you retain the copyright (see Chapter 7, "Consider This: Frequently Asked Questions about Copyright" for more on copyright and works for hire).

Ethics and the Use of Presentation Apps

The heavy use of bulleted lists in PowerPoint and similar apps can help you structure your story or argument and help keep you organized when you present. But some critics have argued that the content outline provided by slide templates can, if the writer is not careful, oversimplify highly complex issues and that an endless list of bullets or

Presentation software and complex technical topics

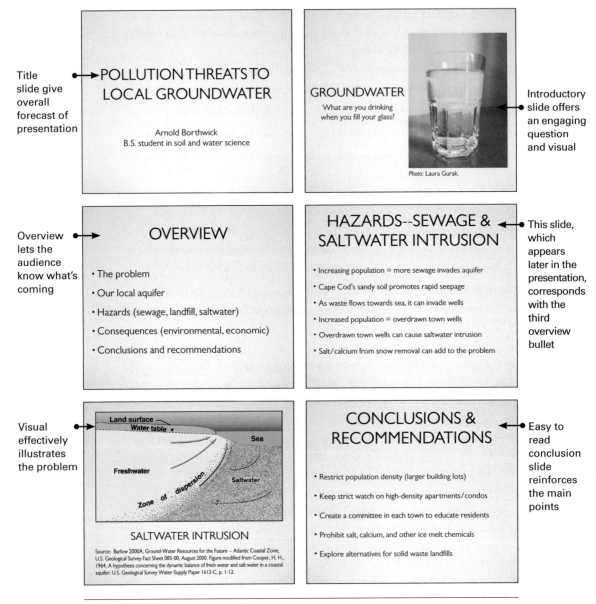

Figure 23.4 **Sample Presentation** These six slides are from a longer presentation (13 slides total).

animations, colors, and sounds can distract from the deeper message. Also, people's learning styles differ, and presentation software may be more suited for those who prefer a clear sequence versus those who learn best when the discussion is more open (Weimer).

In the end, presentation apps are tools. They are not a substitute for the facts, ideas, examples, numbers, and interpretations that make up the clear and complete message audiences expect. Make sure your presentation is well-researched and accurate. If there are subtle but important points that just don't stand out well in a slide format, make these points clear, and use other media (see Figure 23.3) to enhance your discussion. Follow up with a summary email or a short report highlighting the key take-aways from your presentation. Remember: you, not the app, are responsible for making sure your audience understands the key issues, especially when these issues involve safety, major financial implications, or other critical points.

Case

PowerPoint and the Space Shuttle Columbia Disaster

On February 1, 2003, the space shuttle Columbia burned up upon reentering the Earth's atmosphere. The Columbia had suffered damage during launch when a piece of insulating foam had broken off the shuttle and damaged the wing.

While Columbia was in orbit, NASA personnel tried to assess the damage and to recommend a course of action. It was decided that the damage did not seem serious enough to pose a significant threat, and reentry proceeded on schedule. (Lower-level suggestions that the shuttle fly close to a satellite that could have photographed the damage, for a clearer assessment, were overlooked and ultimately ignored by the decision makers.)

The Columbia Accident Investigation Board concluded that a PowerPoint presentation to NASA officials had played a role in the disaster: Engineers presented their findings in a series of confusing and misleading slides that obscured errors in their own engineering analysis. Design expert Edward Tufte points out that one especially crucial slide was so crammed with data and bullet points and so lacking in analysis that it was impossible to decipher accurately (8–9).

> The Board's findings:
>
> As information gets passed up an organization's hierarchy, from people who do analysis to mid-level managers to high-level leadership, key explanations and supporting information are filtered out. In this context, it is easy to understand how a senior manager might read this PowerPoint slide and not realize that it addresses a life-threatening situation.
>
> At many points during its investigation, the Board was surprised to receive similar presentation slides from NASA officials in place of technical reports. The Board views the endemic use of PowerPoint briefing slides instead of technical papers as an illustration of the problematic methods of technical communication at NASA (*Columbia* Accident, *Report* 191).

Guidelines

for Using Presentation Apps

➤ **Don't let the tools do the thinking.** Use your own research, sense of audience and purpose, and other original ideas to shape the presentation. Then use the app to help make the material accessible and interesting. The shape and content should come from you, not from the app.

➤ **Have a backup plan in case the technology fails.** Bring handouts to the presentation and be prepared to give the presentation without the software. Don't distribute handouts until you are ready to discuss them; otherwise, people will start reading the handouts instead of paying attention to you.

➤ **Start with an overview slide.** Orient your audience by showing an opening slide that indicates what you plan to cover and in what order.

➤ **Find a balance between text and visuals.** Excessive text is boring; excessive visuals can be confusing. Each visual should serve a purpose: to summarize information, to add emphasis, to set a tone, and so on. Avoid over-crowding the slides: In general, include no more than seven to nine lines per slide (including the heading) with no more than six to nine words per bulleted item.

➤ **Avoid overcrowding the slide.** In general, include no more than one key point per slide. If you are the kind of person who likes to embellish and enjoys speaking to a crowd, use fewer slides.

➤ **Avoid merely reciting the slides.** Instead, discuss each slide, with specific examples and details that round out the idea—but try not to digress or ramble.

➤ **Don't let the medium obscure the message.** The audience should be focused on what you have to say, and not on the slide. Avoid colors and backgrounds that distract from the content. Avoid the whooshing and whiz-bang and other sounds unless absolutely necessary. Be conservative with any design and display feature.

➤ **Keep it simple but not simplistic.** Spice things up with a light dose of imported digital photos, charts, graphs, or diagrams, but avoid images that look so complex that they require detailed study.

➤ **Keep viewers oriented.** Don't show a slide until you are ready to discuss it. Present one topic per slide, bringing bullets (subtopics) on one at a time as you discuss them. Let your audience digest the slide data *while* you speak, and not before or after.

➤ **End with a "conclusions" or "questions" slide.** Give your audience a sense of having come full circle. On a "conclusions" slide, summarize the key points. A "questions" slide can simply provide a heading and a visual. Allow time at the end of your presentation for questions and/or comments.

➤ **If appropriate for your workplace, make slides available via email or on a shared folder.** Wait until after your presentation to do this, unless you need to send slides in advance to those who are connecting via webinar.

Delivering Your Presentation

23.6 Deliver an effective oral presentation

You have planned and prepared carefully. Now consider the following simple steps to make your actual presentation enjoyable instead of terrifying.

Rehearse Your Delivery

Hold ample practice sessions to become comfortable with the organization and flow of your presentation. Try to rehearse at least once in front of friends or use a full-length mirror and a recorder (the camera app on your phone, for example). Assess your delivery from listener comments or from your recorded voice (which will sound high to you). Use "Checklist: Oral Presentations" at the end of this chapter as a guide.

If at all possible, rehearse using the actual equipment (computer projector, your laptop, and so on) in the actual setting, to ensure that you have all you need and that everything works. Rehearsing a computer-projected presentation is essential.

Check the Room and Setting Beforehand

Make sure you have enough space, electrical outlets, and tables for your equipment. If you will be addressing a large audience by microphone and plan to point to features on your visuals, be sure the microphone is movable. Pay careful attention to lighting, especially for chalkboards, flip charts, and posters. Don't forget a pointer if you need one.

Cultivate the Human Landscape

A successful presentation involves relationship building with the audience.

GET TO KNOW YOUR AUDIENCE. Try to meet some audience members before your presentation. We all feel more comfortable with people we know. If some attendees will join via video link, you could send them a quick email letting them know you appreciate them taking the time to join in.

BE REASONABLE. Don't make your point at someone else's expense. If your topic is controversial (layoffs, policy changes, downsizing), decide how to speak candidly and persuasively with the least chance of offending anyone. For example, in your presentation about groundwater pollution (see "Outline Your Presentation" earlier in this chapter), you don't want to attack the building developers, since the building trade is a major producer of jobs, second only to tourism, on Cape Cod. Avoid personal attacks.

DISPLAY ENTHUSIASM AND CONFIDENCE. Nobody likes a speaker who seems half dead. Clean up verbal tics ("er," "ah," "uuh"). Overcome your shyness; research indicates that shy people are seen as less credible, trustworthy, likable, attractive, and knowledgeable. Don't be afraid to smile. Breath deeply and slowly for a few minutes right before the presentation. Take a drink of water if you need to during your talk.

DON'T PREACH. Speak like a person talking—not someone giving a sermon or the Gettysburg Address. Use *we, you, your, our,* to establish commonality with the audience. Avoid jokes or wisecracks.

Keep Your Listeners Oriented

Help your listeners focus their attention and organize their understanding. Give them a clear sense of the presentation's shape and main points, some guidance, and highlights.

OPEN WITH A CLEAR AND ENGAGING INTRODUCTION. The introduction to a presentation is your chance to set the stage. For most presentations, you have three main tasks:

1. Show the listeners how your presentation has meaning for them. Show how your topic affects listeners personally by telling a quick story, asking a question, or referring to a current event or something else the audience cares about.
2. Establish your credibility by stating your credentials or explaining where you obtained your information.
3. Preview your presentation by listing the main points and the overall conclusion.

An introduction following this format might sound something like this:

An appeal to listeners' concerns

A presentation preview

> Do you know what you are drinking when you turn on the tap and fill a glass? The quality of Cape Cod's drinking water is seriously threatened by rapid population growth. My name is Arnold Borthwick, and I've been researching this topic for a term project. Today, I'd like to share my findings with you by discussing three main points: specific causes of the problem, the foreseeable consequences, and what we can do to avoid disaster.

Outline your main points by using an overview slide.

GIVE CONCRETE EXAMPLES. Good examples are informative and persuasive, such as the following:

A concrete example

> Overdraw from town wells in Maloket and Tanford (two of our most rapidly growing towns) has resulted in measurable salt infusion at a yearly rate of 0.1 mg per liter.

Use examples that focus on listener concerns.

PROVIDE EXPLICIT TRANSITIONS. Alert listeners whenever you are switching gears, as in these two examples:

Explicit transitions

> For my next point. . . .
>
> Turning now to my second point . . .

Repeat key points or terms to keep them fresh in listeners' minds.

REVIEW AND INTERPRET. Last things are best remembered. Help listeners remember the main points, such as in the following example:

A review of main points

> To summarize the dangers to our groundwater, . . .

Also, be clear about what this material means. Be emphatic about what listeners should be doing, thinking, or feeling, for example:

An emphatic conclusion

> The conclusion is obvious: If the Cape is to survive, we must. . . .

Try to conclude with a forceful answer to this implied question from each listener: "What does all this mean to me personally?"

Plan for How You Will Use Any Noncomputer Visual Aids

If you plan to work with a whiteboard, chalkboard, flip chart, poster, or other aid, either alone or in addition to your slides, you need to prepare and organize. Presenting material effectively in these formats is a matter of good timing and careful management.

PREPARE. If you plan to draw something complicated, on a chalkboard or poster, do the drawings beforehand (in multicolors). Otherwise, listeners will be sitting idly while you draw away. However, a quick sketch in front of an audience can be an effective tool in helping listeners see how things work.

Prepare handouts if you want listeners to remember or study certain material. See previous section on Handouts.

ORGANIZE. Make sure you organize your media materials and the physical layout beforehand, to avoid fumbling during the presentation. Check your visual sequence against your storyboard.

AVOID LISTENER DISTRACTION. Make your visuals part of a seamless presentation. Avoid listener distraction, confusion, and frustration by observing the following suggestions.

- Try not to begin with a visual.
- Try not to display a visual until you are ready to discuss it.
- Tell viewers what they should be looking for in the visual.
- Point to what is important.
- Stand aside when discussing a visual, so everyone can see it.
- Don't turn your back on the audience.
- After discussing the visual, remove it promptly.
- Switch off equipment that is not in use.
- Try not to end with a visual.

Manage Your Presentation Style

Think about how you are moving, how you are speaking, and where you are looking. These are all elements of your personal style.

USE NATURAL MOVEMENTS AND REASONABLE POSTURES. Move and gesture as you normally would in conversation, and maintain reasonable postures. Avoid foot shuffling, pencil tapping, swaying, slumping, or fidgeting.

ADJUST VOLUME, PRONUNCIATION, AND RATE. With a microphone, don't speak too loudly. Without one, don't speak too softly. Be sure you can be heard clearly without shattering eardrums. Ask your audience about the sound and speed of your delivery after a few sentences.

Nervousness causes speakers to gallop along and mispronounce words. Slow down and pronounce clearly. Usually, a rate that seems a bit slow to you will be just right for listeners.

MAINTAIN EYE CONTACT. Look directly into listeners' eyes. With a small audience, eye contact is one of your best connectors. As you speak, establish eye contact with as many listeners as possible. With a large group, maintain eye contact with those in the first rows. Establish eye contact immediately—before you even begin to speak—by looking around.

Manage Your Speaking Situation

Do everything you can to keep things running smoothly.

BE RESPONSIVE TO LISTENER FEEDBACK.   Assess listener feedback continually and make adjustments as needed. If you are laboring through a long list of facts or figures and people begin to doze or fidget, you might summarize. Likewise, if frowns, raised eyebrows, or questioning looks indicate confusion, skepticism, or indignation, you can backtrack with a specific example or explanation. By tuning in to your audience's reactions, you can keep listeners on your side.

Consider This

Cross-Cultural Audiences May Have Specific Expectations

Imagine you've been assigned to represent your company at an international conference or before international clients. As you plan and prepare your presentation, remain sensitive to various cultural expectations.

For example, some cultures might be offended by a presentation that gets right to the point without first observing formalities of politeness, well wishes, and the like. Certain communication styles are welcomed in some cultures but are considered offensive in others. In southern Europe and the Middle East, people may expect direct and prolonged eye contact as a way of showing honesty and respect. In Southeast Asia, this may be taken as a sign of aggression or disrespect (Gesteland 24). A sampling of the questions to consider:

➤ *Should I smile a lot or look serious? (Hulbert, "Overcoming" 42)*
➤ *Should I rely on expressive gestures and facial expressions?*
➤ *How loudly or softly, rapidly or slowly should I speak?*
➤ *Should I come out from behind the podium and approach the audience or keep my distance?*
➤ *Should I get right to the point or take plenty of time to lead into and discuss the matter thoroughly?*
➤ *Should I focus only on the key facts or on all the details and various interpretations?*
➤ *Should I be assertive in offering interpretations and conclusions, or should I allow listeners to reach their own conclusions?*
➤ *Which types of visuals and which media might or might not work?*
➤ *Should I invite questions from this audience, or would this be offensive?*

To account for language differences, prepare a handout of your entire presentation, including a list of key words and ideas that may not be familiar to your audience. See Chapter 5 for more tips on technical communication and international/intercultural audiences.

STICK TO YOUR PLAN. Say what you came to say, then summarize and close—politely and on time. Don't punctuate your speech with digressions that pop into your head. Unless a specific anecdote was part of your original plan to clarify a point or increase interest, avoid excursions. We often tend to be more interested in what we have to say than our listeners are! Don't exceed your time limit.

LEAVE LISTENERS WITH SOMETHING TO REMEMBER. Before ending, take a moment to summarize the major points and reemphasize anything of special importance. Are listeners supposed to remember something, have a different attitude, take a specific action? Let them know! As you conclude, thank your listeners. Don't forget about those people who attended via video or webinar. Thank them as well and let them know that the video feed will now end.

ALLOW TIME FOR QUESTIONS AND ANSWERS (Q & A). At the very beginning, tell your listeners that a question-and-answer period will follow. Make sure you include the online attendees in the Q & A session.

Guidelines

for Delivering an Oral Presentation and Managing Listener Questions

Delivering an Oral Presentation

➤ **Be rehearsed and prepared.** The best way to calm your nerves is to remind yourself of the preparation and research you have done. Practice your delivery so that it is professional and appropriate but natural for you. Relax, breath, and be yourself.

➤ **Memorize a brief introduction.** If you begin your presentation smoothly and confidently, you won't be as nervous going forward. Do not open by saying "I'm a little bit nervous today" or "I have a slight cold. Can everyone hear me in the back?" Just begin. A memorized introduction will help you avoid having to ad lib (potentially badly) when you are most anxious.

➤ **Dress for success.** Wear clothes that suggest professionalism and confidence.

➤ **Stand tall and use eye contact.** Good posture and frequent eye contact convey a sense of poise, balance, and confidence. Practice in front of a mirror. If looking people directly in the eye makes you nervous, aim just above people's heads. Or find a friendly face or two and look at those people first. As you gain confidence, be sure to cast your gaze around the entire room or conference table.

➤ **Take charge.** If you get interrupted but don't want to take questions until the end, remember that you are in control. Be polite but firm: "Thank you for that good question. I'll jot it down and take it and any other questions at the end of my presentation."

➤ **Gesture naturally.** Don't force yourself to move around or be theatrical if this is not your style. But do not act like a robot, either. Unless you are speaking from a podium or lectern, move around just a bit. And when it's time to take questions, consider moving closer to the audience. In a conference room setting, if everyone else is seated, you should stand.

Managing Listener Questions

➤ **Announce a specific time limit for the question period.**
➤ **Listen carefully to each question.**
➤ **If you can't understand a question, ask that it be rephrased.**
➤ **Repeat every question, to ensure that everyone hears it.**
➤ **Be brief in your answers.**
➤ **If you need extra time for an answer, arrange for it after the presentation.**
➤ **If anyone attempts lengthy debate, offer to continue after the presentation.**
➤ **If you can't answer a question, say so and move on.**
➤ **Reach out to those connected via webinar or video link.**
➤ **End the session with a clear signal.** Say something such as, "We have time for one more question."

Video Conferencing

23.7 Determine how and when to use video conferencing

When audience members are at different job sites (around the country or in different countries), they may not be able to attend your presentation in person and will need to connect via video. There are many kinds of video conferencing including the following:

Video conferencing options

- **Webinars** allow you to deliver a presentation via the Internet. Audience members are invited to connect to the webinar via their computer and software such as WebEx, GoToMeeting, or Adobe Connect. Most webinars display the presenter's slides on the screen while the presenter talks through each slide and controls when to change slides. You can also display video of the speaker and/or other conference participants. Audience questions are typically handled by a live chat feature or by allowing participants to speak when called on.

- **Skype** allows people to connect via real-time audio and video from a computer, phone, or tablet. Skype software needs to be downloaded in advance, and each person needs to set up a user name and password. Skype has become quite popular, in part because it is free (and easy to use). But the sound and image quality may not always be adequate for complex technical and business presentations. Also, most video streams only allow the camera to point at one area, which is typically the speaker, not the slides. For this reason, most organizations prefer a webinar or secure in-house video conferencing approach that allows slides and presenter to be shown at the same time.

Time considerations and video conferencing

For video conferencing to be effective, everyone must be available at the same time. If you are on the east coast of the United States, for example, and some of your audience is in Europe, there could be anywhere from a 5- to 8-hour time difference. Video conferencing with audiences in Asia requires even more careful time calculations. Depending on the location of team members, it may be impossible to get everyone to connect at the same time. In these cases, you may be able to record a session and allow people to view your presentation, and your slides, later.

Guidelines

for Video Conferencing

➤ **Prepare any slides and other materials well in advance.**

➤ **Test your video connection.** Ask one or two colleagues to sign in the day before the presentation for a trial run.

➤ **Post your slides on the company's server (intranet) or to the video service.** In this way, people unable to attend the video conference will still be able to view your presentation.

➤ **Determine if you should use a webinar, Skype, or other video.** Companies often prefer you to use a secure in-house video connection although most commercial services now offer secure connections as well.

➤ **Keep international audiences in mind.** Remember that 9:00 A.M. your time may not work for people who live on the other side of the globe.

➤ **If more than one person plans to connect via video, make a plan in advance about how to give each person time to contribute.** Since those who are present in the room usually jump in quickly, you may need to encourage those on the video connection to speak up.

Checklist

Oral Presentations

Use the following Checklist to evaluate an oral presentation.

Presentation Evaluation for (*name/topic*) _____

Content Comments

❏ Stated a clear purpose _____

❏ Created interest in the topic _____

❏ Showed command of the material _____

❏ Supported assertions with evidence _____

❏ Used adequate and appropriate visuals _____

❏ Used technology (computer; PowerPoint or other) effectively _____

❏ Used material suited to this audience's needs, knowledge, concerns, and interests

_____

❏ Acknowledged opposing views _____

❏ Gave the right amount of information _____

Organization

❏ Began with a clear overview _____

❏ Presented a clear line of reasoning _____

❏ Moved from point to point effectively _____

❏ Stayed on course _____

❏ Used transitions effectively _____

❏ Avoided needless digressions _____

❏ Summarized before concluding _____

❑ Was clear about what the listeners should think or do _____

❑ PowerPoint (or other) slides were clear and
    easy to follow _____

Style

❑ Dressed appropriately _____

❑ Seemed confident, relaxed, and likable _____

❑ Seemed in control of the speaking situation _____

❑ Showed appropriate enthusiasm _____

❑ Pronounced, enunciated, and spoke well _____

❑ Used no slang whatsoever _____

❑ Used appropriate gestures, tone, volume, and delivery rate _____

❑ Had good posture and eye contact _____

❑ Interacted with the audience _____

❑ Kept the audience actively involved _____

❑ Kept focus on audience; did not just read
    from slides or visuals _____

❑ Answered questions concisely and convincingly _____

❑ **Overall professionalism: Superior** _____ **Acceptable** _____ **Needs work** _____

❑ **Evaluator's signature:** _____

Projects

For all projects, check with your instructor about whether to present your findings in class, bring drafts to class for discussion, upload your project to the class learning management system (LMS), and/or use the LMS forum or discussion boards to collaborate and review each activity below.

General

1. In a memo to your instructor, identify and discuss the kinds of oral reporting duties you expect to encounter in your career.

2. Prepare an oral presentation for your class, based on your written long report. Develop a sentence outline and a storyboard that includes at least three visuals. If your instructor requests, create one or more of your presentation visuals using PowerPoint or another presentation app. Practice your presentation by recording yourself or asking a friend to listen. Use "Checklist: Oral Presentations" in this chapter to assess and refine your delivery.

3. Observe a lecture or speech and evaluate it according to the Checklist. Write a memo to your instructor (without naming the speaker), identifying strong and weak areas and suggesting improvements.

Team

In today's work world, presentations are often prepared by, and sometimes delivered by, more than one person. In groups of 3–4 people, come up with a simple presentation topic that everyone can work on. (You might want to base this topic on a group writing activity from earlier in the semester.) Discuss how, as a team, you will approach the research, planning, and creation of the presentation. For example, will one person be responsible for the research? Will another person design and create the slides? Create a brief presentation for your instructor and classmates (using presentation software), describing the process you will take and the decisions your team made.

Digital and Social Media

In the same group of 3–4 people as above, investigate how you would create a presentation if each of you worked in a different location (too remote from each other to have a face-to-face meeting). If everyone has a Google account, try creating a shared Google presentation (create a new Google Slides file from within Google Drive). Or, use PowerPoint or Keynote (Apple) and share the file in your class learning management system or via email. In class, give a short presentation on the pros and cons of working on a presentation collaboratively and at a distance from each other. End your presentation with tips to help others in your situation.

Global

Imagine you've been assigned to represent your company at an international conference or before international clients. As you plan and prepare your presentation, what can you do to remain sensitive to various cultural expectations? For example, some cultures might be offended by a presentation that gets right to the point without first observing formalities of politeness, well wishes, and the like. Online, research issues about presentation etiquette for a particular country or culture (choose one). Then create a short presentation on this topic for the class.

Chapter 24
Blogs, Wikis, and Web Pages

❝My job involves working on instructions and specifications for our technical racing bikes. The information that I research and write needs to be adapted to our company Web site, our external blog (for customers), and an internal wiki where we keep track of technical changes. I work with a great team of Web developers, and we stay in close touch to be sure all of our digital content is accurate, easy to read, quick to access, and up to date. We make sure that our Web presence is closely aligned with our social media marketing strategy, too.❞

—*Lori Huberman, Senior Technical Writer, international racing cycle company*

Learning Objectives

24.1 Analyze the audience and purpose of blogs, wikis, and Web pages

24.2 Identify the uses of internal and external blogs

24.3 Identify the uses of internal and external wikis

24.4 Explain the purposes of Web pages

24.5 Write and design reader-friendly Web pages

24.6 Storyboard and create a Web page

24.7 Describe the global, ethical, and legal issues of writing for blogs, wikis, and Web pages

Blogs, wikis, and Web pages are important forms of workplace communication for companies and other organizations. Blogs began as forums for people to write about and discuss political topics, hobbies, and personal interests. But today, blogs are also used to help employees network with each other (internal blogs) and stay in touch with customers and clients (external blogs). Wikis, both internal and external, allow organizations to keep technical and scientific content updated and current, based on input from individual contributors. Web pages are essential for almost all organizations, large or small and are used in combination with social media to advertise, maintain a visible presence, and provide information and product updates to customers.

Advantages of blogs, wikis, and Web pages

Considering Audience and Purpose

24.1 Analyze the audience and purpose of blogs, wikis, and Web pages

Audience considerations

Blogs and wikis may be written for internal or external audiences. Web pages, on the other hand, are typically written for external audiences (such as customers who want to learn more about a product or download the user manual). Because these three forms of digital communication may be accessed by thousands of readers, you need to think carefully about your intended audience and purpose. Who are the primary readers? Are they potential customers or existing customers? Are they people with questions about a medical condition? Do they simply want to read the information, or do they want to interact by posting their own thoughts and adding to the site's content?

Using an Audience and Purpose Profile Sheet

Consider your audience carefully by filling out an Audience and Purpose Profile Sheet (Chapter 3). Keep in mind both primary and secondary audiences. Web pages in particular allow you to take a layered approach; you may have one primary audience in mind, but you can write and structure information for a variety of readers. For instance, Figure 24.1 shows a Web page from the U.S. Food and Drug Administration

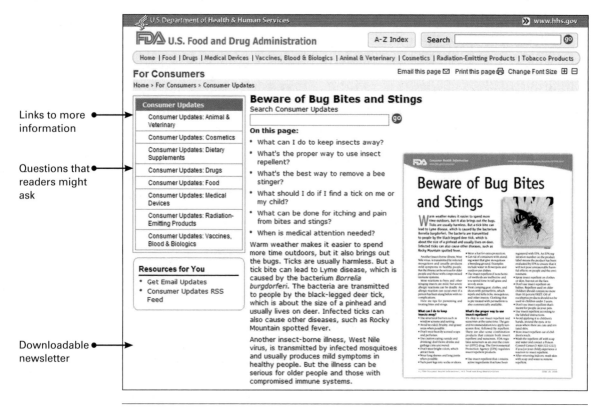

Links to more information

Questions that readers might ask

Downloadable newsletter

Figure 24.1 An Interactive Web Page

Source: U.S. Food and Drug Administration from http://www.fda.gov/ForConsumers/ConsumerUpdates/ucm048022.htm.

on bug bites and bee stings. The primary audience is general readers who want to learn more about the topic. But other readers, such as physicians or nurses, might also access the site to look for new treatment information or links to scientific studies. Researchers might use the site as a starting point to connect with the latest studies on the topic; teachers might use the site as part of a lesson plan. The various links on the page make the content readily available to a variety of readers.

Blogs

24.2 Identify the uses of internal and external blogs

Blogs began as tools to keep track of email discussions. Rather than search through old email, people could view messages in reverse chronological order on a blog. Soon, blogs became popular for discussions on political topics, hobbies, and other mutual interests. Organizations saw the benefits of using blogs—both internally and externally—to share information and provide spaces for discussion. Companies use blogs to help employees share information. These internal blogs, accessible only to those within the company, are focused on audiences such as engineers, managers, technical writers, and others. Companies also create external blogs, designed for more general audiences outside the company.

Definition of blogs

Internal Blogs

Internal blogs enhance workflow and morale. In large organizations, blogs can provide an alternative to email for routine in-house communication. Anyone in the network can post a message or comment on other messages. In the blogging environment, meetings can be conducted without the time and location constraints of face-to-face meetings. Employee training can be delivered, and updates about company developments can be circulated. Blogs are especially useful for collaborating. For example, someone in a company's engineering department can create a forum to discuss various solutions to a technical problem.

How a blog can enhance a company's internal conversation

External Blogs

While internal blogs are used for in-house communication, external blogs are written with a broader audience in mind. External blogs help organizations stay in touch with the public by providing updates on products, services, research findings, and other current and newsworthy events. Blogs offer a space for large organizations to show a more personal side, to respond amiably and quickly to customer concerns, and to allow readers to provide ideas and feedback. Tone, of course, is critical on an external blog; the writing needs to sound friendly, welcoming, and sincere.

How a blog can enhance a company's public conversation

Increasingly, the content on a blog will often mirror what gets posted to the organization's Facebook page and Twitter feed, but blogs offer additional options such as a search function, a sign-in feature for receiving email alerts, and an RSS feed option (RSS stands for "really simple syndication" and is also known as a Web feed or a news feed.) Government and nonprofit organizations, such as the Centers for Disease Control and Prevention (CDC) and the National Wildlife Federation, as well as corporations both large and small, use external blogs for a variety of topics.

Wikis

24.3 Identify the uses of internal and external wikis

Definition of wikis

A wiki (from the Hawaiian phrase *wiki wiki*, meaning "quick") is a type of Web site used to collect information and keep it updated. Most people are familiar with Wikipedia, the online encyclopedia that allows anyone to write and revise its content. Companies and other organizations also use wikis to update content in a wide variety of technical and scientific areas. As with blogs, organizations maintain both internal and external wikis, depending on the project and need. But unlike blogs, which typically have only a few people serving as authors, wikis encourage everyone to contribute to the site's content.

Internal Wikis

How internal wikis enhance access to technical information

In the workplace, internal wikis can be used as a "knowledge management" tool—a way to keep important technical information updated and provide one-stop access for employees who need the latest data, specifications, guidelines, and so forth. To keep the data current, each person with access may log on to the wiki and make changes to content, overwriting previous versions. Despite the potential for error created by such alterations, a well-managed wiki can quickly become a collaborative digital space where content experts contribute to the broader knowledge base. Many companies, from banks to information technology firms to engineering organizations, are finding that wikis are a valuable resource among trustworthy contributors. To ensure that readers can see the history of how changes were made, and by whom, copies of the original entry along with each subsequent edit are saved for later reference.

External Wikis

External wikis related to your field of study

While some wikis are used only internally by a company or organization, others are used by external audiences in specific scientific or technical areas. Because these wikis are open to countless people with knowledge in the fields, these sites can be excellent external sources of information. But as with Wikipedia, the lack of gatekeeping by a central editor also creates the potential for inaccuracy. A well-managed wiki, one that balances open access with editorial control, can be a valuable resource. If you search for a wiki related to your undergraduate major (nursing, chemistry, engineering, technical writing), you may find some useful resources. Or you may find that a Wikipedia entry on this topic contains a link to a more specific external wiki.

Guidelines

for Writing and Using Blogs and Wikis

➤ **Use standard software.** Readers expect blogs and wikis to have a certain look and feel. Tools such as Blogger and Wikispaces can be adapted to confidential workplace environments.

➤ **Keep content brief, to the point, and concise.** Avoid wordiness and repetition. Use hyperlinks to direct readers to additional information.

➤ **Write differently for internal versus external audiences.** Internal audiences do not require as much background. External audiences may require more explanation.

➤ **Think carefully before posting comments.** An internal blog is considered workplace communication; an external blog represents the organization. In both cases, keep your ideas factual and tone professional.

➤ **For an external blog, focus on the customer's priorities and needs.** A friendly, encouraging tone goes a long way in good customer relations.

➤ **Check blog and wiki entries for credibility.** Check the content to see when it was last updated (on a wiki, this information is usually under the "history" tab).

➤ **Before editing or adding to a wiki, get your facts and your tone straight.** If you spot some type of error, be diplomatic in your correction—and be sure you aren't creating some other type of error.

Web Pages

24.4 Explain the purposes of Web pages

Most organizations, whether private or public, large or small, have a Web site. These sites serve many purposes, including to advertise and market products and services; to provide downloadable user manuals, reports, technical data, and software updates; to educate and inform via up-to-date information; and to provide links to related topics. The Web offers a level of interactivity that print documents do not. For instance, the FDA Web site shown in Figure 24.1 contains a list of consumer updates, questions readers can click on for more information, a search box, a resources section, and a PDF fact sheet.

Purposes of Web pages

But both print documents and Web pages must follow the same basic principles of understandable structure (Chapter 10), readable style (Chapter 11), audience-centered visuals (Chapter 12), and a reader-friendly design (Chapter 13). Because Web pages and print pages are read differently, some of these elements play out differently in the online environment. As you read through this section, refer to Figure 24.2, a user-friendly Web page, and note how structure, style, and visual and design elements both resemble and differ from print pages.

> *NOTE* Web site *refers to an entire site with all related pages;* Web page *refers to one single page.*

NPS name and logo are clearly displayed at top of page

A search engine, pull-down menus, and other features allow for quick and easy navigation

"News" and "Events" sections take readers directly to these two popular features

The "Photos and Multimedia" feature is clearly divided into helpful subcategories

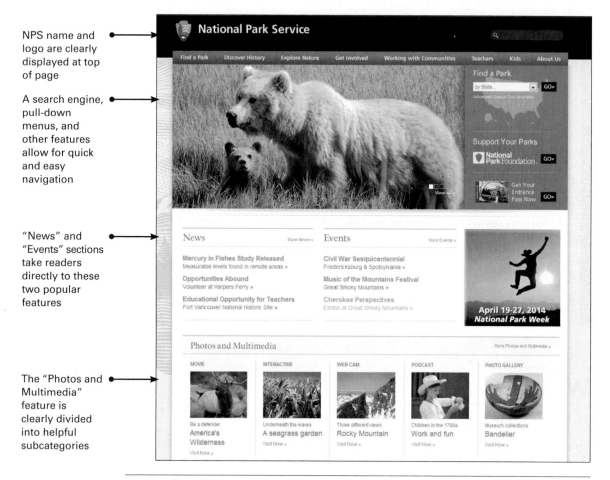

Figure 24.2 A User-Friendly Web Page

Source: From the National Park Service, www.nps.gov/inded.htm.

How People Read Web Pages

When people read a Web page, they typically want to find information quickly and easily. Readers expect Web pages to be accessible and typically share the following expectations.

What readers expect from Web pages

- **Accessibility.** Web pages should be easy to enter, navigate, and exit. Instead of reading word-for-word, readers tend to skim, looking for key material without having to scroll through pages of text. They look for chances to interact (for example, links to click on), and they want to download material quickly. For readers who have accessibility needs related to visual impairments, Web pages must be written and designed so they work properly with computer screen read-ers. The most common app of this kind is called Job Access with Speech, or JAWS. You would work with a specialist on your Web design team to ensure JAWS com-patibility and any other accessibility requirements.

- **Worthwhile Content.** Web pages should contain all the information readers need and want. Content (such as product and price updates) should be accurate and up to date. Readers also look for links to other high-quality sites as indicators of credibility. They look for a "search" tool and for contact information that is easy to find.

- **Sensible Arrangement.** Readers want to know where they are and where they are going. They expect a reasonable design and layout, with links easily navigated forward or backward. They look for navigation features to be labeled ("Company Information," "Ordering," "Job Openings," and so on).

- **Clean, Crisp Page Design.** Readers want a page design that is easy to navigate quickly, with plenty of white space and a balance of text, visuals, and color. A poorly designed Web page is a sure way to turn off readers and makes your orga-nization look unprofessional and untrustworthy.

- **Good Use of Visuals and Special Effects.** Readers expect high-quality visuals (photographs, charts and graphs, company logos) used in a balanced manner (not too many on the page). Special effects, such as fonts that blink, can be annoying and announce to readers that yours is not a serious, professional company or project.

One person rarely tackles all the areas described above. Web pages are typically cre-ated by a team of writers, designers, programmers (coders), networking specialists, marketing experts, accessibility specialists, user experience engineers, and others. Technical communicators are involved in the initial audience and purpose analysis and then play a central role in researching and writing the content. They may also be involved in page design (along with the graphic designer) and with user experience (usability) testing.

Writing and Designing for the Web

24.5 Write and design reader-friendly Web pages

When writing content for Web pages, you should remember that readers will be busy. Especially when reading online, people lack the patience to wade through long pas-sages of prose. Content must be concise, clear, and accurate. Sentences should be kept short and written without too much technical wording. Text should be placed near the visual it describes (see the example in Figure 24.2) The "Guidelines for Writing Web Pages" box offers additional suggestions for writing Web content for busy readers.

Web writing should be concise and clear

An effective Web page, one that people will want to read and explore, must be written clearly and use a clean, attractive, uncluttered design. The page should strike a good balance between text and visuals, offer inviting and complementary colors, and use white space effectively. Ample margins, consistent use of fonts, and clear headings all contribute to the design. Figure 24.2 illustrates a Web page that is visually appealing and makes good use of photographs, white space, and limited amounts of text.

Use an "F-shaped" layout pattern

Even more than with print documents, readers of Web pages skim quickly, employing what one expert calls the "F-shaped" reading pattern (Nielsen). People look across the page a few times (horizontally), then skim down the left margin, moving their eyes back and forth, from left to right (see Figure 24.3). (Note that this reading style applies mainly to cultures that read from left to right.)

The "Guidelines for Designing Web Pages" box offers additional suggestions for designing Web pages.

Figure 24.3 The F-Shaped Reading Pattern Most people follow this pattern as they skim Web pages.

Guidelines
for Writing Web Pages

➤ **Chunk the information.** Break long paragraphs into shorter passages that are easy to acces and quick to read. Chunking is also used in paper documents (see Chapter 13), but is especially important for Web pages. Chunking can also be used to break information into sections that address different audiences.

➤ **Write with a readable style.** Write clear, concise, and fluent sentences. Use a friendly but professional tone. Avoid abbreviations and technical terms that some audience members might not understand.

➤ **Keep sentences short.** Long sentences not only bog down the reader, but may also display poorly on a Web page.

➤ **Keep paragraphs short.** Long paragraphs can make a Web page look prose heavy. Make your online text at least 50 percent shorter than what you would write in a print document.

➤ **Catch reader attention in the first two paragraphs.** One expert suggests that "the first two paragraphs must state the most important information" (Nielsen).

➤ **Write in a factual, neutral tone.** Even on overtly political Web sites (such as a site for a political candidate), readers prefer writing that is fact-based and maintains a neutral tone.

➤ **Choose words that are meaningful.** Start headings, subheads, and bulleted items with "information-carrying words" (Nielsen) that have immediate meaning for readers. Instead of a word like "Introduction," you could use the phrase "Explore nature," as in Figure 24.2.

➤ **Write with interactive features in mind.** Use hyperlinks to provide more information about a technical term or a concept; think about when to link to other Web pages (within and outside your site). When you do create outside links, consider the ethical and legal implications (see "Global, Ethical, and Legal Considerations" later in this chapter).

➤ **Remember that most Web sites can be viewed by people across the globe.** Avoid confusing readers for whom English is not a first language. Avoid violating cultural expectations. For more on writing for global audiences, see "Guidelines for Global, Ethical, and Legal Considerations on the Web" later in this chapter.

Guidelines

for Designing Web Pages

➤ **Keep the F-shaped reading pattern in mind.** Use the top two paragraphs or sections for the most important information or visuals.

➤ **Use plenty of white space.** Cluttered Web pages are frustrating. White space gives the page an open feel and allows the eye to skim the page more quickly. Figure 24.2 uses white space effectively.

➤ **Provide ample margins.** Margins keep your text from blurring at the edges of the computer screen.

➤ **Use an unjustified right margin.** An unjustified margin makes for easier reading.

➤ **Use hyperlinks to direct readers to other information.** In Figure 24.2, links provide quick access to more detail. Don't overuse hyperlinks.

➤ **Use a consistent font style and size.** Figure 24.2 uses the same font style and size for body text and a slightly larger font size—but identical style—for headings. Don't mix and match typefaces randomly.

➤ **Don't use underlining for emphasis.** On a Web page, underlining is only used to indicate a hyperlink.

➤ **Use ample headings.** People skimming a Web page look for headings as guides to the content areas they seek.

➤ **Use visuals (charts, graphs, photographs) effectively.** Excessive visuals confuse readers. Inadequate visuals cause readers to avoid the page. Figure 24.2 nicely balances text with visuals.

➤ **Use a balanced color palette.** Color can make a Web page attractive and easier to navigate. In Figure 24.3, the colors of text and visuals reflect the Web page's theme (e.g., the use of greens, tan, blues, and browns mirror the earth-tone theme of a national park). For more on color, see Chapter 12.

Techniques and Technologies for Creating Web Sites

24.6 Storyboard and create a Web page

Creating a Web site typically involves two major steps. First, you need to plan your overall approach to the site. To do this, you and the team would sketch out a plan for the site, using a storyboard or other approach. After this high-level planning comes the detail work involved in writing and designing each individual Web page on the site. Storyboarding, teamwork, and using the right tools and technologies are all essential to the process.

Planning Web Sites Using Storyboarding

Use storyboarding to get started

When planning a print document, you might create an outline (see Chapter 10 for more on outlining). But to plan a Web site, use a storyboard—a sketch of the site, as in Figure 24.4. This type of storyboard for a Web site is often called a site map. The site map gives you (and others on the design team) an overall view of what the site will contain. (For more on storyboarding, see Chapter 5.)

Teamwork When Creating Web Sites

Get the team started via meetings and brainstorming

As discussed earlier, workplace Web sites are typically developed by a team whose members need to connect their different areas of expertise. Meetings, especially during the early planning stages, allow everyone to have input. Chapter 5 offers advice for conducting meetings. Early in the process, use creative thinking, especially brainstorming and storyboarding, to develop a unified vision for the site.

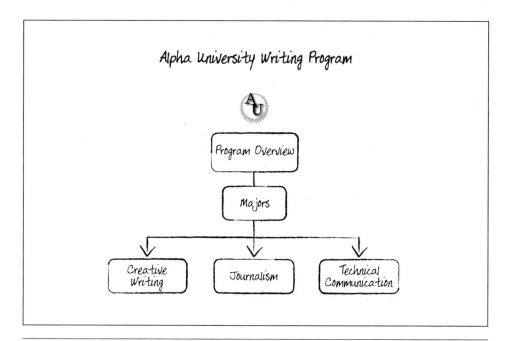

Figure 24.4 A Storyboard Map for a Web Site for Alpha University's Writing Program Each node represents a different Web page. The top node (Program Overview) represents the main Web page (home page).

Tools for Creating Web Pages

Simple Web pages can be created with everyday software, such as word-processing programs. These programs typically allow you to export a Word (or similar) document to a Web page. For more complicated Web sites, programs such as Adobe Dreamweaver and NVU (an open source program) are the norm. Professional Web designers and programmers use these tools to create complex sites that include interactive features, animation, online shopping links, and more. To learn about Web design, look for a workshop or short course through your school, community education center, or local library.

Software for creating Web sites

Web pages should always be tested for usability and the overall user experience (UX). Your team can run some basic tests on early versions of the new Web page to ensure that customers have an outstanding experience, find what they are looking for, don't get frustrated, and so forth. For more on usability and UX, see Chapter 19.

User experience (UX) and Web pages

Global, Ethical, and Legal Considerations

24.7 Describe the global, ethical, and legal issues of writing for blogs, wikis, and Web pages

Blogs, wikis, and Web pages carry with them global, ethical, and legal considerations. Unless entirely internal, all three formats can usually be accessed by readers across the globe. From an ethical perspective, blogs, wikis, and Web pages should always be used to convey accurate, factual ideas but they can be easily written and designed to look "real" but be based on biased, untruthful information. Legally, all three formats raise concerns about copyright and privacy.

Global Considerations

Readers of Web-based online content may come from a variety of countries and cultures. Keep these global audiences in mind as you write and design your site. Word choice, visuals, tone, and even color choices can have different meanings for different cultures. In addition to global considerations, the power of a Web page, external blog, or external wiki to convey information instantaneously and worldwide increases the writer's or designer's need for sound ethical and legal judgment. See Chapters 5 ("Teamwork and Global Considerations") and 4 ("Weighing the Ethical Issues") as well as the "Guidelines for Global, Ethical, and Legal Considerations on the Web" box later in this chapter for more information on each topic.

Most Web pages have global audiences

> NOTE *Some countries censor the Internet, but even in these countries, many Web sites are still available.*

Ethical Considerations

Consider the speed of the Internet, its global reach, and the ability to combine sound, color, images, text, and interactivity. These features create the potential for wide reach

How a Web page can be unethical

of information to thousands of readers but also allow for easy manipulation and distortion. Imagine a Web site for an herbal remedy that some people feel is helpful for anxiety. This remedy may not have FDA approval and may carry risks for harmful side effects. But a site promoting this product could easily, and at little cost, be set up to look scientific and factual. Fancy logos from quasi-scientific organizations might convey a sense of professional credibility. Statistics, charts, and links to other sites might create the appearance of a valid medical site. As a communicator, you need to question the possible outcome for readers of the site and the overall risks to society of setting up such a site.

Legal Considerations

Copyright issues

Legal considerations for Web sites, blogs, and wikis include copyright issues and privacy considerations. Verify the ownership of everything you include. If you have located a bar chart elsewhere on the Web that you'd like to insert, it is technically very easy to cut and paste or download the bar chart to your page. But is this use legal? Even if the chart has no copyright symbol ©, in the United States all expressions fixed in a tangible medium are copyrighted. The © symbol is not absolutely necessary. Therefore, downloading and including the chart in your document could be a copyright infringement. When in doubt, use copyright-free visuals or obtain written permission, even if you are only linking from your site to another site. See Chapter 7 for more information about copyright.

Privacy considerations

Privacy is also an issue, especially in regard to gathering the personal information of site visitors. Many sites offer privacy statements, which let readers know how information will be used and what rights readers have. Privacy statements should be readily available on a site, and they should be written in a language that anyone can understand. As a technical communicator, you may be asked to help write such a privacy statement.

Guidelines

for Global, Ethical, and Legal Considerations on the Web

➤ **Write in clear, simple English, in a way that makes translation easy.** Many commercial Web pages originate in the United States and are written in English, but not every reader speaks English as a first language. Follow the suggestions in Chapter 5, "Guidelines for Communicating on a Global Team."

➤ **Avoid cultural references and humor.** These references are not only hard to translate but can also be offensive to some cultures. See Chapter 5, "Global Considerations when Working in Teams."

➤ **Offer different language options.** For a global audience, Web sites should include links to information in different languages.

➤ **Use colors and visuals appropriately.** If your Web site is aimed at specific cultural groups, find out whether certain colors or images will be offensive. See Chapter 12, "Social Media and Cultural Considerations," for more on cultural considerations when presenting visual information.

➤ **Keep ethical issues in mind.** Make sure your content is factual and credible. It's easy to create a Web site that looks "real" but is based on lies and distorted statistics. Check your content with a subject matter expert and, where appropriate, provide citations and links back to the original source.

➤ **Be sure you are not violating copyright when choosing images for your Web page.** Look for copyright free images or images that use Creative Commons or other licenses specific and appropriate for your use. See Chapter 12, "Guidelines for Obtaining and Citing Visual Material" for more on copyright and Creative Commons.

➤ **Create a clear privacy statement so readers know what, if any, information you are collecting about them and how this information will be used.** Allow readers to opt out of any advertising, email lists, and other commercial uses of their data.

Checklist

Writing and Designing for Blogs, Wikis, and the Web

Use the following Checklist when writing and designing blogs, wikis, and Web pages:

Audience and Purpose

❏ Is the page easy to enter, navigate, and exit? (See "How People Read Web Pages" in this chapter.)

❏ Is the content useful and worthwhile? (See "How People Read Web Pages" in this chapter.)

❏ Is the page at the right level of technicality for its primary audience? (See "Considering Audience and Purpose" in this chapter.)

Writing

❏ Is the most important information contained in the early paragraphs or upper section of the page? (See "Guidelines for Writing Web Pages" in this chapter.)

❏ Is the information chunked for easy access and quick reading? (See "Guidelines for Writing Web Pages" in this chapter.)

❏ Are sentences and paragraphs short and to the point? (See "Guidelines for Writing Web Pages" in this chapter.)

❏ Is the content written in a factual, neutral tone? (See "Guidelines for Writing Web Pages" in this chapter.)

Design

❏ Does the page take into account the "F-shaped" reading pattern? (See "Guidelines for Designing Web Pages" in this chapter.)

❏ Are text and images well balanced? (See "Guidelines for Designing Web Pages" in this chapter.)

❏ Is there plenty of white space and margins to help guide the eye? (See "Guidelines for Designing Web Pages" in this chapter.)

❏ Are hyperlinks used to direct readers to other information? (See "Guidelines for Designing Web Pages" in this chapter.)

❏ Is the page easy for the eye to scan? (See "How People Read Web Pages" in this chapter.)

❏ Is key material highlighted by headings, color, and so on? (See "Guidelines for Designing Web Pages" in this chapter.)

❏ Are visuals used effectively? (See "Guidelines for Designing Web Pages" in this chapter.)

❏ For Web sites, is the entire site sketched out using a storyboard? (See "Planning Web Sites Using Storyboarding" in this chapter.)

Teamwork

❏ Is there a meeting schedule for the project? (See Chapter 5, "Guidelines for Running a Meeting.")

❏ Has the team used brainstorming to get started? (See Chapter 5, "Thinking Creatively.")

❏ Has the team determined what role (writer, designer, programmer, researcher) each person will play? (See Chapter 5, "Guidelines for Running a Meeting.")

Global Issues

❏ Is the writing clear enough for nonnative speakers of English to understand and can it be easily translated? (See "Guidelines for Global, Ethical, and Legal Considerations on the Web" in this chapter.)

❏ Does the page avoid cultural references and humor? (See "Guidelines for Global, Ethical, and Legal Considerations on the Web" in this chapter.)

❏ If appropriate, does the page offer different language options? (See "Guidelines for Global, Ethical, and Legal Considerations on the Web" in this chapter.)

❏ Are colors and visuals used appropriately? (See "Guidelines for Global, Ethical, and Legal Considerations on the Web" in this chapter.)

Ethical and Legal Issues

❏ Are text and visuals truthful and not distorted or manipulative? (See "Ethical Considerations" in this chapter.)

❏ Has permission been granted to use items on the page where copyright is a concern? (See "Legal Considerations" and "Guidelines for Global, Ethical, and Legal Considerations on the Web" in this chapter.)

❏ If appropriate, does the page link to the organization's privacy policy? (See "Legal Considerations" and "Guidelines for Global, Ethical, and Legal Considerations on the Web" in this chapter.)

Projects

For all projects, check with your instructor about whether to present your findings in class, bring drafts to class for discussion, upload your project to the class learning management system (LMS), and/or use the LMS forum or discussion boards to collaborate and review each activity below.

General

1. Do some research to develop your own "top 10" list of effective Web writing principles. Hint: Look online for Web writing guidelines by experts such as the Nielsen Norman Group, college and university sites, and top Web design firms. Based on what you find, assemble your list. Using your list as a guide, find three Web sites that exemplify the principles of effective writing for the Web. Prepare a short presentation on your exemplary sites, and explain why you selected them as examples.

2. In teams of three people, create a storyboard (see Figure 24.4) to map out a Web site you might create for a club, sports team, or other organization on campus.

Team

In teams of two, locate one or two blogs related to technical communication or to your major. Follow the blog for a few days, then write a brief analysis of what you observed related to the blog's intended audience, content areas, and value to members of the profession. Pay attention to the writing style: Does the blog maintain a friendly yet professional tone? Would you consider this blog a good source of career information? What features (writing style, use of visuals, credentials of the author or contributors) influence your view of this blog? Write a short summary of your thoughts to share with the class.

Digital and Social Media

Visit a site such as MediaWiki or see if your campus (through the library or digital media center) offers students access to wiki software. Consider setting up a wiki that would become a knowledge base for a topic you researched for your formal report, proposal, or other assignment for this class. For instance, a report on careers in technical marketing (see the formal analytical report in Chapter 21, "Feasibility Analysis of a Career in Technical Marketing") might become the basis for a wiki containing technical marketing job information, employment trends, schools that offer such a major, and more. Whom would your audience be for this wiki? How would you control who can contribute and edit the entries? Write a few entries following the "Guidelines for Writing and Using Blogs and Wikis" in this chapter. Review your work with another student in class.

Global

Find a blog, wiki, or Web page based in a different country. Besides differences in language, note any other differences between non-U.S. Web sites and those that originate in the United States. Look for issues such as appropriate uses of color, legal considerations regarding online privacy, and options to read the page in a different language. Compare these findings to similar issues on a U.S.-based Web site for a similar product or service. Write a memo to your instructor, indicating the top five issues Web designers need to consider when writing and designing content for a blog, wiki, or Web page in each country.

Chapter 25
Social Media

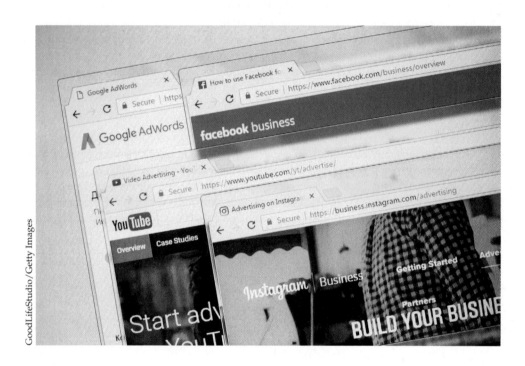

❝As social media director for a national non-profit, my job is to maintain a coordinated media strategy to raise awareness about environmental issues. We use social media—especially Facebook, Twitter, and YouTube—for fundraising, job postings, news items, updates, and links to our reports and studies. Our team worked hard to develop a set of guidelines on how to write for social media; the guidelines ensure a professional, concise, fact-based approach to all of our online communication.**❞**

—Geoffrey Wersan, Director of Social Media, environmental advocacy group

Considering Audience and Purpose

Using Social Media for Technical Communication

Instructional Videos for Social Media

Guidelines for Creating Instructional Videos for Social Media

Credibility and Legal Issues

Guidelines for Writing and Using Social Media

Checklist: Social Media **Projects**

 ## Learning Objectives

25.1 Analyze the audience and purpose of social media

25.2 Use Facebook, Twitter, and other social media for technical communication

25.3 Use social media such as YouTube and Vimeo to create video instructions

25.4 Identify credibility and legal issues related to social media in the workplace

Social media platforms allow people to network and share information, updates, photos, videos, and more. The most popular social media sites for workplace communication include the following:

- Customer review sites, such as Yelp and TripAdvisor, which allow individuals to post reviews of restaurants, hotels, businesses, doctors, and more
- Facebook, originally developed for college students but now used by friends, family, professional associates, businesses, nonprofits, and other organizations
- Google+, a network that includes email, friends and workplace associates, document and photo sharing, and more
- Instagram, originally used for uploading and sharing personal photos but increasingly used by businesses to share photos and coupons, promote new product lines or sales, and connect with customers and others
- LinkedIn and other job sites, used for professional networking, job postings, job searches, and résumé and job application materials
- Twitter, used for sending short messages (tweets) to people who subscribe to a particular Twitter feed
- YouTube, where videos including online instructions and videos about how to perform a task or how to understand a technical concept or idea are posted for viewing and public comment

Blogs and wikis, and even some Web pages, might also be considered social media, depending on how they are used. This chapter focuses on the more obvious forms of social media, listed above. See Chapter 24 for information about workplace blogs, wikis, and Web pages.

As you read this chapter, keep in mind that your interactions with social media in the workplace can take two forms: as a consumer and user of such sites (using LinkedIn for the job search, for instance) and as a creator and writer of such media (writing and maintaining your organization's Facebook site or Twitter feed).

Social media in the workplace

Social media are playing an increasingly important role at work. In technical communication, social media can serve as a resource for staying in touch with colleagues, acquiring information from technical experts, learning more about customer preferences, sharing updated information with customers, and keeping professionals connected. On the job, you might use social media to research a report or proposal, to post a job announcement, or to learn more about the professional background of a job candidate. You might also be asked to help create, maintain, update, and post to these sites.

Social media can help organizations disseminate cutting-edge technical and scientific research—and get feedback from people who may not be experts but who have firsthand experience. Public health organizations, for example, use Twitter and Facebook to get the word out about an influenza outbreak or a food safety recall. Companies might use Twitter or Instagram to announce a new product update or an upcoming seminar. These short social media posts usually contain links directing readers to a Web site (or an online document, such as a long report) with more detailed information.

Jobs in social media

Most organizations with any form of social media presence will end up hiring social media managers; whose responsibilities are to determine the most strategic uses of social media to reach customers and others. Social media management involves not only writing and posting content but also monitoring posted responses and comments. Social media managers require strong writing abilities, an ability to understand the complex nature of audience and purpose in online settings, and skills with digital technology. Technical communicators, engineers, medical professionals, and others with specialized expertise may at one time or another find themselves being asked to contribute to an organization's social media sites and strategies.

Considering Audience and Purpose

25.1 Analyze the audience and purpose of social media

Audience and purpose for social media

Most organizations maintain a social media site to highlight or promote a particular product or service or to provide updates and keep in touch with customers. Initially, a Facebook or Instagram site may be designed with one particular audience in mind. But when set up as open to the public, Facebook and other social media sites can be viewed by thousands (if not more) people. So, your audience and purpose may soon be far more expansive than originally envisioned. For example, the Centers for Disease Control and Prevention (CDC) reports that its initial Facebook page was set up for an audience interested in health and safety updates. But as interest grew, the organization had to expand its Facebook presence, eventually supporting "multiple Facebook profiles connecting users with information on a range of CDC health and safety topics" ("CDC Social Media Guidelines" 1).

For most large organizations, it has become imperative not only to have a social media presence but also to have a social media *policy*. From government agencies to colleges and universities to businesses, social media policies describe everything from appropriate use of visuals to the tone, style, and word count best suited when writing for that organization's social media site. Simon Fraser University (see Figure 25.1), for

Figure 25.1 **A Sample Social Media Policy**
Source: Courtesy of Simon Fraser University.

example, suggests limiting a Facebook post to 35 words or less and keeping voice and tone professional and consistent, no matter how many people are maintaining and writing for the site (Simon Fraser University).

Audience as Contributor

Impact of audience participation

External audience members can be important contributors to social media sites. For instance, organizations of all types, from retail businesses to technical services to restaurants, encourage their customers and clients to provide feedback on a review site, such as Yelp or TripAdvisor. Increasingly, product instructions and user documentation (see Chapter 13) are posted online, with a comment section available so readers can report on or update the documentation if they notice problems or errors. When audience members become contributors, there are more chances to collect important information. But there is also a greater need for fact-checking.

As one technical communication expert noted:

> The challenge for the professional communicator is to contribute at a higher level, welcoming community involvement and building on it. For the technical writer, that may mean doing less writing and more curating of content—for example, identifying and promoting the best wiki contributions, correcting errors, clarifying unclear language, adding illustrations, and improving organization. (Carr)

As discussed previously, social media managers, who are often trained in technical communication, play an important role in encouraging community participation and checking the accuracy of what gets posted.

Personal versus Workplace Uses of Social Media

Personal uses

Social media posts in workplace settings need to adhere to the kind of guidelines discussed throughout this text: words and images should be professional, concise, accurate, and clear. Yet social media is a world where professional and personal lives often intersect. Your personal Facebook page, Twitter, or Instagram account may contain images and posts that are suitable for friends and family: photos of a party or vacation; political opinions; off-the-cuff remarks about a news item or a new technology. When writing these posts as a student and an individual, you probably envision your audience as only those people you have chosen to see your posts (your friends or a custom list).

Workplace uses

But it is important to keep in mind that social media posts can travel far beyond your initial audience, and they can last for years. For example, many people have come to regret a Facebook posting or a tweet when it comes time for a job interview. Remember that although you may control your "friends list," it is very easy for people with enough time on their hands (and some easy-to-use software) to dig through your online history and compile a digital profile of you and your personal opinions and habits.

Using Social Media for Technical Communication

25.2 Use Facebook, Twitter, and other social media for technical communication

The most common social media sites for technical communication are discussed below.

Customer Review Sites

For companies such as restaurants, hotels, and entertainment venues, word of mouth can be the most important—or the most damaging—form of advertising. Customer review sites (Yelp and TripAdvisor are two of the most popular) allow registered members to post feedback in the form of short reviews. People can also rate the business using a number of stars. In some cases, companies are able to write a follow-up response, which can be especially important if the customer review is not favorable.

Business uses of Yelp and other customer review sites

Facebook

Facebook is used by government agencies, nonprofit organizations, and companies both large and small, across the United States and, increasingly, around the globe. Facebook provides a way to keep readers informed about news, updates, products and services, and so forth. Like customer review sites, Facebook also allows readers to participate on the site, posting questions and comments. Social media directors and others who manage the site often need to sift through these comments to ensure accuracy and keep "trolling" and negative information to a minimum.

Organizational uses of Facebook

Facebook recently launched a platform designed for use at work. This service looks and functions similarly to "regular" Facebook but includes more features for businesses, including new kinds of instant messaging and live video streaming designed less for one-to-one communication and more for groups or teams. Facebook and other companies are competing for the growing business market, where employees who grew up with personal social media apps now want and expect the same functionality at work.

Workplace team uses of Facebook

Google+

Google+ is a suite of apps that includes email, document collaboration and sharing, and a network of friends and workplace associates that you can use to stay in touch and share information. Google+ allows you to create special lists of people you want to communicate with: for example, business associates, friends, acquaintances, and so on. In this way, you can customize the audience for your message. Google+ also offers real-time ways to interact with your audience, such as Google Hangouts and Google Chat.

Workplace uses of Google+

Instagram

Workplace uses
of Instagram

Instagram is an app that features the use of photos and other visuals. Instagram was originally developed as a quick and easy phone-based app for people to share photos with friends, but its use has expanded as organizations notice Instagram's potential to reach customers. Businesses with a focus on visual information in particular use Instagram to share photos and coupons, promote new product lines or sales, and connect with existing and new customers.

LinkedIn and Other Job Sites

Employment
uses of *LinkedIn*

Social networking sites can help you stay connected, discover job openings, and advertise yourself. One popular professional networking site is LinkedIn (see Figure 25.2). On this and similar sites, you can keep your profile (work experience, references, and so forth) updated and make this profile available to professional contacts, potential employers, and job recruiters. Importantly, you can connect with former colleagues and classmates. These people are often your best bet for learning about job openings. Don't be shy about contacting people you don't know firsthand but with whom you have some mutual connection through a third party (a former coworker, for example). See Chapter 16 for more on using LinkedIn for networking and the job search.

Figure 25.2 LinkedIn, a Professional Networking Site LinkedIn and similar sites let you connect with potential employers as well as with other people in your field.
Source: Linkedin © 2015.

Twitter is also becoming a popular tool for employment-related communication. Many companies use Twitter to announce job openings. By using Twitter, companies can reach hundreds of potential employees quickly and inexpensively. See Chapter 16 for more information on using LinkedIn and Twitter as part of the job search process.

Twitter as an employment tool

Twitter

While Twitter can be used for employment information, it also provides real-time postings and updates in a concise format. Individuals as well as companies, government agencies, and other organizations maintain Twitter feeds to keep friends, business associates, customers, and citizens informed and updated. Called "tweets," the form of writing on Twitter is concise, with a limit of 280 characters per tweet. Tweets often contain a Web address to take readers elsewhere for more information.

Other uses of Twitter

Twitter can be especially effective in situations that require a rapid response. For example, when a major weather event interrupts air travel, airlines become inundated with calls from customers. A Twitter feed (Figure 25.3) can be a very effective method for providing customers with up-to-the-minute information on flight cancellations, contingency plans, and resources for more information.

PineTree Airways	**PT Airways** @PTAirways @ LauraG Sorry you can't get through by phone. Blizzard has slowed things down. Please rebook online if you can bit.ly/LGjEcl/11 ^CF
PineTree Airways	**PT Airways** @PTAirways @ NitaR International connections are still being rebooked. Check with a gate agent when you get in. ^CF
PineTree Airways	**PT Airways** @PTAirways @ AnetaD Please call us again at the new 800 no. we just posted on the website. Have your confirmation no. handy. ^MC
PineTree Airways	**PT Airways** @PTAirways @ greg_s Greg, did you check your flight status on our website? Here's the link bit.ly/XYZGEx ^MC
PineTree Airways	**PT Airways** @PTAirways Dear customers, as storm conditions improve, we will be updating all flight statuses so please stay tuned to Twitter and our website.

Figure 25.3 Twitter Feed During a Weather Emergency Posts are written with a friendly yet professional style and tone.

YouTube

Using YouTube to post instructions

For workplace communication, YouTube offers an easy-to-access site where customers can watch videos that help explain a concept or demonstrate how to perform a task. Many organizations create video instructions in addition to print or online user guides. Viewers can post comments and ideas to the video's Web page; these comments provide valuable feedback on the product or idea featured and give audience members a chance to offer their input directly to the company or organization. See below for more on using YouTube for video instructions.

Instructional Videos for Social Media

25.3 Use social media such as YouTube and Vimeo to create video instructions

Videos show a full-motion view

Until recently, workplace videos were typically used only for training and safety purposes and were filmed by professional videographers. Today, inexpensive video cameras (including the video feature on most phones and tablets) and easy-to-access video sites (like YouTube and Vimeo) make it simple for organizations large and small to create videos in addition to, or sometimes in place of, documents such as user manuals and instructions.

Popularity of online instructional video

YouTube (and other video sites, such as Vimeo) are often the first place people search when looking for instructions on how to assemble something, make a repair, change their oil, and many other instructional tasks. Even when people go directly to a company Web site to look for instructional material (for example, instructions for assembling a new gas grill, which came with the grill but got lost when the customers accidentally threw these out with the packing material!), the online user manual may be supplemented with a video. If you go to YouTube, most companies that offer extensive product lines have entire YouTube channels just for that company and its products.

Cautions of using online instructional video

Like all forms of social media, online instructional videos on sites like YouTube allow readers to become active participants in the content. Customers can post comments, corrections, and experiences about the video to the YouTube site and can rate the site using a thumbs up or down icon. Regular people, not just company experts, can make their own videos and upload these. When searching on instructional videos, remember to check the source. Keep in mind that bad or inaccurate online video instructions might perpetuate similar videos in a kind of copycat phenomenon. There is little in the way of user testing on these sites, and while the thumbs up or down ratings are helpful, they are no substitute for instructional videos that have been tested for accuracy, ease of use, and safety. With hundreds of videos available on the same topic, start with videos endorsed by the company.

If you are part of a team assigned to create online video instructions, start by considering items listed in Chapter 19 under "Elements of Effective Instructions," such as a clear title, informed and accurate content, and a level of technicality suitable for your audience. Work with a team that includes a trained videographer to be sure the camera angles, lighting, and image quality is suitable for viewing online.

Creating online video instructions

Instead of an outline, for video instructions you would write a script as a way to plan and sketch out what you intend to create. The script outlines the narration, images, motion, and other features that will comprise the complete video. For an instructional video, begin by locating the most recent set of print or PDF instructions. If none exist, write these first. Then adapt the print instructions to a video format: decide on background and foreground details, determine camera placement, decide where and when to include music or text, and write out the full narrative to be delivered.

Start with a script

Script the video (and then edit) to provide separate segments that orient the viewer, give a list of parts, supply step-by-step instructions, and offer a conclusion (such as a shot of the assembled product in action and a closing remark). Remove any unnecessary background or foreground clutter. Position the camera to keep the object or procedure at the center of the frame, at a distance that allows viewers to see clearly. Keep music to a minimum, usually only at the beginning and end. If text is included, make it easy to read on the screen—concise, clear, and to the point. Finally, accompany each step with narration, spoken clearly and slowly, with transitions between each step, and concise information.

Additional steps

As you would with written instructions, test the video instructions with a small group of representative audience members using the usability and user experience ideas in Chapter 19. You might find that a think-aloud evaluation (See Chapter 19, "Approaches for Evaluating a Document's Usability") is especially useful: You can watch people trying to assemble or repair the product and have them talk out loud in places where they appear to get confused or stuck. If you conduct your usability test using a YouTube site, you can ask participants to leave feedback and rate the video, too.

Importance of usability testing

Guidelines

for Creating Instructional Videos for Social Media

➤ **Provide a sense of scale.** Try to show both the object and a person using it (or a ruler or a hand).
➤ **In showing a procedure, simulate the angle of vision of the person actually performing each step.** In other words, allow the viewer to "look over the person's shoulder."
➤ **Show only what the viewer needs to see.** For example, in a video of a long procedure, focus only on the part of the procedure that is most relevant.
➤ **Edit out needless detail.** If you have editing software, shorten the video to include only the essentials.

> ➤ **Avoid excess office or background noise when recording sound.**
> ➤ **Encourage viewers to comment on your video.** You will get good feedback from the thousands of viewers on YouTube and similar sites.
> ➤ **Keep track of viewer feedback and respond quickly and diplomatically.** If people find mistakes or make suggestions, thank them. These customers are saving you countless hours of user testing and may also find potential safety issues before problems occur.

Credibililty and Legal Issues

25.4 Identify credibility and legal issues related to using social media in the workplace

Social media and credibility

Social media platforms provide powerful ways to influence human understanding. For instance, consider how social media posts can affect perceptions of credibility, truth, and accuracy. We all know of situations where a Facebook post or tweet has gone viral, circulating thousands (if not more) times, only to be discovered later to be untrue, inaccurate, or distorted.

Two factors contribute to this phenomenon. First, social media posts are typically shared by others from within a trusted network, for example, friends, family members, valued colleagues, and trusted organizations. As humans, we tend to view as credible what we hear from trusted sources. Second, social media posts combine visuals and text in short, speedy bursts. The visuals in particular play an important role in our response: Images are processed quickly and appeal to the more primitive areas of our brains—the "fight or flight" parts that want to respond (and repost) quickly. Social media images might be shared hundreds, thousands, or even more times, across numerous platforms, reinforcing their messages as the images are repeated over and over.

Because of these features, social media can be highly effective for technical and workplace communication, but the potential for problems also looms large. For example, if companies make a mistake with a social media post (say, an inaccurate claim about a product or an incorrect interpretation about a disease epidemic or medication recall), tens of thousands of Facebook, Twitter, and Yelp users can turn the mistake into a public relations nightmare (Baker 48–50) or even worse, a major health and safety situation. Or, consider the idea of "stealth marketing," in which people who publish supposedly objective product reviews fail to disclose the free merchandise or cash payments they received for their flattering portrayals. These reviews may circulate on social media appearing to be objective, factual discussions when in fact they are a form of marketing hidden behind a credible looking and sounding post.

Social media and legal issues

In addition to credibility concerns, it is important to consider workplace uses of social media in relation to privacy and other legal matters. Most social media sites have privacy policies, but these documents are often difficult to find on the site and are dense and hard to understand. An accessible and easy-to-read privacy policy is important to help inform readers of what data the site is collecting and how the site uses this data.

Employees within the organization must maintain a professional and appropriate tone and approach when posting. Those who discuss proprietary matters or post defamatory comments on social media can face serious legal consequences—not to mention job termination. Also, like email, social media sites create the potential for violations of copyright and confidentiality (see Chapter 18 for more on email and copyright).

To reinforce legal and other standards, organizations such as IBM, the American Red Cross, Dell, Intel, and many others have created social media guidelines and policies. Recognizing the particular challenge of protecting patient privacy, the American Nurses Association and the National Council of State Boards of Nursing created special guidelines on social media and networking specific for nurses ("News Release"). These types of guidelines and policies will become increasingly important as new social media apps come online and businesses and organizations work to adapt to the changing technological environments.

Guidelines

for Writing and Using Social Media

➤ **Create a social media policy that includes guidance on writing for social media.** See the example in Figure 25.1; writers are reminded to keep tweets to 120 characters and to limit Facebook posts to 35 words or less.

➤ **Choose the most appropriate format for your message.** A short tweet can reach thousands of readers quickly, but you'll need a link to another site (Facebook, your Web site, a PDF document) to provide more specific information.

➤ **Write with a friendly yet professional tone.** It's good to be friendly and sound like a real person, but a professional tone will make your message more credible.

➤ **Consider ways to include your readers but make plans for fact-checking.** Social media is interactive; use your postings to encourage customers and others to provide input. (See "Audience as Contributor" earlier in this chapter.) This feedback can be very valuable, but someone in the organization needs to fact-check and review all input.

➤ **Keep ideas focused and specific.** You can always insert a link if you want readers to access more detailed information or download a longer document.

➤ **Think and check before posting.** It's easy to write a quick tweet when an exciting event takes place or when a new product is about to be announced. But you need to coordinate these communications with the larger marketing and social media policies of the company. An unprofessional or poorly prepared message (on Twitter, Facebook, or any similar platform) can easily go "viral" and spread around the world in seconds.

➤ **If using photos or video, check the licensing agreement and other copyright information.** On photo sharing sites like Flickr, people typically list this information. Check to see if you need to request permission or if the photographer has granted some limited rights. See Chapter 12 for more about copyright and using visuals.

➤ **Be discreet about personal uses.** Your personal Facebook page may become a liability when you go on the job market, so be sure that what you post does not come back to haunt you. Also, be careful that what you say on Twitter as a private individual does not cross over into commentary about work.

➤ **Keep global readers in mind.** Social media is often viewed by readers from across many countries and cultures. Keep the content accessible to a wide audience (see Chapter 5 for more on global audiences).

Checklist

Social Media

Use the following Checklist when writing for social media.

❏ Is the message suitable for the specific social media site (Twitter, Facebook, YouTube, other)? (See "Using Social Media for Technical Communication" in this chapter.)

❏ Is the information written in a way that is most appropriate for the audience and purpose? (See "Considering Audience and Purpose" in this chapter.)

❏ Does the organization have a social media policy and social media writing guidelines? (See "Considering Audience and Purpose" in this chapter.)

❏ Does my posting adhere to this policy? (See "Credibility and Legal Issues" in this chapter.)

❏ Are customers and others allowed to contribute to the discussion? If so, is someone on the team responsible for fact-checking and editing? (See "Guidelines for Writing and Using Social Media" in this chapter.)

❏ Does the posting's tone and style seem friendly, yet professional? (See "Guidelines for Writing and Using Social Media" in this chapter.)

❏ Are ideas focused and specific? (See "Guidelines for Writing and Using Social Media" in this chapter.)

❏ For postings that refer to a longer topic, is a link included? (See "Guidelines for Writing and Using Social Media" in this chapter.)

❏ Is the message suitable for a global audience? (See "Guidelines for Writing and Using Social Media" in this chapter.)

❏ Does the message satisfy ethical and legal standards? (See "Credibility and Legal Issues" and "Guidelines for Writing and Using Social Media" in this chapter.)

❏ If video or photographs are included, has permission been granted to use this material? (See "Guidelines for Writing and Using Social Media" in this chapter.)

❏ For personal uses (such as Facebook or personal Twitter accounts), does the information portray you in a way you would want a potential employer to see? (See "Guidelines for Writing and Using Social Media" in this chapter.)

❏ When creating an online video, have you provided a clear and accurate title? (See "Instructional Videos for Social Media" in this chapter.)

❏ When creating an online video, have you started with a script? (See "Instructional Videos for Social Media" in this chapter.)

❏ When creating an online video, have you tested the video for usability? (See "Instructional Videos for Social Media" in this chapter.)

Projects

For all projects, check with your instructor about whether to present your findings in class, bring drafts to class for discussion, upload your project to the class learning management system (LMS), and/or use the LMS forum or discussion boards to collaborate and review each activity below.

General

Online, locate the social media policies of 3–6 organizations (colleges, universities, non-profits, companies). Review these policies, making a list of items they have in common as well as differences. Pay special attention to issues of word choice, tone, and privacy. Use what you learn to draft a set of social media guidelines for a campus organization (a club or other). Share your findings in a brief class presentation.

Team

In teams of 2–3 students, select a formal technical report that one of you wrote for this class or another class. Imagine yourselves as the social media team for an organization that sponsored the report. (For instance, if your report is about solar and wind energy, imagine that you work for a renewable energy company.) Consider a social media strategy that would make your report accessible to a wide audience (people interested in the latest in renewable energy, say). Write a memo to your manager outlining a social media strategy for the report, including ways you would encourage, but monitor, input from readers.

Digital and Social Media

Take a set of instructions you wrote for chapter 19 and make a plan to turn these instructions into a video, suitable for posting on a company web site and on YouTube. Use the print (pdf) instructions as a basis for writing a script. Make a rough version of the video using your phone camera, then test this first draft with several classmates. How could you improve your video? How would you monitor any posted feedback and respond to it?

Global

Customer review sites are written and read by people around the globe. If you are a social media manager, you'll need to read these reviews and write professional, thoughtful responses. Look up a service, restaurant, hotel, or other business using Yelp or TripAdvisor. Find one that has a mix of reviews (some good, some not so good). Did anyone from the company write a response? If so, is the response appropriate in tone and style for a global audience? Keeping a global audience in mind, write your own response to one or several of these (do not post it), and discuss the response with a group of three to four students.

Part 5

Resources for Technical Writers

Appendix A: A Quick Guide to Documentation

Appendix B: A Quick Guide to Grammar, Usage, and Mechanics

Appendix A
A Quick Guide to Documentation

Taking Notes

Ways to take notes

Researchers take notes in many ways, including by hand (on notecards or legal pads) or on a computer, tablet, or phone. While many people use a word-processing program (Microsoft Word, Google Docs, or other) to take notes, citation software, such as EndNote, Mendeley, and Zotero, are increasingly popular. These programs allow you to create an individual entry for each citation, then sort and retrieve your citations based on author, title, date, or key words. No matter what app you use to take notes, consider storing your data on the cloud, for easy access from any location or device. Make sure your notes are easy to organize and reorganize.

Guidelines
for Taking Notes

➤ **Make a separate bibliography listing for each work you consult.** Record that work's complete entry (Figure A.1), using the citation format that will appear in your document. (See "MLA Works Cited Entries" and "APA Reference List Entries" later in this chapter for sample entries.) Record the information accurately so that you won't have to relocate a source at the last minute.

Pinsky, Mark A. *The EMF Book: What You Should Know about Electromagnetic Fields, Electromagnetic Radiation, and Your Health.* Warner Books, 1995.

◀ ● Record each bibliographic citation exactly as it will appear in your final report

Figure A.1   Recording a Bibliographic Citation

When searching online, you can often print out a work's full bibliographic record or save it to your computer, to ensure an accurate citation.

➤ **Skim the entire work to locate relevant material.** Look over the table of contents and the index. Check the introduction for an overview or thesis. Look for informative headings.

➤ **Go back and decide what to record.** Use a separate entry for each item.

➤ **Be selective.** Don't copy or paraphrase every word. (See Chapter 9, "Guidelines for Summarizing Information.")

➤ **Record the item as a quotation or paraphrase.** When quoting others directly, be sure to record words and punctuation accurately. When restating material in your own words, preserve the original meaning and emphasis.

➤ **Highlight all quoted material.** Place quotation marks around all directly quoted material so that you don't lose track of exactly what is being quoted.

Quoting the Work of Others

You must place quotation marks around all exact wording you borrow, whether the words were written, spoken (as in an interview or presentation), or appeared in electronic form. Even a single borrowed sentence or phrase, or a single word used in a special way, needs quotation marks, with the exact source properly cited. These sources include people with whom you collaborate.

Plagiarism is often unintentional

If your notes don't identify quoted material accurately, you might forget to credit the source. Even when this omission is unintentional, you face the charge of *plagiarism* (misrepresenting the words or ideas of someone else as your own). Possible consequences of plagiarism include expulsion from school, loss of a job, and a lawsuit.

The perils of buying plagiarized work online

It's no secret that any cheater can purchase reports, term papers, and other documents on the Web. But antiplagiarism Web sites, such as turnitin.com, now enable professors to cross-reference a suspicious paper against Web material as well as millions of student papers, flagging and identifying each plagiarized source. In the end, plagiarizing negates the very reason for attending college: a student who turns in a plagiarized paper has missed an important opportunity for growth and learning.

Research writing is a process of independent thinking in which you work with the ideas of others in order to reach your own conclusions; unless the author's exact wording is essential, try to paraphrase, instead of quoting, borrowed material.

Guidelines

for Quoting

➤ **Use a direct quotation only when absolutely necessary.** Sometimes a direct quotation is the only way to do justice to the author's own words:

"Writing is a way to end up thinking something you couldn't have started out thinking" (Elbow 15).

Think of the topic sentence as "the one sentence you would keep if you could keep only one" (USAF Academy 11).

Expressions that warrant direct quotation

Consider quoting directly for these purposes:

- to preserve special phrasing or emphasis
- to preserve precise meaning
- to preserve the original line of reasoning
- to preserve an especially striking or colorful example
- to convey the authority and complexity of expert opinion
- to convey the original's voice, sincerity, or emotional intensity

Reasons for quoting directly

➤ **Ensure accuracy.** Copy the selection word for word; record the exact page numbers; and double-check that you haven't altered the original expression in any way (Figure A.2).

Pinsky, Mark A. pp. 29–30.

"Neither electromagnetic fields nor electromagnetic radiation cause cancer per se, most researchers agree. What they may do is promote cancer. Cancer is a multistage process that requires an 'initiator' that makes a cell or group of cells abnormal. Everyone has cancerous cells in his or her body. Cancer— the disease as we think of it—occurs when these cancerous cells grow uncontrollably."

Place quotation marks around all directly quoted material

Figure A.2 Recording a Quotation

➤ **Keep the quotation as brief as possible.** For conciseness and emphasis, use *ellipses:* Use three spaced periods (. . .) to indicate each omission within a single sentence. Add a period before the ellipsis to indicate each omission that includes the end of a sentence or multi-sentence sections of text.

> Use three . . . periods to indicate each omission within a single sentence. Add a fourth period to indicate . . . the end of a sentence. . . .

Ellipses within and between sentences

The elliptical passage must be grammatical and must not distort the original meaning. (For additional guidelines, see Appendix B, "Ellipses.")

➤ **Use square brackets to insert your own clarifying comments or to add transitions between various parts of the original.**

> "Job stress [in aircraft ground control] can lead to disaster."

Brackets setting off the added words within a quotation

➤ **Embed quoted material in your sentences clearly and grammatically.** Introduce integrated quotations with phrases such as "Jones argues that," or "Gomez concludes that." More importantly, use a transitional phrase to show the relationship between the quoted idea and the sentence that precedes it:

> One investigation of age discrimination at select Fortune 500 companies found that "middle managers over age 45 are an endangered species" (Jablonski 69).

An introduction that unifies a quotation with the discussion

Your integrated sentence should be grammatical:

Quoted material integrated grammatically with the writer's words	| "The present farming crisis," Marx argues, "is a direct result of rampant land speculation" (41). (For additional guidelines, see Appendix B, "Quotation Marks.") ➤ **Quote passages longer than four lines in block form.** Avoid relying on long quotations except in these instances: • to provide an extended example, definition, or analogy • to analyze or discuss an idea or concept
Reasons for quoting a long passage	Double-space a block quotation and indent the entire block five spaces (or 1/2 inch from the left margin). Do not indent the first line of the passage but do indent first lines of subsequent paragraphs. Do not use quotation marks. ➤ **Introduce the quotation and discuss its significance.**
An introduction to quoted material	| Here is a corporate executive's description of some audiences you can expect to address: ➤ **Cite the source of each quoted passage after the quotation's closing punctuation mark.**

Paraphrasing the Work of Others

Paraphrasing means more than changing or shuffling a few words; it means restating the original idea in your own words—sometimes in a clearer, more direct, and emphatic way—and giving full credit to the source.

Faulty paraphrasing is a form of plagiarism

To borrow or adapt someone else's ideas or reasoning without properly documenting the source is plagiarism. To offer as a paraphrase an original passage that is only slightly altered—even when you document the source—also is plagiarism. Equally unethical is offering a paraphrase, although documented, that distorts the original meaning.

Guidelines

for Paraphrasing

➤ **Refer to the author early in the paraphrase,** to indicate the beginning of the borrowed passage.
➤ **Retain key words from the original,** to preserve its meaning.
➤ **Restructure and combine original sentences** for emphasis and fluency.
➤ **Delete needless words from the original,** for conciseness.
➤ **Use your own words and phrases** to clarify the author's ideas.
➤ **Cite (in parentheses) the exact source,** to mark the end of the borrowed passage and to give full credit (Weinstein 3).
➤ **Be sure to preserve the author's original intent.**

Figure A.3 shows an entry paraphrased from Figure A.2. Paraphrased material is not enclosed within quotation marks, but it is documented to acknowledge your debt to the source. The paraphrase in the figure is adapted from the quote in Figure A.2.

Pinsky, Mark A.

Pinsky explains that electromagnetic waves probably do not directly cause cancer. However, they might contribute to the uncontrollable growth of those cancer cells normally present—but controlled—in the human body (29–30).

Signal the beginning of the paraphrase by citing the author, and the end by citing the source

Figure A.3 **Recording a Paraphrase**

What You Should Document

Document any insight, assertion, fact, finding, interpretation, judgment, or other "appropriated material that readers might otherwise mistake for your own" (Gibaldi and Achtert 155). Whether the material appears in published form or not, you must document these sources:

- any source from which you use exact wording
- any source from which you adapt material in your own words
- any visual illustration: charts, graphs, drawings, or the like (see Chapter 12 for documenting visuals)

Sources that require documentation

In some instances, you might have reason to preserve the anonymity of unpublished sources: for example, to allow people to respond candidly without fear of reprisal (as with employee criticism of the company), or to protect their privacy (as with certain material from email inquiries or electronic discussion groups). You must still document the fact that you are not the originator of this material. Do this by providing a general acknowledgment in the text such as "A number of employees expressed frustration with. . ." along with a general citation in your list of references or works cited such as "Interviews with Polex employees, May 2013."

How to document a confidential source

You don't need to document anything considered *common knowledge:* material that appears repeatedly in general sources. In medicine, for instance, it has become common knowledge that foods containing animal fat contribute to higher blood cholesterol levels. So in a report on fatty diets and heart disease, you probably would not need to document that well-known fact. But you would document information about how the fat/cholesterol connection was discovered, what subsequent studies have found (say, the role of saturated versus unsaturated fats), and any information for which some other person could claim specific credit. If the borrowed material can be found in only one specific source, not in multiple sources, document it. When in doubt, document the source.

Common knowledge need not be documented

How You Should Document

Documentation practices vary widely, but all systems work almost identically: a brief reference in the text names the source and refers readers to the complete citation, which allows readers to retrieve the source.

Many disciplines, institutions, and organizations publish their own style guides or documentation manuals. This quick guide illustrates citations and entries for two styles widely used for documenting sources in their respective disciplines:

- Modern Language Association (MLA) style, for the humanities
- American Psychological Association (APA) style, for the social sciences

Other citation styles, specific to certain technical and scientific fields, also exist. Use the style required by your instructor or employer. Whatever style you use, apply only that one style consistently throughout the document.

MLA Documentation Style

Cite a source briefly in text and fully at the end

Most writers in English and other humanities fields follow the Modern Language Association's *MLA Handbook*, 8th ed. (2016), which requires you to cite a source briefly within your text (called parenthetical references or in-text citations) and fully at the end of your document (in a Works Cited list). The in-text citation usually includes the author's surname and the exact page number where the borrowed material can be found:

Parenthetical reference in the text

> One notable study indicates an elevated risk of leukemia for children exposed to certain types of electromagnetic fields (Bowman et al. 59).

Readers seeking the complete citation for Bowman can refer easily to the Works Cited section, listed alphabetically by author:

Full citation at document's end

> Bowman, J. D., et al. "Hypothesis: The Risk of Childhood Leukemia Is Related to Combinations of Power-Frequency and Static Magnetic Fields." *Bioelectromagnetics*, vol. 16, no. 1, 1995, pp. 48–59.

This complete citation includes page numbers for the entire article.

MLA Parenthetical References

For clear and informative parenthetical references, observe these rules:

How to cite briefly in text

- If your discussion names the author, do not repeat the name in your parenthetical reference; simply give the page number(s):

Citing page numbers only

> Bowman et al. explain how their study indicates an elevated risk of leukemia for children exposed to certain types of electromagnetic fields (59).

Three works in a single reference

- If you cite two or more works in a single parenthetical reference, separate the citations with semicolons:

> (Jones 32; Leduc 41; Gomez 293-94)

- If you cite two or more authors with the same surnames, include the first initial in your parenthetical reference to each author:

    | (R. Jones 32)

    | (S. Jones 14-15)

Two authors
with identical
surnames

- If you cite two or more works by the same author, include the first significant word from each work's title, or a shortened version:

    | (Lamont, *Biomedicine* 100-01)

    | (Lamont, *Diagnostic Tests* 81)

Two works by
one author

- If the work is by an institutional or corporate author, abbreviate where this is commonly done (such as the abbreviation "dept." for the word "department"). If the Works Cited entry for the source includes the names of administrative units, separated by commas, include all of these names in the in-text citation.

    | (United States, Dept. of the Interior 19)

Institutional,
corporate, or
anonymous
author

Keep parenthetical references brief; when possible, name the source in your discussion and place only the page number(s) in parentheses.

For a paraphrase, place the parenthetical reference *before* the closing punctuation mark. For a quotation that runs into the text, place the reference *between* the final quotation mark and the closing punctuation mark. For a long quotation (longer than four lines) that is set off (indented) from the text, place the reference *after* the closing punctuation mark.

Where to place
a parenthetical
reference

MLA Works Cited Entries

The Works Cited list includes each source that you have paraphrased or quoted. Place your Works Cited list on a separate page at the end of the document. Arrange entries alphabetically by author's surname. When the author is unknown, list the title alphabetically according to its first word (excluding introductory articles such as *A, An, The*). For a title that begins with a numeral, alphabetize the entry as if the number were spelled out. Type the first line of each entry flush with the left margin. Indent the second and subsequent lines by 1/2 inch. Double-space within and between entries. Use one character space after all concluding punctuation marks (period, question mark).

How to format
the Works Cited
list

Certain "core elements" of publication information, according to the *MLA Handbook*, 8th ed., should be included in every source citation, no matter what the delivery method (print or digital). These elements include the author and title of the source, the title of the larger entity (which MLA calls a container; for example, a journal, newspaper, website, or anthology where the source resides), the names of additional contributors when relevant, the publisher and publication date, and the page numbers or Web address (what MLA terms the *location*).

How to cite
individual items

Table A.1 Index to Sample MLA Works Cited Entries

BOOKS

1. Book, single author
2. Book, two authors
3. Book, three or more authors
4. Book, anonymous author
5. Multiple books, same author
6. Book, one or more editors
7. Book, indirect source
8. Anthology selection or book chapter

PERIODICALS

9. Article, magazine
10. Article, journal with new pagination for each issue
11. Article, journal with continuous pagination
12. Article, newspaper

OTHER SOURCES

13. Encyclopedia, dictionary, other alphabetical reference
14. Report
15. Conference presentation
16. Interview, personally conducted
17. Interview, published
18. Letter or memo, unpublished

19. Questionnaire
20. Brochure or pamphlet
21. Lecture, speech, address, or reading
22. Government document
23. Document with corporate or foundation authorship
24. Map or other visual
25. Unpublished dissertation, report, or miscellaneous items

DIGITAL SOURCES

26. Online abstract
27. Online scholarly journal
28. Online newspaper article
29. Article in a reference database
30. Online reference or encyclopedia
31. Computer software
32. Email discussion group or posting
33. Personal email
34. Wiki
35. Blog
36. Podcast
37. Online video
38. Comment on a Web page
39. Tweet
40. Web sites and other digital formats

Table A.1 provides an overview of typical citation types; the rest of this section offers examples based on the MLA guidelines. The introductory section on digital sources later in this section provides more detail on what the MLA calls the "core elements" of MLA citation style.

MLA WORKS CITED ENTRIES FOR BOOKS.   Any citation for a book should contain the following information: author, title, editor or translator, edition, volume number, and publisher, date, and location (for books accessed online provide the site and URL).

What to include in an MLA citation for a book

1. Book, Single Author—MLA

Kerzin-Fontana, Jane B. *Technology Management: A Handbook*. 3rd ed., American Management Association, 2013.

Parenthetical reference: (Kerzin-Fontana 3-4)

2. Book, Two Authors—MLA

Aronson, Linda, and Roger Katz. *Toxic Waste Disposal Methods*. Yale UP, 2010.

Parenthetical reference: (Aronson and Katz 121-23)

Use complete publisher names but omit business words such as *Inc.* and *Corp.* Use the abbreviation UP for university presses ("Yale UP"). For page numbers with more than two digits, give only the final two digits for the second number when the preceding digits in the first number are identical.

3. **Book, Three or More Authors—MLA**

> Santos, Ruth J., et al. *Environmental Crises in Developing Countries.* HarperCollins Publishers, 2017.
>
> *Parenthetical reference:* (Santos et al. 9)

The abbreviation "et al." is a shortened version of the Latin "et alia," meaning "and others."

4. **Book, Anonymous Author—MLA**

> *Structured Programming.* Meredith Press, 2010.
>
> *Parenthetical reference:* (*Structured* 67)

5. **Multiple Books, Same Author—MLA**

> Chang, John W. *Biophysics.* Little, Brown, 2010.
>
> ---. *Diagnostic Techniques.* Radon, 1999.
>
> *Parenthetical reference:* (Chang, *Biophysics* 123-26), (Chang, *Diagnostic* 87)

When citing more than one work by the same author, do not repeat the author's name; simply type three hyphens followed by a period. List the works alphabetically.

6. **Book, One or More Editors—MLA**

> Morris, A. J., and Louise B. Pardin-Walker, editors. *Handbook of New Information Technology.*
>
> HarperCollins Publishers, 2012.
>
> *Parenthetical reference:* (Morris and Pardin-Walker 34)

For more than two editors, name only the first, followed by "et al."

7. **Book, Indirect Source—MLA**

> Kline, Thomas. *Automated Systems.* Rhodes, 2015.
>
> Stubbs, John. *White-Collar Productivity.* Harris, 2010.
>
> *Parenthetical reference:* (qtd. in Stubbs 116)

When your source (as in Stubbs, above) has quoted or cited another source, list each source in its alphabetical place on your Works Cited page. Use the name of the original source (here, Kline) in your text and precede your parenthetical reference with "qtd. in," or "cited in" for a paraphrase.

8. **Anthology Selection or Book Chapter—MLA**

> Bowman, Joel P. "Electronic Conferencing." *Communication and Technology: Today and Tomorrow,*
>
> edited by Al Williams, Association for Business Communication, 1994, pp. 123–42.
>
> *Parenthetical reference:* (Bowman 129)

The page numbers are for the selection cited from the anthology.

What to include in an MLA citation for a periodical

MLA WORKS CITED ENTRIES FOR PERIODICALS. Give all available information in this order: author, article title, periodical title, volume or number (or both), date (day, month, year), and page numbers for the entire article—not just pages cited.

9. Article, Magazine—MLA

DesMarteau, Kathleen. "Study Links Sewing Machine Use to Alzheimer's Disease." *Bobbin*, Oct. 1994, pp. 36–38.

Parenthetical reference: (DesMarteau 36)

If no author is given, list all other information:

"Video Games for the New Decade." *Power Technology Magazine*, 16 Oct. 2009, pp. 18+.

Parenthetical reference: ("Video Games" 18)

This article begins on page 18 and continues on page 21. When an article does not appear on consecutive pages, give only the number of the first page, followed immediately by a plus sign. Use a three-letter abbreviation for any month spelled with five or more letters.

10. Article, Journal with New Pagination for Each Issue—MLA

Thackman-White, Joan R. "Computer-Assisted Research." *American Library Journal*, vol. 51, no. 1, 2010, pp. 3–9.

Parenthetical reference: (Thackman-White 4-5)

Because each issue for a given year will have page numbers beginning with "1," readers need the number of this issue. The "51" denotes the volume number; "1" denotes the issue number.

11. Article, Journal with Continuous Pagination—MLA

Barnstead, Marion H. "The Writing Crisis." *Journal of Writing Theory*, vol. 12, no. 1, 2009, pp. 415–33.

Parenthetical reference: (Barnstead 415-16)

How to cite an abstract

If, instead of the complete work, you are citing merely an abstract found in a bound collection of abstracts and not the full article, include the information on the abstracting service right after the information on the original article.

Barnstead, Marion H. "The Writing Crisis." *Journal of Writing Theory*, vol. 12, no. 1, 2008, pp. 415-33. *Rhetoric Abstracts 67*, 2009, item 1354.

If citing an abstract that appears with the printed article, add "Abstract," followed by a period, immediately after the original work's page number(s).

12. Article, Newspaper—MLA

Baranski, Vida H. "Errors in Medical Diagnosis." *The Boston Times*, evening ed., 15 Jan. 2010, pp. B3+.

Parenthetical reference: (Baranski 3)

When a daily newspaper has more than one edition, cite the edition after the title of the newspaper. If no author is given, list all other information. If a locally published newspaper's name does not include the city of publication, insert it, using brackets; for example, *Sippican Sentinel* [Marion].

MLA WORKS CITED ENTRIES FOR OTHER KINDS OF MATERIALS.   Miscellaneous sources range from unsigned encyclopedia entries to conference presentations to government publications. Give this information (as available): author, title, publisher, date, and page numbers. Some sources will also require a description (Personal interview, Lecture, etc.).

What to include in an MLA citation for a miscellaneous source

13. Encyclopedia, Dictionary, Other Alphabetical Reference—MLA

"Communication." *The Business Reference Book*, Business Resources Press, 2010, pp. 215–16.

Parenthetical reference: ("Communication" 215)

Begin a signed entry with the author's name.

14. Report—MLA

Epidemiologic Studies of Electric Utility Employees. Electrical Power Research Institute (EPRI),

Nov. 1994. Report No. RP2964.5.

Parenthetical reference: (*Epidemiologic* 27)

If a source is created by an organization or corporation, then list that group as the author at the beginning of your citation. But if an organization is both the author and the publisher of a source, as in the example above, then start the citation with the title of the work and list the organization only as the publisher. Do not include *The* as part of the name of any organization in a Works Cited entry.

For any report or other document with group authorship, include the group's abbreviated name (along with its full name) in your first parenthetical reference, and then use only that abbreviation in any subsequent reference.

15. Conference Presentation—MLA

Smith, Abelard A. "Radon Concentrations in Molded Concrete." *First British Symposium in*

Environmental Engineering, London, 11–13 Oct. 2009, edited by Anne Hodkins, Harrison

Press, 2010, pp. 106–21.

Parenthetical reference: (Smith 109)

This citation is for a presentation that has been included in the published proceedings of a conference. For an unpublished presentation, include the presenter's name, the title of the presentation, and the conference title, series or sponsor, date, venue, and city (if city isn't included in venue name), but do not italicize the conference information.

16. Interview, Personally Conducted—MLA

Nasser, Gamel. Personal interview. 2 Apr. 2013.

Parenthetical reference: (Nasser)

17. Interview, Published—MLA

Lescault, James. "The Future of Graphics." Interviewed by Carol Jable. *Executive Views*

of Automation, edited by Karen Prell, Haber Press, 2013, pp. 216–31.

Parenthetical reference: (Lescault 218)

The interviewee's name is placed in the entry's author slot.

18. Letter or Memo, Unpublished—MLA

Rogers, Leonard. Letter to the author. 15 May 2013.

Parenthetical reference: (Rogers)

19. Questionnaire—MLA

Taylor, Lynne. Questionnaire sent to 612 Massachusetts business executives. 14 Feb. 2010.

Parenthetical reference: (Taylor)

20. Brochure or Pamphlet—MLA

Investment Strategies for the 21st Century. Blount Economics Association, 2010.

Parenthetical reference: (*Investment* 3)

If the work is signed, begin with its author.

21. Lecture, Speech, Address, or Reading—MLA

Arrigo, Anthony F. "Augmented Reality." English 368: Internet Communications and Culture, 15

Oct. 2015, U of Massachusetts at Dartmouth. Lecture.

Parenthetical reference: (Arrigo)

If the lecture title is not known, write Address, Lecture, or Reading—without quotation marks—in place of the title (and not at the end of the entry). Include the sponsor, venue, and the location (if the city isn't included in venue name).

22. Government Document—MLA

Virginia State, Department of Transportation. *Standards for Bridge Maintenance.* 2010.

Parenthetical reference: (*Virginia State, Dept. of Transportation* 49)

Include the name of the government, name(s) of any departmental/organizational units (from largest to smallest), name of agency, document title, publisher, and date.

For any congressional document, begin with the name of the government (United States), then Congress and the specific house of Congress (if appropriate), and then the department or committee. At the end of the citation, you can opt to include the number and session of Congress and document or report number.

United States, Congress, House, Armed Services Committee. *Funding for the Military Academies.*

Government Publishing Office, 2010. 108th Congress, 2nd session, House Report 807.

Parenthetical reference: (United States, Congress, House, Armed Services Committee 41)

Here is an example of an entry for a source from the *Congressional Record*:

Congressional Record, vol. 150, no. 30, 10 Mar. 2004, pp. H933–42.

Parenthetical reference: (*Congressional* H939)

23. Document with Corporate or Foundation Authorship—MLA

Hermitage Foundation. *Global Warming Scenarios for the Year 2030.* National Research

Council, 2009.

Parenthetical reference: (Hermitage Foundation 123)

24. Map or Other Visual—MLA

"Deaths Caused by Breast Cancer, by County." *Scientific American*, Oct. 1995, p. 32D. Map.

Parenthetical reference: ("*Deaths Caused*")

If the creator of the visual is listed, give that name first. If the visual has no title, use a description such as Map, Graph, Table, Diagram—without quotation marks—in place of the title (and not at the end of the entry).

25. Unpublished Dissertation, Report, or Miscellaneous Items—MLA

Author (if known). Title (in quotation marks or italicized, as appropriate). Sponsoring organization

or publisher, date.

For any work that has group authorship (corporation, committee, task force), cite the name of the group or agency in place of the author's name.

MLA WORKS CITED ENTRIES FOR DIGITAL SOURCES. MLA guidelines provide flexibility for citation formats based on the medium, but as with the print source examples shown above, they should include as much information as you think your reader will need to locate your original source. *(What to include in an MLA citation for a digital source)*

Digital sources include not only the ones listed here but also other media formats such as an Instagram posting; a discussion from a Facebook page; a YouTube or other online video; a posting to a user forum or discussion group. With so many new apps and formats coming into being every year, the following guidelines and examples provide a basis for creating source citations for new and existing media formats.

As described more generally below and illustrated in the previous examples, MLA guidelines suggest that Works Cited entries contain certain "core elements," including some or all of the following elements, as appropriate:

> Author. Title of source. Title of container, other contributors, version, number, publisher, publication date, location.

Two of these elements are especially important for documenting digital sources. Container refers to the place where the source resides, such as a book in a collection, a journal or magazine, a blog or Web page, or an e-book. Location refers to page numbers, Web address (URL), a digital library, or digital object identifier (DOI). So, for example, you might want to cite one of the Internal Revenue Services's Fact Sheets, located on their Web site. Following the core elements listed above, and the print example shown in item 22 above, you would add information about the digital source in the "location" field. The citation would appear as follows (note that http://is not required):

> Internal Revenue Service. "IRS Identity Theft Victim Assistance: How it Works." IRS Fact Sheet FS-2016-3, Jan. 2016, www.irs.gov/uac/Newsroom/IRSIdentity-Theft-Victim-Assistance-How-It-Works.

See the following examples for more illustrations of MLA style for digital sources; also, see style.mla.org for other examples.

26. Online Abstract—MLA

> Exbrayat, Jean-Marie, et al. "Harmful Effects of Nanoparticles on Animals." *Journal of Nanotechnology*, vol. 2015, doi: 10.1155/2015/861092. Abstract.
>
> *Parenthetical reference:* (Exbrayat et al.)

27. Online Scholarly Journal—MLA

For scholarly journal articles accessed online, use the citation format for an article from a print journal. However, for an online article without page numbers include the URL or DOI for the article at the end of the citation.

> Oswal, Sushil K., and Lisa Meloncon. "Paying Attention to Accessibility When Designing Online Courses in Technical and Professional Communication." *Journal of Business and Technical Communication*, vol. 28, no. 3, July 2014, doi: 10.1177/1050651914524780.
>
> *Parenthetical reference:* (Oswal and Meloncon)

28. Online Newspaper Article—MLAS

List the author (if available), the title of the article, and the publication date as they appear online. Be careful because sometimes the print version of a newspaper article will have a different title or publication date than the online version of the same article. Also, some newspapers (like *The New York Times*) offer permalinks for their articles; if a source has a permalink, use that in your entry, rather than a URL copied from your browser.

> Jabr, Ferris. "Could Ancient Remedies Hold the Answer to the Looming Antibiotics Crisis?" *The New York Times*, 14 Sept. 2016, nyti.ms/2clsr2l.
>
> *Parenthetical reference:* (Jabr)

29. Article in a Reference Database—MLA

> Sahl, J. D. "Power Lines, Viruses, and Childhood Leukemia." *Cancer Causes Control*, vol. 6,
>
> no. 1, Jan. 1995. PubMed Abstracts, PMID 7718739.
>
> *Parenthetical reference:* (Sahl 83)

In the above example, the abstract was published as part of a journal called *Cancer Causes Control* but was accessed through a different "container," the PubMed database.

For abstracts or database entries that were not originally published in print, follow this example, providing enough information for readers to locate the source if necessary:

> Argent, Roger R. "An Analysis of International Exchange Rates for 2009." Accu-Data, Dow Jones
>
> News Service, 10 Jan. 2009, online.djnews.org/ID99783.
>
> *Parenthetical reference:* (Argent)

If the author is not known, begin with the work's title. If the digital document has page numbers, include them in your entry and in your parenthetical reference.

30. Online Reference or Encyclopedia—MLA

If the article or entry has no author, start with the title. If the online reference is updated often, then include the date that the site was most recently updated and also include an access date at the end of your citation.

> "Suspension Bridge." *Encyclopædia Britannica*, 17 June 2016, www.britannica.com/technology/
>
> suspension-bridge. Accessed 5 Oct. 2016.
>
> *Parenthetical reference:* ("Suspension")

31. Computer Software—MLA

> *Virtual Collaboration*. Pearson, 2013. Software.
>
> *Parenthetical reference:* (*Virtual*)

Begin with the author's name, if known.

32. Email Discussion Group or Posting—MLA

> Eimers, Dave. "International dissertation prize for Dr. Anouk Smeekes." *Ercomer*, 19 Apr.
>
> 2015, www.ercomer.eu/2015/04/19/international-dissertation-prize-for-dr-anouk-
>
> smeekes/.
>
> *Parenthetical reference:* (Eimers)

Begin with the author's name (if known), followed by the title of the posting (in quotation marks and with standardized capitalization), the title of the discussion group or site (italicized), the date (and time, if available) of the message, and the URL for the specific message.

33. Personal Email—MLA

> Wallin, John Luther. "Re: Frog Reveries." Received by Virginia Bryant, 12 Oct. 2013.
>
> *Parenthetical reference:* (Wallin)

Use the subject line of the email as the title in your citation. Place it in quotation marks and standardize the capitalization as with a title.

Wikis, blogs, and podcasts, if needed as source material, should also be cited. Begin with the name of the communicator and topic title (in quotation marks); name of forum blog, wiki, or podcast (in italics); followed by the posting date; and the direct URL for the posting.

34. Wiki—MLA

> "Printing Press." *Wikipedia: The Free Encyclopedia*, 6 Sept. 2016, en.wikipedia.org/w/index.
>
> php?title=Printing_press&oldid=738048735. Accessed 16 Sept. 2016.
>
> *Parenthetical reference:* ("Printing")

Insert the full name of the wiki, the date that the entry was last updated, and the URL for the entry. *Wikipedia* offers a permanent link for each of its entries and you should use that link in your citation. Also, because wikis are updated frequently, it is a good idea to include an access date at the end of your citation.

35. Blog—MLA

> Heaven, Douglas. "Bionic Hands Enter the App Age." *New Scientist*, 18 Apr. 2013, www.newsci-
>
> entist.com/blogs/shortsharpscience/2013/04/bionic-hand.html.
>
> *Parenthetical reference:* (Heaven)

36. Podcast—MLA

> "Is the Internet Being Ruined?" *Freakonomics Radio*, narrated by Stephen J. Dubner, WYNC
>
> Studios/Dubner Productions, 13 July 2016, freakonomics.com/podcast/internet/.
>
> *Parenthetical reference:* ("Is the Internet")

37. Online Video—MLA

List the author or creator of the video, the title of the video, the site where you accessed it, the date the video was posted or published, and, if necessary to locate the video, its URL. In the example below, if the video had been accessed on NASA's own website, then the entry would start with the title of the video and NASA would be listed as the site name (instead of *YouTube*). For *YouTube* citations, if a video is uploaded by someone other than the person or organization that created it, then list that information before the date; for example: uploaded by National Aeronautics and Space Administration.

> United States, National Aeronautics and Space Administration (NASA). "Orion Trial by Fire."
>
> *YouTube*, 3 Dec. 2014, youtu.be/gjglwMPvzVo.
>
> *Parenthetical reference:* (United States, NASA)

38. Comment on a Web page—MLA

If citing a comment posted about an article or a blog posting, include the name of the author of the comment (as much as is given), the phrase "Comment on" followed by the title of the article or post being commented on, the name of the site, the date and time of the comment, and the URL for the comment (if a specific URL is available).

> Rodney. Comment on "Bionic Hands Enter the App Age." *New Scientist*, 20 Apr. 2013, 3:55
>
> p.m., www.newscientist.com/blogs/hortsharpscience/2013/04/bionic-hand.html.

Parenthetical reference: (Rodney)

39. Tweet—MLA

If known, insert full name of the author in parentheses after the online username. Use the full text of the tweet (without changes) as the title of the source and set it in quotation marks. Include the time of posting after the date.

> @ProfGurak (Laura Gurak). "Galileo would be proud and amazed—in orbit around
>
> Jupiter! #Juno." *Twitter*, 4 July 2016, 11:10 p.m., twitter.com/ProfGurak/
>
> status/750180000796545024.

Parenthetical reference: (@ProfGurak)

40. Web Sites and Other Digital Formats—MLA

For Web sites and other digital formats such as Facebook postings, online discussion forums, and so forth, follow the "core elements" format shown earlier in this section, choosing items as appropriate to your citation. If the author is unknown, start with the title of source. If the source is easy to locate, you can leave out the URL.

> "Calcium." University of Maryland Medical Center Medical Reference Guide, University
>
> of Maryland, 26 June 2014.

Parenthetical reference: ("Calcium")

> Samantha0212. "Taking Calcium Worked for Me." WebMed Online Discussion Forum, WebMed
>
> Online, 7 July 2016. webmed.org/forums/calcium/posting 9934.

Parenthetical reference: (Samantha0212)

MLA Sample Works Cited Pages

On a separate page at the document's end, arrange entries alphabetically by author's surname. When the author is unknown, list the title alphabetically according to its first word (excluding introductory articles). For a title that begins with a digit ("5," "6," etc.), alphabetize the entry as if the digit were spelled out.

The list of works cited in Figure A.4 accompanies the report on electromagnetic fields in Chapter 21. In the left margin, numbers refer to the elements discussed below. Bracketed labels identify different types of sources cited.

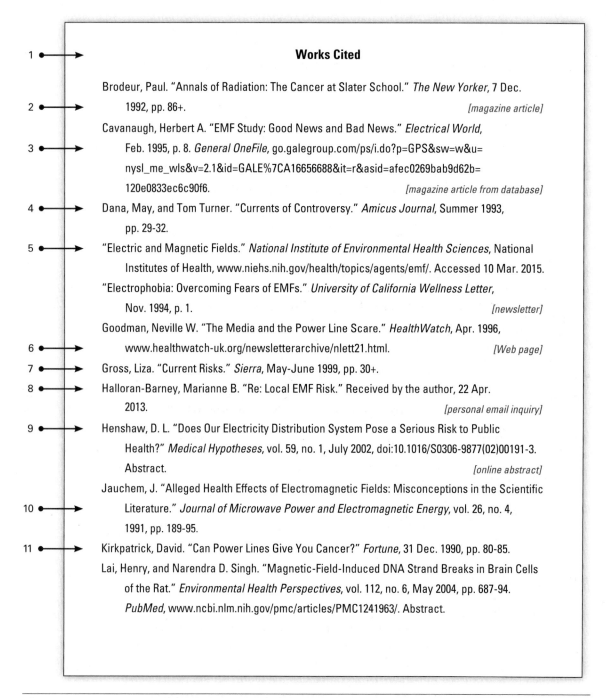

Figure A.4   **A List of Works Cited (MLA Style)**

"Magnetic Field Exposure and Cancer." *National Cancer Institute,* National Institutes of Health, 21 Apr. 2005, www.cancer.gov/cancertopics/factsheet/Risk/magnetic-fields. Accessed 7 Mar. 2015.

Miltane, John. Personal interview. 5 Apr. 2013.

Monmonier, Mark. *Cartographies of Danger: Mapping Hazards in America.* U of Chicago P, 1997.   *[book–one author]*

Moore, Taylor. "EMF Health Risks: The Story in Brief." *EPRI Journal,* Mar.-Apr. 1995, pp. 7-17.

Palfreman, John. "Apocalypse Not." *Technology Review,* 24 Apr. 1996, pp. 24-33.

Pinsky, Mark. A. *The EMF Book: What You Should Know about Electromagnetic Fields, Electromagnetic Radiation, and Your Health.* Warner Books, 1995.

Raloff, Janet. "Electromagnetic Fields May Trigger Enzymes." *Science News,* vol. 153, no. 8, 1998, p. 199.

---. "EMFs' Biological Influences." *Science News,* vol. 153, no. 2, 1998, pp. 29-31.   ← ● 12

Sivitz, Laura B. "Cells Proliferate in Magnetic Fields." *Science News,* vol. 158, no. 18, 2000, pp. 196-97.

Stix, Gary. "Are Power Lines a Dead Issue?" *Scientific American,* Mar. 1998, pp. 33-34.

"Strong Electric Fields Indicated in Major Leukemia Risk for Workers." *Microwave News,* vol. XX, no. 3, May-June 2000, microwavenews.com/news/backissues/ m-j00issue.pdf.   *[article from online journal]*

Taubes, Gary. "Fields of Fear." *Atlantic Monthly,* Nov. 1994, pp. 94-108.

United States, Environmental Protection Agency, Office of Radiation and Indoor Air. *EMF in Your Environment.* Government Printing Office, 1992.

---, National Institutes of Health, National Institute of Environmental Health Sciences. *EMF Magnetic and Electric Fields Associated with the Use of Electric Power.* Government Printing Office, 2002.

---, ---, ---. *Health Effects from Exposure to Power Line Frequency Electric and Magnetic Fields.* NIH Publication No. 99-4493, 1999, www.niehs.nih.gov/health/assets/docs_f_o/report_ powerline_electric_mg_predates_508.pdf.   *[gov. report posted online]*

Wartenburg, Daniel. "Solid Scientific Evidence Supporting an EMF-Childhood Leukemia Connection." *Microwave News,* vol. XXVII. no. 1, 19 Jan. 2007, microwavenews.com/ docs/mwn.1-07.pdf. Transcript.   *[published transcript in an online journal]*

Figure A.4 **Continued**

Discussion of Figure A.4

1. Center Works Cited title at the page top. Double-space entries. Number the Works Cited pages consecutively with text pages.

2. Indent five spaces for the second and subsequent lines of an entry. Place quotation marks around article titles. Italicize periodical or book titles. Capitalize the first letter of key words in all titles. Also capitalize articles, prepositions, and conjunctions only when they are the first or last word in a title. When an article skips pages in a publication, give only the first page number followed by a plus sign.

3. For an article accessed in an online database, include the database name, and the URL, DOI, or other identifier assigned to that article in the database.

4. For additional perspective beyond "establishment" viewpoints, examine alternative but well-researched and credible publications (such as the *Amicus Journal* and *Mother Jones*). Keep in mind that to be credible, even alternative-type publications need to adhere to accepted journalistic standards.

5. For government reports posted on a Web site, name the government agency that the website belongs to, the sponsoring agency and include all available information for retrieving the document.

6. For online sources, include as much information as you think your reader will need to locate your original source (for instance, include the URL if the source is obscure or will be difficult to locate with a simple search).

7. Use a period and one space to separate a citation's three major items (author, title, publication data). Skip one space after a comma or colon.

8. Alphabetize hyphenated surnames according to the name that appears first. Use the subject line of an email message as the title in your citation, set in quotation marks, and standardize capitalization as with a title.

9. When citing an abstract instead of the complete article, indicate this by inserting "Abstract" at the end of your entry.

10. For a journal, include the volume and issue numbers, and the month or season (if available) and year of publication. For page numbers of more than two digits, give only the final differing digits in the second number (but never less than two digits).

11. Use three-letter abbreviations for months with five or more letters.

12. When the same author is listed for two or more works, use three dashes to denote the author's name in any entries after the first one.

APA Documentation Style

Another common citation style is one published in the *Publication Manual of the American Psychological Association*, 6th ed. (Washington: American Psychological Association, 2009). Called APA for short, this method emphasizes the date. APA style (or some similar author-date style) is preferred in the sciences and social sciences, where information quickly becomes outdated. A parenthetical reference in the text briefly identifies the source, date, and page number(s):

In one study, mice continuously exposed to an electromagnetic field tended to die earlier than mice in the control group (de Jager & de Bruyn, 1994, p. 224).

Reference cited in the text

The full citation then appears in the alphabetical listing of "References," at the report's end:

de Jager, L., & de Bruyn, L. (1994). Long-term effects of a 50 Hz electric field on the life-expectancy of mice. *Review of Environmental Health, 10*(3–4), 221–224.

Full citation at document's end

APA Parenthetical References

APA's parenthetical references differ from MLA's as follows: The APA citation includes the publication date. A comma separates each item in the reference; and "p." or "pp." precedes the page number. When a subsequent reference to a work follows closely after the initial reference, the date need not be included. Here are specific guidelines:

How APA and MLA parenthetical references differ

- If your discussion names the author, do not repeat the name in your parenthetical reference; simply give the date and page numbers:

    Researchers de Jager and de Bruyn (1994) explain that experimental mice exposed to an electromagnetic field tended to die earlier than mice in the control group (p. 224).

Author named in the text

When two authors of a work are named in the text, their names are connected by "and," but in a parenthetical reference, their names are connected by an ampersand, "&."

- If you cite two or more works in a single reference, list the authors in alphabetical order and separate the citations with semicolons:

    (Jones, 2007; Gomez, 2005; Leduc, 2002)

Two or more works in a single reference

- If you cite a work with three to five authors, try to name them in your text, to avoid an excessively long parenthetical reference.

    Franks, Oblesky, Ryan, Jablar, and Perkins (2008) studied the role of electromagnetic fields in tumor formation.

A work with three to five authors

In any subsequent references to this work, name only the first author, followed by "et al." (Latin abbreviation for "and others").

- If you cite two or more works by the same author published in the same year, assign a different letter to each work:

    (Lamont, 2009a, p. 137)
    (Lamont, 2009b, pp. 67-68)

Two or more works by the same author in the same year

Other examples of parenthetical references appear with their corresponding entries in the following discussion of the reference list entries.

APA Reference List Entries

The APA reference list includes each source you have cited in your document. Type the first line of each entry flush with the left margin. Indent the second and subsequent lines five character spaces (one-half inch). Skip one character space after any period, comma, or colon. Double-space within and between each entry.

How to space and indent entries

Following are examples of complete citations as they would appear in the References section of your document. Shown immediately below each entry is its corresponding parenthetical reference as it would appear in the text. Note the capitalization, abbreviation, spacing, and punctuation in the sample entries.

Table A.2 provides an overview of typical citation types; the rest of this section offers examples based on the APA guidelines. The introductory section on electronic sources provides an overview of the commonalities of citing these types of sources in APA style.

What to include in an APA citation for a book

APA ENTRIES FOR BOOKS. Book citations should contain all applicable information in the following order: author, date, title, editor or translator, edition, volume number, and facts about publication (city, state, and publisher).

1. Book, Single Author—APA

Kerzin-Fontana, J. B. (2013). *Technology management: A handbook* (3rd ed.). Delmar, NY:

American Management Association.

Parenthetical reference: (Kerzin-Fontana, 2013, pp. 3–4)

Use only initials for an author's first and middle name. Capitalize only the first word of a book's title and subtitle and any proper names. Identify a later edition in parentheses between the title and the period.

Table A.2 Index to Sample Entries for APA References

BOOKS

1. Book, single author
2. Book, two to seven authors
3. Book, eight or more authors
4. Book, anonymous author
5. Multiple books, same author
6. Book, one to five editors
7. Book, indirect source
8. Anthology selection or book chapter

PERIODICALS

9. Article, magazine
10. Article, journal with new pagination for each issue
11. Article, journal with continuous pagination
12. Article, newspaper

OTHER SOURCES

13. Encyclopedia, dictionary, alphabetical reference
14. Report
15. Conference presentation

16. Interview, personally conducted
17. Interview, published
18. Personal correspondence
19. Brochure or pamphlet
20. Lecture
21. Government document
22. Miscellaneous items

DIGITAL SOURCES

23. Online abstract
24. Print article posted online
25. Book or article available only online
26. Journal article with DOI
27. Online encyclopedia, dictionary, or handbook
28. Personal email
29. Blog posting
30. Newsgroup, discussion list, or online forum
31. Wiki
32. Facebook and Twitter
33. Press release
34. Technical or research report

2. Book, Two to Seven Authors—APA

> Aronson, L., Katz, R., & Moustafa, C. (2010). *Toxic waste disposal methods.* New Haven, CT.
>
> Yale University Press.
>
> *Parenthetical reference:* (Aronson, Katz, & Moustafa, 2010)

Use an ampersand (&) before the name of the final author listed in an entry. As an alternative parenthetical reference, name the authors in your text and include date (and page numbers, if appropriate) in parentheses.

Give the publisher's full name (as in "Yale University Press") but omit the words "Publisher," "Company," and "Inc."

3. Book, Eight or More Authors—APA

> Fogle, S. T., Gates, R., Hanes, P., Johns, B., Nin, K., Sarkis, P. . . . Yale, B. (2009). *Hyperspace*
>
> *technology.* Boston, MA: Little, Brown.
>
> *Parenthetical reference:* (Fogle et al., 2009, p. 34)

List the first six authors' names, insert an ellipsis, then add the last author's name. "Et al." is the Latin abbreviation for "et alia," meaning "and others."

4. Book, Anonymous Author—APA

> Structured programming. (2010). Boston, MA: Meredith Press.
>
> *Parenthetical reference:* (Structured Programming, 2010, p. 67)

In your list of references, place an anonymous work alphabetically by the first key word (not *The, A,* or *An*) in its title. In your parenthetical reference, capitalize all key words in a book, article, or journal title.

5. Multiple Books, Sa me Author—APA

> Chang, J. W. (2010a). Biophysics. Boston, MA: Little, Brown.
>
> Chang, J. W. (2010b). MindQuest. Chicago, IL: John Pressler.
>
> *Parenthetical references:* (Chang, 2010a) (Chang, 2010b)

Two or more works by the same author not published in the same year are distinguished by their respective dates alone, without the added letter.

6. Book, One to Five Editors—APA

> Morris, A. J., & Pardin-Walker, L. B. (Eds.). (2012). *Handbook of new information technology.*
>
> New York, NY: HarperCollins.
>
> *Parenthetical reference:* (Morris & Pardin-Walker, 2012, p. 79)

For more than five editors, name only the first, followed by "et al."

7. Book, Indirect Source—APA

> Stubbs, J. (2010). White-collar productivity. Miami, FL: Harris.
>
> *Parenthetical reference:* (cited in Stubbs, 2010, p. 47)

When your source (as in Stubbs, above) has cited another source, list only your source in the References section, but name the original source in the text: "Kline's study (cited in Stubbs, 2010, p. 47) supports this conclusion."

8. Anthology Selection or Book Chapter—APA

Bowman, J. (1994). Electronic conferencing. In A. Williams (Ed.), *Communication and technology:*

Today and tomorrow (pp. 123–142). Denton, TX: Association for Business Communication.

Parenthetical reference: (Bowman, 1994, p. 126)

The page numbers in the complete reference are for the selection cited from the anthology.

What to include in an APA citation for a periodical

APA ENTRIES FOR PERIODICALS.   Give this information (as available), in order: author, publication date, article title (no quotation marks), periodical title, volume or number (or both), and page numbers for the entire article—not just page(s) cited.

9. Article, Magazine—APA

DesMarteau, K. (1994, October). Study links sewing machine use to Alzheimer's disease.

Bobbin, 36, 36–38.

Parenthetical reference: (DesMarteau, 1994, p. 36)

If no author is given, provide all other information. Capitalize the first word in an article's title and subtitle, and any proper nouns. Capitalize all key words in a periodical title. Italicize the periodical title, volume number, and commas (as shown above).

10. Article, Journal with New Pagination for Each Issue—APA

Thackman-White, J. R. (2010). Computer-assisted research. *American Library Journal, 51*(1), 3–9.

Parenthetical reference: (Thackman-White, 2010, pp. 4–5)

Because each issue for a given year has page numbers that begin at "1," readers need the issue number (in this instance, "1"). The "51" denotes the volume number, which is italicized.

11. Article, Journal with Continuous Pagination—APA

Barnstead, M. H. (2009). The writing crisis. *Journal of Writing Theory, 12*, 415–433.

Parenthetical reference: (Barnstead, 2009, pp. 415–416)

The "12" denotes the volume number. When page numbers continue from issue to issue for the full year, readers won't need the issue number. (You can include the issue number if you think it will help readers retrieve the article more easily.)

12. Article, Newspaper—APA

Baranski, V. H. (2010, January 15). Errors in technology assessment. *The Boston Times*, p. B3.

Parenthetical reference: (Baranski, 2010, p. B3)

In addition to year of publication, include month and day. If the newspaper's name begins with "The," include it. Include "p." or "pp." before page numbers. For an article on nonconsecutive pages, list each page, separated by a comma.

APA ENTRIES FOR OTHER SOURCES.   Miscellaneous sources range from unsigned encyclopedia entries to conference presentations to government documents. Give this information (as available): author, publication date, work title (and report or series number), page numbers (if applicable), city, and publisher.

What to include in an APA citation for a miscellaneous source

13. Encyclopedia, Dictionary, Alphabetical Reference—APA

> Communication. (2010). In *The business reference book.* Boston, MA: Business Resources Press.
>
> *Parenthetical reference:* ("Communication," 2010)

For an entry that is signed, begin with the author's name and publication date.

14. Report—APA

> Electrical Power Research Institute. (1994). *Epidemiologic studies of electric utility employees*
>
> (Report No. RP2964.5). Palo Alto, CA: Author.
>
> *Parenthetical reference:* (Electrical Power Research Institute [EPRI], 1994, p. 12)

If authors are named, list them first, followed by publication date. When citing a group author, as above, include the group's abbreviated name in your first parenthetical reference, and use only that abbreviation in subsequent references. When the organization and publisher are the same, list "Author" in the publisher's slot.

15. Conference Presentation—APA

> Smith, A. A. (2009, March). Radon concentrations in molded concrete. In A. Hodkins (Ed.), First British
>
> Symposium on Environmental Engineering (pp. 106–121). London, UK: Harrison Press, 2010.
>
> *Parenthetical reference:* (Smith, 2009, p. 109)

In parentheses is the date of the presentation. The symposium's name is proper and so is capitalized. Following the publisher's name is the date of publication.

For an unpublished presentation, include the presenter's name, year and month, presentation title (italicized), and all available information about the conference: "Symposium held at. . . ." Do not italicize this last information.

16. Interview, Personally Conducted—APA

> *Parenthetical reference:* (G. Nasser, personal interview, April 2, 2013)

This material is considered a nonrecoverable source, and so is cited in the text only, as a parenthetical reference. If you name the respondent in text, do not repeat the name in the citation.

17. Interview, Published—APA

> Jable, C. K. (2009). The future of graphics [Interview with James Lescault]. In K. Prell (Ed.),
>
> *Executive views of automation* (pp. 216–231). Miami, FL: Haber Press, 2010.
>
> *Parenthetical reference:* (Jable, 2009, pp. 218–223)

Begin with the interviewer's name, followed by interview date and title (if available), the designation (in brackets), and publication information, including the date.

18. Personal Correspondence—APA

Parenthetical reference: (L. Rogers, personal correspondence, May 15, 2013)

This material is considered nonrecoverable data, and so is cited in the text only, as a parenthetical reference. If you name the correspondent in your discussion, do not repeat the name in the citation.

19. Brochure or Pamphlet—APA

This material follows the citation format for a book entry. After the title of the work, include the designation "Brochure" in brackets.

20. Lecture—APA

Dumont, R. A. (2010, January 15). *Managing natural gas*. Lecture presented at the University of

Massachusetts at Dartmouth.

Parenthetical reference: (Dumont, 2010)

If you name the lecturer in your discussion, do not repeat the name in the citation.

21. Government Document—APA

Virginia Highway Department. (2010). Standards for bridge maintenance. Richmond, VA: Author.

Parenthetical reference: (Virginia Highway Department, 2010, p. 49)

If the author is unknown, present the information in this order: name of the issuing agency, publication date, document title, place, and publisher. When the issuing agency is both author and publisher, list "Author" in the publisher's slot.

For any congressional document, identify the house of Congress (Senate or House of Representatives) before the date.

U.S. House Armed Services Committee. (2010). *Funding for the military academies*. Washington,

DC: U.S. Government Printing Office.

Parenthetical reference: (U.S. House, 2010, p. 41)

22. Miscellaneous Items (Unpublished Manuscripts, Dissertations, and so on)—APA

Author (if known). (Date of publication.) *Title of work*. Sponsoring organization or publisher.

For any work that has group authorship (corporation, committee, and so on), cite the name of the group or agency in place of the author's name.

What to include in an APA citation for an electronic source

APA ENTRIES FOR DIGITAL SOURCES. The *Publication Manual of the American Psychological Association*, 6th edition (2009) provides instructions for citation of print and electronic sources. For digital sources in particular, the APA notes that "in general. . . include the same elements, in the same order, as you would for a reference to a fixed-media source and add as much electronic retrieval information as needed for others to locate the sources you cited" (187). Including the Web address (URL) is still recommended. However, one new feature is the use of digital object identifiers (DOIs). These are unique identifiers designed to last longer than Web addresses, which often disappear or get changed when Web pages are moved or renamed.

Identify the original source (printed or electronic) and give readers a path for retrieving the material. Provide all available information in the following order.

1. Author, editor, creator, or sponsoring organization.
2. Date the item was published or was created electronically. For magazines and newspapers, include the month and day as well as the year. If the date of an electronic publication is not available, use *n.d.* in place of the date.
3. Publication information of the original printed version (as in previous entries), if such a version exists. Follow this by designating the electronic medium [CD-ROM] or the type of work [Abstract], [Brochure]—unless this designation is named in the work's title (as in "Inpatient brochure").
4. Database names. Do not list database names (unless the database is obscure or the material hard to find), but do include the Web address (or DOI, discussed below).
5. Web addresses and DOIs. Provide the full electronic address. For Internet sources, only provide the Web address if the source would be impossible to locate without it. APA recommends only using the home page Web address. For CD-ROM and database sources, give the document's retrieval number (see entry 23, below). Start the Web address with http://, but do not underline, italicize, use angle brackets, or add a period at the end of a Web address. When a Web address continues from one line to the next, break it only after a slash or other punctuation (except for http://, which should not be broken).

The APA now recommends using the DOI (Digital Object Identifier), when available, in place of a Web address in references to electronic texts. DOI numbers are found on some recent scholarly journals, especially in the sciences and social sciences. Here is a sample reference for a journal article with a DOI assigned:

Schmidt, D., et al. (2009). Advances in psychotropic medication. *Boston Journal of*

Psychotherapy, 81(3), 398-413. doi: 10.1037/0555-9467.79.3.483

Parenthetical reference: (Schmidt, 2009)

23. Online Abstract—APA

Stevens, R. L. (2010). Cell phones and cancer rates. *Oncology Journal, 57*(2), 41–43. [Abstract].

Retrieved from http://nim.mh.gov/medlineplus. (MEDLINE Item: AY 24598).

Parenthetical reference: (Stevens, 2010)

Ordinarily an APA entry ends with a period. Entries with a DOI, however, omit the period at the end of the electronic address. If you are citing the entire article retrieved from a full-text database, delete [Abstract] from the citation.

24. Print Article Posted Online—APA

Alley, R. A. (2009, January). Ergonomic influences on worker satisfaction. *Industrial Psychology,*

5(12), 672–678. Retrieved from http://www.psycharchives

Parenthetical reference: (Alley, 2009)

If you were confident that the document's electronic and print versions were identical, you could omit the Web address and insert "[Electronic version]" between the end of the article title and the period.

25. Book or Article Available Only Online (no DOI)—APA

Kelly, W. (2013). *Early graveyards of New England.* Retrieved from http://www.onlinebooks.com

Parenthetical reference: (Kelly, 2013)

This source exists only in electronic format.

26. Journal Article with DOI

Tijen, D. (2009). Recent developments in understanding salinity tolerance. *Environmental & Experimental Botany. 67*(1), 2–9. doi:10.1016/j.envexpbot.2009.05.008

Parenthetical reference: (Tijen, 2009)

27. Online Encyclopedia, Dictionary, or Handbook—APA

Ecoterrorism. (2009). *Ecological encyclopedia.* Washington, DC: Redwood. Retrieved May 1, 2010, from http://www.eco.floridastate.edu

Parenthetical reference: ("Ecoterrorism," 2009)

Include the retrieval date for works that are routinely updated. If a work on CD-ROM has a print equivalent, cite it in its printed form.

28. Personal Email—APA

Parenthetical reference: Fred Flynn (personal communication, May 10, 2013) provided these statistics.

Instead of being included in the list of references, personal email (considered a nonretrievable source) is cited fully in the text.

29. Blog Posting—APA

Owens, P. (2010, June 1). How to stabilize a large travel trailer. Message posted to http://rvblogs.com

Parenthetical reference: (Owens, 2010)

30. Newsgroup, Discussion List, or Online Forum—APA

LaBarge, V. S. (2013, October 20). A cure for computer viruses. Message posted to http://www.srb/forums

Parenthetical reference: (LaBarge, 2013)

Although email should not be included in the list of references, postings from blogs, newsgroups, and online forums, considered more retrievable, should be included.

31. Wiki—APA

Skull-base tumors. (n.d.). Retrieved June 10, 2009, from the Oncology Wiki:http://oncology.wikia.com

Parenthetical reference: ("Skull-Base," n.d.)

Notice the "n.d." ("no date") designation for this collaborative Web page that can be written or edited by anyone with access.

32. Facebook and Twitter

> NASA. (2013, January 8). Astronomers have made a 3D weather map of a brown dwarf using
>
> > NASA's Spitzer and Hubble Space telescopes! [Facebook update]. Retrieved from www.
> >
> > facebook.com/NASA/posts/128611970637320
>
> *Parenthetical reference:* (NASA, 2013)

APA ENTRIES FOR GRAY LITERATURE. Gray literature is material that is not peer reviewed but according to the APA can play an important role in research and publication. Examples of gray literature include annual reports, fact sheets, consumer brochures, press releases, and technical reports (each type is so named in the title, in brackets, or elsewhere in the citation). In the sample citation in entry 32, the type of item is identified in brackets; in entry 33, it is part of the titling information.

What to include in an APA citation for gray literature

33. Press Release—APA

> American Natural Foods Association. (2009, January 20). *Newest food additive poses special*
>
> > *threat to children, according to the upcoming issue of* Eating for Health [Press release].
> >
> > Retrieved from American Natural Foods Association website: http://www.anfha.org
>
> *Parenthetical reference:* (American Natural Foods Association, 2009)

34. Technical or Research Report—APA

> Gunderson, H., et al. (2007). *Declining birthrates in rural areas: Results from the 2005 National*
>
> > *Census Bureau Survey* (Report No. 7864 NCB 2005–171). Retrieved from the National
> >
> > Center for Population Statistics: http://ncps.gov
>
> *Parenthetical reference:* (Gunderson, 2007)

Notice that the report number, if available, is given after the title.

APA Sample Reference List

APA's References section is an alphabetical listing (by author) equivalent to MLA's Works Cited section. Like Works Cited, the reference list includes only those works actually cited. (A bibliography usually would include background works or works consulted as well.) Unlike MLA style, APA style calls for only "recoverable" sources to appear in the reference list. Therefore, personal interviews, email messages, and other unpublished materials are cited in the text only.

The list of references in Figure A.5 accompanies the report on a technical marketing career in Chapter 21. In the left margin, numbers denote elements of Figure A.5 discussed below. Bracketed labels on the right identify different types of sources.

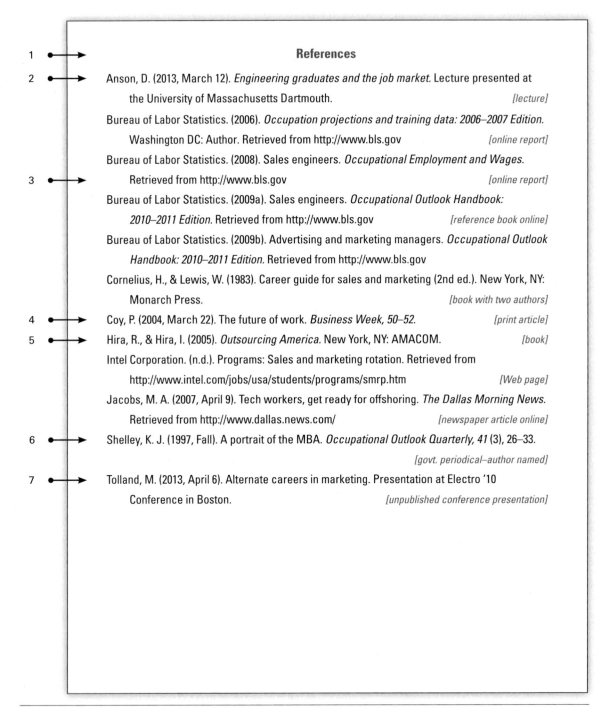

1   **References**

2   Anson, D. (2013, March 12). *Engineering graduates and the job market.* Lecture presented at
the University of Massachusetts Dartmouth.   *[lecture]*

Bureau of Labor Statistics. (2006). *Occupation projections and training data: 2006–2007 Edition.*
Washington DC: Author. Retrieved from http://www.bls.gov   *[online report]*

Bureau of Labor Statistics. (2008). Sales engineers. *Occupational Employment and Wages.*
3   Retrieved from http://www.bls.gov   *[online report]*

Bureau of Labor Statistics. (2009a). Sales engineers. *Occupational Outlook Handbook:
2010–2011 Edition.* Retrieved from http://www.bls.gov   *[reference book online]*

Bureau of Labor Statistics. (2009b). Advertising and marketing managers. *Occupational Outlook
Handbook: 2010–2011 Edition.* Retrieved from http://www.bls.gov

Cornelius, H., & Lewis, W. (1983). Career guide for sales and marketing (2nd ed.). New York, NY:
Monarch Press.   *[book with two authors]*

4   Coy, P. (2004, March 22). The future of work. *Business Week, 50–52.*   *[print article]*

5   Hira, R., & Hira, I. (2005). *Outsourcing America.* New York, NY: AMACOM.   *[book]*

Intel Corporation. (n.d.). Programs: Sales and marketing rotation. Retrieved from
http://www.intel.com/jobs/usa/students/programs/smrp.htm   *[Web page]*

Jacobs, M. A. (2007, April 9). Tech workers, get ready for offshoring. *The Dallas Morning News.*
Retrieved from http://www.dallas.news.com/   *[newspaper article online]*

6   Shelley, K. J. (1997, Fall). A portrait of the MBA. *Occupational Outlook Quarterly, 41* (3), 26–33.
[govt. periodical–author named]

7   Tolland, M. (2013, April 6). Alternate careers in marketing. Presentation at Electro '10
Conference in Boston.   *[unpublished conference presentation]*

Figure A.5   **A References List (APA Style)**

Discussion of Figure A.5

1. Center the "References" title at the top of page. Use one-inch margins. Number reference pages consecutively with text pages. Include only recoverable data (material that readers could retrieve for themselves); cite personal interviews, email, and other personal correspondence parenthetically in the text only. See also item 7 in this list.

2. Double-space entries and order them alphabetically by author's last name (excluding *A, An,* or *The*). List initials only for authors' first and middle names. Write out names of all months. In student papers, indent the second and subsequent lines of an entry five spaces. In papers submitted for publication in an APA journal, the first line is indented instead.

3. Omit punctuation from the end of an electronic address.

4. Do not enclose article titles in quotation marks. Italicize periodical titles. Capitalize the first word in article or book titles and subtitles, and any proper nouns. Capitalize all key words in magazine or journal titles.

5. For more than one author or editor, use ampersands instead of spelling out "and."

6. Use italics for a journal's name, volume number, and the comma. Give the issue number in parentheses only if each issue begins on page 1. Do not include "p." or "pp." before journal page numbers (only before page numbers from a newspaper).

7. Treat an unpublished conference presentation as a "recoverable source"; include it in your list of references instead of only citing it parenthetically in your text.

Appendix B
A Quick Guide to Grammar, Usage, and Mechanics

Grammar

The following common grammatical errors are easy to repair.

Sentence Fragments

A sentence fragment is a grammatically incomplete sentence. A grammatically complete sentence consists of at least one subject-verb combination and expresses a complete thought. It might include more than one subject-verb combination, and it might include other words or phrases as well. The following are all complete sentences:

Complete sentences

| This book summarizes recent criminal psychology research.

| The smudge tool creates soft effects.

| My dog, Zorro, ate my paper.

Even though the following example contains a subject-verb combination, it is a sentence fragment, because it doesn't express a complete thought:

Sentence fragment

| Although the report was not yet complete.

Although and other words like it, including *because, if, as, while, since, when,* and *unless,* are called subordinating conjunctions. Any of these words combined with a subject-verb combination produces a subordinate clause (a clause that expresses an incomplete idea). Subordinating conjunctions leave readers waiting for something to complete the thought. The thought can be completed only if another subject-verb combination that does express a complete idea is added:

Additional subject-verb completes the thought

| Although the report was not yet complete, I began editing.

This next group of words is a fragment because it contains no verb:

Fragment with no verb

| DesignPro, a new desktop publishing program.

Simply add a verb to turn the fragment into a complete sentence:

| DesignPro, a new desktop publishing program, will be available soon.

or

| DesignPro is a new desktop publishing program.

Avoid sentences that seem to contain a subject-verb combination but actually do not. Gerunds (verb forms ending in -*ing* that act like nouns, such as *being* in the first sentence below) and participles (verb forms ending in -*ing* that act like adjectives, such as *barking* in the second sentence below) look like verbs but actually are not. The first sentence is a gerund fragment and the second is a participle fragment:

| Dale being a document design expert
| The barking dog

These fragments can be turned into complete sentences by substituting a verb for the gerund (in the first correction below), or adding a verb to the participle (in the second correction):

| Dale is a document design expert.
| The barking dog finally stopped.

Run-On Sentences

A run-on sentence crams grammatically complete sentences together, as in the following example:

| For emergencies, we dial 911 for other questions, we dial 088.

This sentence can be repaired by dividing it into two sentences:

| For emergencies, we dial 911. For other questions, we dial 088.

Another possibility is to join the two parts of the sentence with a semicolon, as in the sentence below. This option indicates a break that is not quite as strong as the period, and therefore signals to the reader that the two items are closely related:

| For emergencies, we dial 911; for other questions, we dial 088.

Another possibility is to add a comma followed by a coordinating conjunction (*for, and, nor, but, or, yet, so*):

| For emergencies, we dial 911, but for other questions, we dial 088.

Comma Splices

In a comma splice, two complete ideas (independent clauses) that should be *separated* by a period or a semicolon are incorrectly *joined* by a comma, as in the following example:

| Sarah did a great job, she was promoted.

One option for correcting comma splices is to create two separate sentences:

| Sarah did a great job. She was promoted.

Another option is a semicolon to show a relationship between the two items:

Semicolon	| Sarah did a great job; she was promoted.

A third option is a semicolon with a conjunctive adverb (an adverb, ending in -*ly*, that shows a relationship between the items, such as *consequently*):

Semicolon with conjunctive adverb	| Sarah did a great job; consequently, she was promoted.

A fourth option is a coordinating conjunction: (*for, and, nor, but, or, yet, so*)—to create an equal relationship between items:

Coordinating conjunction	| Sarah did a great job and was promoted.

Finally, you can use a subordinating conjunction (a conjunction that creates a dependent relationship between items—see "Faulty Subordination" later in this appendix):

Subordinating conjunction	| Because Sarah did a great job, she was promoted.

Faulty Agreement—Subject and Verb

The subject must agree in number with the verb. But when subject and verb are separated by other words, we might lose track of the subject-verb relationship, as in the following sentence with faulty subject-verb agreement:

Faulty subject-verb agreement	| The lion's share of diesels are sold in Europe.

Make the verb agree with its subject (*share*), not with a word that comes between the subject and the verb (in this case, the plural noun *diesels*). In the following correction, the subject and verb now agree:

Correct subject-verb agreement	| The lion's share of diesels is sold in Europe.

Treat compound subjects connected by *and* as plural:

Compound subject takes plural verb	| Terry and Julie enjoy collaborating on writing projects.

With compound subjects connected by *or* or *nor*, the verb is singular if both subjects are singular and plural if both subjects are plural. If one subject is singular and one is plural, the verb agrees with the one closer to the verb. In the following sentence, the verb agrees with the closer subject, *students*:

Verb agrees with closer subject in compound subject	| Neither the professor nor the students were able to see what was going on.

Also, treat most indefinite pronouns (*anybody, each, everybody*, etc.) as singular subjects:

Indefinite pronoun agrees with singular verb	| Almost everybody who registered for the class was there on the first day.

Treat collective subjects (*team, family, group, committee*, etc.) as singular unless the meaning is clearly plural. Both of the following sentences are correct. In the first sentence, the *group* is understood as a singular subject (the group is a single entity) and agrees with the singular verb *respects*, whereas in the second sentence, the *board* is understood as plural (as indicated by the plural word *authors*) and agrees with the plural verb *are*:

Collective subjects agree with singular or plural verbs, depending on meaning	| The group respects its leader. | The editorial board are all published authors.

Faulty Agreement—Pronoun and Referent

A pronoun must refer to a specific noun (its *referent* or *antecedent*), with which it must agree in gender and number. Faulty pronoun-referent agreement is easy to spot when the gender and number clearly don't match (e.g., "He should proceed at their own pace"). However, when an indefinite pronoun such as *each, everyone, anybody, someone,* or *none* is the referent, the pronoun is always singular, as both of the sentences below demonstrate:

> Everybody should proceed at his or her own pace.
> None of the candidates described her career plans in detail.

Indefinite referents agree with singular pronouns

Dangling and Misplaced Modifiers

Problems with ambiguity occur when a modifying phrase has no word to modify, as in the following sentence containing a dangling modifier:

> **Writing an email,** the cat ran out the open door.

Dangling modifier

The cat obviously did not use the computer, but because the modifier **Writing an email** has no word to modify, the noun beginning the main clause (*cat*) seems to name the one who used the computer. Without any word to join itself to, the modifier *dangles*. Inserting a subject, as in the following example in which the subject **Judy** is added, repairs this absurd message.

> As **Judy** wrote the email, the cat ran out the open door.

Correct

A dangling modifier can also obscure your meaning, as in this next sentence in which it's unclear who completes the form—the student or the financial aid office:

> **After completing the student financial aid application form,** the Financial Aid Office will forward it to the appropriate state agency.

Dangling modifier

Here are two more examples of sentences with dangling modifiers that obscure the message, followed by their corrected versions, in which subjects are added for clarification:

> **While walking,** a cold chill ran through my body.

Dangling modifier

> While **I** walked, a cold chill ran through my body.

Correct

> Impurities have entered our bodies **by eating chemically processed foods.**

Dangling modifier

> Impurities have entered our bodies by **our** eating chemically processed foods.

Correct

The order of adjectives and adverbs also affects meaning.

> I **often** remind myself of the need to balance my checkbook.

> I remind myself of the need to balance my checkbook **often.**

Position modifiers to reflect your meaning. In the following two examples, the first sentence features a confusingly placed modifier. However, by repositioning the modifiers in two possible ways, the problem is solved:

Misplaced modifier	Joe typed another memo on our computer **that was useless.**
	(Was the computer or the memo useless?)
	Joe typed another useless memo on our computer.
	or
Correct	Joe typed another memo on our useless computer.
	Mary volunteered **immediately** to deliver the radioactive shipment.
Misplaced modifier	(Volunteering immediately, or delivering immediately?)
	Mary immediately volunteered to deliver . . .
	or
Correct	Mary volunteered to deliver immediately . . .

Faulty Parallelism

To reflect relationships among items of equal importance, express them in identical grammatical form, as in the following famous quotation from Abraham Lincoln:

Correct	We here highly resolve . . . that government **of the people, by the people, for the people** shall not perish from the earth.

Otherwise, the message would be garbled, like this:

Faulty	We here highly resolve . . . that government **of the people, which the people created and maintain, serving the people** shall not perish from the earth.

If you begin the series with a noun, use nouns throughout the series; likewise for adjectives, adverbs, and specific types of clauses and phrases. In the following pairs of sentences, notice how the consistent use of subjunctive complements, adjectives, infinitive phrases, or verbs makes the corrected versions of each sentence easier to understand and remember:

Faulty	The new apprentice is **enthusiastic, skilled,** and **you can depend on her.**
Correct	The new apprentice is **enthusiastic, skilled,** and **dependable.**
	(all subjective complements)
Faulty	In his new job, he felt **lonely** and **without a friend.**
Correct	In his new job, he felt **lonely** and **friendless.**
	(both adjectives)
Faulty	She plans **to study** all this month and **on scoring well** in her licensing examination.
Correct	She plans **to study** all this month and **to score well** in her licensing examination.
	(both infinitive phrases)
Faulty	She **sleeps** well and **jogs** daily, **as well as eating** high-protein foods.
Correct	She **sleeps** well, **jogs** daily, and **eats** high-protein foods.
	(all verbs)

Faulty Coordination

Give equal emphasis to ideas of equal importance by joining them with coordinating conjunctions: **and, but, or, nor, for, so,** and **yet**, as in the following examples:

| This course is difficult **but** worthwhile.

| My horse is old **and** gray.

| We must decide to support **or** reject the dean's proposal.

But do not confound your meaning by coordinating excessively, as in the following difficult to understand sentence:

The climax in jogging comes after a few miles **and** I can no longer feel stride after stride **and** it seems as if I am floating **and** jogging becomes almost a reflex **and** my arms **and** legs continue to move **and** my mind no longer has to control their actions. | Excessive coordination

Excessive coordination can be repaired by breaking a sentence with multiple coordinating conjunctions into several sentences, as in the following revision of the previous sentence:

The climax in jogging comes after a few miles when I can no longer feel stride after stride. By then I am jogging almost by reflex, nearly floating, my arms and legs still moving, my mind no longer having to control their actions. | Revised

Notice how the meaning becomes clear when the less important ideas (**nearly floating, arms and legs still moving, my mind no longer having**) are shown as dependent on, rather than equal to, the most important idea (**jogging almost by reflex**)—the idea that contains the lesser ones.

Avoid coordinating two or more ideas that cannot be sensibly connected. The following faulty sentence implies that the simple act of John's lateness wrecked his car:

| John was late for work and wrecked his car. | Faulty

The problem can be repaired by adding more information:

Because John was late for work, he backed his car out of the driveway into oncoming traffic, wrecking his car. | Revised

Finally, instead of *try and*, use *try to*. In the following example, *try and* implies that *try* and *help* are not necessarily connected in the first sentences, whereas *try to* makes the connection clear in the second sentence:

| I will try and help you.

| I will try to help you.

Faulty Subordination

Proper subordination shows that a less important idea is dependent on a more important idea. A dependent (or subordinate) clause in a sentence is signaled by a subordinating conjunction: **because, so, if, unless, after, until, since, while, as,** and **although.** Consider these complete ideas:

| Joe studies hard. He has severe math anxiety.

Because these ideas are expressed as simple sentences, they appear coordinate (equal in importance). But if you wanted to convey an opinion about Joe's chances of succeeding in math, you would need a third sentence: **His disability probably will prevent him from succeeding,** or **His willpower will help him succeed.** To communicate the intended meaning concisely, combine the two ideas. Subordinate the one that deserves less emphasis and place the idea you want emphasized in the independent (main) clause:

> Despite his severe math anxiety (*subordinate idea*), Joe studies hard (*independent idea*).

This first version suggests that Joe will succeed. In the next example, the subordination suggests the opposite meaning:

> Despite his diligent studying (*subordinate idea*), Joe has severe math anxiety (*independent idea*).

Do not coordinate when you should subordinate, as in the following weak sentence:

Weak
> Television viewers can relate to an athlete they idolize and they feel obliged to buy the product endorsed by their hero.

Of the two ideas in the sentence above, one is the cause, the other the effect. Emphasize this relationship through subordination, as in the following revision:

Revised
> Because television viewers can relate to an athlete they idolize, they feel obliged to buy the product endorsed by their hero.

When combining several ideas within a sentence, decide which is most important, and subordinate the other ideas to it. Do not merely coordinate, as in the faulty example below, but subordinate, in in the revised version that follows:

Faulty
> This employee is often late for work, and he writes illogical reports, and he is a poor manager, and he should be fired.

Revised
> Because this employee is often late for work, writes illogical reports, and has poor management skills, **he should be fired**. (*This last clause is independent.*)

Faulty Pronoun Case

A pronoun's case (nominative, objective, or possessive) is determined by its role in the sentence: as subject, object, or indicator of possession.

If the pronoun serves as the subject of a sentence (*I, we, you, she, he, it, they, who*), or follows a version of the linking verb *to be*, its case is *nominative*, as in the following three examples:

Nominative
pronoun case
> She completed her graduate program in record time.
>
> Who broke the chair?
>
> The chemist who perfected this distillation process is he.

If the pronoun serves as the object of a verb or a preposition (*me, us, you, her, him, it, them, whom*), its case is *objective*, as the two sentences below illustrate:

Objective
pronoun case
> The employees gave her a parting gift.
>
> To whom do you wish to complain?

If a pronoun indicates possession (*my, mine, our, ours, your, yours, his, her, hers, its, their, whose*), its case is *possessive*, as in the following two examples:

| The brown briefcase is mine.

| Whose opinion do you value most?

Possessive pronoun case

The incorrect sentences below are examples of faulty pronoun case based on subject-object position, followed by corrected versions in parentheses:

| Whom is responsible to who? (Who is responsible to whom?)

| The debate was between Marsha and I. (The debate was between Marsha and me.)

Faulty pronoun case

Punctuation

Punctuation marks are like road signs and traffic signals. They govern reading speed and provide clues for navigation through a network of ideas. The three marks of end punctuation—period, question mark, and exclamation point—work like a red traffic light by signaling a complete stop.

Period

A period ends a declarative sentence, as in the following example:

| I see that you've all completed the essay.

It is also used as the final mark in some abbreviations, such as "Dr." and "Inc." and as a decimal point such as "$18.43" and "26.2%."

Period ends declarative statement

Question Mark

A question mark follows a direct question, as in the following sentence:

| Have you all completed the essay?

Do not use a question mark to end an indirect question. The first sentence below incorrectly does so, while the second sentence correctly ends with a period:

| Professor Grim asked if all students had completed the essay?

| Professor Grim asked if all students had completed the essay.

Question mark ends direct question

Incorrect

Period ends indirect question

Exclamation Point

Use an exclamation point only when expression of strong feeling is appropriate, as in the following sentence in which emphasis is appropriate:

| I can't believe you finished the essay so fast!

Exclamation point provides emphasis

Semicolon

Like a blinking red traffic light at an intersection, a semicolon signals a brief but definite stop. Semicolons have several uses.

TO SEPARATE INDEPENDENT CLAUSES.   A semicolon can separate independent clauses (logically complete ideas) whose contents are closely related and are not already connected by a comma and a coordinating conjunction (*and, or, but*, etc.):

Semicolon
separates
independent
clauses

| The project was finally completed; we had done a good week's work.

TO ACCOMPANY CONJUNCTIVE ADVERBS.   A semicolon must accompany a conjunctive adverb such as *besides, otherwise, still, however, furthermore, moreover, consequently, therefore, on the other hand, in contrast*, or *in fact*:

Semicolon
accompanies
conjunctive
adverbs

| The job is filled; however, we will keep your résumé on file.
| Your background is impressive; in fact, it is the best among our applicants.

TO SEPARATE ITEMS IN A SERIES.   When items in a series contain internal commas, semicolons provide clear separation between items:

Semicolon
separates items
in a series

| I am applying for summer jobs in Santa Fe, New Mexico; Albany, New York; Montgomery, Alabama; and Moscow, Idaho.

Colon

Like a flare in the road, a colon signals you to stop and then proceed, paying attention to the situation ahead. Colons have several uses.

TO SIGNAL A FOLLOW-UP EXPLANATION.   Use a colon when a complete introductory statement requires a follow-up explanation:

Colon signals
a follow-up
explanation

| She is an ideal colleague: honest, reliable, and competent.

Do not use a colon if the introductory statement is incomplete, such as in the following incorrect sentence:

Incorrect

| My plans include: finishing college, traveling for two years, and settling down in Santa Fe.

TO REPLACE A SEMICOLON.   A colon can replace a semicolon between two related, complete statements when the second one explains or amplifies the first:

Colon replaces a
semicolon

| Pam's reason for accepting the lowest-paying job offer was simple: she had always wanted to live in the Northwest.

TO INTRODUCE A QUOTATION.   Colons can introduce quotations:

Colon introduces
a quotation

| The supervisor's message was clear enough: "You're fired."

TO FOLLOW SALUTATIONS.   Colons follow salutations in formal correspondence (e.g., Dear Ms. Jones:).

Comma

The comma is the most frequently used—and abused—punctuation mark. It works like a blinking yellow light, for which you slow down briefly without stopping. Never use a comma to signal a *break* between independent ideas, only a brief slow down. Use commas only in the following situations.

TO PAUSE BETWEEN COMPLETE IDEAS.   In a compound sentence in which a coordinating conjunction (*and, or, nor, for, but*) connects equal (independent) statements, a comma usually precedes the conjunction:

| This is an excellent course, but the work is difficult.

> Comma to pause between complete ideas

TO PAUSE BETWEEN AN INCOMPLETE AND A COMPLETE IDEA.   A comma is usually placed between a complete and an incomplete statement in a complex sentence when the incomplete statement comes first, as the following example demonstrates:

| Because he is a fat cat, Jack diets often.

When the order is reversed (complete statement followed by an incomplete one), the comma is usually omitted:

| Jack diets often because he is a fat cat.

> Comma between an incomplete and complete idea
> No comma needed

TO SEPARATE ITEMS (WORDS, PHRASES, OR CLAUSES) IN A SERIES.   Use commas after items in a series, including the next-to-last item, as the two sentences below demonstrate:

| Helen, Joe, Marsha, and John are joining us on the term project.
| The new employee complained that the hours were long, the pay was low, the work was boring, and the supervisor was paranoid.

> Comma separates items in a series

Use no commas if *or* or *and* appears between all items in a series:

| She is willing to study in San Francisco or Seattle or even in Anchorage.

> No commas needed if *or* or *and* appears between all items

TO SET OFF INTRODUCTORY PHRASES.   Introductory phrases include infinitive phrases (*to* plus a simple form of the verb), prepositional phrases (beginning with *at, of, in, on,* etc.), participial phrases (beginning with an *-ing* or *-ed* form of a verb), and interjections (emotional words that have no connection with the rest of the sentence). These phrases are set off from the remainder of the sentence by a comma. Following are examples of each:

| To be or not to be, that is the question. (infinitive phrase)
| In the event of an emergency, use the fire exit. (prepositional phrase)
| Being an old cat, Jack was slow at catching mice. (participial phrase)
| Oh, is that the verdict? (interjection)

> Commas to set off introductory phrases

TO SET OFF NONRESTRICTIVE PHRASES AND CLAUSES.   A *restrictive* phrase or clause modifies or defines the subject in such a way that deleting the modifier would change the meaning of the sentence. In the following sentence, "who have work experience" *restricts* the subject by limiting the category from all students to just those with work experience. Because this phrase is essential to the sentence's meaning it is *not* set off by commas:

Restrictive phrase (no comma)

| All students who have work experience will receive preference.

A *nonrestrictive* phrase or clause could be deleted without changing the sentence's meaning ("Our new manager is highly competent") and *is* therefore set off by commas:

Nonrestrictive phrase (use comma)

| Our new manager, who has only six weeks' experience, is highly competent.

TO SET OFF PARENTHETICAL ELEMENTS. Elements that interrupt the flow of a sentence (such as *of course*, *as a result*, *as I recall*, and *however*) are considered parenthetical and are enclosed by commas. These items may denote emphasis, afterthought, or clarification. Following are examples of each:

Commas to set off parenthetical elements

| This deluxe model, of course, is more expensive. (emphasis)
| Your essay, by the way, was excellent. (afterthought)
| The loss of my job was, in a way, a blessing. (clarification)

A direct address also interrupts a sentence and is set off by commas:

Commas to set off direct address

| Listen, my children, and you shall hear my story.

A parenthetical element at the beginning or the end of a sentence, as in the two examples below, is also set off by a comma:

Commas to set off parenthetical elements at beginning and end

| Naturally, we will expect a full guarantee.
| You've done a good job, I think.

TO SET OFF QUOTED MATERIAL. Quoted items within a sentence are set off by commas:

Comma to set off quoted material

| The customer said, "I'll take it," as soon as he laid eyes on our new model.

TO SET OFF APPOSITIVES. An *appositive*, a word or words explaining a noun and placed immediately after it, is set off by commas when the appositive is nonrestrictive, as in the two examples below:

Comma to set off appositives

| Martha Jones, our new president, is overhauling all personnel policies.
| Alpha waves, the most prominent of the brain waves, are typically recorded in a waking subject whose eyes are closed.

OTHER USES. Commas are used to set off the day of the month from the year in a date (May 10, 2014), to set off numbers in three-digit intervals (6,463,657), to separate city and state in an address (Albany, Iowa), to set off parts of an address in a sentence (J.B. Smith, 18 Sea Street, Albany, Iowa 51642), to set off day and year in a sentence (June 15, 2013, is my graduation date), and to set off degrees and titles from proper names (Roger P. Cayer, M.D. or Gordon Browne, Jr.).

COMMAS USED INCORRECTLY.   Avoid needless or inappropriate commas. Read a sentence aloud to identify inappropriate pauses. All of the sentences below are faulty for the reasons indicated in parentheses:

| The instructor told me, that I was late.   Faulty

(separates the indirect from the direct object)

| The most universal symptom of the suicide impulse, is depression.

(separates the subject from its verb)

| This has been a long, difficult, semester.

(second comma separates the final adjective from its noun)

| John, Bill, and Sally, are joining us on the trip home.

(third comma separates the final subject from its verb)

| An employee, who expects rapid promotion, must quickly prove his or her worth.

(separates a modifier that should be restrictive)

| I spoke by phone with John, and Marsha.

(separates two nouns, linked by a coordinating conjunction)

| The room was, 18 feet long.

(separates the linking verb from the subjective complement)

| We painted the room, red.

(separates the object from its complement)

Apostrophe

Apostrophes indicate the possessive, a contraction, and the plural of numbers, letters, and figures.

TO INDICATE THE POSSESSIVE.   At the end of a singular word, or of a plural word that does not end in *-s*, add an apostrophe plus *-s* to indicate the possessive. Single-syllable nouns that end in *-s* take the apostrophe before an added *-s*. Following are examples of each:

| The people's candidate won.   Apostrophe
| I borrowed Chris's book.   to indicate

For words that already end in *-s* and have more than one syllable, add an *-s* after the   possessives
apostrophe:

| Aristophanes's death

Do not use an apostrophe to indicate the possessive form of either singular or plural pronouns, as the following two examples demonstrate:

| The book was hers.
| Ours is the best school in the county.

At the end of a plural word that ends in -*s*, add an apostrophe only, as shown in the two examples below:

| the cows' water supply

| the Jacksons' wine cellar

At the end of a compound noun, add an apostrophe plus -*s*:

| my father-in-law's false teeth

At the end of the last word in nouns of joint possession, add an apostrophe plus -*s* if both own one item:

| Joe and Sam's lakefront cottage

Add an apostrophe plus -*s* to both nouns if each owns specific items:

| Joe's and Sam's passports

TO INDICATE A CONTRACTION.   An apostrophe shows that you have omitted one or more letters in a phrase that is usually a combination of a pronoun and a verb, as in these three contractions:

Apostrophes to show contractions

| I'm
| they're
| you'd

Faulty contractions

Avoid faulty contractions: For example, *they're* (short for "they are" and often confused with the possessive *their* and the adverb *there*); *it's* (short for "it is" and often confused with the possessive *its*); *who's* (short for "who is" and often confused with the possessive *whose*); and *you're* (short for "you are" and often confused with the possessive *your*). See Table B.1 in the "Usage" section of this appendix for more examples.

TO INDICATE THE PLURALS OF NUMBERS, LETTERS, AND FIGURES.   For example:

Apostrophes to pluralize numbers, letters, and figures

| The 6's on this new printer look like smudged *G*'s, 9's are illegible, and the %'s are unclear.

Quotation Marks

Quotation marks have a variety of uses:

TO SET OFF THE EXACT WORDS BORROWED FROM ANOTHER SPEAKER OR WRITER.   The period or comma at the end is placed within the quotation marks, as both of the following sentences demonstrate:

Quotation marks to set off a speaker's exact words

| "Hurry up," Jack whispered.
| Jack told Felicia, "I'm depressed."

A colon or semicolon is always placed outside quotation marks:

| Our student handbook clearly defines "core requirements"; however, it does not list all the courses that fulfill the requirements.

When a question mark or exclamation point is part of a quotation, it belongs within the quotation marks, replacing the comma or period, as in the following two examples:

| "Help!" he screamed.

| Marsha asked John, "Can't we agree about anything?"

But if the question mark or exclamation point pertains to the attitude of the person quoting instead of the one being quoted, it belongs outside the quotation mark:

| Why did Boris wink and whisper, "It's a big secret"?

TO INDICATE TITLES.   Use quotation marks for titles of articles, book chapters, and poems (but italicize titles of books, journals, paintings, or newspapers instead):

| The enclosed article, "The Job Market for College Graduates," should provide some helpful insights.

Quotation marks to indicate titles

TO INDICATE IRONY.   Finally, use quotation marks (with restraint) to indicate irony:

| She is some "friend"!

Quotation marks to indicate irony

Ellipses

Three dots . . . indicate that you have omitted material from a quotation. If the omitted words include the end of a sentence, a fourth dot indicates the period. (Also Appendix A, "Guidelines for Quoting the Work of Others.")

| "Three dots . . . indicate . . . omitted . . . material . . . . A fourth dot indicates the period."

Ellipses to indicate omitted material from a quotation

Brackets

Brackets are used within quotations to set off material that was not in the original quotation but is needed for clarification, as in the following two examples:

| "She [Amy] was the outstanding candidate for the scholarship."

| "It was in early spring [April 2, to be exact] that the tornado hit."

Use *sic* (Latin for "thus," or "so") in brackets when quoting an error from the original source:

| The assistant's comment was clear: "He don't [sic] want any."

Brackets to indicate material added to a quotation

Brackets to indicate an error in a quotation

Italics

Use italics or underlining for titles of books, periodicals, films, newspapers, and plays; for the names of ships; and for foreign words or technical terms. Also, you may use italics *sparingly* for special emphasis. Following are four uses of italics:

| The *Oxford English Dictionary* is a handy reference tool.

| My only advice is *caveat emptor.*

Various uses of italics

> *Bacillus anthracis* is a highly virulent organism.

> *Do not* inhale these fumes under any circumstances!

Parentheses

Material between parentheses, like all other parenthetical material discussed earlier, can be deleted without harming the logical and grammatical structure of the sentence. Use parentheses to enclose material that defines or explains the statement that precedes it, as in both of the following sentences:

Parentheses to define or explain preceding material

> An anaerobic (airless) environment must be maintained for the cultivation of this organism.

> The cost of running our college has increased by 15 percent in one year (see Appendix for full cost breakdown).

Dashes

Dashes can be effective to set off parenthetical material if, like parentheses, they are not overused. Parentheses deemphasize the enclosed material, while dashes emphasize it, as shown in the two sentences that follow:

Dashes to set off and emphasize material

> Have a good vacation—but watch out for sandfleas.

> Mary—a true friend—spent hours helping me rehearse.

On most word-processing programs, a dash is created by two hyphens, which the software will typically convert to a dash.

Mechanics

The mechanical aspects of writing a document include abbreviation, hyphenation, capitalization, use of numbers, and spelling. (Keep in mind that not all of these rules are hard and fast; some may depend on style guides used in your field.)

Abbreviation

The following should *always* be abbreviated:

Always abbreviate

- Titles such as *Ms., Mr., Dr.,* and *Jr.,* when they are used before or after a proper name.
- Specific time designations (*400* B.C.E., *5:15* A.M.).

The following should *never* be abbreviated:

Never abbreviate

- Military, religious, academic, or political titles (*Reverend, President*).
- Nonspecific time designations (*Sarah arrived early in the morning*—not *early in the* A.M.).

Avoid abbreviations whose meanings might not be clear to all readers. Units of measurement (for example, *mm* for *millimeter*) can be abbreviated if they appear often in the document. However, spell out a unit of measurement the first time it is used. Avoid abbreviations in visual aids unless saving space is essential.

Hyphenation

Hyphens divide words at line breaks, and join two or more words used as a single adjective if they precede the noun (but not if they follow it). Following are examples of each use:

| Com-puter (*at a line break*)

| An all-too-human error (*but* "The error was all too human")

Other commonly hyphenated words include the following:

Correct use of hyphens

- Most words that begin with the prefix *self-* (*self-reliance, self-discipline*—see your dictionary for exceptions).

- Combinations that might be ambiguous (*re-creation* versus *recreation*).

- Words that begin with *ex* when *ex* means "past" (*ex-faculty member* but *excommunicate*).

- All fractions, along with ratios that are used as adjectives and that precede the noun, and compound numbers from twenty-one through ninety-nine (*a two-thirds majority, thirty-eight windows*).

Other uses of hyphens

Capitalization

Use capitalization in the following situations:

- The first words of all sentences (*This is a good idea.*)
- Titles of people if the title precedes the person's name, but not after (*Senator Barbara Boxer* but *Barbara Boxer, U.S. senator*)
- Titles of books, films, magazines, newspapers, operas, and other longer works. In addition to capitalizing the first word, also capitalize all other words within the title (*A Long Day's Journey into Night*) except articles, short prepositions, and coordinating conjunctions (*and, but, for, or, nor, yet*).
- Parts of a longer work (*Chapter 25, Opus 23*)
- Languages (*French, Urdu*)
- Days of the week (*Saturday*)
- Months (*November*)
- Holidays (*Thanksgiving*)
- Names of organizations or groups (*World Health Organization*)
- Races and nationalities (*Asian American, Australian*)
- Historical events (*War of 1812*)
- Important documents (*Declaration of Independence*)

Uses of capitalization

- Names of structures or vehicles (*Empire State Building*, the *Queen Mary*)
- Adjectives derived from proper nouns (*Chaucerian English*)
- Words such as *street, road, corporation, university,* and *college* only when they accompany a proper noun (*High Street, Rand Corporation, Stanford University*).
- The words *north, south, east,* and *west* when they denote specific regions (*the South, the Northwest*) but not when they are simply directions (*turn east at the light*)

Do not capitalize the seasons (*spring, winter*) or general groups (*the younger generation, the leisure class*).

Numbers and Numerals

Numbers expressed in one or two words can be written out or written as numerals. Use numerals to express larger numbers, decimals, fractions, precise technical figures, or any other exact measurements, as in the following examples:

Uses of numerals

| 543
| 2,800,357
| 3.25
| 15 pounds of pressure
| 50 kilowatts
| 4,000 rpm

Use numerals for dates, census figures, addresses, page numbers, exact units of measurement, percentages, times with A.M. or P.M. designations, and monetary and mileage figures, as shown below.

Additional uses of numerals

| page 14
| 1:15 P.M.
| 18.4 pounds
| 9 feet
| 12 gallons
| $15

Do not begin a sentence with a numeral. If the figure needs more than two words, revise your word order. The following two examples show the uses of numbers in sentences:

Uses of numbers in sentences

| Six hundred students applied for the 102 available jobs. The 102 available jobs attracted 600 applicants.

Do not use numerals to express approximate figures, time not designated as A.M. or P.M., or streets named by numbers less than 100, as shown below:

When not to use numerals

| About seven hundred fifty
| Four fifteen
| 108 East Forty-Second Street

In contracts and other documents in which precision is vital, a number can be stated both in numerals and in words, as in the following sentence:

| The tenant agrees to pay a rental fee of eight hundred and seventy-five dollars ($875.00) monthly.

Spelling

Always use the spell-check function in your word-processing software. However, don't rely on it exclusively. Take the time to use a dictionary for all writing assignments. If you are a poor speller, ask someone else to proofread every document before you present the final version.

Usage

Be aware of the pairs of words (and sometimes groups of three words) that are often confused. Refer to Table B.1 for a list of the most commonly confused words.

Table B.1 Commonly Confused Words

SIMILAR WORDS	USED CORRECTLY IN A SENTENCE
Accept means "to receive willingly."	She *accepted* his business proposal.
Except means "otherwise than."	They all agreed, *except* Bob.
Affect means "to have an influence on."	Meditation *affects* concentration in a positive way.
Affect can also mean "to pretend."	Boris likes to *affect* a French accent.
Effect used as a noun means "a result."	Meditation has a positive *effect* on concentration.
Effect used as a verb means "to make happen" or "to bring about."	Meditation can *effect* an improvement in concentration.
Already means "before this time."	Our new laptops are *already* sold out.
All ready means "prepared."	We are *all ready* for the summer tourist season.
Among refers to three or more.	The prize was divided *among* the four winners.
Between refers to two.	The prize was divided *between* the two winners.
Cite means "to document."	You must always *cite* your sources in research.
Sight means "vision."	Margarita seems to have the gift of second *sight*.
Site means "a location."	Have the surveyors inspected the *site* yet?
Continual means "repeated at intervals."	Our lower field floods *continually* during the rainy season.
Continuous means "without interruption."	His headache has been *continuous* for three days.
Council means "a body of elected people."	I plan to run for student *council*.
Counsel means "to offer advice."	Since you have experience, I suggest you *counsel* Jim on the project as it moves along.

Table B.1 Commonly Confused Words (*Continued*)

SIMILAR WORDS	USED CORRECTLY IN A SENTENCE
Differ from refers to unlike things.	This plan *differs* greatly from our earlier one.
Differ with means "to disagree."	Mary *differs with* John about the plan.
Disinterested means "unbiased" or "impartial."	Good science calls for *disinterested* analysis of research findings.
Uninterested means "not caring."	Boris is *uninterested* in science.
Eminent means "famous" or "distinguished."	Dr. Ostroff, the *eminent* physicist, is lecturing today.
Imminent means "about to happen."	A nuclear meltdown seemed *imminent*.
Farther refers to physical distance (a measurable quantity).	The station is 20 miles *farther*.
Further refers to extent (not measurable).	*Further* discussion of this issue is vital.
Fewer refers to things that can be counted.	*Fewer* than fifty students responded to our survey.
Less refers to things that can't be counted.	This survey had *less* of a response than our earlier one.
Imply means "to insinuate."	This report *implies* that a crime occurred.
Infer means "to reason from evidence."	From this report, we can *infer* that a crime occurred.
It's stands for "it is."	*It's* a good time for a department meeting.
Its stands for "belonging to it."	The cost of the project has exceeded *its* budget.
Lay means "to set something down."	Please *lay* the blueprints on the desk.
Lie means "to recline." It takes no direct object.	This patient needs to *lie* on his right side all night.
(Note that the past tense of *lie* is *lay*.)	The patient *lay* on his right side all night.
Precede means "to come before."	Audience analysis should *precede* a written report.
Proceed means "to go forward."	If you must wake the cobra, *proceed* carefully.
Principle is always a noun that means "basic rule or standard."	Ethical *principles* should govern all our communications.
Principal, used as a noun, means "the major person(s)."	All *principals* in this purchase must sign the contract.
Principal, used as an adjective, means "leading."	Martha was the *principal* negotiator for this contract.
Stationary means "not moving."	The desk is *stationary*.
Stationery means "writing supplies."	The supply cabinet needs *stationery*.
Their means "belonging to them."	They all want to have *their* cake and eat it too.
There means "at that location."	The new copy machine is over *there*.
They're means "they are."	*They're* not the only ones who disagree.

Transitions

You can choose from three techniques to achieve smooth transitions within and between paragraphs.

Use Transitional Expressions

Use words such as *again, furthermore, in addition, meanwhile, however, also, although, for example, specifically, in particular, as a result, in other words, certainly, accordingly, because,* and *therefore.* Such words serve as bridges between ideas.

Transitional expressions

Repeat Key Words and Phrases

To help link ideas, repeat key words or phrases or rephrase them in different ways, as in this next paragraph (emphasis added):

> Whales are among the most *intelligent* of all mammals. Scientists rank whale *intelligence* with that of higher primates because of *whales' sophisticated* group behavior. These *bright creatures* have been seen teaching and disciplining *their* young, helping *their* wounded comrades, engaging in elaborate courtship rituals, and playing in definite *gamelike patterns*. *They* are able to coordinate such *complex cognitive activities* through *their* highly effective communication system of sonar clicks and pings. Such remarkable social organization apparently stems from the *humanlike* devotion that whales seem to display toward one another.

Repeated key words and phrases

The key word *intelligent* in the above topic statement reappears as *intelligence* in the second sentence. Synonyms describing intelligent behavior (*sophisticated, bright, humanlike*) reinforce and advance the main idea throughout.

Use Forecasting Statements

Forecasting statements tell your readers where you are going next. Following are three examples of forecasting statements:

| The next step is to further examine the costs of this plan.

| Of course, we can also consider other options.

| This plan should be reconsidered for several reasons.

Forecasting statements

Lists

Listed items can be presented in one of two ways: running in as part of the sentence (embedded lists) or displayed with each item on a new line (vertical lists).

Embedded Lists

An embedded list integrates a series of items into a sentence. To number an embedded list, use parentheses around the numerals and either commas or semicolons between the items:

Embedded list

> In order to complete express check-in for your outpatient surgery, you must (1) go to the registration office, (2) sign in and obtain your registration number, (3) receive and wear your red armband, and (4) give your check-in slip to the volunteer, who will escort you to your room.

Vertical Lists

Embedded lists are appropriate for listing only a few short items. Vertical lists are preferable for multiple items. If the items belong in a particular sequence, use numerals or letters; if the sequence of items is unimportant, use bullets.

There are a number of ways to introduce vertical lists. You can use a sentence that closes with "the following" or "as follows" and ends with a colon:

> All applicants for the design internship must submit the following:

Vertical list using *following* and a colon

> - Personal statement
> - Résumé
> - Three letters of reference
> - Portfolio

You can also use a sentence that closes with a noun and ends with a colon:

> All applicants for the design internship must submit four items:

Vertical list using a noun and colon

> 1. Personal statement
> 2. Résumé
> 3. Three letters of reference
> 4. Portfolio

Finally, you can introduce a vertical list with a sentence that is grammatically incomplete without the list items:

> To register as a new student:

Vertical list using grammatical incompleteness

> 1. Take the placement test at the Campus Test Center.
> 2. Attend a new student orientation.
> 3. Register for classes online or by telephone.
> 4. Pay tuition and fees by the due date.

Do not use a colon with an introductory sentence that ends with a verb (as in the first incorrect sentence below), a preposition (as in the second incorrect sentence below), or an infinitive (as in the third incorrect sentence below):

> All applicants for the design internship must submit:

Incorrect because the introductory sentence ends with a verb

> - A personal statement
> - A résumé
> - Three letters of reference
> - A portfolio

| All applicants for the design internship need to:

| Submit a personal statement and résumé.

| Forward three letters of reference.

| Provide a portfolio.

Incorrect because the introductory sentence ends with a preposition

| All applicants for the design internship need to submit:

| A personal statement

| A résumé

| Three letters of reference

| A portfolio

Incorrect because the introductory sentence ends with an infinitive

If the sentence that introduces the list is followed by another sentence, use periods after both sentences. Do not use a colon to introduce the list:

| The next step is to configure the following fields. Consult Chapter 3 for more information on each field.

| IP address

| Network location

| User ID

| Encryption status

Vertical list without a colon

Note that some of the preceding examples use a period after each list item and some do not. Use a period after each list item if any of the items contains a complete sentence. Otherwise do not use a period. Also note that items included in a list should be grammatically parallel. For more on parallelism, see "Faulty Parallelism" earlier in this appendix.

Works Cited

Adams, Gerald R., and Jay D. Schvaneveldt. *Understanding Research Methods*. Longman, 1985.

"Add Useful Headings." Plain Language Action and Information Network, plainlanguage.gov, 2018.

Aldus Guide to Basic Design, The Aldus Corporation, 1988.

American Psychological Association *(APA). Publication Manual of the American Psychological Association*. 6th ed. Washington, DC: APA, 2009.

Anson, Chris M., and Robert A. Schwegler. *The Longman Handbook for Writers and Readers*. 2nd ed., Longman, 2000.

Archee, Raymond K. "Online Intercultural Communication." Intercom Sept./Oct. 2003: 40–41.

"Are We in the Middle of a Cancer Epidemic?" *University of California at Berkeley Wellness Letter,* vol. 10, no. 9, 1994, pp. 4–5.

Armstrong, William H. "Learning to Listen." *American Educator,* Winter 1997–98, pp. 24+.

Baker, Stephen. "Beware Social Media Snake Oil." *Bloomberg Businessweek*, 14 Dec. 2009, pp. 48–50.

Baldelomar, Raquel. "Where is the Line between Ethical and Legal?" *Forbes*, 21 July 2016.

Ball, Charles. "Figuring the Risks of Closer Runways." *Technology Review*, Aug.-Sept. 1996, pp. 12–13.

Barbour, Ian. *Ethics in an Age of Technology*. HarperCollins Publishers, 1993.

Barck, Jonas. "View of the 2020 Talent Market." *Universum*, 2 Mar. 2015, universumglobal.com/articles/2015/03/view-2020-talent-market/.

Barnett, Arnold. "How Numbers Can Trick You." *Technology Review*, Oct. 1994, pp. 38–45.

Baumann, K. E., et al. "Three Mass Media Campaigns to Prevent Adolescent Cigarette Smoking." *Preventive Medicine*, vol. 17, no. 5, Sept. 1988, pp. 510–30.

Bedford, Marilyn S., and F. Cole Stearns. "The Technical Writer's Responsibility for Safety." *IEEE Transactions on Professional Communication*, vol. 30, no. 3, 1987, pp. 127–32.

Bernstein, Peter L. *Against the Gods: The Remarkable Story of Risk*. John Wiley & Sons, 1998.

Bogert, Judith, and David Butt. "Opportunities Lost, Challenges Met: Understanding and Applying Group Dynamics in Writing Projects." *Bulletin of the Association for Business Communication*, vol. 53, no. 2, 1990, pp. 51–53.

Bosker, Bianca. "The Binge Breaker." *The Atlantic*, Nov. 2016.

Bosley, Deborah. "International Graphics: A Search for Neutral Territory." *INTERCOM*, Aug.–Sept. 1996, pp. 4–7.

Brill, Alida. "Saving Face. Moving forward boldly despite a drug's betrayal." *Psychology Today*, 29 Nov. 2010.

Brownell, Judi, and Michael Fitzgerald. "Teaching Ethics in Business Communication: The Effective/Ethical Balancing Scale." *Bulletin of the Association for Business Communication*, vol. 55, no. 3, 1992, pp. 15–18.

Bryan, John. "Down the Slippery Slope: Ethics and the Technical Writer as Marketer." *Technical Communication Quarterly*, vol. 1, no. 1, 1992, pp. 73–88.

Burghardt, M. David. *Introduction to the Engineering Profession*. HarperCollins Publishers, 1991.

Caher, John M. "Technical Documentation and Legal Liability." *Journal of Technical Writing and Communication*, vol. 25, no. 1, 1995, pp. 5–10.

CareerBuilder. "Number of Employees Using Social Media to Screen Candidates at All-Time High, Finds Latest CareerBuilder Study." CareerBuilder/Cision PR Newswire, 15 June 2017.

Carliner, Saul. "Demonstrating Effectiveness and Value: A Process for Evaluating Technical Communication Products and Services." *Technical Communication*, vol. 44, no. 3, 1997, pp. 252–65.

---. "Physical, cognitive, and affective: a three-part framework for information design," *Technical Communication*, vol. 47, no. 4, 2000, pp. 561–576.

Carr, David F. "How Social Media Changes Technical Communication." *InformationWeek*, 4 Jan. 2012.

Caswell-Coward, Nancy. "Cross-Cultural Communication: Is It Greek to You?" *Technical Communication*, vol. 39, no. 2, 1992, pp. 264–66.

Centers for Disease Control and Prevention (CDC). *Social Media Guidelines and Best Practices*. United States Department of Health and Human Services, 16 May 2012.

Chauncey, C. "The Art of Typography in the Information Age." *Technology Review*, Feb.–Mar. 1986, pp. 26+.

Christians, C. G., et al. *Media Ethics: Cases and Moral Reasoning*. 2nd ed., Longman, 1978.

Cialdini, Robert B. "The Science of Persuasion." *Scientific American*, Feb. 2001, pp. 76–81.

Claiborne, Robert. *Our Marvelous Native Tongue: The Life and Times of the English Language*. Times Books, 1983.

Clark, Gregory. "Ethics in Technical Communication: A Rhetorical Perspective." *IEEE Transactions on Professional Communication*, vol. 30, no. 3, 1987, pp. 190–95.

Clark, Thomas. "Teaching Students How to Write to Avoid Legal Liability." *Business Communication Quarterly*, vol. 60, no. 3, 1997, pp. 71–77.

Cochran, Jeffrey K., et al. "Guidelines for Evaluating Graphical Designs." *Technical Communication*, vol. 36, no. 1, 1989, pp. 25–32.

Coe, Marlana. *Human Factors for Technical Communicators.* John Wiley & Sons, 1996.

---. "Writing for Other Cultures: Ten Problem Areas." *INTERCOM*, Jan. 1997, pp. 17–19.

Cohn, Victor. "Coping with Statistics." *A Field Guide for Science Writers: The Official Guide of the National Association of Science Writers,* edited by Deborah Blum and Mary Knudson, Oxford UP, 1997, pp. 102–09.

Cole-Gomolski, B. "Users Loathe to Share Their Know-How." *Computer world,* 17 Nov. 1997, p. 6.

Collier, Mary Jane. "Intercultural Communication Competence: Continuing Challenges and Critical Directions." *International Journal of Intercultural Relations,* vol. 48, 2015, pp. 9–11.

Congressional Research Report. "Dioxins." Washington, DC: GPO, 1990.

"Consequences of Whistle Blowing in Scientific Misconduct Reported." *Professional Ethics Report,* vol. IX, no. 1, Winter 1996, American Association for the Advancement of Science, p. 5.

Cooper, Lyn O. "Listening Competency in the Workplace: A Model for Training." *Business Communication Quarterly,* vol. 60, no. 4, Dec. 1997, pp. 75–84.

Corbett, Edward P. J. *Classical Rhetoric for the Modern Student.* 3rd ed., Oxford UP, 1990.

Cotton, Robert, editor. *The New Guide to Graphic Design.* Chartwell Books, 1990.

Crosby, Olivia. *Résumés, Applications, and Cover Letters.* United States Department of Labor, 1999.

Cross, Mary. "Aristotle and Business Writing: Why We Need to Teach Persuasion." *Bulletin of the Association for Business Communication,* vol. 54, no. 1, 1991, pp. 3–6.

Crossen, Cynthia. *Tainted Truth: The Manipulation of Fact in America.* Simon and Schuster, 1994.

Davenport, Thomas H. *Information Ecology.* Oxford UP, 1997.

Debs, Mary Beth, "Collaborative Writing in Industry." *Technical Writing: Theory and Practice,* edited by Bertie E. Fearing and W. Keats Sparrow, Modern Language Association, 1989, pp. 33–42.

Devlin, Keith. *Infosense: Turning Information into Knowledge.* W. H. Freeman, 1999.

Dombrowski, Paul M. "*Challenger* and the Social Contingency of Meaning: Two Lessons for the Technical Communication Classroom." *Technical Communication Quarterly,* vol. 1, no. 3, 1992, pp. 73–86.

Dörner, Dietrich. *The Logic of Failure: Recognizing and Avoiding Error in Complex Situations.* 1989. Translated by Rita Kimber and Robert Kimber, Metropolitan Books, 1996.

Duin, Ann Hill. "Terms and Tools: A Theory and Research-Based Approach to Collaborative Writing." *Bulletin of the Association for Business Communication,* vol. 53, no. 2, 1990, pp. 45–50.

Dumont, Raymond A., and John M. Lannon. *Business Communications.* 3rd ed., Scott Foresman & Co., 1990.

Easton, Thomas, and Stephan Herrara. "J&J's Dirty Little Secret." *Forbes,* 12 Jan. 1998, pp. 42–44.

Elbow, Peter. *Writing without Teachers.* Oxford UP, 1973.

Evans, James. "Legal Briefs." *Internet World,* Feb. 1998, p. 22.

"Fact Sheet 7: Workplace Privacy and Employee Monitoring." *Privacy Rights Clearinghouse,* Feb. 2013, www.privacyrights.org/workplace-privacy-and-employee-monitoring. Accessed 30 Mar. 2013.

Farnham, Alan. "How Safe Are Your Secrets?" *Fortune,* 8 Sept. 1997, pp. 114–20.

Felker, Daniel B., et al. *Guidelines for Document Designers.* American Institutes for Research, 1981.

Fineman, Howard. "The Power of Talk." *Newsweek,* 8 Feb. 1993, pp. 24–28.

Fisher, Anne. "Is My Team Leader a Plagiarist? . . . How Do I Deal With an Office Lothario?" *Fortune,* 27 Oct. 1997, pp. 291–92.

---. "My Company Just Announced I May Be Laid Off. Now What?" *Fortune,* 3 Mar. 2003, p. 184.

---. "Truth and Consequences." *Fortune,* 29 May 2000, p. 292.

Fogg, B. J. "Stanford Guidelines for Web Credibility." Stanford Persuasive Technology Lab, 2002.

Friedland, Andrew J., and Carol L. Folt. *Writing Successful Science Proposals.* Yale UP, 2000.

Gallo, Amy. "How to Deliver Bad News to Your Employees." *Harvard Business Review,* 30 Mar. 2015.

Gallo, Carmine. "Jeff Bezos Banned PowerPoint in Meetings. His Replacement is Brilliant." *Inc.,* 25 April 2018.

Gartaganis, Arthur. "Lasers." *Occupational Outlook Quarterly,* Winter 1984, pp. 22–26.

Gensler, Arthur. "Trust is the Most Powerful Currency in Business." *Fortune,* 28 July 2015.

Gesteland, Richard R. "Cross-Cultural Compromises." *Sky,* May 1993, pp. 20+.

Gibaldi, Joseph, and Walter S. Achtert. *MLA Handbook for Writers of Research Papers.* 3rd ed., Modern Language Association, 1988.

Gilbert, Nick. "1-800-ETHIC." *Financial World,* 16 Aug. 1994, pp. 20+.

Gilsdorf, Jeanette W. "Executives' and Academics' Perception of the Need for Instruction in Written Persuasion." *Journal of Business Communication,* vol. 23, no. 4, 1986, pp. 55–68.

---. "Write Me Your Best Case for . . . ." *Bulletin of the Association for Business Communication,* vol. 54, no. 1, 1991, pp. 7–12.

Girill, T. R. "Technical Communication and Art." *Technical Communication,* vol. 31, no. 2, 1984, p. 35.

---. "Technical Communication and Law." *Technical Communication,* vol. 32, no. 3, 1985, p. 37.

Glidden, H. K. *Reports, Technical Writing and Specifications.* McGraw-Hill, 1964.

Goby, Valerie P., and Justus Helen Lewis. "The Key Role of Listening in Business: A Study of the Singapore Insurance Industry." *Business Communication Quarterly,* vol. 63, no. 2, June 2000, pp. 41–51.

Goman, Carol Kinsey. "How Culture Controls Communication." *Forbes*, 28 Nov. 2011.

Gribbons, William M. "Organization by Design: Some Implications for Structuring Information." *Journal of Technical Writing and Communication*, vol. 22, no. 1, 1992, pp. 57–74.

Grice, Roger A. "Focus on Usability: Shazam!" *Technical Communication*, vol. 42, no. 1, 1995, pp. 131–33.

Griffin, Robert J. "Using Systematic Thinking to Choose and Evaluate Evidence." *Communicating Uncertainty: Media Coverage of New and Controversial Science*, edited by Sharon Friedman et al., Lawrence Erlbaum Associates, 1999, pp. 225–48.

Guidelines: How to Write and Report about People with Disabilities. 8th ed., U of Kansas Research and Training Center on Independent Living, 2015.

Gurak, Laura J. Cyberliteracy: *Navigating the Internet with Awareness*. Yale UP, 2001.

Harcourt, Jules. "Teaching the Legal Aspects of Business Communication." *Bulletin of the Association for Business Communication*, vol. 53, no. 3, 1990, pp. 63–64.

Harris, Richard F. "Toxics and Risk Reporting." *A Field Guide for Science Writers: The Official Guide of the National Association of Science Writers*, edited by Deborah Blum and Mary Knudson, Oxford UP, 1997, pp. 166–72.

Haskin, David. "Meetings without Walls." *Internet World*, Oct. 1997, pp. 53–60.

Hauser, Gerald. *Introduction to Rhetorical Theory*. HarperCollins Publishers, 1986.

Hayakawa, S. I. *Language in Thought and Action*. 3rd ed., Harcourt Brace Jovanovich, 1972.

Hein, Robert G. "Culture and Communication." *Technical Communication*, vol. 38, no. 1, 1991, pp. 125–26.

Hernandez, Ivan, and Jesse Lee Preston. "Disfluency disrupts the confirmation bias." *Journal of Experimental Social Psychology*, vol. 49, no. 1, 2013, pp. 178-182.

Huff, Darrell. *How to Lie with Statistics*. W. W. Norton, 1954.

Hulbert, Jack E. "Developing Collaborative Insights and Skills." *Bulletin of the Association for Business Communication*, vol. 57, no. 2, 1994, pp. 53–56.

---. "Overcoming Intercultural Communication Barriers." *Bulletin of the Association for Business Communication*, vol. 57, no. 2, 1994, pp. 41–44.

Janis, Irving L. *Victims of Groupthink: A Psychological Study of Foreign Policy Decisions and Fiascos*. Houghton Mifflin, 1972.

Johannesen, Richard L. *Ethics in Human Communication*. 2nd ed., Waveland Press, 1983.

Kane, Kate. "Can You Perform under Pressure?" *Fast Company*, Oct.–Nov. 1997, pp. 54+.

Kelley-Reardon, Kathleen. *They Don't Get It Do They? Communication in the Workplace—Closing the Gap between Women and Men*. Little, Brown, 1995.

Kelman, Herbert C. "Compliance, Identification, and Internalization: Three Processes of Attitude Change." *Journal of Conflict Resolution*, vol. 2, no. 1, Mar. 1958, pp. 51–60.

Kerr, Dara. "Six States Outlaw Employer Snooping on Facebook." *CNET*, 2 Jan. 2013, www.cnet.com/news/six-states-outlaw-employer-snooping-on-facebook/.

Kiely, Thomas. "The Idea Makers." *Technology Review*, Jan. 1993, pp. 33–40.

King, Ralph T. "Medical Journals Rarely Disclose Researchers' Ties." *The Wall Street Journal*, 2 Feb. 1999, pp. B1+.

Kipnis, David, and Stuart Schmidt. "The Language of Persuasion." *Understanding Persuasion*, edited by Raymond S. Ross, 3rd ed., Prentice Hall, 1990.

Kohl, John R. "Improving Translatability and Readability with Syntactic Cues." *Technical Communication*, May 1999, pp. 149–166.

Kremers, Marshall. "Teaching Ethical Thinking in a Technical Writing Course." *IEEE Transactions on Professional Communication*, vol. 32, no. 2, 1989, pp. 58–61.

Ladegaard, Hans. J. "'Doing Power' at Work: Responding to Male and Female Management Styles in a Global Business Corporation." *Journal of Pragmatics*, vol. 43, no. 1, 2011, pp. 4–19.

Lang, Thomas A., and Michelle Secic. *How to Report Statistics in Medicine*. American College of Physicians, 1997.

Larson, Charles U. *Persuasion: Perception and Responsibility*. 7th ed., Wadsworth Publishing Company, 1995.

Lavin, Michael R. *Business Information: How to Find It, How to Use It*. 2nd ed., Oryx Press, 1992.

Leber, Jessica. "The Immortal Life of the Enron E-mails." *MIT Technology Review*, 2 July 2013, www.technologyreview.com/s/515801/the-immortal-life-of-the-enron-e-mails/.

Leki, Ilona. "The Technical Editor and the Non-native Speaker of English." *Technical Communication*, vol. 37, no. 2, 1990, pp. 148–52.

Lemonick, Michael. "The Evils of Milk?" *Time*, 15 June 1998, p. 85.

Lenzer, Robert, and Carrie Shook. "Whose Rolodex Is It Anyway?" *Forbes*, 23 Feb. 1998, pp. 100–04.

Lewis, Howard L. "Penetrating the Riddle of Heart Attack." *Technology Review*, Aug.–Sept. 1997, pp. 39–44.

Liu, Ziming. "Reading Behavior in the Digital Environment: Changes in Reading Behavior over the Past Ten Years." *Journal of Documentation*, vol. 61, no. 6, 2005, pp. 700–12.

MacKenzie, Nancy. Unpublished review of *Technical Writing*, by John M. Lannon. May 1991.

Mackin, John. "Surmounting the Barrier between Japanese and English Technical Documents." *Technical Communication*, vol. 36, no. 4, 1989, pp. 346–51.

Manning, Michael. "Hazard Communication 101." *INTERCOM*, June 1998, pp. 12–15.

Martin, Judith N. "Revisiting intercultural communication competence: Where to go from here." *International Journal of Intercultural Relations*, vol. 48, 2015, pp. 6–8.

Matson, Eric. "The Seven Sins of Deadly Meetings." *Fast Company*, Oct.–Nov. 1997, pp. 27–31.

McGuire, Gene. "Shared Minds: A Model of Collaboration." *Technical Communication*, vol. 39, no. 3, 1992, pp. 467–68.

Menczer, Filippo. "Misinformation on social media: Can technology save us?" *The Conversation*, 27 Nov. 2016.

Meyer, Benjamin D. "The ABCs of New-Look Publications." *Technical Communication*, vol. 33, no. 1, 1986, pp. 13–20.

Meyer, Erin. "Getting to Si, Ja, Oui, Hai, and Da." *Harvard Business Review*, December 2015.

Monastersky, Richard. "Courting Reliable Science." *Science News*, vol. 153, no. 16, 1998, pp. 249–51.

Monmonier, Mark. *Cartographies of Danger: Mapping Hazards in America*. U of Chicago P, 1997.

Morgan, Meg. "Patterns of Composing: Connections between Classroom and Workplace Collaborations." *Technical Communication*, vol. 38, no. 4, 1991, pp. 540–42.

Murphy, Kate. "Separating Ballyhoo from Breakthrough." *Business Week*, 13 July 1998, p. 143.

Nakache, Patricia. "Is It Time to Start Bragging about Yourself?" *Fortune*, 27 Oct. 1997, pp. 287–88.

Naragon, Kristin. 29. "Consumers are Still Email Obsessed, but They're Finding More Balance." *Adobe Blog*, 29 Aug. 2017.

National Institutes of Health (NIH). "NIH-led Study Finds Genetic Test Results Do Not Trigger Increased Use of Health Services." 17 May 2012.

Nelson, Sandra J., and Douglas C. Smith. "Maximizing Cohesion and Minimizing Conflict in Collaborative Writing Groups." *Bulletin of the Association for Business Communication*, vol. 53, no. 2, 1990, pp. 59–62.

Nielsen, Jakob. "F-Shaped Pattern for Reading Web Content." *Nielsen Norman Group*, 17 Apr. 2006, www.nngroup.com/articles/f-shaped-pattern-reading-web-content/.

Nordenberg, Tamar. "Direct to You: TV Drug Ads That Make Sense." *FDA Consumer*, Jan.-Feb. 1998, pp. 7–10.

Notkins, Abner L. "New Predictors of Disease." *Scientific American*, March 2007, pp. 72–79.

Oetzel, John, et al. "Face and Facework in Conflict: A Cross-Cultural Comparison of China, Germany, Japan, and the United States." *Communication Monographs*, vol. 68, no. 3, Sept. 2001, pp. 235–258.

Ornatowski, Cezar M. "Between Efficiency and Politics: Rhetoric and Ethics in Technical Writing." *Technical Communication Quarterly*, vol. 1, no. 1, 1992, pp. 91–103.

Oxfeld, Jesse. "Analyze This." *Brill's Content*, Mar. 2000, pp. 105–06.

Parish, Steve. "The Profit Potential in Running an Ethical Business." *Forbes* 4 Feb. 2016.

Pearce, C. Glenn, et al. "Enhancing the Student Listening Skills and Environment." *Business Communication Quarterly*, vol. 58, no. 4, Dec. 1995, pp. 28–33.

"Performance Appraisal—Discrimination." *The Employee Problem Solver*. Alexander Hamilton Institute, 2000.

Perloff, Richard M. *The Dynamics of Persuasion*. Lawrence Erlbaum Associates, 1993.

Pinelli, Thomas E., et al. "A Survey of Typography, Graphic Design, and Physical Media in Technical Reports." *Technical Communication*, vol. 32, no. 2, 1986, pp. 75–80.

"Plain Language." *Bureau of Land Management*, United States Department of the Interior, 10 Apr. 2004.

Plumb, Carolyn, and Jan H. Spyridakis. "Survey Research in Technical Communication: Designing and Administering Questionnaires." *Technical Communication*, vol. 39, no. 4, 1992, pp. 625–38.

Porter, James E. "Truth in Technical Advertising: A Case Study." *IEEE Transactions on Professional Communication*, vol. 33, no. 3, 1987, pp. 182–89.

Powell, Corey S. "Science in Court." *Scientific American*, Oct. 1997, pp. 32+.

Publication Manual of the American Psychological Association. 6th ed., American Psychological Association, 2009.

Raloff, Janet. "Chocolate Hearts: Yummy and Good Medicine?" *Science News*, vol. 157, no. 12, 2000, pp. 188–89.

Read Me First!: A Style Guide for the Computer Industry. Sun Microsystems Press, 2003.

Redish, Janice C., et al. "Making Information Accessible to Readers." *Writing in Nonacademic Settings*, edited by Lee Odell and Dixie Goswami, Guilford Press, 1985, pp. 129–153.

Rokeach, Milton. *The Nature of Human Values*. Free Press, 1973.

Rottenberg, Annette T. *Elements of Argument*. 3rd ed., St. Martin's Press, 1991.

Ruggiero, Vincent R. *The Art of Thinking*. 3rd ed., HarperCollins Publishers, 1991.

Savan, Leslie. "Truth in Advertising?" *Brill's Content*, March 2000, pp. 62+.

Sabath, Ann Marie. *Business Etiquette: 101 Ways to Conduct Business with Charm and Savvy*. Career Press, 1998.

Schafer, Sarah. "Is Your Data Safe?" *Inc.*, Feb. 1997, pp. 93–97.

Scheffler, Israel. *Reason and Teaching*. New York: Bobbs-Merrill, 1973.

Schein, Edgar H. "How Can Organizations Learn Faster? The Chalenge of Entering the Green Room." *Strategies for Success: Core Capabilities for Today's Managers*, Sloan Management Review Association, 1996, pp. 34–39.

Schrage, Michael. "Time for Face Time." *Fast Company*, Oct.-Nov. 1997, p. 232.

Seglin, Jeffrey L. "Would You Lie to Save Your Company?" *Inc.*, July 1998, pp. 53+.

Seligman, Dan. "Gender Mender." *Forbes*, 6 Apr. 1998, pp. 72+.

Senge, Peter M. "The Leader's New York: Building Learning Organizations." *Sloan Management Review*, vol. 32, no. 1, Fall 1990, pp. 1–17.

Shedroff, Nathan. "Information Interaction Design: A Unified Field Theory of Design." *Information Design*, edited by Robert Jacobson, MIT P, 2000, pp. 267–92.

Sheehy, Kelsey. "Four Degrees That Are Better to Earn at a Community College." *U.S. News & World Report*, 3 Dec. 2014.

Silverstein, Ken. "Enron, Ethics and Today's Corporate Values." *Forbes*, 14 May 2013.

Sittenfeld, Curtis. "Good Ways to Deliver Bad News." *Fast Company*, Apr. 1999, pp. 88+.

Smith, Gary. "Eleven Commandments for Business Meeting Etiquette." *INTERCOM*, Feb. 2000, p. 29.

"The Social Media Marketing Blog." *Scott Monty*, 30 June 2009, www.scottmonty.com. Accessed 3 Mar. 2013.

Sowell, Thomas. "Magic Numbers." *Forbes*, 20 Oct. 1997, p. 120.

Spencer, SueAnn. "Use Self-Help to Improve Document Usability." *Technical Communication*, vol. 43, no. 1, 1996, pp. 73–77.

Sproull, Lee, and Sara Kiesler. *Connections: New Ways of Working in the Networked Organization*. MIT P, 2001.

Spyridakis, Jan H., and Michael J. Wenger. "Writing for Human Performance: Relating Reading Research to Document Design." *Technical Communication*, vol. 39, no. 2, 1992, pp. 202–15.

Stanton, Mike. "Fiber Optics." *Occupational Outlook Quarterly*, Winter 1984, pp. 27–30.

Stevenson, Richard W. "Workers Who Turn in Bosses Use Law to Seek Big Rewards." *The New York Times*, 10 July 1989, p. A7.

Stix, Gary. "Plant Matters: How Do You Regulate an Herb?" *Scientific American*, Feb. 1998, pp. 30+.

"Study to Prove How Healthy Younger Adults Make Use of Genetic Tests." *National Institutes of Health*, United States Department of Health and Human Services, 3 May 2007, www.nih.gov/news-events. Accessed 4 May 2007.

Sturges, David L. "Internationalizing the Business Communication Curriculum." *Bulletin of the Association for Business Communication*, vol. 55, no. 1, 1992, pp. 30–39.

Taubes, Gary. "Telling Time by the Second Hand." *MIT Technology Review*, May–June 1998, pp. 76–78.

Thrush, Emily A. "Bridging the Gap: Technical Communication in an Intercultural and Multicultural Society." *Technical Communication Quarterly*, vol. 2, no. 3, 1993, pp. 271–83.

Timmerman, Peter D., and Wayne Harrison. "The Discretionary Use of Electronic Media: Four Considerations for Bad-News Bearers." *Journal of Business Communication*, vol. 42, no. 4, 2005, pp. 379–89.

Trafford Abigail. "Critical Coverage of Public Health and Government." *A Field Guide for Science Writers: The Official Guide of the National Association of Science Writers*, edited by Deborah Blum and Mary Knudson, Oxford UP, 1997, pp. 131–41.

Tufte, Edward R. *The Cognitive Style of PowerPoint*. Graphics Press, 2003.

Unger, Stephen H. *Controlling Technology: Ethics and the Responsible Engineer*. Holt Rinehart, 1982.

United States Air Force Academy. *Executive Writing Course*. Government Printing Office, 1981.

United States, Congress, Office of Technology Assessment. *Harmful Non-Indigenous Species in the United States*. Government Printing Office, 1993.

---. National Aeronautics and Space Administration. *Columbia Accident Investigation Board Report*. Vol. 1, Government Printing Office, 2003.

University of Kansas. "Guidelines: How to Write and Report about People with Disabilities." University of Kansas Research and Training Center on Independent Living, 2013.

University of Nottingham. "Covering Letters: UK Conventions." University of Nottingham Careers and Employability Service, 2015.

van der Meij, Hans, and John M. Carroll. "Principles and Heuristics for Designing Minimalist Instruction." *Technical Communication*, vol. 42, no. 2, 1995, pp. 243–61.

Van Pelt, William. Unpublished review of *Technical Writing*, by John M. Lannon. June 1983.

Varchaver, Nicholas. "The Perils of E-mail." *Fortune*, 17 Feb. 2003, pp. 96–102.

Vaughan, David K. "Abstracts and Summaries: Some Clarifying Distinctions." *Technical Writing Teacher*, vol. 18, no. 2, 1991, pp. 132–41.

Victor, David A. *International Business Communication*. HarperCollins Publishers, 1992.

"Walking to Health." *Harvard Men's Watch*, vol. 2, no. 12, 1998, pp. 3–4.

Walter, Charles, and Thomas F. Marsteller. "Liability for the Dissemination of Defective Information." *IEEE Transactions on Professional Communication*, vol. 30, no. 3, 1987, pp. 164–67.

Wang, Linda. "Veggies Prevent Cancer through Key Protein." *Science News*, vol. 159, no. 12, 2001, p. 182.

Warshaw, Michael. "Have You Been House-Trained?" *Fast Company*, Oct. 1998, pp. 46+.

Weimer, Maryellen. "Does PowerPoint Help or Hinder Learning?" *Faculty Focus*, 1 Aug. 2012, www.facultyfocus.com/articles/teaching-professor-blog/does-powerpoint-help-or-hinder-learning/.

Weinstein, Edith K. Unpublished review of *Technical Writing*, by John M. Lannon. May 1991.

Weymouth, L. C. "Establishing Quality Standards and Trade Regulations for Technical Writing in World Trade." *Technical Communication*, vol. 37, no. 2, 1990, pp. 143–47.

White, Jan. *Color for the Electronic Age*. Watson-Guptill, 1990.

---. *Editing by Design*. 2nd ed., RR Bowker Company, 1982.

---. *Great Pages*. Serif Publishing, 1990.

---. *Visual Design for the Electronic Age*. Watson-Guptill, 1988.

Wickens, Christopher D. *Engineering Psychology and Human Performance*. 3rd ed., Pearson, 1999.

Wight, Eleanor. "How Creativity Turns Facts into Usable Information." *Technical Communication*, vol. 32, no. 1, 1985, pp. 9–12.

Wojahn, Patricia G. "Computer-Mediated Communication: The Great Equalizer between Men and Women?" *Technical Communication*, vol. 41, no. 4, 1994, pp. 747–51.

Writing Reader-Friendly Documents. Plain Language Action and Information Network, 2007.

Yoos, George. "A Revision of the Concept of Ethical Appeal." *Philosophy and Rhetoric*, vol. 12, no. 4, 1979, pp. 41–58.

Zibell, Kristin J. "Usable Information through User-Centered Design." *INTERCOM*, Dec. 1999, pp. 12–14.

Index

I

icons, 240, 265–266
identification, persuasion and, 38
illustrations, 240
 typeface, 293
immediate audience, 18
imperative mood, instructional documents, 450
implicit persuasion, 34–35
imply/infer, 676
impromptu oral presentation, 573, 574
Indeed, 369
indexes, 131
indirect approach
 in letters, 347–348
 in memos, 332
infographics, 240, 266–267
informal proposals, 535
informal reports, 469–470
 informational, 470–471, 474–475, 476–479
informal tone, 226
informational interviews, 134–137
informational reports, 470–471, 471–474
 meeting minutes, 478–479
 periodic activity, 474–475
 progress reports, 471–474
 trip reports, 476–478
information documents, 10
information gathering, 105–106
informative abstracts, 173–174
informative interview, request, inquiry letter, 354
informative oral presentation, 572
InfoTrac, 127
inquiry letters, 350, 352–353, 354
 guidelines, 355
Instagram, 613, 618
instructional documents, 10–11, 438
 assembly guides, 439
 audience considerations, 439
 background, 445–446
 cautions, 449
 content accuracy, 443
 danger notice, 449
 design, 452
 detail, 444–448, 446, 447
 effective, 443–456
 examples, 446
 formats, 439–440
 guidelines, 452, 465–466
 hazard notices, 448–449
 introduction-body-conclusion structure,
 453–456

legal implications, 442
 manuals, 439, 440
 notes, 448–449
 online help, 441, 457
 outlines, 453
 quick reference materials, 439, 440
 readability, 450–452
 social media, 457–459
 steps, 448
 technicality, 444–448
 title, 443
 troubleshooting, 448
 usability, 462–466
 visuals, 443–444, 445
 warnings, 449
 web-based instructions, 441
 wordless instructions, 444
instructional videos, 619–622
 guidelines, 621
intercultural communication, 5
interests, résumés, 373
internalization, persuasion and, 38
Internet Archive, 127
Internet Public Library, 127
interpersonal issues
 email and, 314–315
 teams and, 85–86
interpreting findings, 149
 assumptions, 150–151
 causal reasoning, 153–155
 certainty, 150
 considering other interpretations, 151
 generalization, 152–153
 guidelines, 161–162
 personal bias and, 151
 standards of proof, 152
 statistical analysis, 155–159
interviews, 134–137
 APA style, 651
 job, 385–388
 MLA style, 638
 request letter (inquiry style), 354
introduction
 formal analytical reports, 501–502
 proposals, 547–549
introduction-body-conclusion structure,
 182, 183
 instructional documents,
 453–456
introductory phrases, 667
irony, quotation marks and, 671

X-Y-Z